# SPECIALITY CHEMICALS

# INNOVATIONS IN INDUSTRIAL SYNTHESIS AND APPLICATIONS

# SPECIALITY CHEMICALS

# INNOVATIONS IN INDUSTRIAL SYNTHESIS AND APPLICATIONS

*Edited by*

**BRIAN PEARSON**

Spring Innovations Limited, Stockport, Cheshire, UK

ELSEVIER APPLIED SCIENCE

LONDON AND NEW YORK

ELSEVIER SCIENCE PUBLISHERS LTD
Crown House, Linton Road, Barking, Essex IG11 8JU, England

*Sole distributor in the USA and Canada*
ELSEVIER SCIENCE PUBLISHING CO., INC.
655 Avenue of the Americas, New York, NY 10010, USA

WITH 78 TABLES AND 550 ILLUSTRATIONS



**British Library Cataloguing in Publication Data**

Speciality chemicals
  1. Industrial chemicals
I. Pearson, Brian 1939–
661 ISBN 1-85166-646-X

**Library of Congress Cataloging-in-Publication Data**

Speciality chemicals: innovations in industrial synthesis and
applications / edited by Brian Pearson.
p. cm.
"Compendium of articles based on papers presented at conferences
on speciality chemicals organized by Spring Innovations Ltd.
throughout Europe and the USA"—P.
Includes bibliographical references and index.
1. Chemicals—Congresses. I. Pearson, Brian, 1939–.
II. Spring Innovations, Ltd.
TP200.S62 1991
660—dc20

91-15420
CIP



Typeset and printed by The Universities Press (Belfast) Ltd.

# Preface

There are many definitions of speciality chemicals but the most commonly accepted subdivide these added value products into two main groups: (i) fine chemicals and intermediates and (ii) performance products. The first group is concerned with distinct chemicals which can be identified by chemical formulae and are generally sold against analytical specification which confirm the purity of the product. In the second group the chemical identity is often a commercial secret carefully guarded by the supplier and the product sold against a performance specification. In this book we concentrate mainly on the first group.

It is not always possible to translate a laboratory route for a specific chemical to a plant production process. The successful fine chemical manufacturer devotes much of his research and development effort into identifying the nuances of this essential step, whereas the main development effort of the performance chemical manufacturer is directed to technical service to meet his customer's needs.

This book gives an insight into those routes that have been adapted to industrially important processes and advice on chemicals that are now becoming commercially available and will affect the manufacturing routes in the future. It is not a comprehensive survey of all available techniques but is intended to provide information on developments in synthetic routes. It is a compendium of articles based on papers presented at conferences on speciality chemicals organised by Spring Innovations Limited throughout Europe and the United States over the last few years, updated to ensure that the information is the latest available. These papers have been specifically selected to cover as wide a range of topics as possible and bring together for the first time the industrialist's view rather than the academic's on practical innovations in synthesis. Consequently they are not solely technical articles but many also contain useful commercial and market information to demonstrate the practical importance of the chemicals and processes.

The papers have been grouped into sections on similar topics. The scene setter is a commercial paper by Enrico Polastro of Arthur D. Little which provides an overview and forward predictions for this sector of the industry.

The section on the halogens comprehensively covers the use of fluorine, chlorine and bromine in speciality chemicals manufacturing. Fluoro derivatives are a particularly important development of the last decade and much research effort is devoted to these compounds. Consequently three papers have been included on this topic which demonstrate the range of approaches that can be taken by different manufacturers. Chlorine-containing products have come under increased environmental pressure but the importance of this element is reflected in the two articles on this topic. The paper on the bromine derivatives completes the picture and demonstrates that each halogen has its own specific use in speciality chemicals synthesis, and the application of the end product.

Sulphur-, nitrogen- and phosphorus-containing products find use both as performance chemicals in rubber, plastics, etc., and intermediates for pharmaceuticals and agrochemicals.

Heterocyclics are featured in the previous section but two papers have been separated to specifically highlight the latest developments in the industrial synthesis of this important class of compounds.

Classical organic chemistry places great emphasis on progressing up the chain. A means of introducing another carbon atom into the molecule is considered to be a most useful synthetic tool. The papers in the building blocks section demonstrate that building up molecules by the introduction of functional groups opens up a wide range of synthetic routes to industrially important compounds. There are two papers on hydrogen cyanide and its derivatives which together provide a most comprehensive study of the wide-ranging possibilities of these products in synthetic chemistry. The value of phosgene as a building block is demonstrated in the paper by G. Sennyey of SNPE, which highlights that, although mainly used in the manufacture of polyisocyanates for polyurethanes its use in fine chemicals production provides useful routes to many important compounds. *t*-Butyl acetoacetate has recently been developed as an alternative to diketene and is detailed in a paper from Eastman Kodak. The section is completed with a paper on the use of substituted ethylene diamines.

Organometallic compounds are widely used in both laboratory and industrial synthesis. The use of magnesium and aluminium in this context is widely reported and this section concentrates on the newer areas of sodium, lithium and cesium. The sodium paper devotes much of the discussion to the use of the metal itself which again is an innovation in industrial synthesis. The paper on lithium provides what is probably the most comprehensive review of the uses of lithium derivatives to be published in recent years. Cesium is not widely used in organic synthesis but can solve some problems, particularly in esterification reactions, to result in improved yields over other routes.

Oxidation and reduction reactions are used as steps in the manufacturing routes of a wide range of fine chemicals. The papers in this section have been selected to provide information on new developments in this area and alternatives to the classical techniques.

Often the key to economic viability of a particular route is the correct choice of catalyst. The following section deals with some developments in this area, particularly with thoughts also on protecting the environment by re-use of catalysts or presenting them in a form that reduces the waste generation. A range of process innovations is also given which provide alternative routes to the manufacture of compounds and hence affect the competitiveness of the traditional routes. The economics of existing routes can often be undermined by use of completely different technology. The use of sonochemistry and electrochemistry can change the reaction kinetics and produce compounds in much higher yields and purities than classical synthetic routes. Papers on these techniques complete this section.

Biotransformations have become important in the manufacture of chiral intermediates for agrochemical and pharmaceuticals and four papers have been devoted to this important topic.

The final section is a miscellany of interesting developments in the use of speciality chemicals. This section does not concentrate on the synthesis of products but on the more interesting applications areas and illustrates how products can be modified to achieve performance characteristics. It demonstrates the problem of definition of speciality chemicals as the products in this section are distinct, identifiable chemicals but are sold for their performance.

The book should provide you with a most important reference source for the industrial synthesis of speciality chemicals and hopefully be enjoyable to read.

# Contents

## OXIDATIONS AND REDUCTIONS



## NEW PROCESSES AND CATALYSTS



## BIOTRANSFORMATIONS

## APPLICATION INNOVATIONS

# List of Contributors

**B. PEARSON:** *Editor*
Present address: Peer Chemicals Limited, 24 Waterloo Road, Bramhall, Stockport, Cheshire SK7 2NX, UK

**H. ASADA**
UBE Industries Limited, 102 Jermyn Street, London SW1Y 6EE, UK

**P. ASHWORTH**
The Associated Octel Company Limited, PO Box 17, Oil Sites Road, Ellesmere Port, South Wirral L65 4HF, UK

**A. BADGER**
Robinson Brothers Limited, Phoenix Street, West Bromwich, West Midlands B70 0AH, UK

**T. D. BAILEY**
Reilly Industries, Inc., 1500 South Tibbs Avenue, Indianapolis, IN 46241, USA

**A. J. BANISTER**
Department of Chemistry, University Science Laboratory, University of Durham, South Road, Durham DH1 3LE, UK

**T. W. BASTOCK**
Contract Chemicals Limited, Penrhyn Road, Knowsley Industrial Park, Prescot, Merseyside L34 9HY, UK

**C. BIRR**
Orpegen GmbH, Czerny-Ring 22, D-6900 Heidelberg, FRG

**N. W. BOAZ**
Corporate Research Laboratories, Eastman Kodak Company, Rochester, New York, NY 14650-2115, USA

**M. J. BRAITHWAITE**
Allied Colloids Limited, PO Box 38, Low Moor, Bradford, West Yorkshire BD12 0JZ, UK

**J. CHETLAND**
The Associated Octel Company Limited, PO Box 17, Oil Sites Road, Ellesmere Port, South Wirral L65 4HF, UK

**J. H. CLARK**
Department of Chemistry, University of York, Heslington, York Y01 5DD, UK

**D. A. CORNFORTH**
Anchor Chemical (UK) Limited, Clayton Lane, Clayton, Manchester M11 4SR, UK

**K. M. DEAR**
Interox Research & Development, PO Box 51, Moorfield Road, Widnes, Cheshire WA8 0FE, UK

**H. DEGER**
Advanced Technology Group, Hoechst Celanese Corporation, 51 John F. Kennedy Parkway, Short Hills, NJ 07078, USA

**L. DOLBY-GLOVER**
Rhône-Poulenc Chemicals, ISC Division, St Andrews Road, Avonmouth, Bristol BS11 9YF, UK

**J. H. EDWARDS**
W. R. Grace Limited, Rexolin Product Line, Seal Sands, Middlesbrough, Cleveland TS2 1UD, UK

**P. G. S. FIELD**
P.G.F. Associates, 7 Hall Bank, Buxton, Derbyshire SK17 6EW, UK

**G. L. GOE**
Reilly Industries, Inc., 1500 South Tibbs Avenue, Indianapolis, IN 46241, USA

**T. GOODWIN**
AEA Technology, Harwell, Oxon OX11 0RA, UK
*Present address*: Britsh Steel, Welsh Laboratory, Port Talbot, West Glamorgan SA13 2NG, UK

**J. F. HAMLIN**
A. H. Marks & Company Limited, Wyke, Bradford, West Yorkshire BD12 9EJ, UK

**W. HILLENBACH**
ANGUS Chemie GmbH, Postfach 40 03, AM Uffelner Moor 30, D-4530 Ibbenburen, FRG

**G. F. HILLSTROM**
Reilly Industries, Inc., 1500 South Tibbs Avenue, Indianapolis, IN 46241, USA

**A. HOWELL**
Anchor Chemical (UK) Limited, Clayton Lane, Clayton, Manchester M11 4SR, UK

**K. HUTHMACHER**
Degussa AG, Forschung Chemie, Postfach 1345, D-6450 Hanau 1, FRG

**M. S. IRVINE**
Nipa Laboratories Limited, Llantwit Fardre, Pontypridd, Mid Glamorgan, Wales CF38 2SN, UK

**A. JACKSON**
Fine Organics Limited, Teeside Site, Seal Sands, Middlesbrough, Cleveland TS2 1UB, UK

**M. A. JOHNSON**
Radcure Specialities SA, Anderlecht Str 33, Drogness B-1620, Brussels, Belgium

**T. S. JOHNSON**
Advanced Technology Group, Hoechst Celanese Corporation, 51 John F. Kennedy Parkway, Short Hills, NJ 07078, USA

**A. C. JONES**
Epichem Limited, Power Road, Bromborough, Wirral L62 3QF, UK

**J. G. KEAY**
Reilly Industries, Inc., 1500 South Tibbs Avenue, Indianapolis, IN 46241, USA

**R. R. KEMMERER**
Radcure Specialties SA, Anderlecht Str 33, Drogness B-1620, Brussels, Belgium

**H. KLENK**
Degussa AG, Forschung Chemie, Postfach 1345, D-6450 Hanau 1, FRG

**W. KLOP**
DSM Research, PO Box 18, 6160 MD Geleen, The Netherlands

**K. LAUMEN**
Corporate Research Laboratories, Eastman Kodak Company, Rochester, New York, NY 14650-2115, USA

**A. B. LEESE**
Epichem Limited, Power Road, Bromborough, Wirral L62 3QF, UK

**D. MARSHALL**
Allied Colloids Limited, PO Box 38, Low Moor, Bradford, West Yorkshire BD12 0JZ, UK

**C. R. MARSTON**
Reilly Industries, Inc., 1500 South Tibbs Avenue, Indianapolis, IN 46241, USA

**P. MARTIN**
AEA Technology, Harwell, Oxon OX11 0RA, UK

**D. W. McQUIGG**
Reilly Industries, Inc., 1500 South Tibbs Avenue, Indianapolis, IN 46241, USA

**I. M. McROBBIE**
A. H. Marks & Company Limited, Wyke, Bradford, West Yorkshire BD12 9EJ, UK

**O. METH-COHN**
*Present address*: Department of Pharmaceutcal Chemistry, Sunderland Polytechnical College, Chester Road, Sunderland SR1 3SD, UK

**W. D. NOTTINGHAM**
Eastman Chemicals Division, Eastman Kodak Company, PO Box 1972, Kingsport, TN 37662, USA

**J. A. PARR**
Nipa Laboratories Limited, Llantwit Fardre, Pontypridd, Mid Glamorgan, Wales CF38 2SN, UK

**L. C. PAYNE**
Nipa Laboratories Limited, Llantwit Fardre, Pontypridd, Mid Glamorgan, Wales CF38 2SN, UK

**E. POLASTRO**
Arthur D. Little International, Inc., Boulevard de la Woluwe 2, B-1150 Brussels, Belgium

**H. PRINZ**
Chemetall GmbH, Reuterweg 14, Postfach 10 15 01, D-6000, Frankfurt 1, FRG

**T. L. RATHMAN**
FMC Corporation, Lithium Division, 449 North Cox Road, Box 3925, Gastonia, NC 28053, USA

**J. REGAN**
La Porte Organic Chemicals Division, Teeside Site, Seal Sands, Middlesbrough, Cleveland TS2 1UB, UK.
*Present address*: Contrachem Limited, 3 The Courtyard, Ochrelands, Hexham, Northumberland NE46 1SB

**E. F. V. SCRIVEN**
Reilly Industries, Inc., 1500 South Tibbs Avenue, Indianapolis, IN 46241, USA

**C. A. SELDON**
Nipa laboratories Limited, Llantwit Fardre, Pontypridd, Mid Glamorgan, Wales CF38 2SN, UK

**G. SENNYEY**
SNPE, Chemical Division, BP2, Le Bouchet, 91710 Vert-le-Petit, France

**R. A. SHELDON**
Andeno BV, PO Box 81, 5900 AB Venlo, The Netherlands
*Present address*: Department of Chemical Engineering, Technical University of Delft, Julianalaan 136, P.O. Box 5045, 2600 GA Delft, The Netherlands

**E. E. SOWERS**
Reilly Industries, Inc., 1500 South Tibbs Avenue, Indianapolis, IN 46241, USA

**M. STOHLMEIER**
Dynamit Nobel AG, Kalkstrasse 218, PB 10 03 40, D-5090 Leverkusen 1, FRG

**E. A. SULLIVAN**
Morton International, 150 Andover Street, Danvers, MA 01923-1428, USA

**K. THEWALT**
Dynamit Nobel AG, Kalkstrasse 218, PB 10 03 40, D-5090 Leverkusen 1, FRG

**J. E. TOOMEY**
Reilly Industries, Inc., 1500 South Tibbs Avenue, Indianapolis, IN 46241, USA

**R. VALENTINE**
La Porte Organic Chemicals Division, Teeside Site, Seal Sands, Middlesbrough, Cleveland TS2 1UB, UK
*Present address*: High Force Research Limited, Mount Join Research Centre, Stockton Road, Durham DH1 3FW, UK

**K. WALL**
P.G.F. Associates, 7 Hall Bank, Buxton, Derbyshire SK17 6EW, UK

**N. WEFERLING**
Hoechst AG, Werk Knapsack, Abteilung Forschung und Entwicklung Chemikalien, Postfach, D-5030 Hürth-Knapsack, FRG

**S. WEST**
Biocatalysts Limited, Main Avenue, Treforest Industrial Estate, Pontypridd, Mid Glamorgan, Wales CE37 5UT, UK

**J. S. WITZEMAN**
Eastman Chemicals Division, Eastman Kodak Company, PO Box 1972, Kingsport, TN 37662, USA

**M. YAMASHITA**
UBE Industries Limited, 102 Jermyn Street, London SW1Y 6EE, UK

# The fine chemicals industry—the challenges at the turn of the century

**E. Polastro**

*Arthur D. Little International, Inc., Brussels, Belgium*

## Introduction

The fine chemicals industry is in a period of fundamental change. Players and prospective new entrants must face key challenges for successful participation in the industry.

## Key points

—Trends in the fine chemicals industry scene are dominated by increasing industry concentration, evolving demand patterns, mounting competitive intensity and tighter environmental scrutiny.

—The challenge for participants in this sizeable, growing and highly fragmented industry is to maintain or develop a profitable position.

—The strategic implications of meeting this challenge vary, depending on the type of industry participant.

Newcomers, typically commodity chemical companies, will have to:

- Develop an approach that is compatible with the peculiar industry dynamics.

- Successfully integrate this new business into a fully fledged division.

Established fine chemical producers will have to:

- Grow to optimal size, balancing portfolio critical mass with organizational responsiveness.
- Attain competitive differentiation.
- Innovatively manage their business, including developing new managerial systems.

Integrated fine chemical users like pharmaceutical companies that are contemplating competing in the fine chemicals arena face the most complex problem: avoid sawing off the branch that they are sitting upon!

## A large and growing industry—with its own dynamics and characteristics

—The fine chemicals industry turnover is about US$16–20 billion. This includes both the merchant and captive market segments, as well as the intermediates and bulk active ingredient segment (such

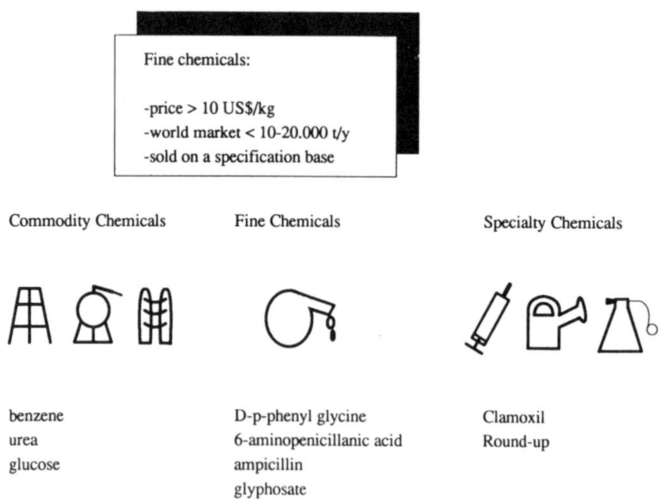


**Fig. 1.** Fine chemicals definition and industry boundaries.

as bulk pharmaceuticals, agrochemicals and aroma chemicals but excluding 'doublecounting'—see Fig. 1).

—Pharmaceutical and agrochem applications account for 50 and 20% of the total market respectively. Other important outlets for fine chemicals are food additives (including vitamins) and intermediates for advanced polymers (see Fig. 2).

—The overall growth outlook for the fine chemicals industry is favourable. Annual growth rates over the next 5 year period are expected to outpace the general GNP evolution.

Growth will be particularly healthy for advanced polymer applications, driven by increasing demand for these products and greater market acceptance.

—The fine chemicals industry offers some unique characteristics and features (see Table 1), when compared with the other chemical sector industries (commodity and speciality chemicals). These include:

- continuous emergence of new products;
- relatively low volume/high value products;
- basis of competition largely driven by price, but with image (e.g. product tree base, technological capacities) and 'customer service' also playing a role—allowing for competitive differentiation;

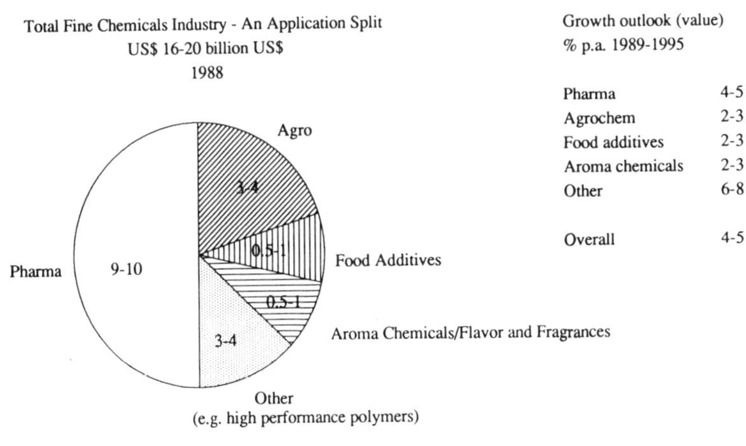


**Fig. 2.** Fine chemicals industry: a tentative market split.

**Table 1.** The chemical sector—main characteristics of the industries

| Industry characteristics | Bulk chemicals | Fine chemicals | Speciality chemicals |
|---|---|---|---|
| Product life cycle[a] | Long | Moderate | Short/moderate |
| Product range (number of products) | 00's | 000's | 0 000's |
| Product volumes[a] | >10 000 t/y | <10 000 t/y | Highly variable |
| Product prices[a] | <5 US$/kg | >5 US$/kg | >5 US$/kg |
| Product differentiation | None | Very low | High |
| Valued added | Low | High | High |
| Capital intensity | High | Moderate | Moderate/low |
| R&D focus | Process improvement | Process development | Application/product |
| Key success factors[b] | | | |
| —cost position | XXX | XX | X |
| —technical service | — | XX | XXX |
| —close links with the customer | — | XXX | XXX |

[a] Typical examples—exceptions may apply.
[b] Relative importance: X low; XX, average; XXX, high importance.

- R & D focus mainly on process development and optimization;
- wide product and customer slate making this industry relatively immune to the gyration of the economy.

## A fragmented, intricate and complex structure

—The fine chemicals industry consists of innumerable segments, with different products, competitor sets and technologies.

Individually, most segments have limited market—often less than US$10 Mio.

—Overall, the industry has more than 10 000 diverse products. Most compounds have sales below US$10 Mio per year (see Fig. 3) and are typically supplied by several competing companies. A monopolistic supply situation rarely occurs.

The product suppliers' insight into production applications is typically very limited. They have no control over the end product demand and are completely dependent on the customer to develop new applications for their products.

—More than twenty end market application segments can be identified for fine chemicals, from pharmaceuticals to food additives and including special polymers. Each of these end markets has a multitude of customers.

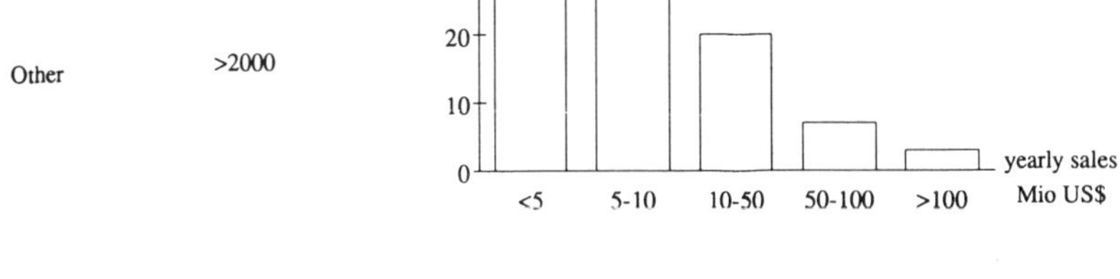


Source : ADL estimates

**Fig. 3.** Fine chemicals product slate.

The range of end markets and the continuing emergence of new product opportunities make the industry relatively attractive for newcomers such as commodity chemicals producers, hoping to reduce their dependence on highly cyclical activities.

—The supplier/producer/customer structure is particularly intricate and complex. Overlaps and unclear boundaries between the various industry participants are common. Of particular interest is the ambiguous customer/competitor situation where the customer often has the option to produce in-house rather than buy its fine chemicals requirements from a third party.

—The competitor set in fine chemicals can be subdivided as follows (Table 2):

- *Specialized fine chemicals suppliers,* like Lonza, Omnichem, Andeno, ISC, that consider fine chemicals to be their main business. These companies' annual sales are typically less than US $100 Mio. They tend to concentrate on selected product/technology industry niches (e.g. fluorinated intermediates, azide chemistry, diketenes). Their level of upstream integration is typically very limited. Most, if not all, fine chemicals producers have no downstream activities. They make a strong point of being independent and not competing with their customers.

  These companies tend to compete on the basis of:

  —specialization (e.g. unique technology combination);
  —customer service and responsiveness to offer tailor-made solutions (e.g. exclusive synthesis).

- *Divisions of large commodity chemicals companies* (e.g. BASF, DSM, Enichem, Monsanto, . . .). Many of these companies started their fine chemicals activities to extract value from by-products derived from mainstream processes (examples include oxime derivatives from caprolactam production, benzaldehyde from toluene conversion to phenol, . . .).

  These competitors usually offer a limited product and service range. However, they are increasingly trying to upgrade their product offering, entering in direct competition with the specialized fine chemicals producers.

  Several commodity chemicals producers have recently expanded their fine chemicals activities, through either in-house

**Table 2.** Industry participants

| | Established fine chemicals producers | Commodity chemicals companies | Fine chemicals users backward integrated |
|---|---|---|---|
| Examples | Lonza, Omnichem, Farchemia | BASF, DSM | Roche, Syntex |
| Turnover | Typically <50 Mio US$ | Variable | Variable |
| Focus | Fine chemicals main business | Fine chemicals not a core business | |
| | Technology/product specialization | Add value to by-products | Load available equipment |

organic growth or acquisitions (e.g. Tessenderlo, Shell, Nobel Chemi, . . .).

This reinforced thrust in fine chemicals was prompted by their strategy to reduce dependence on commodity chemicals and to develop activities in a 'related' but less cyclical business arena.

- *Integrated users of fine chemicals,* including pharma and agrochem companies (e.g. Syntex, Zambon, Roche, . . .). These companies may either produce their fine chemical requirements in-house or rely on outside sources.

  They are potential competitors in the industry arena. Recently, some companies facing market saturation and increasing competition in their pharma/agrochem business have been trying to develop their activities as merchant suppliers of fine chemicals. For example, Syntex has explicitly announced its intention to supply generic producers with bulk Naproxene.

## Facing significant challenges

Dramatic events are currently occurring in the fine chemicals arena that will spell out the future issues and challenges that participants must face to maintain or develop their competitive position.

Among the most noteworthy events are:

—Industry concentration.
—Evolving demand patterns.
—Increasing competitive intensity.
—Tighter environmental scrutiny.
—Accelerating technological pace.

—*Industry concentration.* This trend over the past few years is due to the acquisition of several independent companies mainly by commodity chemicals producers. Examples include:

- Andeno and Plaine, acquired by DSM;
- Farchemia (Tessenderlo-EMC);
- Technochemie, Ward Blankinsop, Synthetic Chemicals Ltd and recently La Mesta, absorbed by Shell Chemicals;
- Blaschim and Profarmaco, acquired by Nobel Chemi of Sweden.

Clusters of companies with broad technology and product portfolio capabilities have emerged. This is a significant change compared to the past where the centre of gravity of the industry was represented by companies competing mainly in selected niches and with limited sales.

The customer base is also concentrating, with mergers taking place among pharmaceutical and agrochem companies. This trend, which is poised to continue and accelerate, has far reaching implications such as:

- a shrinking customer base for chemicals suppliers;
- the high likelihood of available free capacity among the newly merged companies—making the 'buy or make' dilemma even more acute (e.g. although in the past Smith Kline Beckmann farmed out most of its fine chemicals requirements, it will probably rely increasingly on the strong in-house chemical synthesis capabilities of its new partner Beecham, which used to be fully integrated for its fine chemicals requirements).

—*Demand patterns*. These are also changing and evolving towards increased molecular complexity and smaller volumes, driven up by the trend towards higher specific activity among the agrochem and pharma industries. In addition, the product demand cycle is shorter and more volatile, reflecting heavier competition in the end product arena.

The balance between intermediates and bulk active end products is also changing. One of the driving forces behind this evolution is the growing importance of generic producers (which often have no chemical production capacity) in the customer mix as well as the tendency of several customers to rely entirely on an outsider for all their chemical synthesis requirements. This trend is particularly marked among US pharmaceutical companies and is largely driven by their desire to concentrate on R&D and marketing—which they view as core activities.

Key implications of this evolving demand pattern are:

- a higher degree of uncertainty and risk;
- the need for stronger focus and specialization to offer tailor-made solutions to prospective clients.

—*Competitive intensity*. Rising overall in the fine chemicals industry. Overcapacity problems are looming along with pressure on profit margins. This increase in competitive intensity is driven by:

- The entry of new competitors, such as NICs (Newly Industrialized Countries), that are eager to develop their fine chemicals activities, particularly in bulk active molecules, where the marketing requirements are simpler. They compete primarily on the basis of price and often ruin the market. Examples include China and India, whose entry onto the bulk sulphamide market has effectively driven most European producers out of the business.
- Frequent fights between established industry participants that are all chasing the same opportunities. A recent example of this is the rush in fluoroaromatic intermediates, which has led to a price erosion of 25% over the past 6 months and will probably result in severe overcapacity.
- The inroads made by both traditional suppliers and industry customers, that are increasingly trying to develop their fine chemicals activities (the 'historic' Syntex ad: 'we view generic producers not as competitors but as partners').

—*Tighter environmental scrutiny*. The industry is plagued by its poor image and reputation in environmental issues. It is subject to public and governmental scrutiny that often leads to operating hurdles and the need for investment to comply with the tightening environmental standards.

This has resulted in the closure of some old plants and the migration of some polluting/challenging reactions towards countries that take a more relaxed environmental stance (for example of the exodus of the menadione production based on a chromic acid oxidation step, from Italy to Spain and now to South America).

New legislation now calls for extensive toxicological and safety data sheets on all chemicals to be shipped. This burden on the fine chemicals industry will increase, and eventually severely limit the shipment of intermediates from one plant to another, which has, until now, been a very common procedure in the industry. The legislation could redirect the demand towards bulk active products, for which extensive safety data are usually available.

Rather than posing a threat, the 'green trend' could represent an important opportunity for industry participants that are able to position themselves as experts in environmental issues. They could develop and offer a sales proposition which would relieve prospective clients of all environmental related issues. Lonza has

# Multiple choice technology to fluoroaromatic intermediates

**L. Dolby-Glover**

*Rhône-Poulenc Chemicals, Bristol, UK*

## Introduction

Fluoroaromatic intermediates are assuming greater importance as the cost effectiveness of biologically active fluorine containing products and the synthesis value of fluorine substituents becomes more widely acknowledged. 'Fluoroaromatics' are defined here as molecules where the fluorine is attached directly to the aromatic system as opposed to, for example, the very important group of benzotrifluoride intermediates where the manufacturing technology is very different (aliphatic substitution). The effectiveness of fluorine substituents either in subtle drug modifications, enzyme blocking or as strategic synthesis markers is well documented. However, fluoraromatic intermediates do have a high price relative to other ranges of intermediates. That high price is mainly a reflection of the chemistry of fluorine but is also partly due to the relatively small scale of operations. Some of the chemistry of the major industrial routes to fluoroaromatic intermediates will be described to illustrate the advantages and disadvantages of each process.

Substituent stability, substitution pattern and market requirements means that a multiple choice of technology is needed to access the whole range of fluoroaromatic intermediates in the most economic way.

# Technology—Introduction

The techniques for making fluoroaromatic building blocks are significantly different from those used in, for example, classical halogenations, nitrations or sulphonations. The reasons for this are embodied in fluorine's electronic configuration. Table 1 compares some of the physical and chemical properties of fluorine with hydrogen and the other halogens. Fluorine is an extremely reactive gaseous element, categorised as the most electronegative, with a highly stable minus one oxidation state, either as a fluoride ion or in, for example, a carbon–fluorine covalent bond. With a small ionic radius, fluoride ions also have a high charge density: loss of entropy on solvation and high crystal lattice energies resulting in a very low water solubility. For example, calcium chloride is 10 000 times more soluble than calcium fluoride, fluorspar, one of the major sources of fluoride ions. The low water solubility and high electrode potential means that techniques used for chlorine production, for example, cannot be readily translated to fluorine production. Hydrogen fluoride, HF, production is currently the major large scale method of obtaining fluorine in a usable form. Reaction of fluorspar with sulphuric acid at elevated temperatures is the main technique (Fig. 1), production having increased from a few thousand tonnes in the 1940s to almost half a million tonnes per year. Table 2 gives the US consumption of HF and Table 3 illustrates the

**Table 1.** A comparison of physical and chemical properties

| | Hydrogen | Bromine | Chlorine | Fluorine |
|---|---|---|---|---|
| Van der Waal's radius (nm)[1] | 0·13 | 0·195 | 0·18 | 0·135 |
| Pauling electronegativity[2] | 2·1 | 2·8 | 3·0 | 4·0 |
| Lattice energy of sodium salt (kJ per mole)[1] | — | 743 | 781 | 915 |
| Water solubility of calcium salt (moles per 100 g water)[3] | Decomposes | $6{\cdot}25\times10^{-1}$ | $5{\cdot}36\times10^{-1}$ | $2{\cdot}3\times10^{-5}$ |
| Average strength of bond to carbon (kJ per mole)[4] | 413 | 280 | 339 | 450 |
| Standard electrode potential (volts)[5] | 0 | +1·07 | +1·36 | +2·87 |

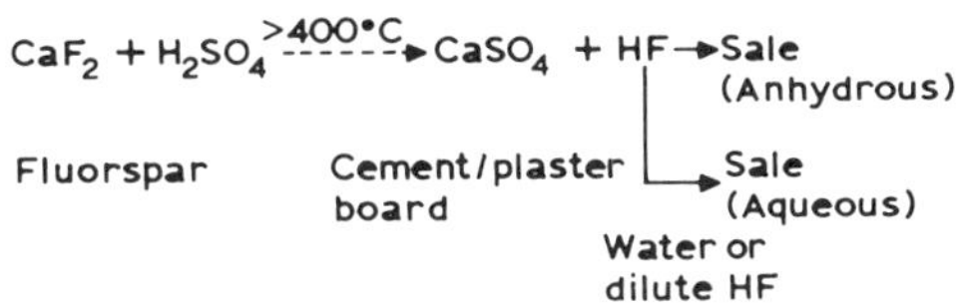


**Fig. 1.** HF production.

difference in price between HF and HCl. These figures were taken from one particular issue of a trade magazine and, although there are many local variations, do indicate the relative costs.

Low volume HF production began at the Avonmouth site in 1939 and was boosted in 1946 by purchase of a second-hand plant from the Pennsylvania Salt Company. HF is used for a wide range of industrial processes, either anhydrous or in aqueous solution. HF boils at 190°C, is highly toxic and reacts with glass and many metals. Special techniques are necessary for safe handling. Figure 2 shows part of an HF plant.

HF is the backbone of the methods used for making fluoroaromatics, either as anhydrous HF, as fluoride salts or as the consumable in fluorine cells. Fluorine is normally produced from a cell with a potassium fluoride/hydrogen fluoride eutectic as the electrolyte. Figure 3 shows the heavy engineering of a fluorine cell. The high electrode potential, equipment and HF costs mean that the production of free fluorine is significantly more expensive than chlorine generation where an aqueous electrolyte, low voltages and sodium chloride can be used. The terrestrial abundance of fluorine is actually three times greater than that of chlorine.

The methods for making fluoroaromatic building blocks can be split into classical reaction pathways: electrophilic substitution, nucleophilic substitution, free radical substitution and additive oxidation with subsequent reduction.

**Table 2.** USA HF consumption (000 tonnes)[6]

| | | |
|---|---|---|
| 1930 | 6 | |
| 1940 | 37 | Production starts at Avonmouth |
| 1950 | 97 | |
| 1960 | 165 | |
| 1970 | 295 | |
| 1980 | 350 | |

**Table 3.** Market prices[7]

| | Aqueous HF (70%) | Aqueous HCl (30%) | Anhydrous HCl | Anhydrous HF |
|---|---|---|---|---|
| Cost per tonne | $1310 | $68 | $300 | $1515 |
| Relative cost per mole | 4·6 | 1 | 1·33 | 3·7 |

These will be examined in more detail with some aspects of downstream chemistry.

## Nucleophilic Substitution

With HF being the main source of usable 'fluorine', methods that can utilise HF or simple salts have a significant cost advantage. However, although fluoride ions are very stable entities, they are poor nucleophiles for organic substrates. They have low solubilities in many

**Fig. 2.** HF plant at Avonmouth.

**Fig. 3.** Fluorine cell—new facility under construction.

solvents and their charge density leads to highly co-ordinated solvation spheres in others. An absence of competitive nucleophiles is normally required and two major approaches are used. The choice of route depends mainly on the ultimate substitution pattern required. The approaches differ in the leaving group, the first being an activated halogen, the second a diazonium ion.

## Halogen Exchange

Halogen exchange using HF has been the route to aliphatic and benzylic fluorine-containing intermediates since the 1930s. Short chain chlorofluorocarbons, hydrofluorocarbons and benzotrifluoride derivatives are all made in this way, the exchange being activated by the proximity of the other halogen substituents. Chlorobenzene with HF gives no usable products but with potassium fluoride the reaction can be forced (Fig. 4). Where the leaving group is a halogen, the reaction is normally termed a halogen exchange or halex reaction. Suitable electron-withdrawing substituents can activate the substrate to attack

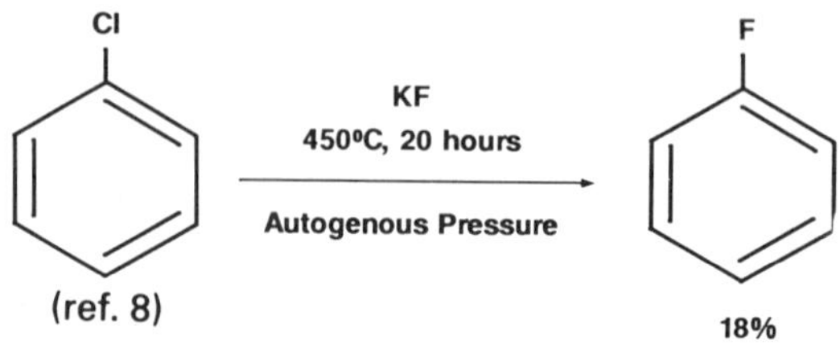


**Fig. 4.** Reaction of chlorobenzene with KF.

by nucleophiles, giving the general reaction depicted in Fig. 5. Apart from exceptionally powerful electron-withdrawing substituents such as diazonium or dinitro groups, elevated temperatures are needed for usable reaction times. Alkali fluorides are the usual source of fluoride ions to minimise pressure requirements: an aprotic high boiling point non-nucleophilic solvent helps to increase solubility. Typical solvents include DMSO, DMF, sulpholane, *N*-methyl pyrrolidone and others. As with nucleophilic substitution in general, inductive and mesomeric effects must be considered, inductive effects decreasing with distance from the withdrawing substituents and mesomeric effects being most powerful when exerted ortho or para (Fig. 6). Stepwise exchange is possible in some cases.

Multiple exchange is relatively easy without too many side reactions. It was first used in the mid 1960s at Avonmouth for making hexafluorobenzene and related products (Fig. 7). In particular, note the substitution pattern. Hexafluorobenzene had potential as an

Z: $NO_2$, Cl, CHO, CN, COX, $CF_3$, HCX

M: Na, K, Cs

Solvent: DMSO, Sulpholane, DMF, DMAC, NMP, DMID

Temperature: 140-240°C

**Fig. 5.** Halogen exchange—general reaction.

**Fig. 3.** Fluorine cell—new facility under construction.

solvents and their charge density leads to highly co-ordinated solvation spheres in others. An absence of competitive nucleophiles is normally required and two major approaches are used. The choice of route depends mainly on the ultimate substitution pattern required. The approaches differ in the leaving group, the first being an activated halogen, the second a diazonium ion.

## Halogen Exchange

Halogen exchange using HF has been the route to aliphatic and benzylic fluorine-containing intermediates since the 1930s. Short chain chlorofluorocarbons, hydrofluorocarbons and benzotrifluoride derivatives are all made in this way, the exchange being activated by the proximity of the other halogen substituents. Chlorobenzene with HF gives no usable products but with potassium fluoride the reaction can be forced (Fig. 4). Where the leaving group is a halogen, the reaction is normally termed a halogen exchange or halex reaction. Suitable electron-withdrawing substituents can activate the substrate to attack

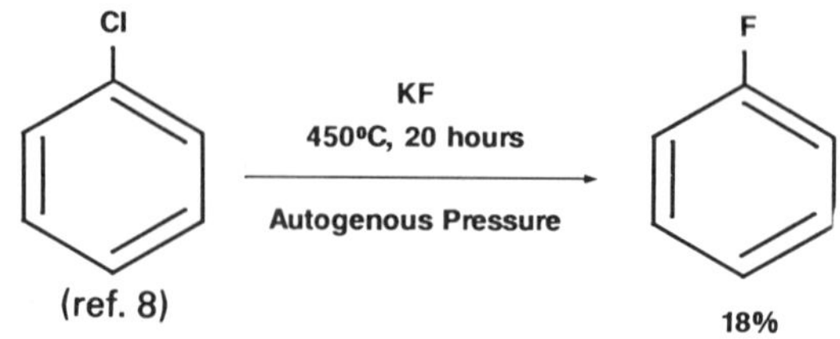


**Fig. 4.** Reaction of chlorobenzene with KF.

by nucleophiles, giving the general reaction depicted in Fig. 5. Apart from exceptionally powerful electron-withdrawing substituents such as diazonium or dinitro groups, elevated temperatures are needed for usable reaction times. Alkali fluorides are the usual source of fluoride ions to minimise pressure requirements: an aprotic high boiling point non-nucleophilic solvent helps to increase solubility. Typical solvents include DMSO, DMF, sulpholane, *N*-methyl pyrrolidone and others. As with nucleophilic substitution in general, inductive and mesomeric effects must be considered, inductive effects decreasing with distance from the withdrawing substituents and mesomeric effects being most powerful when exerted ortho or para (Fig. 6). Stepwise exchange is possible in some cases.

Multiple exchange is relatively easy without too many side reactions. It was first used in the mid 1960s at Avonmouth for making hexafluorobenzene and related products (Fig. 7). In particular, note the substitution pattern. Hexafluorobenzene had potential as an

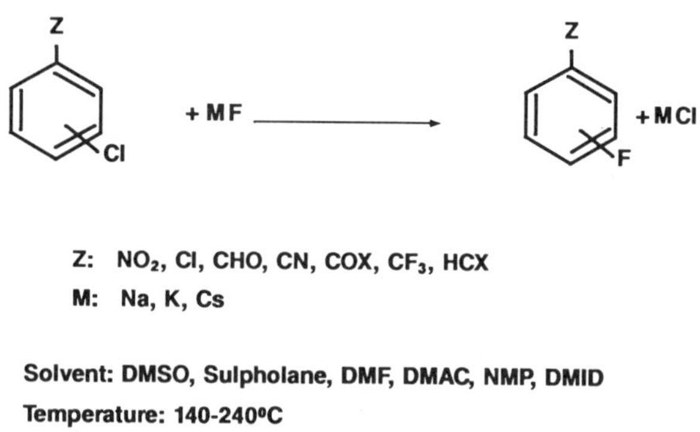


**Fig. 5.** Halogen exchange—general reaction.

successfully implemented such a strategy in its Visp plant, where it has built a comprehensive and efficient waste treatment unit, including an incineration plant. These capabilities have helped Lonza to enhance its image in the industry with customers who are concerned about their image in environmental compliance.

—*Accelerating technological pace.* The acceleration is fuelled by developments in the end user markets (e.g. the trend towards optical purity in pharmaceuticals and agrochemicals as well as the emergence of biotechnology as a production tool) and the increasingly competitive environment that is pushing participants to develop the lowest cost base.

Clear implications of this trend are the risk of process obsolescence and the need to master/obtain complementary technologies to sustain a long term position. A dramatic example was the development in 1985 by Enichem Sintesi in Italy of a very elegant stereoselective catalytic hydrogenation process for the production of L-phenylalanine—made obsolete even before the plant was operational—by a complete fermentative process (Aijinomoto).

## Strategic implications

The events taking place in the fine chemicals industry have fundamental implications for potential entrants and current participants if they are to maintain or develop a successful position in the industry.

For *existing fine chemicals producers* burning issues will be:

—To maintain or, if necessary, develop optimal size large enough to offer a viable product and technology portfolio and to represent a credible partner to the end user industries. However, large size alone will not guarantee success and the successful industry participant must maintain an agile and responsive organization to meet the customers' requests effectively and in time.

—To develop some form of competitive advantage, by offering either unique technological capabilities (e.g. challenging reactions, environmental compliance) or innovative services to the end users, including new partnership options (e.g. risk sharing in fixed assets investments).

—To redeploy corporate assets to maintain or improve competitive position by:

- increasing the project and activity selectivity and permanently reviewing the product portfolio critically;
- taking advantage of the available resources within the organization by, for example, setting up a multifunctional team to evaluate critically new leads.

For *commodity chemical producers* trying to establish a position in the industry the challenges will be:

—To evaluate possible acquisition candidates carefully and avoid paying unrealistically high prices for 'mediocre' companies.

—To integrate acquired companies in their typically more structured and bureaucratic environment, while maintaining the indispensable operating flexibility.

Past experience shows that the successful merger of two often dramatically different cultures is often problematic.

—To develop a coherent vision in fine chemicals activites thus avoiding dispersion of efforts in all directions (DSM, for example, has focused its effort to build on its strong position in D/L phenylglycine by acquiring Andeno and Plaine).

—To develop a clear understanding of the industry and give fine chemicals activities a fully fledged status, rather than treating them as side, low priority activities.

Examples demonstrate that the failure of commodity chemical producers to develop a viable business in fine chemicals is frequently associated with a lack of managerial commitment to this activity, and giving only second class citizenship to the organization responsible for handling these 'tiny–small volume' chemicals.

—To compete, not only on the basis of price but also on customer service, and develop a more customer responsive attitude. Producers must develop a market image of commitment to forge a long term relationship with potential customers.

For *integrated fine chemical users* such as pharmaceutical companies that are contemplating entering the industry, the key will be to avoid sawing off the branch they are sitting upon!

Some of these companies are facing the imminent expiration of patent protection for their key proprietary compounds and/or an overcapacity problem in their chemical production plant. They are increasingly considering supplying bulk active ingredients to generic producers.

These companies must develop an integrated approach to balance both drawbacks and advantages associated with such a venture from a corporate perspective, particularly to avoid cruel disappointments and major conflicts among the various parts of the organization.

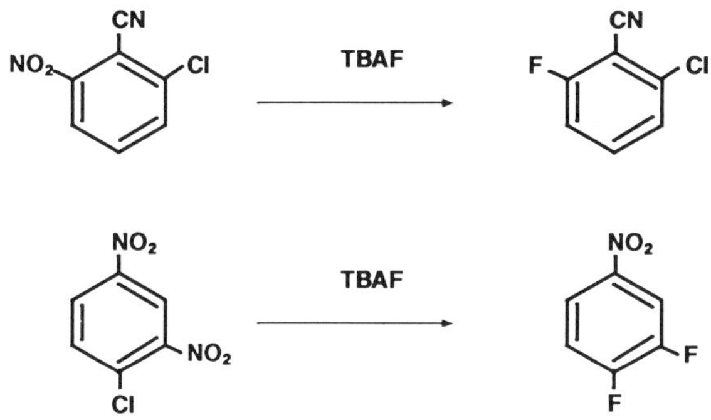


**Fig. 10.** Fluorodenitration.

## Diazotisation

The second major route to fluoroaromatic intermediates involves a diazonium salt. Production of diazonium salts from anilines and sodium nitrite has been a mainstay of the chemical industry since the last century. By using anhydrous hydrogen fluoride as the solvent, or by precipitating with fluoroboric acid, two important salts can be made (Fig. 11).

Thermal decomposition of these salts is the second major route to fluoroaromatics by fluoride substitution of the diazo group with

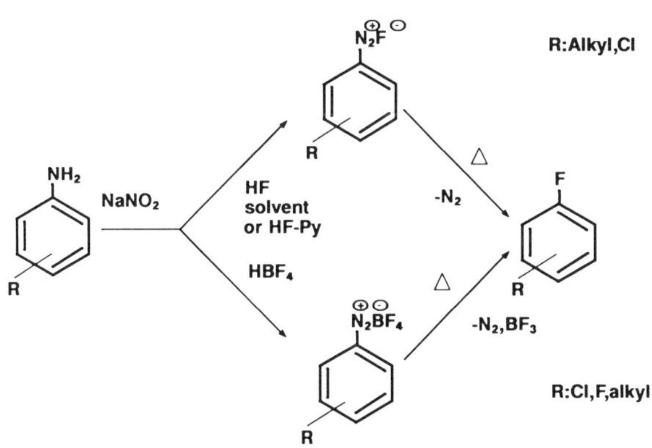


**Fig. 11.** Diazotisation–general reaction.

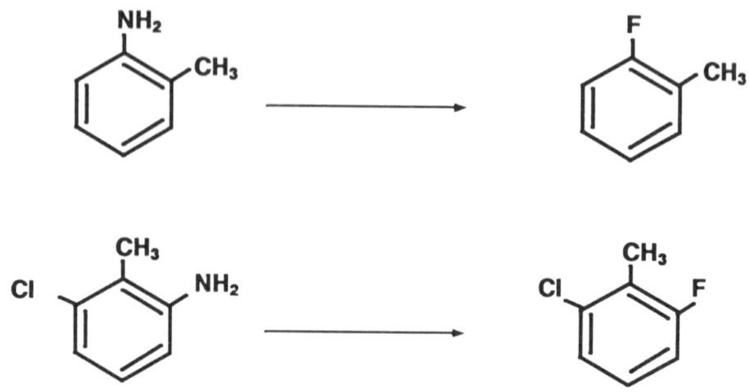

**Fig. 12.** Diazotisation in HF—typical intermediates.

consecutive nitrogen evolution. When carried out in anhydrous HF, the water from the reaction of the sodium nitrite on the aniline wets the HF. The fluoride value can be reclaimed by, for example, incorporating into aqueous HF or fluoroboric acid production. Boron trifluoride liberated during the decomposition of diazonium fluoroborate salts can be absorbed in aqueous HF, also generating fluoroboric acid.

Radical formation leading to coupling is a problem with any diazonium reaction and further problems are also caused when electron-withdrawing substituents are present, in particular, nitro, hydroxy, carboxy and ester groups. Single substitution by the HF solvent technique is particularly useful for making non-activated aromatics such as fluorobenzene, fluorotoluenes, chlorofluorotoluenes, etc. (Fig. 12). The fluoroborate method can be extended to some electron-withdrawing substituents or those insoluble in HF (Fig. 13).

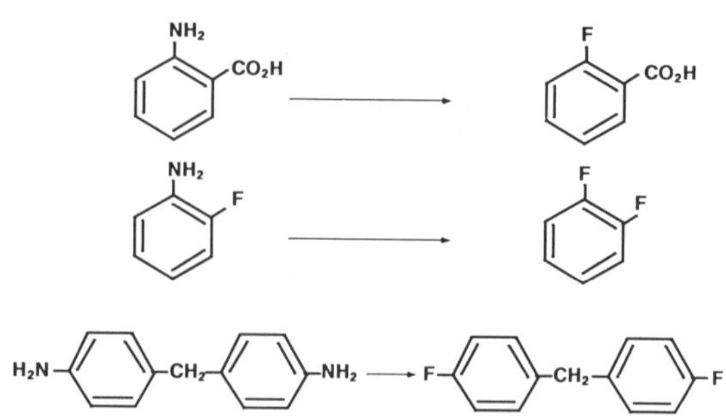

**Fig. 13.** Diazotisation via fluoroborates—typical intermediates.

Consecutive diazotisation and fluorination is also possible. Multiple fluorination in a single stage is difficult by either method, multiple diazotisation giving very poor yields.

A wide variety of inexpensive anilines are available, nitration/reduction being a standard procedure.

Choice of diazotisation method depends mainly on the thermal stability of the diazonium salt relative to the boiling point of the HF/salt mixtures. 4,4-Difluorodiphenylmethane and its oxidation product 4,4-difluorobenzophenone is an important example of this technique. Unfortunately the solids handling involved in fluoroborate production increases their inherent cost relative to HF solution.

HF diazotisation techniques have historically been batch or semi-batch but, recently, continuous plants for diazotisation in HF have been commissioned and further plants are under construction. The world market for fluoroaromatics made by diazotisation techniques is approximately 2500 tonnes with a total current and planned production capacity of 5000 tonnes.

Diazotisation techniques utilising HF as the source of fluoride have a significant raw material cost advantage and intermediates such as fluorobenzene are less expensive than, for example, simple fluoronitrobenzenes made by halogen exchange. However, the major consideration between a halex and a diazo approach is the choice of methods in the downstream chemistry and in particular the orientation of substituents.

Fluorine itself is electron-withdrawing and tends to dominate alkyl substituents. Conversely, fluoronitrobenzenes tend to substitute relative to the nitro group. Halex and diazotisation are, therefore, complementary in terms of substitution pattern.

## Electrophilic Substitution

Because it is the most electronegative element, creating a positive character on fluorine is very difficult. At very low temperatures diluted in a nitrogen stream, fluorine gas can be used and the substitution pattern suggests electrophilic attack (Fig. 14). However, the monofluorination yield is low and the multitude of products are difficult to separate. Additive oxidation is difficult to stop and, occasionally, the mixtures can explode.

Similar substitution patterns can be obtained using *N*-fluoropyridinium triflate (Fig. 15) but with much better yields.

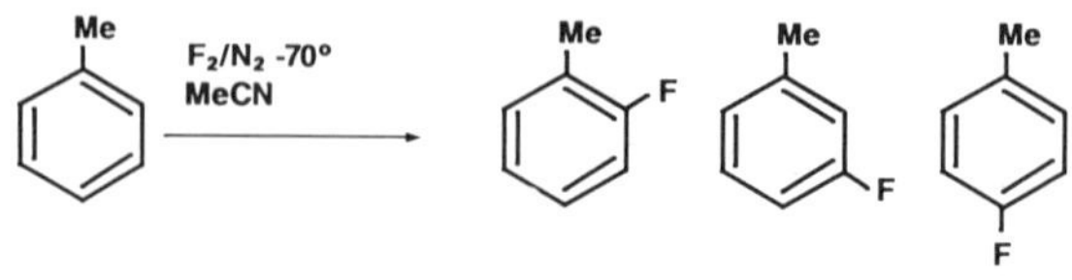


(ref. 7) Ratio 5 : 1 : 4

**Fig. 14.** Electrophilic substitution pattern using fluorine.

Fluorodemercuration of an organo-mercury salt using acetyl hypofluorite or the action of trifluoromethyl hypofluorite on aromatics with electron donating substituents are similar electrophilic fluorination reagents. Electrochemical fluorination has been attempted, but apparently unsuccessfully, the major products being polymers.[12,13]

All of these electrophilic reagents require fluorine for their preparation so that there is a major cost disadvantage over the nucleophilic approaches.

Compare the case of fluorine with chlorine where hundreds of thousands of tonnes of chlorobenzenes are made each year. Isomer separation benefits from the scale of operation and the ability to recycle chlorobenzenes.

Despite their inherent disadvantages, there is great potential for these electrophilic reagents in synthesising difficult substitution patterns or for fluorinating unreactive substrates. They will undoubtedly become more important in future years.

(ref. 9), (ref. 10) (ref. 11)

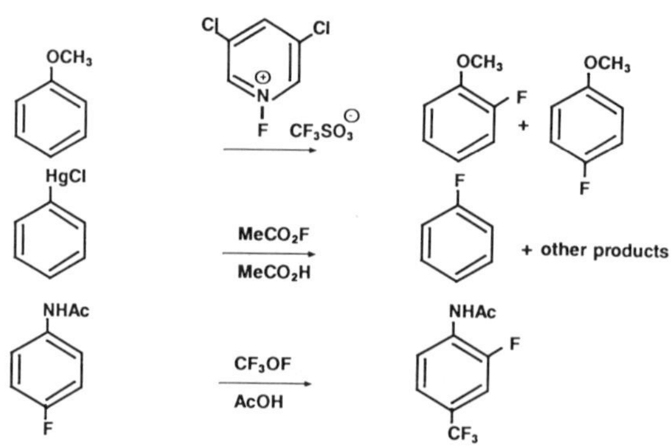


**Fig. 15.** Other electrophilic substitution patterns.

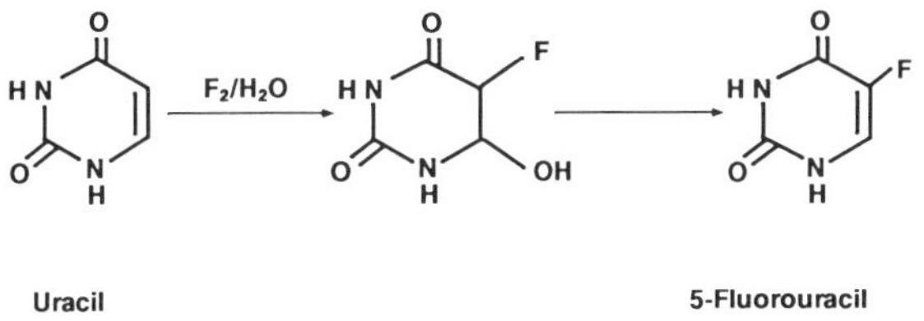


**Fig. 16.** Oxidative addition/reduction of uracil.

## Oxidative Addition

Fluoroaromatics are also accessible by oxidative addition of fluorine with subsequent reduction—some of the electrophilic substitution by-products are of this type. Fluorine itself has been used with pseudoaromatic systems such as uracil (Fig. 16), but with most classical aromatic substrates the reaction is difficult to control and high valence metal fluorides are normally used—silver, manganese or cobalt. The reaction probably involves one electron oxidation[14] followed by fluoride addition. Hydrogen substitution also occurs under most reaction conditions and, at the extreme, perfluorocarbons are formed (Fig. 17). These and less highly fluorinated intermediates can be reduced over iron at 400–600°C to generate highly fluorinated aromatics. However, the process is difficult to control and the multiplicity of products obtained are difficult to separate. The use of silver fluoride at low temperatures has been claimed to give small amounts of fluoroaromatics directly (Fig. 18). Hexafluorobenzene,

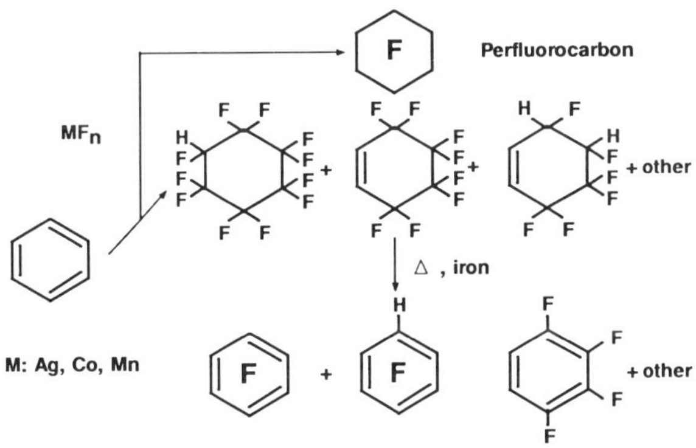


**Fig. 17.** Oxidative addition using high valency metal fluorides.

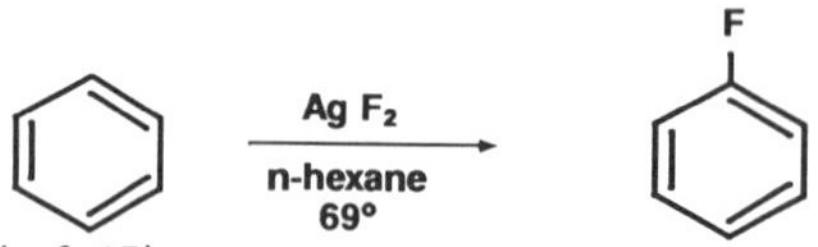


(ref. 15)

**Fig. 18.** Oxidative addition using silver difluoride.

1,2,3,4-tetrafluorobenzene and others were made at Avonmouth in the 1950s and 1960s using this method, but it was superseded by halogen exchange (see earlier) which proved more cost effective. The technique is currently used to make Flutec® perfluorocarbon liquids for vapour phase soldering, coolants and for gas and heat transport.

As with electrophilic substitution, this approach to fluoroaromatics uses fluorine gas and does have a cost disadvantage relative to the nucleophilic methods. Also, some non-carbon substituents are sensitive to the reagents and selectivity to specific intermediates is poor. However, for certain highly fluorinated intermediates, this method is the only viable route and some of the by-products may also have value of their own.

## Radical Substitution

Generating fluorine radicals is incredibly easy and is anomalous for its position in the periodic table. Table 4 shows typical dissociation energies for the halogens and their bond strengths to carbon. The difference between these two figures represents the quantity of heat liberated by these reactions. Fluorine is significantly different from the other halogens and dispersing this heat to control the reaction is the major problem. A variety of approaches have been tried such as jet reactors, substrates absorbed on to molecular sieves and even dissolving fluorine in a perfluorocarbon.[16–18] The molecular sieve work in particular was promising. All these methods, however, are still laboratory curiosities and even if scaled up would be unlikely to

**Table 4.** Halogen bond energies (kJ $mol^{-1}$)

| | $F_2$ | $Cl_2$ | $Br_2$ | $I_2$ |
|---|---|---|---|---|
| Bond dissociation energy | 159 | 243 | 192 | 151 |
| Bond energy to carbon | 450 | 339 | 280 | 226 |
| Difference | 291 | 91 | 88 | 75 |

**Fig. 19.** Other routes to fluoroaromatics—cyclopentadiene addition.

compare with the economics achieved with UV chlorination carried out on a scale of thousands of tonnes per year.

A bold approach and novel substrates would be needed to make the process commercially significant.

## Others

There are a large number of other routes to fluoroaromatics, but problems with preparing reagents, control of the reaction, scale-up costs, toxicology, separating intermediates, etc., make most of them impracticable on an industrial scale. Three routes, however, should be mentioned. These are the pyrolysis of cyclopentadiene with dichlorofluoromethane (Fig. 19), Deacon catalysts (Fig. 20) and the decomposition of arylthallium difluorides (Fig. 21).

The first two routes benefit from their use of inexpensive reagents and the second also has the potential for recycling by-products. The third can access certain difficult substitution patterns. Significant changes in the market would be needed to justify investment in these alternatives.

## Technologies Summary

Fluoride ions are relatively expensive and fluorine more so. The routes to fluoroaromatics rely to a great extent on preserving the 'fluoride value' both on economic and environmental grounds.

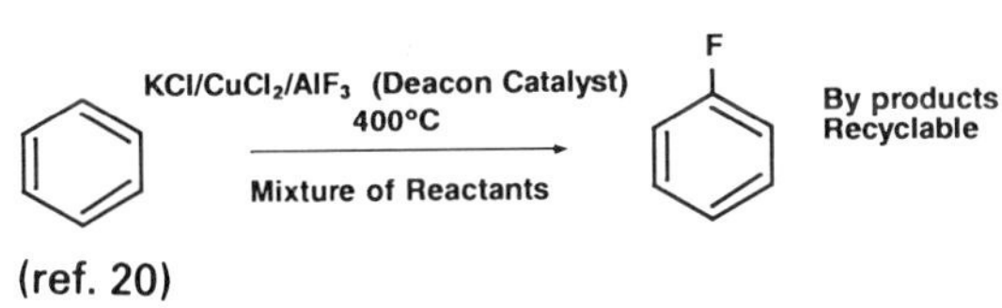


**Fig. 20.** Other routes to fluoroaromatics—Deacon catalysts.

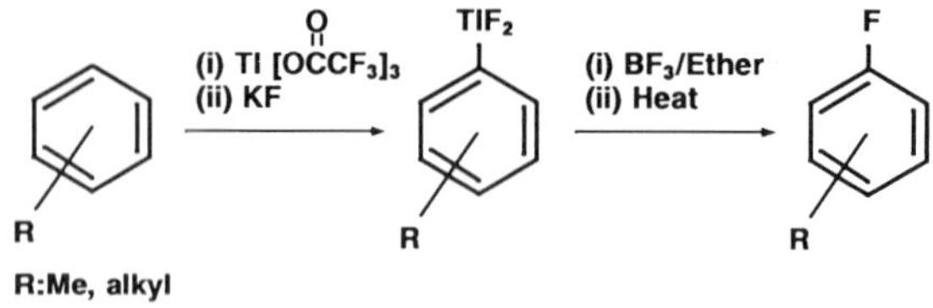


(ref. 21)

**Fig. 21.** Other routes to fluoroaromatics—arylthallium salts.

The nucleophilic approaches to fluoroaromatics: diazotisation (fluoroborate and anhydrous HF) and halogen exchange are the major industrial reaction pathways. Figure 22 shows some of the interactions between these and hydrogen fluoride. Hydrogen fluoride by nucleophilic substitution on aliphatics is used for making benzotrifluoride, hydrofluorocarbons and chlorofluorocarbons, the latter in turn being used for making fluoropolymers. Dilute aqueous HF from these reactions can be scrubbed/recovered to give useful by-products.

Reaction of HF with potassium hydroxide gives potassium fluoride, the main halex reagent. Potassium fluoride is also used in fluorine cells for the melt, HF being the consumable for making fluorine.

HF is used directly or via a $BF_3$ loop in the diazotisation approach. Pure $BF_3$ itself is normally made by reaction of a boron salt with sulphuric acid and fluorspar.

HF is the major primary reagent for making fluoroaromatics.

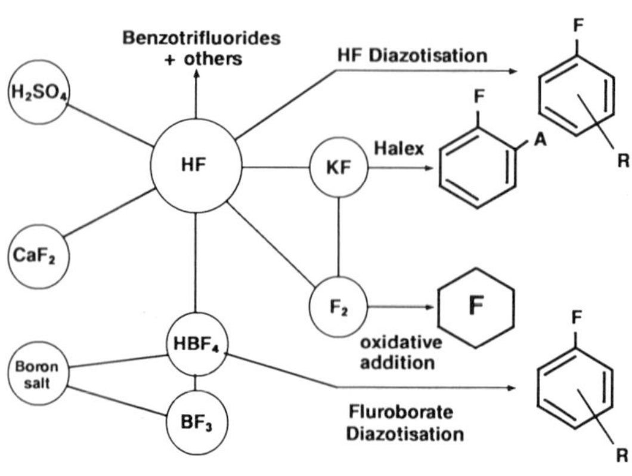


**Fig. 22.** HF in industrial fluoroaromatic processes.

Different technologies are required to access all the possible substitution patterns.

## Derivatisation

Once made, fluoroaromatic building blocks are fairly stable and many derivatives can be made using standard chemistry, but process engineering must always consider the consequences of fluoride ion release.

Fluorine is electron-withdrawing, giving, for example, ortho/para substitution on nitration. Fluoronitrobenzenes can be reduced and diazotised to give a range of products (Fig. 23).

Fluorotoluenes can be oxidised by UV chlorination to benzoic acid derivatives (Fig. 24). Some fluorobenzaldehydes are directly accessible from the chlorobenzaldehyde by halogen exchange, as are benzonitriles.

More recently, certain highly fluorinated intermediates have used a combination of fluorination techniques (Fig. 25).

The most cost effective approach to any particular molecular structure must consider precursor availability; scale of requirement; impurity profile, particularly isomers; substitution pattern for down-

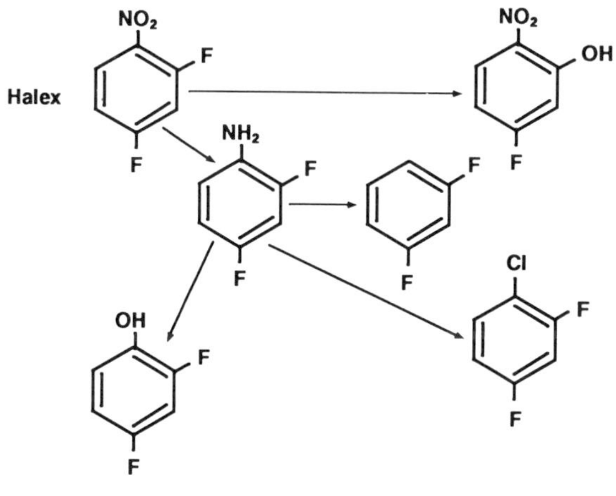


**Fig. 23.** Derivatives of fluoronitrobenzenes.

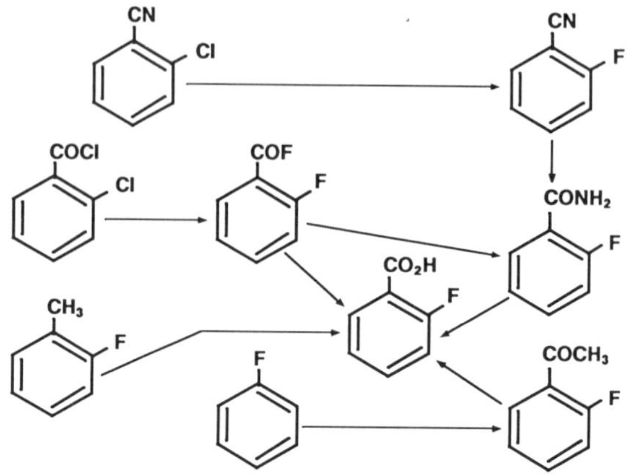


**Fig. 24.** Derivatives—fluorotoluenes.

stream reactions; etc. Fluorophenol isomers, for example, are made by different methods: 3-fluorophenol by fluoroborate diazotisation of 3-aminophenol, 2-fluorophenol by aqueous diazotisation of 2-fluoroaniline via halogen exchange and 4-fluorophenol either from 4-fluoroaniline or 4-bromofluorobenzene (Fig. 26).

In downstream chemistry, the major problem encountered is nucleophilic substitution of the fluorine itself. Although useful in, for example, the manufacture of PEEK polymers or Sanger's Reagent, the concomitant release of fluoride ions creates highly corrosive mixtures. Special equipment and techniques are needed to minimise this effect.

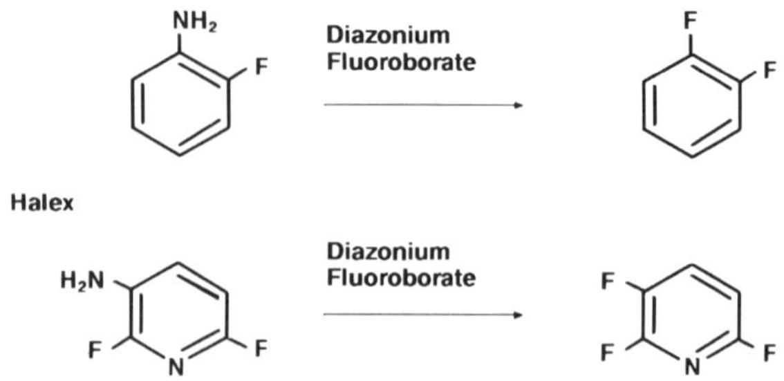


**Fig. 25.** Combinations of fluorination techniques.

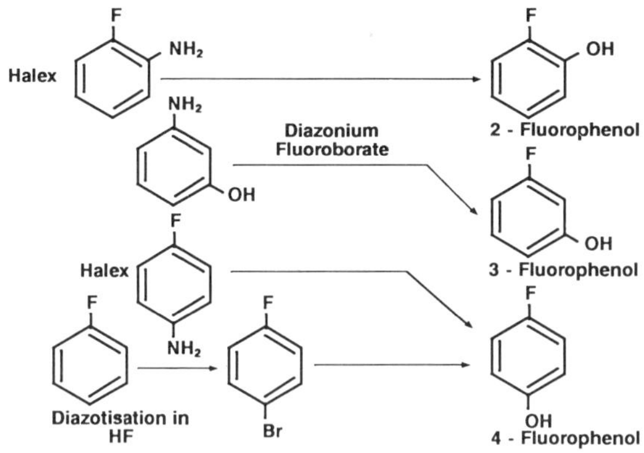


**Fig. 26.** Fluorophenols.

# Summary

Despite the intrinsic high cost of making fluoroaromatics, because of their enhancement of biological activity, resistancc to enzyme breakdown or in synthesis, their use continues to grow: the market is now in the region of 5000–6000 tonnes per year, having grown at an average 15–20% per year over the last 20 years. Increasing biological efficacy of fluorine-containing products, however, and the fact that some are now superseding earlier fluorine-containing products will reduce this growth rate.

HF will continue to be the dominant precursor unless some direct means of utilizing the fluoride value of calcium fluoride can be introduced.

Nucleophilic methods predominate but the electrophilic methods will undoubtedly assume more importance.

Finally, Fig. 27 illustrates various approaches to 2,3,4-trifluoroaniline. Halogen exchange on 2,3,4-trichloronitrobenzene stops at 3-chloro, 2,4-difluoronitrobenzene, which is fortunately a useful intermediate in its own right.

A variety of fluorination techniques coupled with a wide range of standard chemistry gives a multiple choice of routes, each of which has its own particular advantages.

The ISC Division of Rhône-Poulenc Chemicals has over forty years' experience in the development and application of fluorination tech-

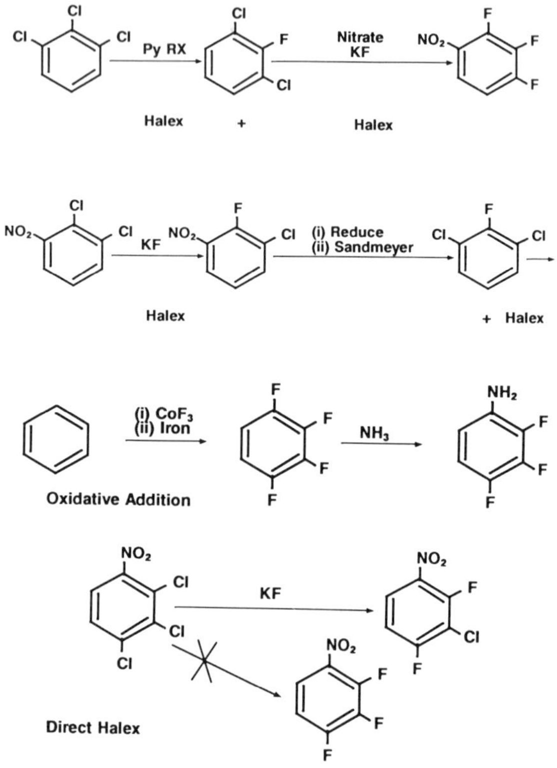


**Fig. 27.** Multiple choice technology routes to trifluoroanilines.

niques. With commercial scale manufacturing capability for anhydrous and aqueous HF, potassium fluoride, boron trifluoride chemicals and elemental fluorine all located at the Avonmouth complex, it is well placed to offer the full range of alternatives discussed, and in a reliable and cost effective way.

## References

1. *International Encyclopaedia of Chemical Science*. Van Nostrand, 1964.
2. Pauling, L., *The Nature of the Chemical Bond*. Oxford University Press, Oxford, 1967.
3. Stephen, H. & Stephen, T., *Solubilities of Inorganic and Organic Compounds*. Pergamon, Oxford, 1963.

4. Cottrell, T. L., *The Strengths of Chemical Bonds*. 1958.
5. Parsons, R., *Handbook of Electrochemical Constants*. Butterworth, 1959.
6. Cacace, F. & Wolf, A. P., *J. Amer. Chem. Soc.*, **100**(11) (1978) 3639–41; and Grakauser, V., *Intra-Science Chem. Rep.*, **5**(1) (1971) 85–104.
7. *Chemical Marketing Reporter*, 5 June 1989.
8. Yakobson, G. G. & Plantonov, V. E. *et al.*, *J. Gen. Chem. USSR*, **36** (1966) 1212124–7.
9. Umemoto, T., Onodera, K. & Tomita, K., 52nd National Meeting of Japanese Chemical Society, Abstract 1415, Kyoto, 1 April 1986.
10. Visser, G. M. & Bakker, C. N. M. *et al.*, *J. Org. Chem.*, **51** (1986) 1886–9.
11. US Patent 4 766 243 to Occidental.
12. Kokhanov, G. N. & Per'kova, S. A., *Electrokhimiye*, **3**(8) (1967) 977–80.
13. Masson, J. P., Devynck, J. & Tremillan, B., *Electroanal. Chem. and Interfac. Electrochem.*, **54** (1974) 232–6.
14. Bardon, Parsons & Tatlow, *Tetrahedron*, **28** 43–52.
15. Zweig, A., Fischer, R. G. & Lancester, J. E., *J. Org. Chem.*, **45** (1980) 3597.
16. Baltenburg, V. K., *Chem. Zeit*, **96**(2) (1972) 84–92.
17. Vasek, A. H. & Sams, L. C., *J. Flourine Chem.*, **2** (1972) 257.
18. Sams, L. C., Reamer, T. A. & Durrance, M. A., *J. Org. Chem.*, **43**(11) (1978) 2273–5.
19. US Patent 3 499 942, 10 March 1970; British Patent 1 130 263, 10 October 1968.
20. US Patent 4 012 453, 15 March 1977.
21. Taylor, E. C., Bigham, E. C., Johnson, D. K. & McKillop, A., *J. Org. Chem.*, **42** (1977) 362.
22. For example Roe, A., *Org. Reaction*, **5** (1949) 193 or Suschitzky, H., *Adv. Fluorine Chem.*, **4** (1965) 1, plus others.

# The industrial application of the Balz–Scheimann reaction

**John Regan***

*La Porte Organic Chemicals Division, Middlesbrough, UK*

The growing interest in the use of aromatic fluorine compounds in industrial applications during the last 20 years has led to increased efforts towards devising improved manufacturing methods.

Many ring-fluorinated aromatics are prepared via the diazotisation of substituted anilines with sodium nitrate in anhydrous hydrogen fluroide, followed by the *in situ* decomposition of the aryl diazonium fluoride, generally at temperatures between 30 and 120°C[1] depending on the stability of the diazonium salt. For example, fluorobenzene is prepared from aniline by this method:[2]

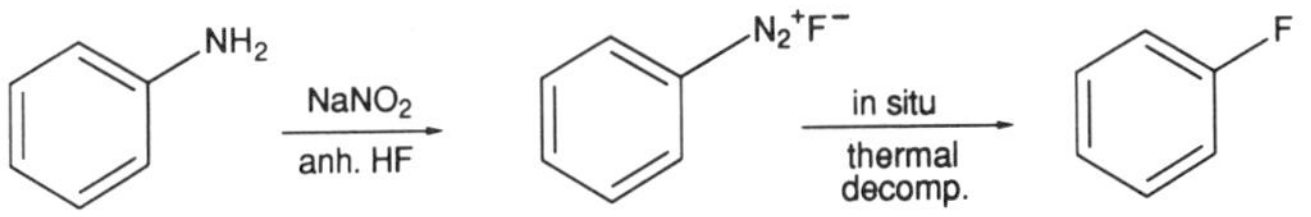


Variations on the above method include carrying out the diazotisation in the presence of ammonium fluoride or bifluroide ($NH_4HF_2$)[3] or with nitrosyl chloride[4] or nitrosyl fluoride[5] instead of sodium nitrite.

Haloheterocycles and haloaromatics which are activated by an

* Present address: Contrachem Ltd, 3 The Courtyard, Ochrelands, Hexham, Northumberland NE46 1SB, UK.

electron-withdrawing substituent in the ortho or para position can be converted into the corresponding fluorides via exchange fluorination by heating with an alkali metal fluoride, generally potassium fluoride. The reaction is normally carried out in a polar aprotic solvent.[6,7] 4-Fluoronitrobenzene has thus been prepared in 92·2% yield using dimethylacetamide as solvent:[7]

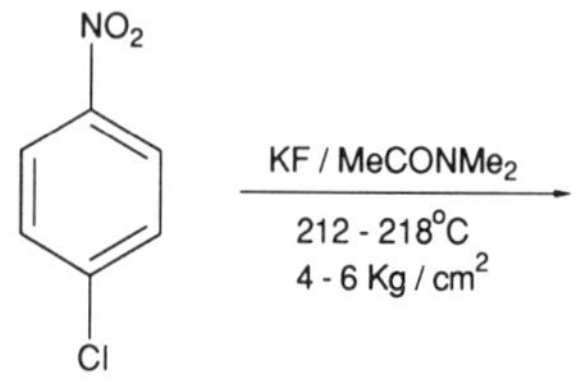


It is also possible, in some cases, to use molten reactants and no solvent, for example in the preparation of 2,4,6-trifluoropyrimidine[8] and in the preparation of 2-fluoropyridine.[9]

Some fluoronucleic bases and fluoronucleosides have been prepared by direct substitutive fluorination; for example 5-fluorouracil has been prepared from uracil and fluorine or trifluoromethyl hypofluorite.[10] When fluorine itself is used it is diluted with an inert carrier gas, usually nitrogen, and bubbled through a solution of uracil in a suitable solvent, for example a 78–81% yield of 5-fluorouracil is obtained by bubbling a 1:1 mixture of fluorine and nitrogen through an aqueous solution of uracil at 35–50°C:[10(b)]

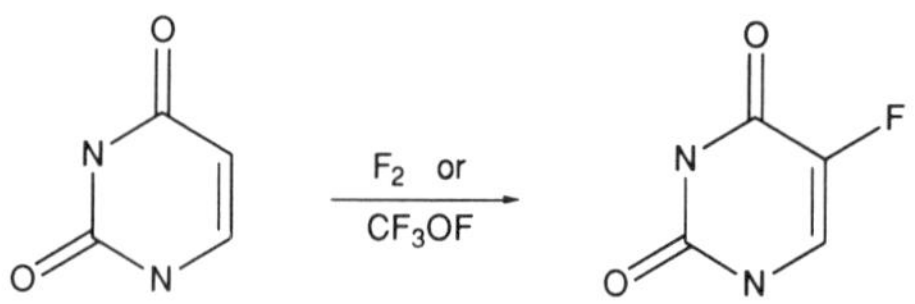


Up until now, apart from the work done at Wendstone Chemicals, only one attempt to adapt the Balz–Scheimann reaction (the isolation and subsequent thermal decomposition of an aryl diazonium tetrafluoroborate) for use on an industrial scale has been reported in the literature:

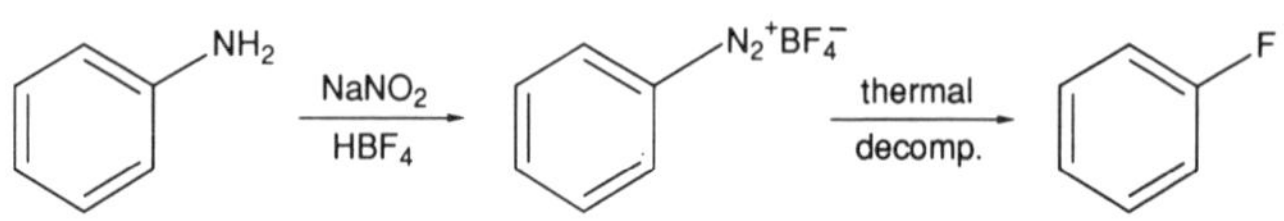

# Synthesis of side-chain fluorinated aromatics and heterocycles

Many industrial processes involve the fluorination of substituted benzotrichlorides using fluoride. Benzotrifluoride itself is manufactured by this method:[12]

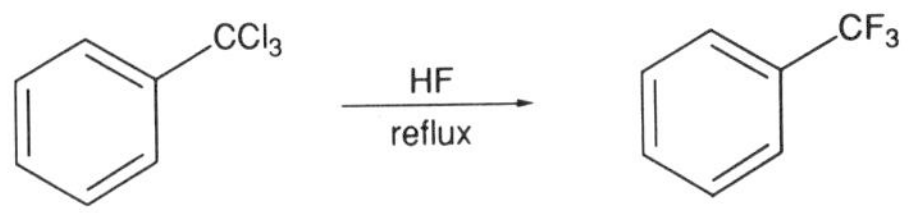


A method that has been employed in the synthesis of the drug 'Tolrestat' is the copper-assisted coupling of perfluoroalkyl iodides with aromatic halides:[13]

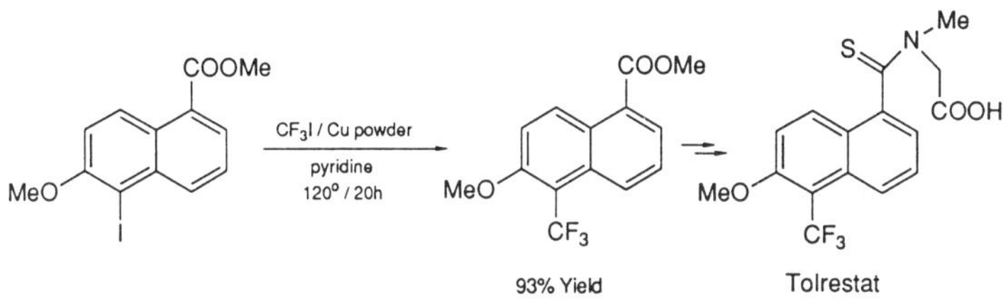


# Industrial uses of ring-fluorinated aromatic compounds and heterocycles drug manufacture

Electrophilic substitution reactions of fluorobenzene lead predominantly to para-substituted products. Acylation with 4-chlorobutyryl chloride gives 4-chloro-4′-fluorobutyrophenone which is a key intermediate in the synthesis of fluorobutyrophenone tranquilisers such as Haloperidol. Haloperidol is one of the four most widely prescribed anti-psychotic agents and is synthesised by the following route:[14]

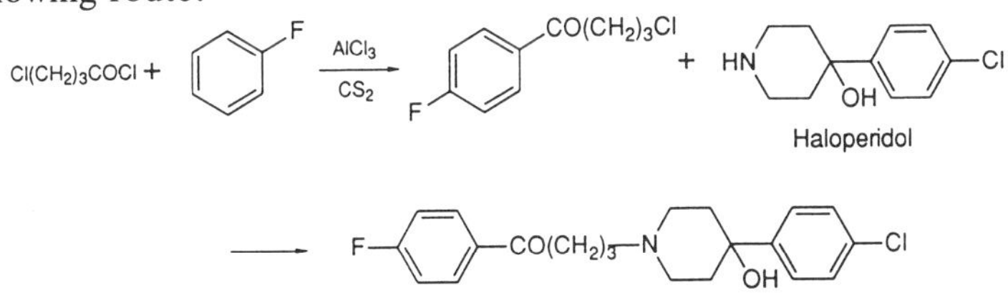

Acetylation with acetyl chloride or acetic anhydride in dichloromethane with aluminium trichloride as a catalyst gives 4-fluoroacetophenone, a principal intermediate in the synthesis of the anti-inflammatory agent Flazalone:[15]

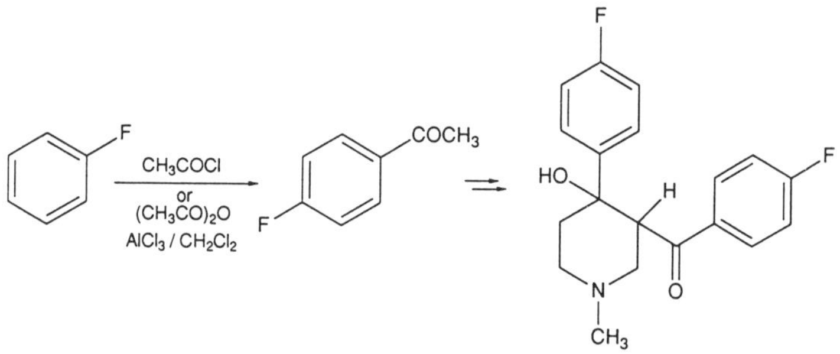


Flazalone

The antibiotic Flucloxacillin,[16] a semi-synthetic penicillin marketed in the United Kingdom under the name of Floxapen® by SmithKline Beecham Pharmaceuticals, is synthesised by the following route:

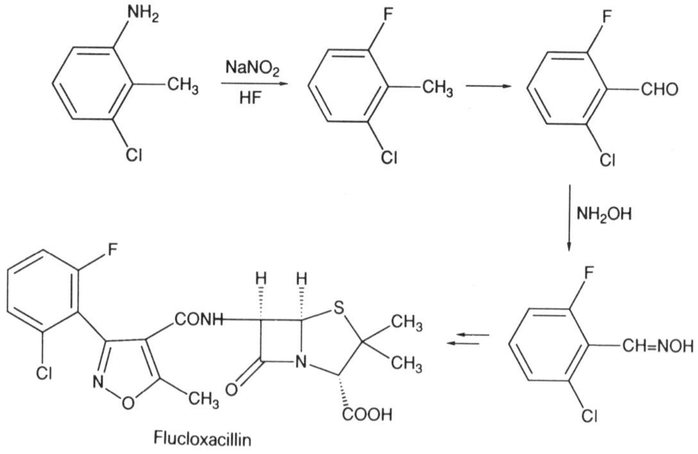


Flucloxacillin

One of the most effective of all hyponotic drugs is Flurazepam hydrochloride[17] (sold as Dalmane® by Hoffmann-La Roche) which is the fluoro-analogue of diazepam, the well-known drug Valium.

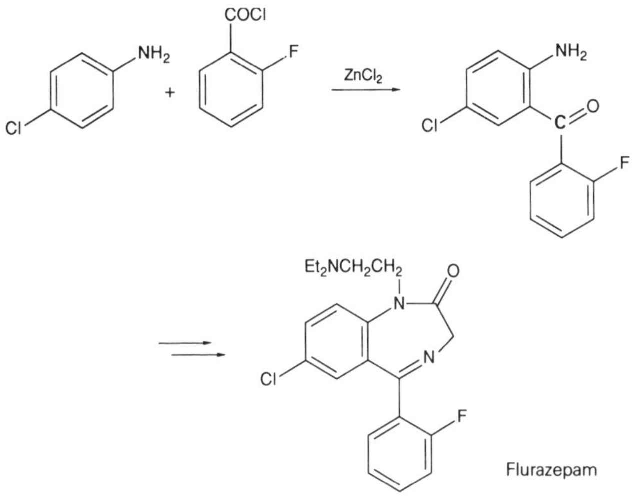


Flurazepam

The starting material, 2-fluorobenzoyl chloride, can be made by the following method:[18]

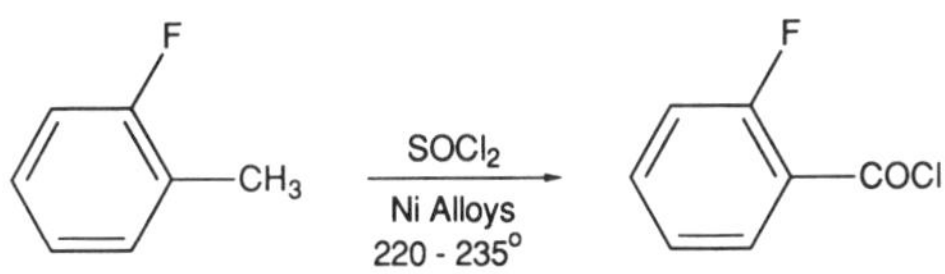


An extremely important analgesic and anti-inflammatory drug is Diflunisal (marketed as Dolobid® by Merck Sharpe and Dohme) which is an analogue of 1,3-difluorobenzene. It is a salicylaldehyde-type drug (like aspirin) but is superior to aspirin, and is widely used as an anti-inflammatory agent for osteoarthritis and post-operative analgesis.[20] It is synthesised by the following route:[21]

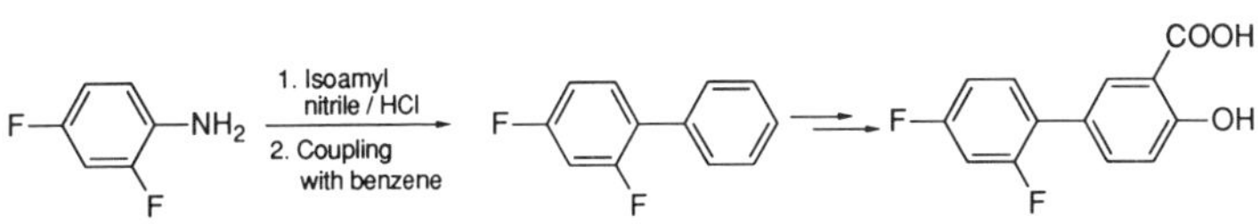

An improved method for the preparation of the starting material 2,4-difluoroaniline was patented in 1979:[22]

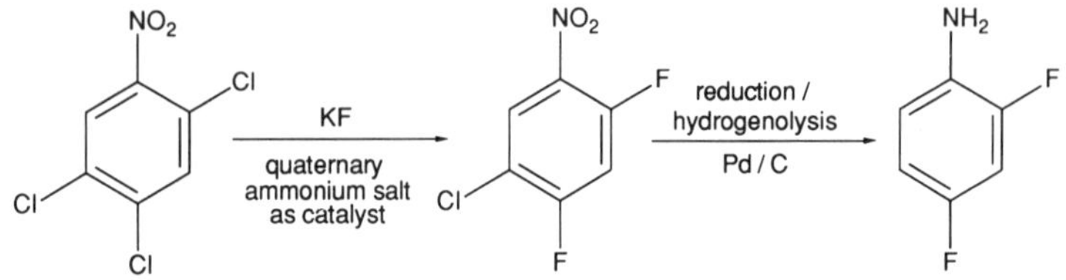


70.5% yield 48% yield

In the exchange-fluorination step only the chlorines in the activated ortho and para positions are substituted. 5-Fluoropyrimidine derivatives are very important in cancer chemotherapy. The most well known is 5-fluorouracil (see Ref. 10).

The Balz–Scheimann method for forming an aromatic fluoride composes of forming diazonium salt, converting this to the tetrafluoroborate salt and then thermally decomposing this. Boron trifluoride and nitrogen are liberated and fluorine goes into the ring. The reaction is represented below:

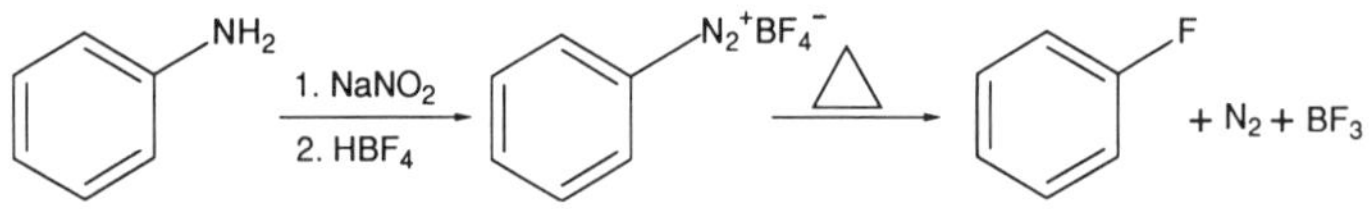


The stoichiometric equation representing the formation of an aromatic fluoride by both Balz–Scheimann reaction and by diazotisation in anhydrous hydorfluoric acid are represented below:

$$\underset{198}{NH_2{\cdot}Ph{\cdot}CH_2{\cdot}Ph{\cdot}NH_2} + \underset{2\times 98}{2H_2SO_4} + \underset{2\times 60}{2NaNO_2} + \underset{2\times 88}{2HBF_4} \rightarrow \underset{204}{F{\cdot}Ph{\cdot}CH_2{\cdot}Ph{\cdot}F} + \underset{2\times 120}{2NaHSO_4} + \underset{2\times 28}{2N_2} + \underset{2\times 68}{2BF_3} + \underset{4\times 18}{4H_2O} \quad (1)$$

Water, 2HF (recycled from $2BF_3$ to $2HBF_4$)

$$\underbrace{\underset{198}{NH_2{\cdot}Ph{\cdot}CH_2{\cdot}Ph{\cdot}NH_2} + \underset{6\times 20}{6HF} + \underset{2\times 69}{2NaNO_2}}_{456} \rightarrow \underbrace{\underset{204}{F{\cdot}Ph{\cdot}CH_2{\cdot}Ph{\cdot}F} + \underset{2\times 28}{2N_2} + \underset{2\times 62}{2NaHF_2} + \underset{4\times 18}{4H_2O}}_{456} \quad (2)$$

It may be seen from these equations that water is liberated in both reaction sequences. In the case of fluoridation in anhydrous HF, the diazonium salt decomposes spontaneously in the hydrofluoric acid medium. Hydrofluoric acid is, therefore, continuously diluted with water as the reaction proceeds, and this must be disposed of as an effluent. In the case of the Balz–Scheimann reaction, the diazonium salt is isolated as the tetrafluoroborate and dried. Boron trifluoride liberated during this decomposition is recycled. The 100% utilisation of the fluorine in the tetrafluoric acid is obtained. The reaction is thus free of difficult effluents, and this represents a major advantage.

## References

1. *Ullmann's Encykl Tech Chem,* 4th edn, Vol. 11. Verlag Chemie, Weinheim, 1976, p. 650.
2. German Patent 600 706, 30 July 1934; P. Osswald & O. Scherer (to Farbenind), Ca **28** (1934) 7260b.
3. US Patent 4 075 252, 21 February 1978; M. M. Boudakian (to Olin Corp.), CA **89** (1978) 6086s.
4. US Patent 2 563 796, 7 August 1951; W. H. Sherk Jr & G. R. Pellon (to Harshaw Chem Co.), CA **46** (1952) 9125b.
5. Belgium Patent 611 545, 29 December 1961 (to Allied Chem. Corp.), CA **57** (1962) 12380e.
6. British Patent Appl. 2 058 067, 8 April 1981; D. E. M. Wotton (to ISC Chemicals Ltd), CA **95** (1981) 150146c.
7. British Patent 1 360 327, 17 July 1974; S. Misaki *et al.* (to Daikin Kogyo) CA **81** (1974) 169273e.
8. US Patent 3 280 124, 18 October 1966; M. M. Boudakian & E. R. Shipkowski (to Olin Corp.), CA **66** (1967) 2582f.
9. (a) UK Patent 3 296 269, 3 January 1967; M. M. Boudakian (to Olin Corp.). (b) Boudakian, M. M., *J. Heterocychc. Chem.*, **4** (1967) 381.
10. (a) German Offen 2 014 216, 14 October 1971; I. L. Knunyants, L. S. German & N. B. Kaz'mine (to Acad. Sci. USSR), CA **76** (1972) 25310h.
    (b) US Patent 3 954 758, 4 May 1976; P. D. Schuman *et al.* (to PCR), CA **85** (1976) 123964e.
    (c) US Patent 3 846 429, 5 November 1974; S. Hillers *et al.*, CA **83** (1975) 97346s.
11. US Patent 2 705 730, 1949; J. D. Head (to Dow Chemical Co.), CA **50** (1956) 2666.
12. US Patent 3 136 822, 9 June 1964; L. J. Frainer (to Olin Corp.), CA **61** (1964) 4263g.
13. (a) Sestanj, K. *et al.*, *J. Med. Chem.*, **27**(3) (1984) 255.
    (b) Sestanj, K. & Bellini, F., Eur. Patent Appl. EP 104 078, 28 March 1984 (to Ayerst), CA **101** (1984) 23645y.
14. British Patent 881 893, 8 November 1961; P. A. J. Janssen, CA **57** (1962) 2199a.
15. US Patent 3 408 445, 28 October 1968, L. Levy & D. A. McClure (to Rexall Drug and Chemical Co.), CA **70** (1969) 47302k.
16. British Patent 978 299, 23 December, 1964; J. H. C. Naylor (to Beecham Group Ltd), CA **62** (1965) 10441b.
17. Archer, G. A. *et al.*, Belgium Patent 629 005, 28 August 1963 (to Hoffmann–La Roche), CA **60** (1964) 15896f; Neth. Patent Appl., 6401 335, 1964 (to Hoffmann-La Roche), CA **62** (1965) 5289a.

18. US Patent 3 681 452, 1 August 1972; C. S. Rondesuedt (to Du Pont), CA **77** (1972) 114 070t.
19. US Patent 3 755 427, 28 August 1973; S. S. Adams (to Boots), CA **79** (1973) 104 952j.
20. Banks, R. E. (ed.), *Organofluorine Chemicals and Their Industrial Applications*. Horwood, Chichester, 1979, p. 137.
21. Hannah, J. *et al.*, *J. Med. Chem.*, **21** (1978). 1093.
22. US Patent 4 140 719, 20 February 1979; R. J. Tull *et al.* (to Merck & Co. Inc.), CA **90** (1979) 186 558e.

## Bibliography

*Kirk-Othmer Encyclopaedia of Chemical Technology*, 3rd edn, Vols 10, 11. Interscience, New York, 1981.

Kleemann & Engel, *Pharmaceutische Wirkstaffe*, 2nd edn. Thieme Verlag, Stuttgart 1982.

*Organofluorine Chemicals and Their Industrial Applications* (Symposium), ed. R. E. Banks. Horwood, Chichester, 1979.

*Ullmann's Encykl Tech Chem*, 4th edn, Vol. 11. Verlag Chemie, Weinheim, 1976.

# Speciality fluoroaliphatic chemicals

**T. S. Johnson & H. Deger**

*Hoechst Celanese Corporation, Short Hills, New Jersey, USA*

## Introduction

The first large scale US research activity in the area of fluoroaliphatics began in the early 1930s. It was soon discovered that certain fluorocarbons make excellent refrigerants, possessing among other properties remarkable chemical inertness and thermal stability. Continuing emphasis is being placed upon research in this field today as the unique and valuable properties which fluorine can impart are being exploited to meet increasing performance demands for chemical products.

Fluorine is the most electronegative, and most chemically reactive, of all the elements. The fluorine atom is relatively small and it forms a strong covalent bond with carbon. In a highly fluorinated compound the fluorine atoms act to shield the carbon skeleton from attack. In addition to chemical and solvent resistance, and thermal stability, fluorination can impart UV stability and corrosion resistance. One of the properties utilized in the development of CFCs is that addition of enough fluorine to a compound renders it nonflammable.

One of the most well known aspects of highly fluorinated compounds is their interesting surface properties (e.g. the 'no-stick' properties of PTFE). Fluorinated compounds have good lubricating properties due to their low coefficient of friction and low surface tension. They are nonwettable by most substances and they promote water, oil, and soil repellancy.

Fluorinated compounds also have interesting electronic and optical properties. They possess low dielectric constants and low indices of refraction.

## Methods of manufacture

Generally speaking, there are four major methods of manufacturing fluoroaliphatic compounds. The method of choice depends upon the availability of raw materials, desired product(s), and economic viability of the process.

Electrolysis of organic substrates (e.g. hydrocarbons, carboxylic acids sulfonic acids, or tertiary amines) in an anhydrous hydrogen fluoride medium results in the exchange of hydrogen by fluorine. This process is commonly referred to as the Simons process. Some by-product formation occurs based on carbon–carbon cleavage and rearrangement of the carbon skeleton. The following equation illustrates the electrochemical fluorination of butyryl fluoride:

$$CH_3CH_2CH_2{-}\overset{\overset{\displaystyle O}{\|}}{C}{-}F \xrightarrow{e/HF} CF_3CF_2CF_2{-}\overset{\overset{\displaystyle O}{\|}}{C}{-}F \qquad (1)$$

The substitution of hydrogen by fluorine can also be accomplished using elemental fluorine. However, this reaction is extremely difficult to control and typically results in perfluorination and scission of the hydrocarbon chain (i.e. formation of $CF_4$). Better control can be obtained by employing fluoride salts of transition metals such as cobalt (III) fluoride. The cobalt (III) fluoride can be regenerated with elemental fluorine.

A second method of producing fluoroaliphatic compounds involves the halogen exchange of either chloride or bromide by fluoride within an organic substrate. One such reaction, known as the Swarts reaction, uses HF and can be carried out in either the liquid or vapor phase. Antimony halides are typical liquid phase catalysts whereas, in the vapor phase, a catalyst such as chromium oxyfluoride is used. The following equation illustrates the formation of chlorodifluoromethane:

$$CHCl_3 \xrightarrow{Sb(III/V)/HF} CHClF_2 \qquad (2)$$

Another way of carrying out the halogen exchange reaction employs fluoride salts of alkali metals (e.g. potassium fluroide, cesium fluoride). This method is more commonly employed with aromatic substrates activated toward nucleophilic substitution (e.g. production of *p*-fluoronitrobenzene).

A third method of producing fluoroaliphatic compounds is through dimerization, oligimerization, and polymerization of lower molecular weight species of fluoroalkanes and fluoroalkenes. A commercial example is the addition of iodopentafluoroethane to tetrafluoroethylene, yielding a homologous series of long-chain 1-iodoperfluoroalkanes:

$$CF_3CF_2I + nCF_2{=}CF_2 \xrightarrow{R^*} CF_3CF_2(CF_2 - CF_2)_nI \qquad (3)$$

The fourth method is the addition of hydrogen fluoride to alkenes. Hydrogen fluoride, like the other hydrogen halides, typically adds in Markovnikov fashion.

## Applications

Fluoroaliphatics find utility in a broad range of end uses. Their chemical and thermal stability combined with good heat transfer capabilities and low dielectric constant lead to applications in the electronics industry. Inert liquids based on perfluorinated polyethers, tertiary amines, or phenanthrene find utility in electronic cooling and testing applications as well as vapor phase soldering of printed circuit boards (e.g. Fluorinert™).

Higher molecular weight perfluoropolyethers find utility as high performance oils and greases (e.g. Krytox®) because of their ability to withstand temperature extremes as well as chemically aggressive environments. One specific application for these oils is in vacuum pumps used in the chemical vapor deposition process of the semiconductor industry.

The oleophobic and hydrophobic nature of fluoroaliphatics is useful in textile and leather treatment applications (e.g. Scotchguard®). In these applications there is typically a spacer between the perfluoro group and the 'anchor' (functional group through which polymerization or reaction with the substrate occurs). The purpose of the spacer is to prevent the perfluoro group from impacting the reactivity of the functional group.

The combination of low intermolecular interactions, high capacity for dissolving oxygen, and physiological inertness lends compounds such as perfluorooctylbromide utility in medical applications such as artificial blood. In addition, the presence of the bromine atom on this compound makes it useful as a contrast agent in various imaging techniques.

Fluoroaliphatics are useful in speciality surfactants in which the perfluoro group acts as a surface tension depressor and strong hydrophobe. Fluoroaliphatics are also used as modifiers for silicones, providing improved high temperature stability and hydrophobicity. The low refractive index of fluoroaliphatics makes them useful in optical fibers.

Other application areas for fluoroaliphatics include aqueous film forming foams, corrosion inhibitors, speciality coatings, release agents, catalysts, antifoams, and paper treatment.

## Hoechst Celanese Fluoroaliphatics

Hoechst Celanese, through its parent company Hoechst AG, offers a diverse line of fluoroaliphatic compounds derived from hexafluoropropylene (HFP) (Table 1) and tetrafluoroethylene (TFE) (Table 2).

From the HFP side, a series of inert liquids based on per-

**Table 1.** Selected HFP/HFPO derivatives and their applications

| Product | End use |
|---|---|
| Hostinert® fluids | Vapor phase soldering<br>Electronic testing (e.g. thermal shock, burn-in)<br>Chemical processing<br>Biotechnology |
| Hexafluoroisopropanol | Polymer optical fiber<br>Pharmaceutical/ag chemicals<br>Analytical reagent |
| R-227 | Niche replacement for CFCs |

**Table 2.** Selected TFE derivatives and their applications

| Product | End use |
|---|---|
| Perfluoroalkylbromide | Blood substitute<br>Medical contrast agent |
| Perfluorooctanoic acid | Emulsion for polymers<br>Leveling agent for photographic films |
| Fluorinated alcohols | Polymer optical fibers<br>Contact lenses |
| Perfluoroalkyl ethylene | Silicone modifiers |

fluoroalkane and perfluoropolyether structure are available. For example, Hostinert® 216 inert liquid has recently been introduced commercially for use in electronic and heat transfer applications. (The suffix refers to the boiling point in degrees centigrade.) Hostinert® is composed of a single, high purity chemical compound. It offers a distinct boiling point and exceptional stability in applications such as vapor phase soldering and electronic testing. Developmental Hostinert® liquids, all offering exceptional stability, are available with boiling points of 130, 175, 245, and 272°C.

Hexafluoroacetone, available as a sesquihydrate, and hexafluoroisopropanol are also available. Hexafluoroacetone is a useful intermediate. As an example of its utility within Hoechst, it is condensed with various activated aromatic compounds and subsequently used to produce a series of high performance fluoropolyimides. Hexafluoroisopropanol is useful as an intermediate for pharmaceuticals and agricultural chemicals, as a component in polymer optical fibers,and as an analytical solvent for polymers.

Hoechst has product R-227 under development. This is a compound with no ozone depletion potential and a useful vapor pressure curve which is expected to capture niche replacement markets for banned CFCs. Examples include refrigeration systems and medical sprays.

From the TFE side, a wide variety of extremely useful intermediates is derived. Under base-catalyzed conditions TFE can be inserted between the hydrogen and oxygen atoms of an alcohol or phenol to form the corresponding ether. An example of this reaction is the insertion of TFE into allyl alcohol to form the 2-hydroperfluoroethylether of allyl alcohol. This product, which is useful as a silicone modifier, is currently available.

In a free radical reaction TFE or HFP is inserted between a carbon and hydrogen atom of methanol to produce either 2,2,3,3-tetrafluoropropanol-1 or 2,2,3,4,4,4-hexafluorobutanol-1. These compounds, useful as components of optical fibers, are also available.

In addition, the reaction of TFE with iodoperfluoroethane, as described earlier, produces a series of telomer iodides as intermediates. From these are made available a series of perfluoroalkyl carboxylic acids (e.g. perfluorooctanoic acid), perfluoroalkyl halides (e.g. perfluorooctyl bromide), perfluoroaíkylethylenes, and perfluoroalkyl ethyl alcohols.

Perfluorooctanoic acid is an example of an available perfluoroalkyl carboxylic acid. The effect of the fluorine substitution in these compounds is an increase in acidity. Perfluorooctanoic acid is useful as a component in photographic emulsions.

Perfluorooctyl bromide, whose uses in medical applications were described earlier, is a developmental product available from Hoechst Celanese.

Perfluorohexylethylene is one example of the perfluorinated olefins currently available. The double bond offers a convenient site for further derivatization. The effect of fluorination of olefins such as these is to make the double bond electron-deficient, causing it to react preferentially with nucleophiles.

Perfluoroalkyl ethyl alcohols form the basis for a series of speciality surfactants and textile and leather treatment chemicals (Fluowet®) offered by the Speciality Chemicals Group of Hoechst Celanese.

In summary, the valuable properties that fluorine can impart are being utilized increasingly to meet tougher performance demands. Hoechst Celanase, through its parent Hoechst AG, offers a versatile array of fluorinated products. We are committed to bringing these products as well as the worldwide resources of Hoechst AG to bear in helping our customers solve unique problems.

## Bibliography

*Chemistry Today, The World Book Encyclopedia of Science,* World Book Inc, Chicago, 1986.

Lovelace, A. M., Rausch, D. A. & Postelnek, W., *Aliphatic Fluorine Compounds,* Reinhold Publishing Corporation, New York, 1958.

Silversten, Michael, Fluorination offers diverse applications for specialties. In *Performance Chemicals,* June 1988, pp. 18–22.

*Ulmann's Encyclopedia of Industrial Chemistry,* 5th edn, Vol. A11, 1988, pp. 349–375, VCH Verlagsgesellshaft, Weinheim, FRG.

# The role of chlorine in organic intermediates

**P. G. S. Field & K. Wall**

*P.G.F. Associates, Buxton, UK*

The synthesis and utilisation of organochlorine compounds is an area of major importance in the manufacture of organic intermediates. Chlorine has many characteristics that make it ideal as a substituent in organic intermediates. It is a cheap, highly reactive reagent which can be used to modify the properties of an end-product or provide a reactive site for subsequent substitution.

Chlorine can be introduced into a molecule in a number of ways, the choice of technique often determining the position of the chlorine in the molecule. In biologically active molecules, chlorine atoms are frequently introduced to impart a specific physiological function. For instance the chlorine atom in Diazepam (Fig. 1(i)) has been shown to be essential for the compound to be an effective tranquilliser. Figure 1 illustrates some examples of well established pharmaceutical and agrochemical end-products containing chlorine.

An interesting modification of the properties of a molecule is obtained by substituting hydroxyl groups in sucrose with chlorine atoms. The product sucralose (1′,4,6′-trichloro-galacto-sucrose) (see Fig. 2) has been developed as a high intensity sweetener. Regulatory approval is currently being sought for the widespread use of this product by Tate & Lyle. The chlorine atom not only contributes to the

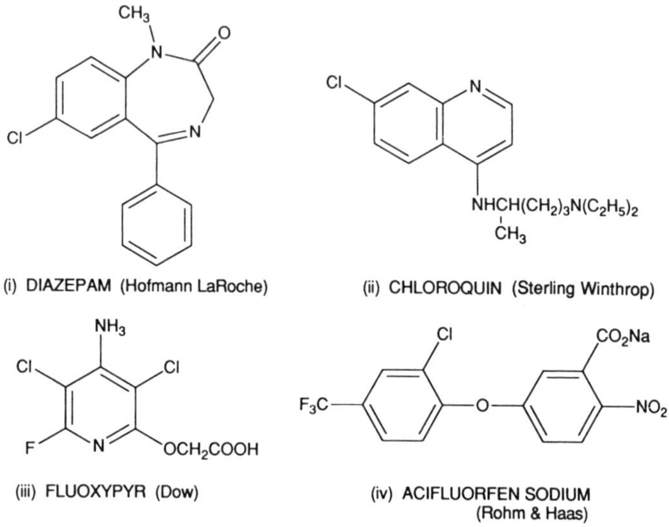


**Fig. 1.** Examples of important chlorine containing biologically active products.

increased sweetness but also prevents breakdown in the body, resulting in a calorie-free product..

Chlorine in both aliphatic and aromatic molecules can be substituted by simple nucleophiles such as —OH, —CN, —$NH_2$ and, with increasing importance, —F. The substitution by fluorine will form a disproportionate part of this paper in view of its implications in 'state of the art' chemistry. More complex moieties such as phenoxide ions and nitrogen containing heterocycles are also used to displace chloride. Two important examples are shown in Fig. 3.

In the anti-inflammatory drug Diclofenac (Voltarol) chlorine features both as a leaving group in the synthesis and appears in the final product. Its synthesis route is shown in Fig. 4. 2,6-Dichloroaniline, an intermediate for Diclofenac represents an interesting

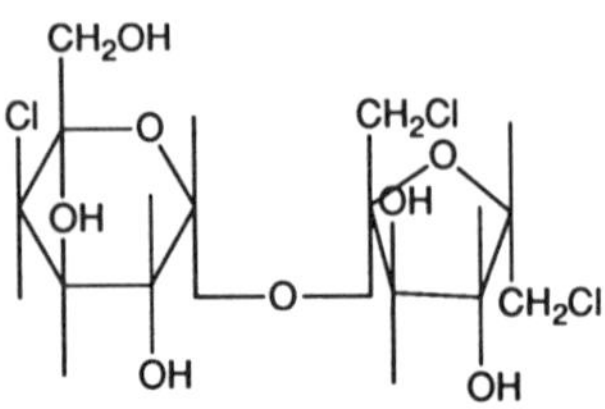

**Fig. 2.** Sucralose.

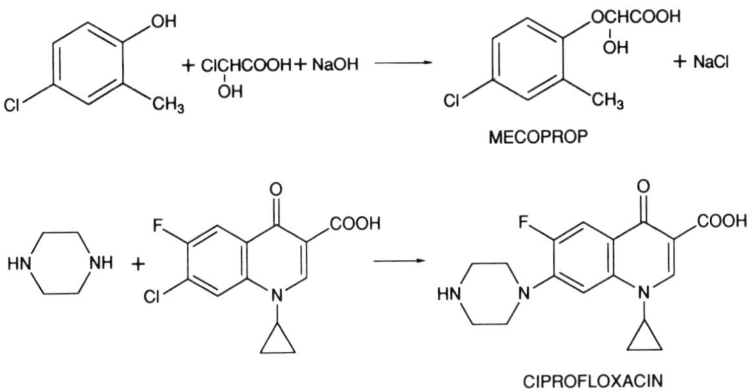


**Fig. 3.** Examples of chloride displacement reactions.

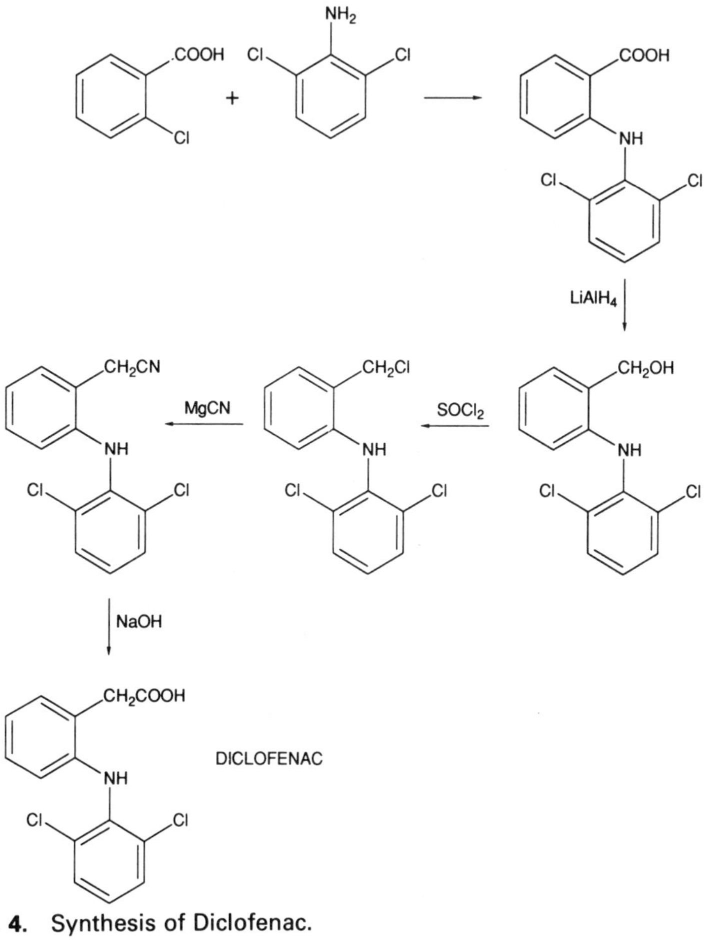


**Fig. 4.** Synthesis of Diclofenac.

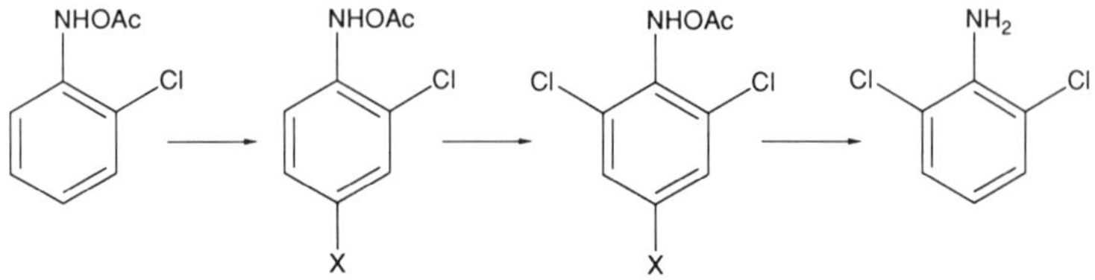


**Fig. 5.** Synthesis of 2,6-dichloroaniline.

example of the indirect synthesis of an aromatic chloroamine, where the required chlorine substitution pattern cannot be obtained by direct chlorination (Fig. 5).

A 'one-pot' process from 2-chloroaniline has been developed in which a blocking group is introduced into the 4-position after protection of the amine function. The blocking group is subsequently removed after chlorination in the 6-position using oxidative chlorination.

The side-chain photochlorination of toluene (Fig. 6) and nuclear chlorinated toluenes to benzyl chlorides, benzal chlorides and benzotrichlorides provides a route to key intermediates for a wide range of useful products. The chloride group in benzyl chlorides can be readily substituted by:

(a) Cyanide ion using sodium cyanide to product benzyl cyanides, the precursor to phenyl acetic acids. Phenyl acetic acids are used as intermediates in the production of Penicillins (Fig. 7).
(b) Hydroxide ion, to produce benzyl alcohols, e.g. technical benzyl alcohol used to produce benzyl esters for the fragrance industry and BP grade product for the manufacture of the artificial sweetener, aspartame. 2,4-Dichlorobenzyl alcohol produced from 2,4-dichlorotoluene is an important biocide (Fig. 8).

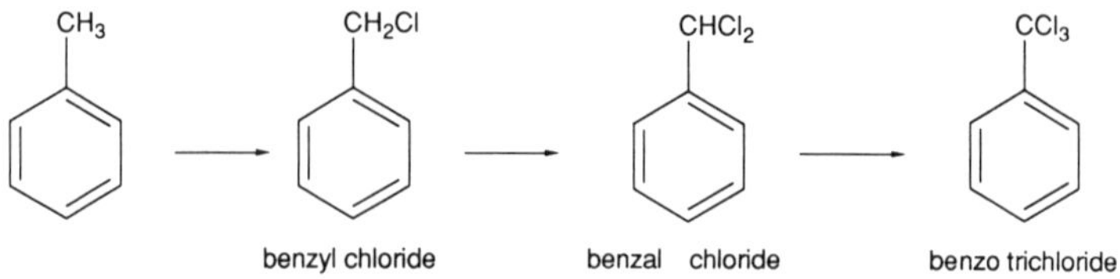


**Fig. 6.** Side-chain photochlorination of toluene.

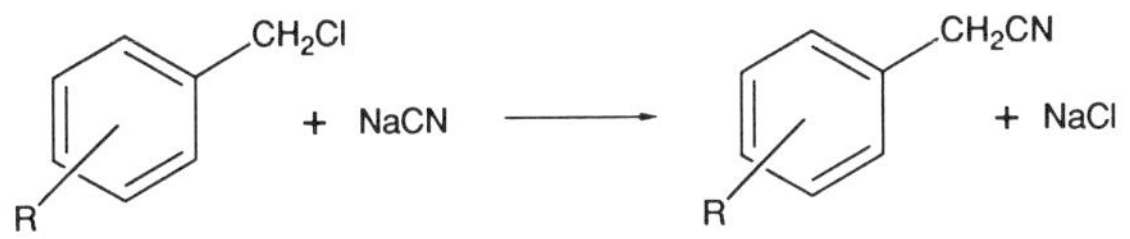


**Fig. 7.** Displacement of chloride ion by cyanide in penicillin G synthesis.

There are many other synthetically useful reactions of benzyl chloride to produce benzyl quaternary ammonium compounds: dibenzyl disulphide, used as an extreme pressure lubricant additive, *n*-benzyl aniline, a pharmaceutical intermediate, etc. The main application for benzal chloride was formerly its hydrolysis to benzaldehyde, but since the removal of the final traces of chlorine are difficult the use of benzaldehyde produced by this route for flavours and fragrances has decreased.

Benzotrichlorides finds two major applications:

(a) The reaction with a carboxylic acid to produce benzoyl chlorides and the corresponding acid chloride (Fig. 9). Acetyl, propionyl and benzoyl chlorides are produced by the reaction of benzotrichloride and the carboxylic acid.
(b) The conversion to benzotrifluorides, which has become extremely important and will be discussed in detail later.

Acid chlorides are widely used to produce industrially important products. Typical processes are:

1. Friedel–Craft processes with aromatic substrates to produce ketones, e.g. benzophenone from benzoyl chloride and benzene 2,4′-difluorobenzophenone from 2-fluorobenzoylchloride and fluorobenzene.

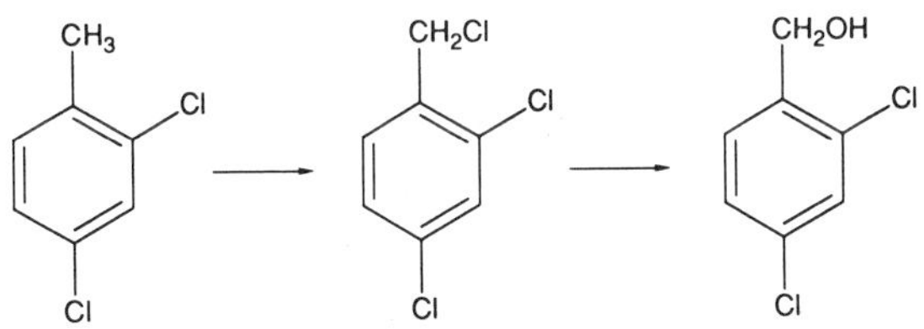


**Fig. 8.** Displacement of chloride ion by hydroxyl in 2,4-dichlorobenzyl alcohol production.

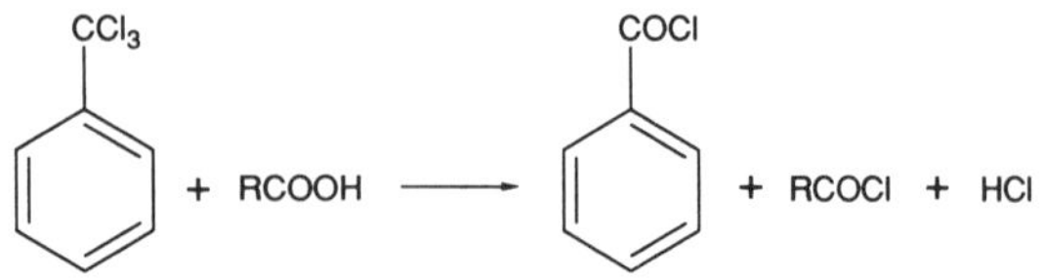

**Fig. 9.** Production of acid chlorides using benzotrichloride.

2. Reaction with hydrogen peroxide derivatives to produce diacyl peroxides and peroxy esters used as initiators in polymerisation processes.

The acid chlorides are reacted with an inorganic peroxide, usually sodium peroxide, to produce a diacyl peroxide:

$$\mathrm{R{-}\overset{\overset{\displaystyle O}{\|}}{C}{-}Cl + Na_2O_2 \longrightarrow R{-}\overset{\overset{\displaystyle O}{\|}}{C}{-}O{-}O{-}\overset{\overset{\displaystyle O}{\|}}{C}{-}R + 2NaCl}$$

or alternatively with the sodium salt of a hydroperoxide to produce a peroxy ester:

$$\mathrm{R{-}\overset{\overset{\displaystyle O}{\|}}{C}{-}Cl + R'{-}OO^{-}Na^{+} \longrightarrow R{-}\overset{\overset{\displaystyle O}{\|}}{C}{-}O{-}OR + NaCl}$$

Acid chlorides find a very large number of synthetic applications in the manufacture of pharmaceuticals and agrochemicals, the diversity which can be judged from Table 1.

**Table 1**

| Acid chloride | End-product |
|---|---|
| Acetyl | Acephate<br>Ibuprofen<br>Naproxen |
| Chloroacetyl | Alachlor<br>Oxadiazon<br>Diazepam<br>Dicyclomine hydrochloride |
| Dichloroacetyl | Diloxamide<br>Diphenamide<br>Chloramphenicol |
| Pivavoyl | Flamprop isopropyl<br>Metalachlor<br>Propachlor<br>Pivampicillin |
| Butyryl | Binapacryl |

Acid chlorides can also be produced by the reaction of the carboxylic acid with phosphorous trichloride, thionyl chloride or phosgene. The relative merits of each technology are often discussed but a consensus view is that the phosgene route, in the hands of experienced phosgenators, provides the purest products at the lowest cost.

Sulphuryl chloride is a specialised chlorinating agent used for regiospecific chlorination of phenols in the 4-position. The utility of this reaction can be judged from the products that can be produced from 4-chlorophenol (Fig. 10). Further chlorination of 4-chlorophenol to produce industrially significant intermediates can be effected using molecular chlorine (Fig. 11).

Sulphuryl chloride chlorination of methylated phenols produces highly effective biocides and disinfectants, the most prominent examples of which are *para*-chloro-*meta*-xylenol (PCMX), dichloro-*meta*-xylenol (DCMX) and *para*-chloro-*meta*-cresol (PCMC).

*N*-Chlorosuccimimide is used as a specific chlorinating agent in the production of intermediates for the manufacture of several pharmaceuticals, e.g. Methacycline, Doxycycline and Clomiphene. The

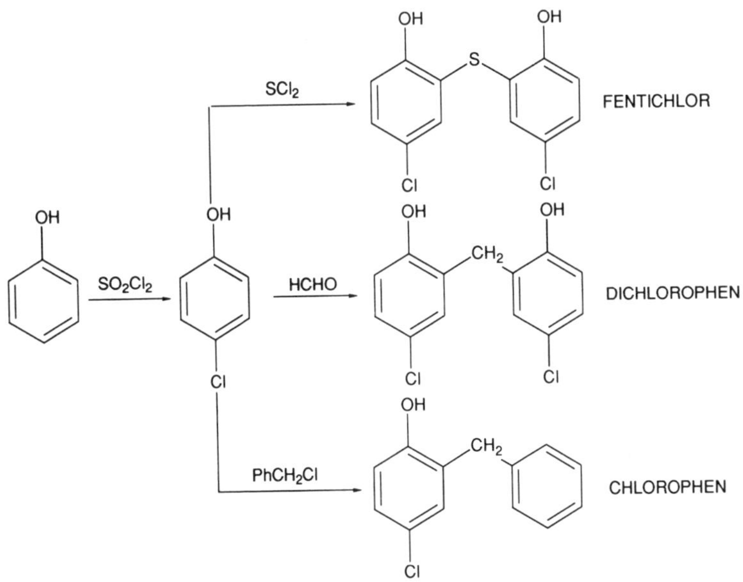


**Fig. 10.** Products derived from 4-chlorophenol.

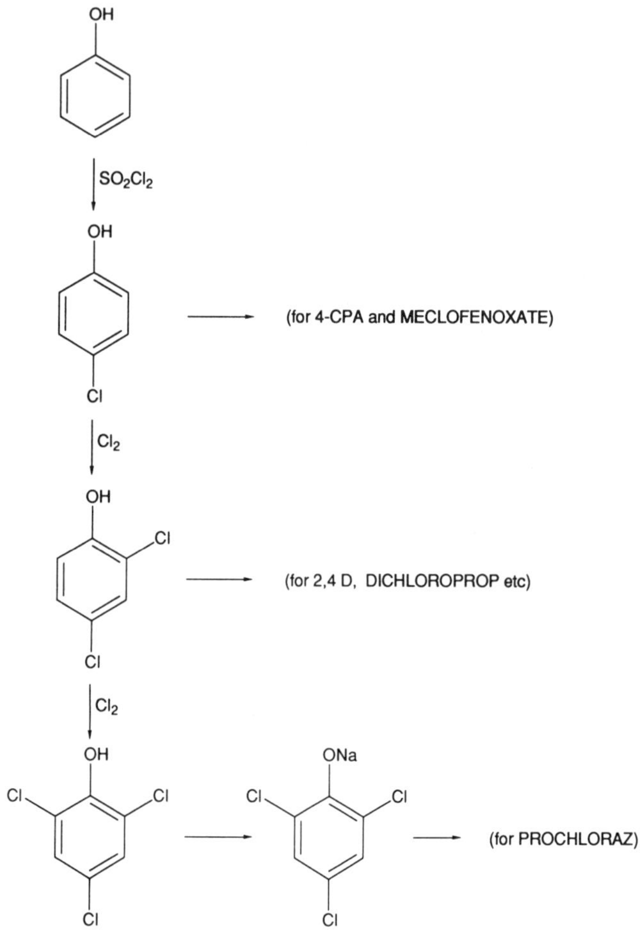


**Fig. 11.** Derivatives of polychlorinated phenols.

synthesis of Clomiphene (Fig. 12) demonstrates the versatility of chlorine in a multistage process.

Chlorine as the sulphonyl chloride group can be introduced into a molecule using chlorosulphonic acid. An example is the process used by Merck Sharpe & Dohme to produce the carbo-anhydrase inhibitor Dichlorphenamide, as shown in Fig. 13.

Alpha-chlorination of propionic acid with elemental chlorine in the presence of phosphorus trichloride (Hell–Volhard–Zelinsky reaction)

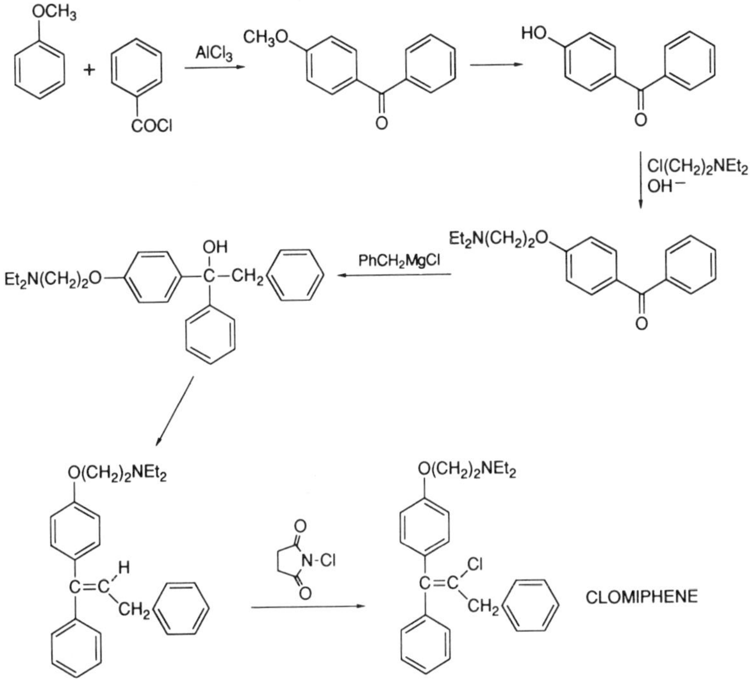


**Fig. 12.** Synthesis of Clomiphene.

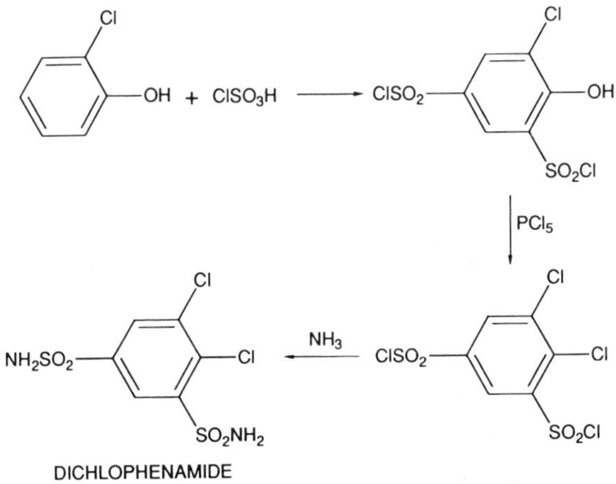


**Fig. 13.** Synthesis of Dichlorphenamide.

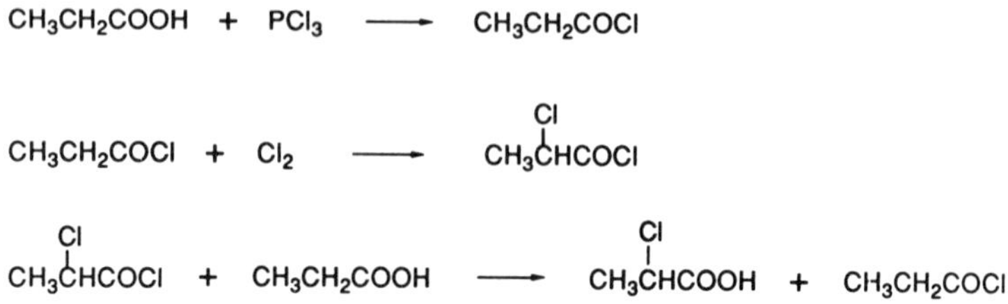

**Fig. 14.** $\alpha$-Chlorination of propionic acid.

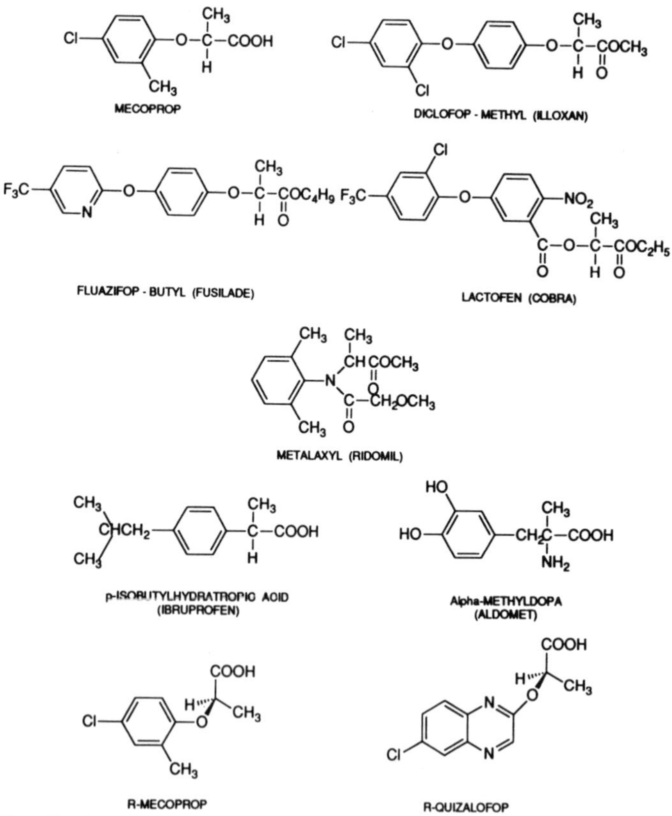


**Fig. 15.** Derivatives of 2-chloropropionic acid.

produces 2-chloropropionic acid (Fig. 14). The reaction produces both chiral forms as a racemic mixture. A number of agrochemical and pharmaceutical products derived from 2-chloropropionic acid has been commercialised (Fig. 15).

In several cases the desired biological activity has been shown to reside in only the derivatives of the (*S*)-2-chloropropionic acid. Optically resolved 2-chloropropionic acid is now commercially available which permits the manufacture of the optically pure active chiral form of several of the products.

## Chlorinated intermediates as precursors to organofluorine compounds

An often neglected topic when discussing the role of chlorine in organic chemistry in its intermediacy in the synthesis of many organofluorine compounds. Much research is currently focused on this subject with emphasis in three main areas:

1. aliphatic halogen exchange
2. aromatic nuclear halogen exchange (Halex reaction)
3. aromatic side-chain halogen exchange

The well known fully halogenated chlorofluorocarbons (CFCs) acquired and lost their reputation largely on the basis of their intrinsic stability. Therefore these products at first sight are hardly likely to stimulate much excitement as intermediates. This conclusion is valid but a number of related products are increasingly used and produced using variations of the technology originally used to produce the CFCs.

CFC-11 and CFC-12, now controlled under the terms of the Montreal Protocol, are made from carbon tetrachloride, whereas the ethane derivatives use either tetrachloroethylene or hexachloroethane as the raw material. The extent of *chlorine exchange* by reaction with HF is controlled by varying the hydrogen fluoride concentration, the contact time or the reaction temperature.

There are two commercially important partially halogenated products, hydrochlorofluorocarbons:

Chlorodifluoromethane—HCFC-22
1-Chloro-1,1-difluoroethane—HCFC-142b

HCFC-22 and HCFC-142b are produced from chloroform $CHCl_3$ and 1,1,1-trichloroethane, respectively, and anhydrous hydrofluoric acid:

$$CHCl_3 \rightarrow CHClF_2$$

$$CCl_3{-}CH_3 + 2HF \rightarrow CClF_2{-}CH_3 + 2HCl$$

either in the liquid phase, using Swarts-type catalysts based on antimony chlorides and fluorides, or in the vapour phase over chromium oxide, ferric chloride or thorium tetrafluoride catalysts.

CFCs and HCFCs serve as raw materials for the production of fluoroalkanes. Fluoroalkanes include the partially fluorinated alkanes or hydrofluoroalkanes (HFAs) and perfluoroalkanes (PFAs). HFAs are currently considered to be the favoured products to eventually replace the CFCs in aerosols, refrigerants, etc.

HCFC-22 is the raw material for tetrafluoroethylene (TFE). It is produced by the pyrolysis of difluorochloromethane (HCFC-22) at 800–1000°C:

$$2CHClF_2 \rightarrow CF_2{=}CF_2 + 2HCl$$

Hexafluoropropylene (HFP), a key monomer for the production of fluoropolymers and elastomers, is also used to produce a range of derivatives including perfluoropolyethers, hexafluoroacetone, fluoro-alcohols and hexafluoropropylene oxide. HFP is produced by the low-pressure pyrolysis of TFE.

The first HFA to be commercialised, HFA-134a, can be produced by the catalytic hydrogenation of CFC-114a, but purification is complex and hence expensive:

$$\underset{\text{CFC-114a}}{CF_3{-}CFCl_2} \rightarrow \underset{\text{HFA-134a}}{CF_3{-}CFH_2}$$

Another important aliphatic fluorine product depending on the availability of an appropriately substituted chloro compound is trifluoroacetic acid (TFA). The most important industrial process for the production of TFA, patented by both Rhône-Poulenc SA and Hoechst AG, is based on the sequence shown in Fig. 16. Trifluoroacetic acid serves as the raw material for the synthesis of its esters and salts and also 2,2,2-trifluoroethanol, an increasingly important product used for the manufacture of the chlorine containing vapour phase anaesthetic

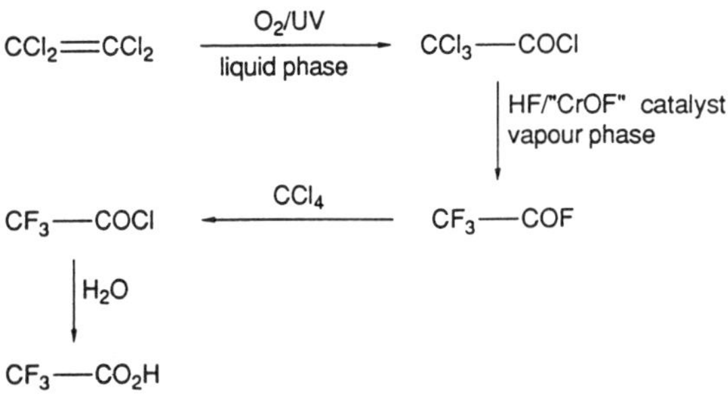


**Fig. 16.** Production of trifluoroacetic acid.

Isofluorane:

$$CF_3CH_2OH \xrightarrow[NaOH]{CHF_2Cl} CF_3CH_2OCHF_2 \xrightarrow{Cl_2} CF_3CHClOCHF_2$$

Many attempts have been made to produce 2,2,2-trifluoroethanol from $CF_3CH_2Cl$ (F-133) which is the key intermediate in the manufacture of Halothane. It is produced by the reaction of trichloroethylene with HF in the liquid phase and subsequent bromination:

$$CCl_2{=}CHCl \xrightarrow{\text{HF liquid phase}} CF_3{-}CH_2Cl \xrightarrow{Br_2} CF_3CHClBr$$

Halothane F-133

The second area of interest in halogen exchange is aromatic nuclear halogen exchange, the Halex reaction. The Halex reaction is a nucleophilic aromatic substitution ($SN_{Ar}$) in which a chlorine or bromine atom, activated by an electron-withdrawing group in the ortho or para position, is displaced by fluorine upon reaction with a metal fluoride in a polar aprotic solvent (Fig. 17). It is a commercially attractive manufacturing process which can be operated in multi-purpose equipment. It is, however, susceptible to certain hazards,

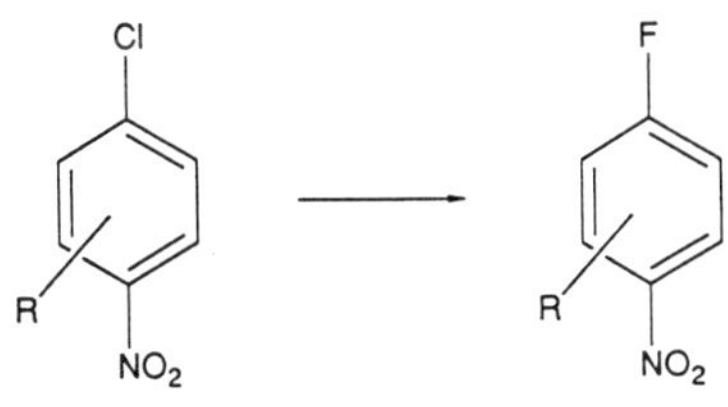


**Fig. 17.** A typical Halex reaction.

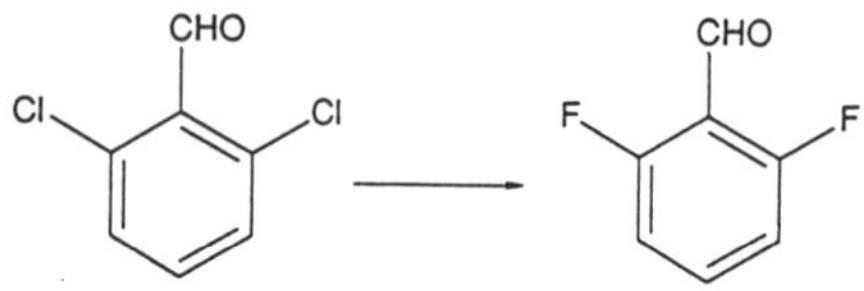


**Fig. 18.** 'Super' Halex synthesis of 2,6-diflurorbenzaldehyde.

particularly a tendency to 'runaway'. The generally severe conditions required have led to a great deal of patent activity in the search for milder reaction conditions.

Potassium fluoride offers the most readily available cost effective source of fluoride ion for the Halex reaction, although its physical form is of vital importance to reaction efficiency. Spray-dried potassium fluoride is much more efficient than the calcined product.

The scope of the Halex reaction is broad, because of the variety of substituents that can activate a halide substituent, ranging from one strongly activating nitro group inducing the exchange of *ortho*- and *para*-chlorines to a number of moderately activating substituents such as —CN, $CF_3$ and the pyridine nitrogen, effecting the replacement of a single chlorine atom.

A recent elaboration of the Halex reaction is the so-called 'super Halex' reaction, in which halogen exchange is effected without the presence of a nitro or nitrile group, to activate a chlorine substituent, but containing an aldehyde or acid chloride in the ortho or para position. Higher temperatures are used than in the normal Halex reaction.

Quaternary ammonium and phosphonium salts are effective catalysts in $S_N$-Ar reactions and lead to commercially attractive yields in, for example, the conversion of 2,6-dichlorobenzaldehyde to its difluoro analogue (Fig. 18). 2-Chlorobenzoyl chloride can be converted to 2-fluorobenzoyl chloride via the corresponding acid fluoride (Fig. 19).

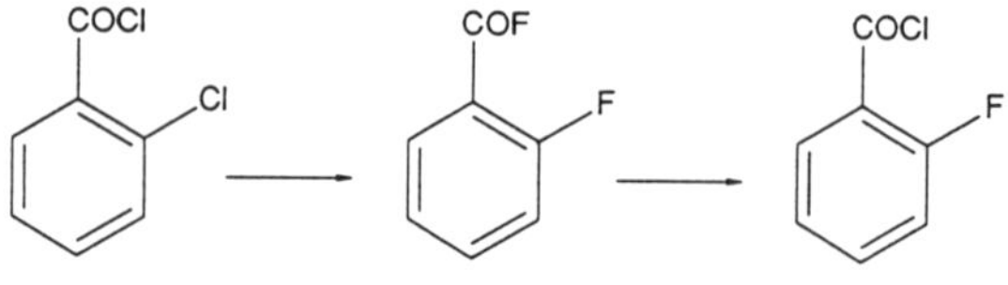


**Fig. 19.** 'Super' Halex synthesis 2-fluorobenzoyl chloride.

The fluorinated 4-quinolone antibacterials represent the most important single product category driving the exploitation of the Halex reaction, in terms of product diversity. However, in volume terms the insecticides based on 2,6-difluorobenzamide, e.g. Diflubenzuron, are much more important. A list of frequently encountered fluorinated products together with their chloro precursors is shown in Table 2.

The third important class of halogen exchange reactions, where the availability of the appropriately substituted chloro precursor is critical to the technology, is aromatic side-chain fluorination. Side-chain fluorination can be effected using very similar technology to that used for aliphatic halogen exchange.

Benzotrichloride can either be fluorinated at atmospheric pressure with gaseous hydrogen fluoride in the presence of a catalyst from the range $MoCl_4$, $MoCl_5$, $TaF_5$, $RhF_5$ $NbCl_5$, fluorosulphonic acid (Fig. 20). More commonly used processes are: liquid hydrogen fluoride

**Table 2.** Some of the more important nuclear fluorinated aromatics produced by the Halex reaction

| Chloroaromatic RM | Fluoroaromatic intermediate | End-product |
|---|---|---|
| 2-Chloronitrobenzene | 2-Fluoroaniline | Flurbiprofen |
| 4-Chloronitrobenzene | 4-Fluoroaniline | Flumequine |
| | | Flumazenil |
| | | Clinoril |
| 3,4-Dichloronitrobenzene | 3-Chloro-4-fluoroaniline | Flamprop isopropyl |
| | | Ciprofloxacin |
| | | Norfloxacin |
| | | Pefloxacin |
| | | Amifloxacin |
| 2,4-Dichloronitrobenzene | 2,4-Difluoroaniline | Diflunisal |
| | | Diflufenican |
| | | Temafloxacin |
| 2,3,5,6-Tetrachloro-phthalonitrile | 2,3,5,6-Tetrafluorophthalo-nitrile | Force |
| 2,6-Dichlorobenzaldehyde | 2-Chloro-6-fluorobenzaldehyde | Flucloxacillin |
| 2,6-Dichlorobenzonitrile | 2,6-Difluorobenzamide | Diflubenzuron |
| | | Teflubenzuron |
| | | Flufenoxuron |
| Pentachloropyridine | 2,4,6-Trifluoro-3,5-dichloro-pyridine | Fluoxypyr |

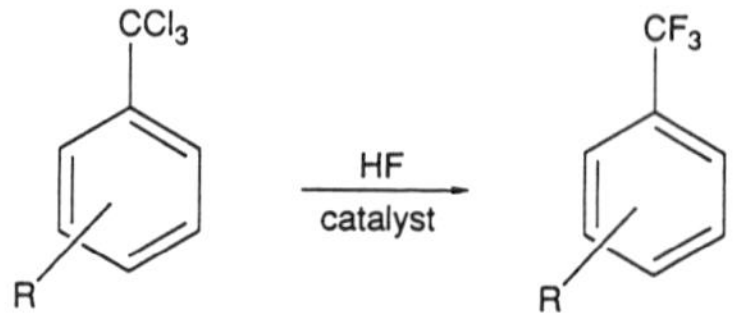

**Fig. 20.** Side-chain fluorination of benzotrichlorides.

under autogenous pressure in a batch system or gaseous hydrogen fluoride and countercurrent liquid substrate in a column heated to 90–120°C operating under pressure of 8–15 bar.

The total world demand for side-chain fluorinated aromatics is much larger than that for their nuclear substituted counterparts. There are, however, only four high volume products: benzotrifluoride, 4-chlorobenzotrifluoride, 3.4-dichlorobenzotrifluoride and trifluoromethylpyridines. These are listed in Table 3 together with their chloroaromatic precursors and end-products.

**Table 3.** Side-chain fluorinated aromatic products and their fluoraromatic precursors

| Side-chain chlorinated precursor | Side-chain fluorinated intermediate | End-product |
|---|---|---|
| Benzotrichloride | Benzotrifluoride | Flumeturon<br>Diflufenican<br>Flufenamic acid<br>Trifluperidol<br>Norflurazon<br>Fluridone<br>Flurochloridone |
| 4-Chlorobenzotrichloride | 4-Chlorobenzotrifluoride | Trifluralin<br>Benfluralin<br>Ethalfluralin<br>Fluchloralin |
| 3,4-Dichlorobenzotrichloride | 3,4-Dichlorobenzotrifluoride | Lactofen<br>Acifluofen<br>Oxyfluorfen<br>Fomesafen |
| 2-Chloro-5-trichloromethylpyridine | 2-Chloro-5-trifluoromethylpyridine | Fluazifop butyl<br>Haloxyfop |

## Conclusion

This short exposé on the role of chlorine in the field of organic chemical intermediates demonstrates that although it is labelled 'environmentally UN-friendly', with our current knowledge, its participation is indispensible in many instances, particularly in the area of organofluorine chemistry. We must then learn to treat it with the respect it deserves.

# The future for chlorophenolic biocides

**J. A. Parr, M. S. Irvine, C. A. Seldon & L. C. Payne**

*Nipa Laboratories Limited, Pontypridd, Wales, UK*

## Introduction and review

### History of Chlorinated Phenolic Biocides (Chlorophenolics)

From the middle of the last century, when Lister's work on the antimicrobial activity of phenol was documented, the activity of this compound and its derivatives has been extensively studied.[1] The halogenation of phenolic compounds and, in particular, the production of chlorinated derivatives results in an increase of antimicrobial efficacy and a reduction in toxicity.

Several microbiocides of commerical importance are currently produced. These include:

*Para*-chloro *meta*-xylenol (PCMX)
Dichloro *meta*-xylenol (DCMX)
*Para*-chloro *meta*-cresol (PCMC)
*Ortho*-benzyl *para*-chlorophenol (Chlorophen)
5,5′-Dichloro-2,2′-dihydroxydiphenyl methane (Dichlorophen)
2,2′-Dihydroxy-5′,5′-dichlorodiphenyl monosulphide (Fentichlor)
Monochloro *ortho*-phenyl phenol (MCOPP)

Additionally, several commercial products are available based on mixtures of benzylated chlorophenols and chloroxylenol solutions in methanol or isopropanol. Chlorophenolics, therefore, are still of great industrial and commercial significance. Over 8000 tonnes per year are produced by 15 manufacturers world wide. This is a market worth £40 million per year as active agent and £200 million when formulated into the 120 million litres per year of disinfectant products.[2]

## Applications of Chlorophenolics

These biocides enjoy the advantage of a long history and the benefits of many years of research concerning activity, toxicity and compatibility with other compounds and are in widespread use throughout the world. Applications range from pharmaceutical, e.g. PCMX as antiseptic and as active agent in bactericidal hand soaps; cosmetic, e.g. PCMX, PCMC, chlorophen as preservatives for toiletries, barrier creams, etc.; industrial, e.g. PCMC as a biocide for metal working fluids, leather finishing; household, hospital and agricultural, e.g. chlorinated xylenols as active ingredients for low cost disinfectant formulations. Specialised products such as Fentichlor are important in water treatment applications including the control of *Legionella* sp. Dichlorophen is extensively used as a industrial preservative and for agricultural/horticultural applications of plants and certain food crops. Chlorophenolics are therefore used both as preservatives, i.e. to maintain the integrity of raw materials or finished products from the effects of microbial biodegradation, or as microbiocides to actively reduce the microbial contamination on surfaces, industrial process machinery and animal tissue through disinfection and antisepsis.

## Antimicrobial Activity of Chlorophenolics

Chlorophenolics have a broad spectrum of antimicrobial activity. Specific compounds exhibit antibacterial activity (both Gram positive and Gram negative bacteria); antifungal activity (including yeast, moulds and dermatophytes) and antiviral activity. Suitably formulated products are useful for the majority of household, institutional and agricultural disinfectant applications.[3]

The key to the successful use of chlorinated phenolics relies on the sensible formulation of individual compounds in reliable finished products suitable for the intended end use. Careful selection of active agents can produce a very broad spectrum of antimicrobial activity at relatively low costs. Certain chlorophenolics, e.g. chlorophen, are generally accepted to be the best actives for control of *Mycobacterium tuberculosis* which remains an important pathogen in the hospital environment and less well developed areas of the world. Mixtures of chlorophenolics can produce synergistic activity and result in an even wider spectrum of activity. Successful products of the Hycolin type, for example, consist of well proven mixtures of phenolic and chlorinated phenolic biocides and are well known for their consistent, broad spectrum activity and hold a significant part of the hospital and laboratory disinfectant market.

## Toxicity of Chlorophenolics

Investigation shows that despite popular belief the majority of these compounds are of relatively low toxicity compared to biocides of other types. Typically, they exhibit $LD_{50}$ (oral rat) values 5 to 10 times lower than quaternary ammonium compounds (source: *Register of Toxic Effects of Chemicals*). PCMX, for example, has an $LD_{50}$ greater than 5000 mg/kg bodyweight compared to dimethyl alkyl benzyl ammonium chloride, which has an $LD_{50}$ of 300 mg/kg.[2] Many formulations based on chlorinated phenolics are registered for use throughout the world, including Europe and the USA.

## Antimicrobial Properties and Applications of Specific Chlorophenolics

### *Chlorophen*

This compound is extensively used as an active agent in disinfectant formulations for application in household, hospital, industrial and farming environments. It is used at 3–10% (as active agent) in hospital disinfectants and disinfectant cleansers which in final use could be further diluted, typically 1:40 to 1:120, with water. Chlorophen is used either alone or in combination with other active agents and has demonstrable activity against *Mycobacterium tuberculosis* and several viruses including *Herpes simplex, Vaccinia,* Adeno type 2 and HIV. It is also active against Avian influenza virus which causes infections of commercial importance in intensive chicken and turkey farming. The US Department of Agriculture has approved a

5% chlorphen based formulation for its Avian Influenza Eradication Program.

Table 1 illustrates the antimicrobial spectrum of chlorphen, and Table 2 summarises its toxicity profile.

More recently, chlorophen has been successfully incorporated as an

**Table 1.** The antimicrobial activity of chlorophen

| Test species | MIC value (ppm) Nipacide BCP | Phenol coefficient | Killing dilution (as 100% active) |
|---|---|---|---|
| Bacteria | | | |
| *Staphylococcus aureus* | 25 | 200–300 | 1:30 000 |
| *Staphylococcus epidermis* | 50 | | |
| *Streptococcus pyogenes* | | 220 | 1:22 000 |
| *Pseudomonas aeruginosa* | 5000 | | 1:1000 |
| *Klebsiella pneumonia* | 2500 | | |
| *Enterobacter cloacae* | 250 | | |
| *Proteus vulgaris* | 10 | | |
| *Serratia marcescens* | 4000 | | |
| *Salmonella enteritidis* | 100 | | |
| *Escherichia coli* | 250 | 160 | 1:16 000 |
| *Salmonella typhi* | | 200 | 1:22 000 |
| *Mycobacteria* | | | |
| *Mycobacterium tuberculosis* | | 160 | 1:16 000 |
| *Mycobacterium smegmatis* | | 220 | 1:22 200 |
| Yeasts | | | |
| *Candida albicans* | 50 | | |
| *Saccharomyces cerevisiae* | 100 | | |
| *Saccharomyces rouxii* | 50 | | |
| Moulds | | | |
| *Mucor racemosis* | 50 | | |
| *Aspergillus niger* | 250 | | 1:20 000 |
| *Aspergillus glaucus* | 100 | | 1:4000 |
| *Penicillium expansum* | 100 | | |
| *Penicillium notatum* | 25 | | |
| *Stachybotrys atra* | 25 | | |
| Virus | | | |
| *Herpes simplex* | | | 1:6400 |
| *Vaccinia* | | | 1:6400 |
| Asian influenza | | | 1:6400 |
| Adeno type 2 | | | 1:6400 |

(These data were collected under different test protocols and are thus not necessarily comparable. Most tests were based on 5% active solutions with 10 minutes exposure times. Stated results indicate levels of 100% active chlorophen.)

**Table 2.** Toxicity summary of chlorophen

| | |
|---|---|
| $LD_{50}$ Acute oral, rat | 4147 mg/kg |
| $LD_{50}$ Acute percutaneous, rat | >2000 mg/kg |
| $LC_{50}$ Dietary, Bob White quail & Mallard duck | >5620 mg/kg |
| $LC_{50}$ Bluegill Sunfish, 96 hour acute | 0·33 mg/l |
| $LC_{50}$ Rainbow trout, 96 hour acute | 0·72 mg/l |
| Mutagenicity | Negative salmonella assay and negative Chinese hamster |

Source: BTP Cocker Chemicals, internal communication.

active agent for glass washing fluids where its low taint and odour is an advantage over other chlorinated phenolics. Typically formulations containing 2% chlorophen plus anionic surfactants have proven reliable and produced approximately 99·99% kill in 30 seconds contact time (Nipa Laboratories, internal communication). Chlorophen has also been used as the basis of hard surface cleansing/sanitising products.

### *Phenosept*

This is an example of a chlorophenolic based formulation specially produced for use in antiseptic liquid soaps, surgical handwash preparations and teat dip products (for the prevention of bovine mastitis). It consists of a synergistic combination of PCMX and phenoxy-2-propanol and illustrates how chlorophenolics may be formulated with other active agents to produce toxicologically acceptable products with rapid and broad spectrum microbial kill.

In tests of their bactericidal activity, liquid soap formulations containing Phenosept were subjected to bactericidal challenge tests. Three portions (20 g) of each sample were inoculated with 0·2 ml pure culture of test bacteria at high initial inoculum levels. The time taken to achieve a total bactericidal effect was determined by standard plate counting techniques.

Table 3 illustrates the results obtained compared with the liquid soap control (with no antiseptic).

The antiseptic activity of Phenosept was also compared to chlorhexidine gluconate, using the method of Lowbury and Lilly.[4] Figure 1 illustrates the mean percentage reduction obtained after one and six washes. The activity of Phenosept against *Streptococcus agalactiae*, one of the causative microorganisms of bovine mastitis, was investigated in microbial challenge tests. Results are shown in Table 4.

**Table 3.** Bactericidal activity—challenge test

| Test species | Initial inoculum level of colony forming units per ml | Time taken to achieve total lethal effect in | |
|---|---|---|---|
| | | Liquid soap + 2% Phenosept | Liquid soap control |
| *Escherichia coli* | $1{\cdot}2 \times 10^7$ | 5 minutes | >28 days |
| *Staphylococcus aureus* | $2{\cdot}3 \times 10^8$ | 30 minutes | >28 days |
| *Streptococcus mutans* | $6{\cdot}7 \times 10^7$ | 30 minutes | >28 days |

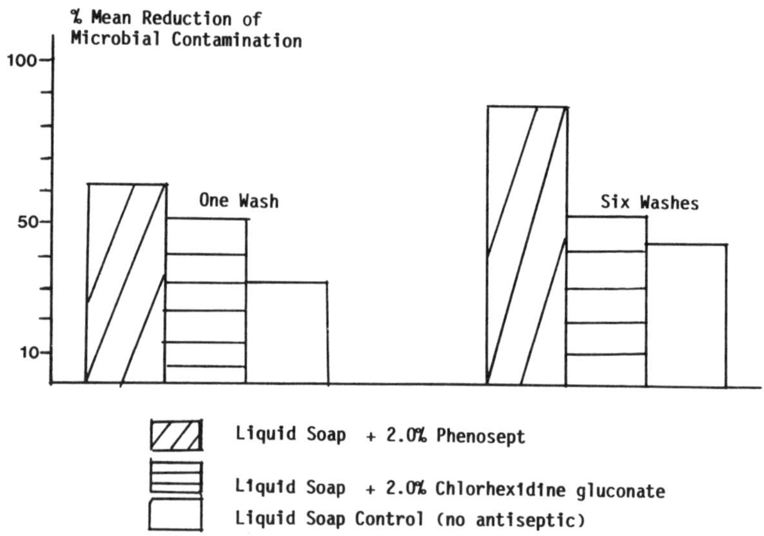


**Fig. 1.** Comparison of antiseptic activity of Phenosept and Chlorhexidine gluconate.

**Table 4.** Activity of Phenosept against the causative organism of bovine mastitis

| Test species | Initial inoculum level | Time taken to achieve total lethal effect in | | | |
|---|---|---|---|---|---|
| | | Teat dip + 0·3% Phenosept | Teat dip + 0·5% Phenosept | Teat dip + 0·75% Phenosept | Control |
| *Streptococcus agalactiae* | $7{\cdot}2 \times 10^7$ | 15 min | 5 min | 30 s | >30 min |

The results of such work illustrate the considerable potential that exists for formulating chlorophenolics with other antimicrobial compounds, thereby extending the traditional applications of the products. Care, of course, must be exercised in selection of the active agents for particular end uses as taint and odour of some chlorophenolics would preclude the use for many food or dairy applications.

Both PCMX and chlorophen, however, have proved acceptable for use in such applications. Sensibly formulated they can provide an effective, low cost alternative to other antimicrobial agents.

## Discussion

The size of the UK market for preservatives, industrial biocides and disinfectants exceeds £100 million per year.[5] Consumption of disinfectants and sanitisers in the UK are estimated to be 66 000 tonnes of formulated product and substituted phenols account for approximately 15% of this market and remain market leaders for hospital and agricultural applications. Consumption of disinfectants and sanitisers in EEC countries for 1989 was forecast at 300 000 tonnes of formulated products and the market is growing, albeit slowly.

Chlorophenolics should continue to play an important role both as industrial biocides and as active agents for disinfectants, sanitisers and antiseptics, especially for the hospital, agricultural and institutional environments. Chlorophenolics have certain advantages over some other active agents including pH tolerance (pH 4–9, in some cases even wider); good activity in the presence of organic matter; broad spectrum activity and ease of formulation. For some markets, however, the disadvantage of taint and smell will outweigh these factors. The long history of chlorophenolics is a great advantage and much information exists to help formulators produce even more effective blends of microbial products to satisfy future demand.

## References

1. Prindle, R. F., from *Disinfection, Sterilization and Preservation,* ed. S. S. Block, 3rd ed. Lea and Febinger, 1983, p. 197.
2. Peacock, J., *Speciality Chemicals,* January 1986.

3. Russell, A. D. & Hugo, W. B., from *Disinfection in Veterinary and Farm Animal Practice* ed. A. H. Linton, W. B. Hugo & A. D. Russell. Blackwell Scientific Publications, 1987, p. 12.
4. Lowbury, E. J. & Lilly, H. A., *British Medical Journal*, **1** (1973) 510.
5. Parr, J. A. & Irvine, M. S., *Manufacturing Chemist*, **59**(9) and **59**(10) (1988) 22–4 and 69–71.

# Developments in the use of organobromine intermediate chemicals

**P. Ashworth & J. Chetland**

*The Associated Octel Company Limited, Ellesmere Port, UK*

## Introduction

Organohalogen intermediates contain one or more halogen atoms directly bonded to carbon. In the application of an intermediate, halogen is displaced, usually by a nucleophilic agent, generating a bond to a new functional group (Fig. 1).

Such halide displacements from carbon (aliphatic and aromatic) constitute one of the most valuable and versatile general processes in speciality organics manufacture and provide new links to oxygen, nitrogen, sulphur, carbon and metals for pharmaceuticals, agrochemicals, surfactants, fuel additives and many other product areas.

## Choice of halogen

Of the halogens available for commercial use (F, Cl, Br, I), only chlorine and bromine are widely employed in intermediate applications.

Fluorine is relatively expensive, not easy to introduce selectively and the high C—F bond energy makes displacement from aliphatic carbon

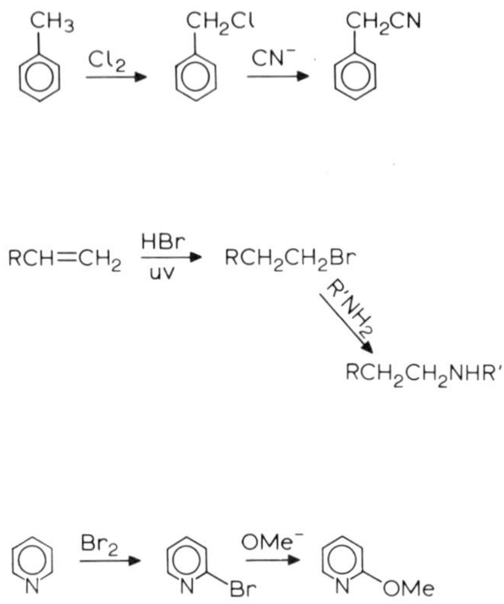


**Fig. 1.** Typical halide displacements.

extremely difficult. Certain nucleophilic substitutions of fluorine from aromatic systems, activated by electron withdrawing groups, have been commercialised for high value specialities, but this type of process is not widely exploited.

Iodine is very readily displaced from aliphatic carbon, but the element is scarce, expensive and heavy. Applications are limited to a small number of low volume specialities.

Chlorine and bromine, on the other hand, are plentiful, relatively cheap and can be incorporaed, often very selectively, into organic molecules in a wide variety of ways. The halo products are stable, easy to purify and handle and the halide anion can generally be displaced under mild conditions by a multitude of important nucleophilic groups.

The annual usage of bromine in intermediate applications is greater than $10^5$ tons, but is nevertheless vastly exceeded by the usage of chlorine. The latter is boosted by a number of mature, high volume products, which were developed from the cheap raw materials of the chlor-alkali industry.

More recently, in the development of higher value speciality chemicals, the direct competition of chlorine versus bromine for application in organic intermediates is more evident.

The choice of halogen for a particular intermediate application is governed by a number of major factors:

Production of organohalogen intermediate
—cost of halogen/halogenating agent
—ease and selectivity of halogenation.
Application of organohalogen intermediate
—ease and efficiency of displacement of halide
—by-products, effluents generated.

## Production—cost of halogen/halogenating agent

1990 prices in the United States of bulk chlorine and bromine are around \$200 and \$950 respectively. The ratio of 2·25 in atomic weights makes bromine ten times more expensive in terms of equivalent use. This is clearly an outstanding factor in the economics of processes requiring halo intermediates.

Table 1 shows percentages of halogen and costs of halogen per ton of organic radical contained in some chlorine and bromine intermediate chemicals. The table illustrates the constant ratio cost penalty in choosing bromine against chlorine for intermediate applications. However, it can be seen that whilst the ratio of halogen cost remains 10:1, the absolute differential decreases with the halogen content of

**Table 1.** Percentage and costs of halogen, per ton of organic radical, in some intermediates

| | Halogen (%) | Cost per ton of organic radical (\$) |
|---|---|---|
| Dichloro methane | 83 | 1000 |
| Dibromo methane | 92 | 10 000 |
| Dichloro decane | 34 | 100 |
| Dibromo decane | 53 | 1000 |
| Chloro cetane | 14 | 30 |
| Bromo cetane | 26 | 300 |
| Chloro benzene | 32 | 90 |
| Bromo benzene | 51 | 900 |
| Chloro *t*-butyl napthalene | 16 | 40 |
| Bromo *t*-butyl napthalene | 30 | 400 |

the intermediate. Indeed, for large and expensive organic radicals, the differential from bromine is small relative to the total cost of producing the downstream product and can often be recovered many-fold by the more efficient processing provided by bromine (discussed in later sections).

## Bromine recycling

The economics of bromine utilisation can be improved by reprocessing bromide ion released in intermediate operations.

Bromide arises in two ways:

(a) In production of bromine intermediates by elemental bromination (H-substitution), one half of the bromine utilised is released as hydrogen bromide. This by-product can be oxidised *in situ* using chlorine or peroxide, but the approach has limitations. Chlorine is often incorporated into the organic product and peroxide is expensive.

Alternatively, hydrogen bromide by-product can often be used, after necessary purification, for on-site hydrobrominations or absorbed to produce inorganic bromides. While being efficient in terms of material usage, this approach can place severe constraints on the flexibility of operations.

(b) Bromide is displaced in the application of organobromine intermediates, usually generating bromide-rich aqueous streams contaminated with organics. Disposal in effluent, after suitable treatment of organics, is the common practice of many intermediates users.

The bromide streams from displacement processes, and those from elemental brominations, can be processed for elemental bromine recovery in dedicated units employing chlorine oxidation. This bromine recycling is illustrated in Fig. 2 for production and subsequent reaction of an aromatic intermediate.

A number of plants currently carry out this process for in-house recovery of generated bromide. One such plant is operated by The Associated Octel Company at its site in the United Kingdom and also serves to recover bromine from streams generated locally by bromide intermediates users.

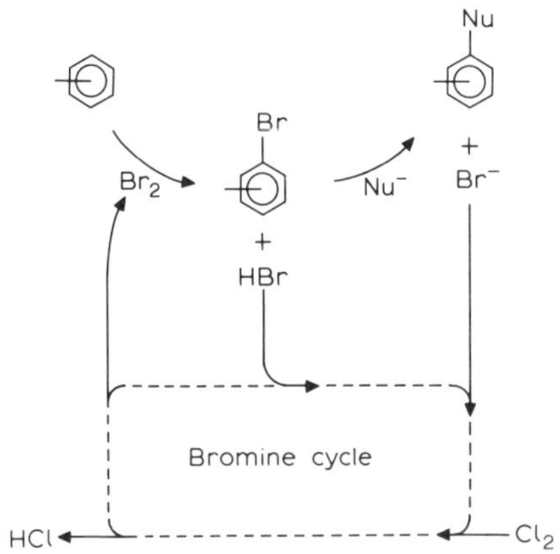


**Fig. 2.** The bromine cycle.

Figure 2 shows a net usage of chlorine only in the completed cycle, although in practice small amounts of bromine are lost in processing. Acid and energy are also consumed in recovering the elemental bromine, but the process provides a significant reduction of the cost differentials quoted in Table 1.

The cycle serves to illustrate a further feature of organobromine operations. The high atomic weight of the element places a burden on producers and users of intermediates in respect of transport and storage particularly. This burden is minimised, in cases of high bromine content chemicals, by performing the halide displacement at the site of manufacture of the intermediate. Containment and transport of generated bromide streams, often high in water content, are likewise reduced to a minimum.

Consideration of halogen cost has highlighted a clear penalty in using bromine in place of chlorine for intermediate applications, although this is seen to be reduced where recycling of bromide is performed. However, the establishment of many commercial bromine intermediate operations is proof of strongly positive features acting to outweigh this penalty in these areas.

An outstanding example from Table 1 is dibromomethane, several thousand t.p.a. being used in the manufacture of the biocide, methylene bisthiocyanate (MBT, Fig. 3). Despite the very low cost of dichloromethane resulting from massive production for other markets,

$$\text{Br–CH}_2\text{–Br} \xrightarrow{2\text{SCN}^-} \underset{\text{(MBT)}}{\text{NCS–CH}_2\text{–SCN}} + 2\text{Br}^-$$

**Fig. 3.** Methylene bisthiocyanate (MBT).

it finds no application for MBT manufacture due to the poor reactivity with thiocyanate.

The positive features acting on behalf of bromine are discussed in the following sections.

# Production—ease and selectivity of halogenation

Where an organohalide is already manufactured in appreciable volume for other markets, its selection for a new application is supported by the cost benefits of this existing production. There are many examples of this support for chlorine (e.g. allyl chloride, ethyl chloride, chlorobenzene) and a number for bromine. For example, the use of dibromoethane as intermediate for introducing an ethylene bridge into certain speciality products owes much to its availability at a low price from lead alkyl anti-knock operations (Fig. 4).

Where no such support exists and the organohalide is manufactured specifically for an application, the choice of halogen can often be influenced or dictated by some special feature of the chemistry developed for introducing bromine and chlorine into organic molecules.

A selection of areas in which bromine demonstrates a clear advantage over chlorine in providing commercially important intermediates is set out below.

## Aromatics

Aromatic halo intermediates are important in the manufacture of speciality phenols, amines, diaryl ethers and arylated aliphatics for a range of applications.

$$\text{Br–CH}_2\text{CH}_2\text{–Br} \xrightarrow[(2)\text{Y}^-]{(1)\text{X}^-} \text{X–CH}_2\text{CH}_2\text{–Y}$$

**Fig. 4.** Ethylene bridge via DBE.

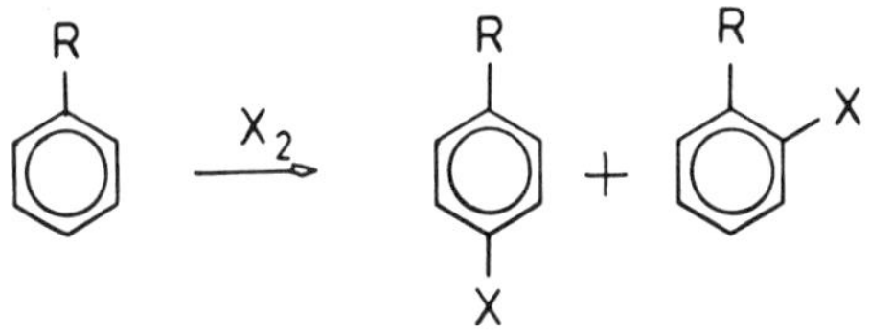

**Fig. 5.** Aromatic halogenation.

Bromine and chlorine are normally introduced into aromatics by electrophilic substitution. Elemental bromine, being less reactive than chlorine in this mechanism and of larger size, is generally more selective under similar conditions. This enhanced selectivity can be particularly valuable in providing high levels of para substitution in activated rings, where the co-production of ortho isomer is a notorious problem throughout aromatic chemistry. Demand for pure isomers usually requires expensive equipment for separation of the o/p mix, and can result in much low value co-product or waste being generated (Fig. 5).

Table 2 gives data for ortho and para substitution of hydrogen by bromine and chlorine in some activated aromatic systems. Both steric and electronic factors are required to account for the enhanced selectivity found in bromination. The commercial implications are clear, particularly in the manufacture of speciality diaryl ethers where the copper catalysed halide displacement occurs much more readily with the bromo derivative.

Further increases in selectivity of bromine, and to a lesser extent of chlorine, are afforded by maintaining low concentrations of halogen in the reaction mixture. This may be achieved by introducing a dilute

**Table 2.** Ortho and para substitution of hydrogen by bromine and chlorine in some aromatics

| Aromatic | Halogen | % ortho | % para |
|---|---|---|---|
| Chlorobenzene | $Cl_2$/Fe | 39 | 55 |
| | $Br_2$/Fe | 10 | 88 |
| Bromobenzene | $Cl_2$/Fe | 41 | 52 |
| | $Br_2$/Fe | 12 | 86 |
| Toluene | $Cl_2$/Fe | 53 | 46 |
| | $Br_2$/Fe | 28 | 71 |
| Phenol | $Cl_2$ | 48 | 52 |
| | $Br_2$ | 9 | 91 |

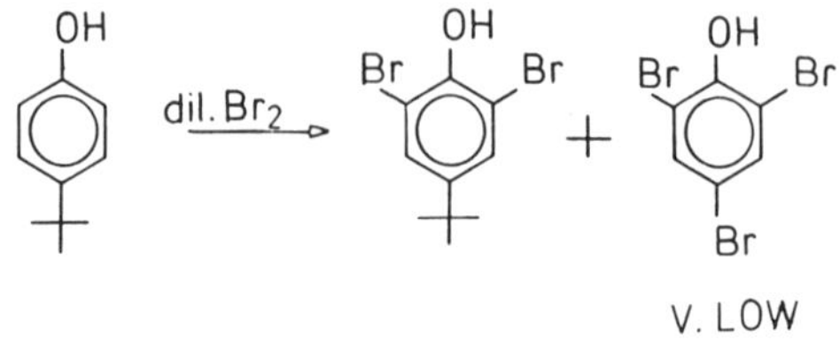


**Fig. 6.** Dibromination of alkyl phenols.

bromine stream or by generation of bromine *in situ*, e.g. with peroxide and bromide.

In a recent example,[1] the unwanted dealkylation during the 2,6-dibromination of 4-alkyl phenols is suppressed by feeding highly diluted bromine to the reaction mixture (Fig. 6).

Current work in our own laboratory on aromatic brominations is identifying new examples of this selectivity. In some cases bromine is so selective for para substitution that the product mix can be used, without expensive isomer separation, for the subsequent displacement reaction.

## Aliphatics

Bromine offers advantages in a number of areas of aliphatic chemistry.

The free radical addition of hydrogen halide to olefins is practised commercially only for hydrogen bromide. The process is thermodynamically unfavourable for hydrogen chloride and is generally unsuccessful.

The radical addition of HBr, initiated by UV light or peroxide, provides $\alpha$-bromo and $\alpha,\omega$-dibromo alkanes from terminal olefins and diolefins (Fig. 7). These products have important applications as intermediates in the manufacture of amines, thiols, ethers, etc., the bromide displacement proceeding under relatively mild conditions.

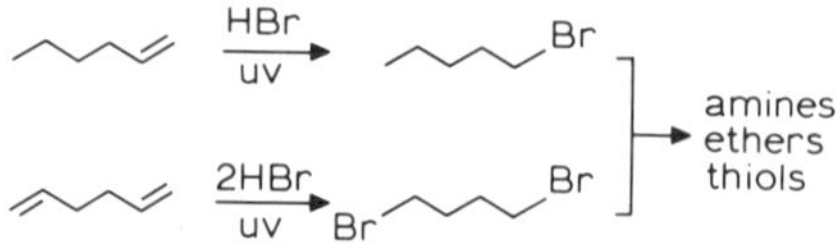


**Fig. 7.** Radical addition of HBr to olefins.

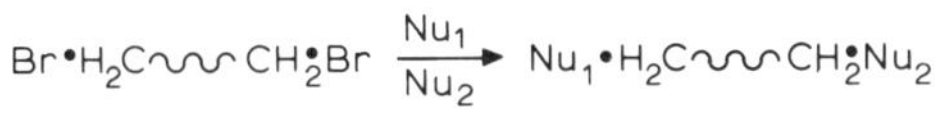

**Fig. 8.** Alkylene bridges from dibromides.

Terminal monobromides are effectively alkylating agents to nucleophiles, while the dibromides provide alkylene 'bridges' linking functional groups (Fig. 8).

$\alpha,\omega$-Bromochloro alkanes are produced by radical hydrobromination of the corresponding chloroalkenes (Fig. 9).

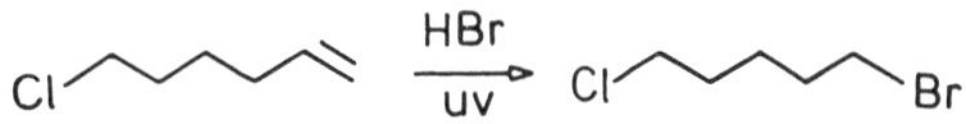

**Fig. 9.** Bromochloro alkanes.

Secondary and tertiary alkyl halides are generated by ionic addition of HX to olefins, a process successful for producing both bromides and chlorides (Fig. 10). Here chlorine is often the halogen of choice on cost grounds, provided the alkyl chloride reacts efficiently in the downstream process.

Alcohols and diols compete with olefins as raw material for alkyl halide production (Fig. 11), and again chlorides compete with bromides for each application. The benefits of cleaner processes for production and displacement of the bromide are weighted against the lower cost of hydrogen chloride. In general, bromine is chosen for complex and higher value products where halogen cost is relatively insignificant.

Thermal or light induced halogenation reactions are of major commercial value for production of polychlorinated aliphatic compounds by hydrogen substitution. The thermodynamics of this process for bromine are generally unfavourable except where the intermediate radical has an allylic or tertiary alkyl structure (Fig. 12). In such cases the substitution reaction is more selective than for chlorine, particularly in limiting the degree of substitution. The photolytic side

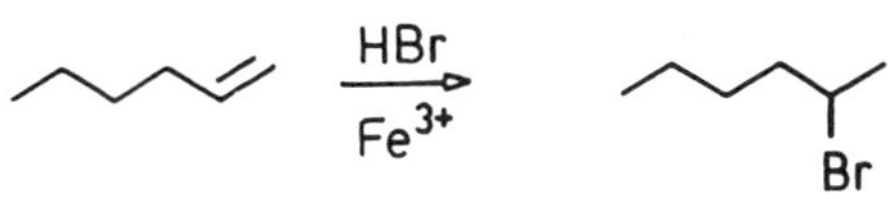

**Fig. 10.** Ionic addition of HBr to olefins.

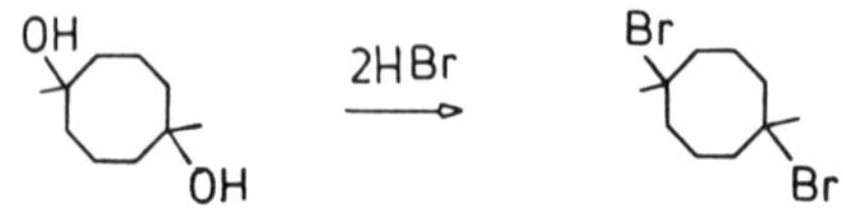


**Fig. 11.** Bromides from alcohols.

chain bromination of ring-substituted toluenes, for example (Fig. 13), tends to cease at the dibromo derivative, providing a route to aldehyde by subsequent hydrolysis.[2]

The smaller and more reactive chlorine atom provides a mixed product containing much trichloro derivative.

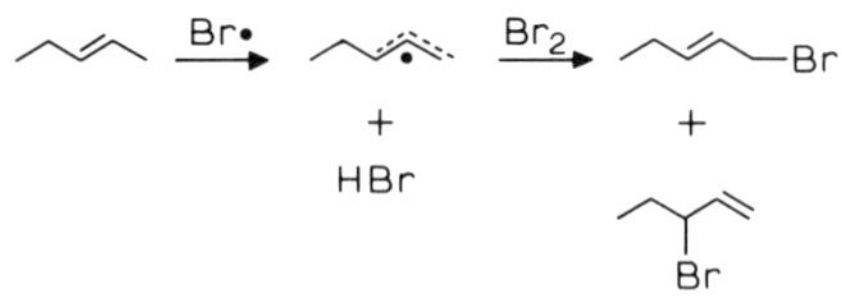


**Fig. 12.** Bromination in Allylic systems.

## Application—ease and efficiency of displacement of halide

Carbon–bromine bonds in both aliphatics and aromatics are weaker than the corresponding carbon–chlorine. As a result, displacement mechanisms that involve substantial bond breakage in the transition state generally proceed more readily for bromine. These include $S_N1$ and $S_N2$ nucleophilic displacements in aliphatics and copper-assisted nucleophilic displacements of aromatic halogen. Both processes have extensive commercial application for bromine and chlorine.

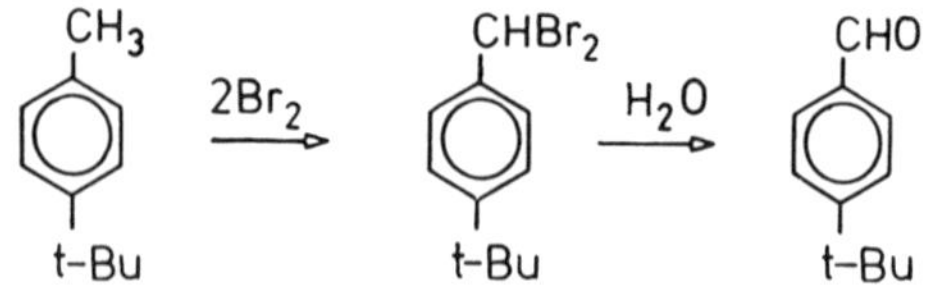


**Fig. 13.** Side chain bromination of toluenes.

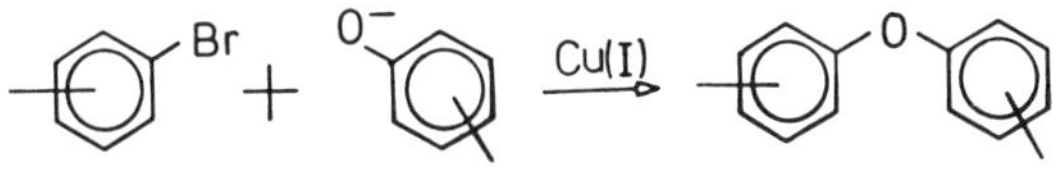


**Fig. 14.** Ullmann ether synthesis.

A reverse order of reactivity is found in uncatalysed aromatic displacements ($S_N$-Ar), where fluorine is generally most readily displaced. Here, fluorine, bromine and chlorine all find application commercially.

## Aromatics

The $S_N$-Ar displacement of halogen by various nucleophiles occurs readily in aromatics activated by strongly electron-withdrawing groups, particularly ortho and para to the halogen. Fluorine has importance where it can be easily incorporated at the required ring position but the major applications involve chlorine and, to a lesser extent, bromine.

Halo anthraquinones are highly activated to displacement and both chlorine and bromine are used to provide amino, hydroxy and alkoxy derivatives of anthraquinone in dyestuff manufacture.

Bromine is more reactive than chlorine in the copper-assisted reactions (Ullmann displacements)[3] and finds major application in producing speciality diaryl ethers. The efficiencies of Ullmann displacements are not generally high, due to competing reactions of reductive dehalogenation and diaryl coupling, but bromine provides consistently better yields of required ether product (Fig. 14).

The pyrethroid intermediate *m*-phenoxybenzaldehyde (Fig. 15) is produced in good yield by reaction of *m*-bromobenzaldehyde with sodium phenoxide in the presence of copper (I) catalyst. The yield with *m*-chlorobenzaldehyde is poor.

Strong electron-withdrawing groups ortho and para to halogen have a marked activating effect on the latter, providing milder process

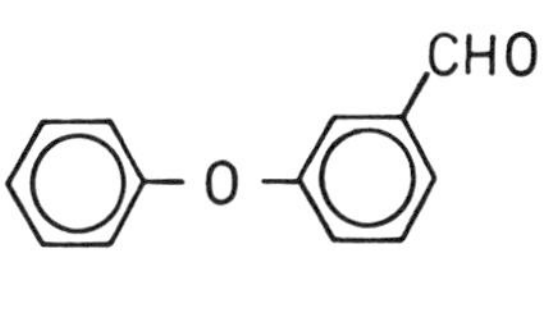


**Fig. 15.** *m*-phenoxy benzaldehyde.

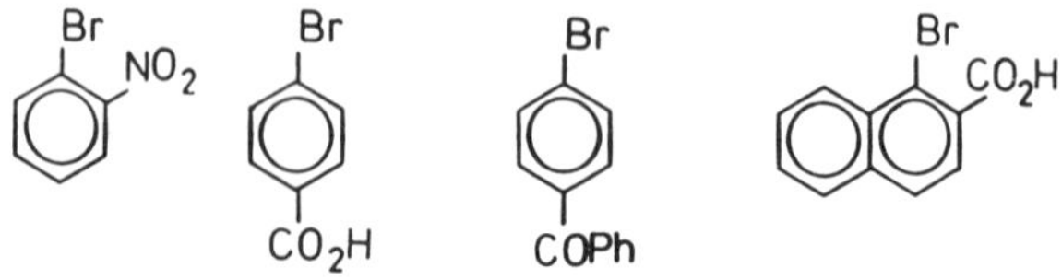

Fig. 16. Aromatic bromide intermediates.

conditions and improved yields. The bromo intermediates of Fig. 16 are very reactive and have potential for constructing useful diaryl ethers by reaction with a range of phenoxides.

The greater reactivity of aromatic bromine is valuable in production of organometallic intermediates of magnesium and lithium for arylation processes. Aromatic chloride is markedly less reactive and presents difficulties to the familiar problem of initiating Grignard formation. Bromoaromatics initiate more readily and provide more controlled reactions under milder process conditions.

Production of the anti-inflammatory drug 'Naproxen' shown in Fig. 17 uses a Grignard coupling reaction of 6-bromo,2-methoxy naphthalene and 2-bromo propionic acid halide.

Arylation of carbonyl compounds via Grignard intermediates finds widespread application in the manufacture of many agrochemicals and pharmaceutical products containing the tertiary alcohol structure of Fig. 17.

## Aliphatics

The nucleophilic displacement of halide from aliphatic carbon is almost invariably easier for bromide than for chloride. However, the reactivity is greatly influenced by other substituents on the halogen-bearing carbon, and the acid halides and $\alpha$-halo acids (Fig. 18) are so reactive to nucleophiles that the cheaper chlorides generally offer excellent yields under mild conditions.

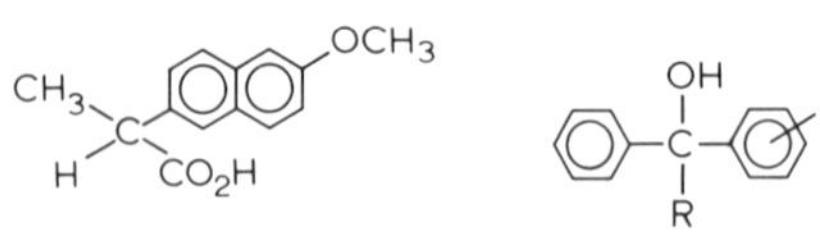

Fig. 17. Products via bromide Grignards.

**Fig. 18.** Highly reactive halides.

Halo alkanes are much less reactive in nucleophilic displacement and competing base-catalysed reactions, particularly dehydrohalogenation, can seriously reduce yields of the required product. Here, the benefits of bromine emerge very clearly in milder process conditions and in less waste product arising from competing reactions.

Gem-dihalides are particularly unreactive to most common nucleophiles. The production of MBT biocide, referred to earlier in Fig. 3, uses dibromomethane despite a very large cost penalty for bromine.

Dichloromethane is so unreactive in the required displacement process with thiocyanate that the major product is resinous material arising from a competing polymerisation reaction.

Monohalo alkanes are moderately reactive to most important nucelophiles and here the competition between bromine and chlorine for application is much keener.

Some recent kinetic work in our own laboratory has compared rates of bromide and chloride displacement from aliphatics using a range of important nucleophilic groups. Where possible, direct competition experiments were performed with both organohalides in the same reaction mixture. The form of trace shown in Fig. 19 is typical of that found for *n*-halo alkane reactions with methoxide, thiophenate and alkylamine, the organobromide reacting some tens of times faster than the chloride.

The reaction of sodium thiophenate, for example, with *l*-halo hexanes proceeds 40–50 times faster with the bromo compound. To compensate for the lower reactivity of the chloride, an increase in

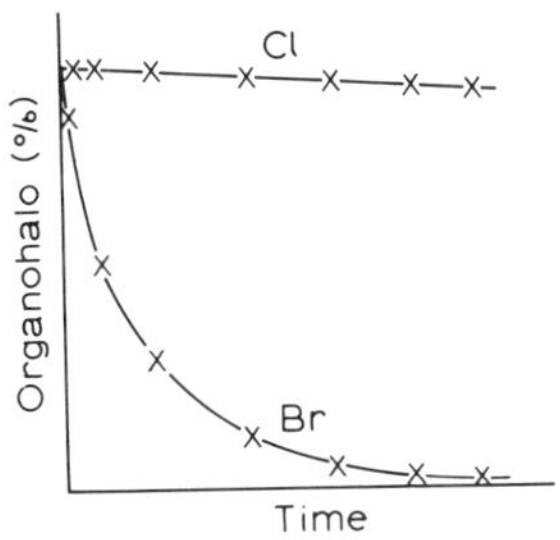


**Fig. 19.** Bromide vs chloride reactivities.

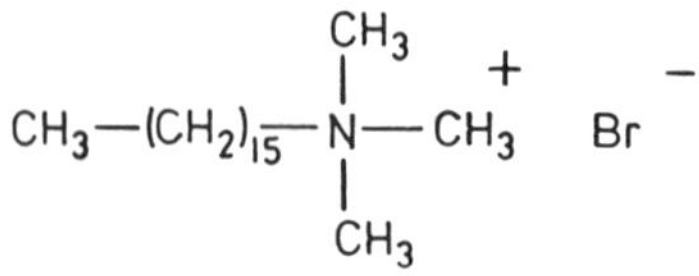

**Fig. 20.** Cetrimide.

processing temperature of approximately 50°C would be required to provide an equivalent reaction rate.

In general, the benefits of easier displacement of bromide, usually with improved yields of product, are more evident with the weaker nucleophiles, e.g. amines. These benefits must be weighed against the higher cost of bromine, the result being that the organobromide is often chosen for manufacture of high value speciality products where yield is of major importance.

The *l*-bromo alkanes find substantial application in alkylations of nitrogen for pharmaceuticals, agrochemicals and other speciality products. For example, cetyl bromide is used in the manufacture of the quaternary bromide 'cetrimide', an important antiseptic (Fig. 20).

The $\alpha,\omega$-dibromo alkanes are valuable 'bridging intermediates', the first two members of the series dibromo methane and dibromo ethane having been cited earlier. The higher members of the series continue to attract a great deal of interest for providing inert polymethylene bridges to link functional units in polymers, surfactants, agrochemicals and pharmaceuticals. The bridge can be tailored to optimise performance of the functional groups in the molecule (Fig. 21).

The alkylene bridge offers opportunities for constructing dimer and polymer forms of alkylated monomer products, for example, metal alkyls and quaternary nitrogen compounds (Fig. 22).

Bromochloro alkanes can be reacted sequentially with two different nucleophilic groups by virtue of the very different ease of displacement of bromide and chloride. Bromochloro methane is used for manufacture of the biocide (Fig. 23, I), while bromochloro propane is

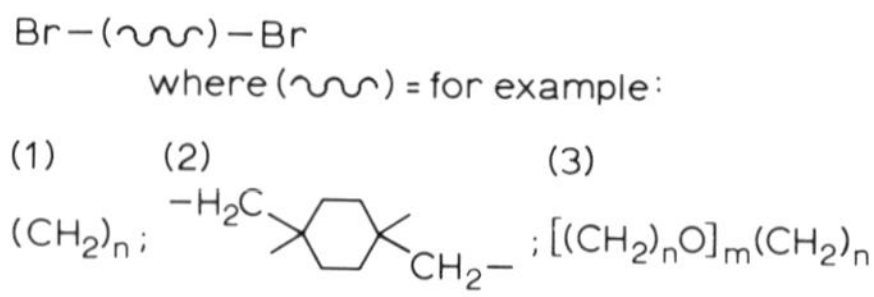


**Fig. 21.** Typical 'bridges' via dibromides.

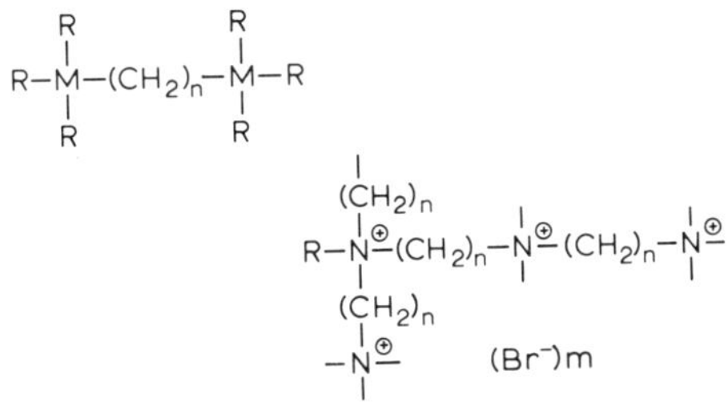

**Fig. 22.** Polymer forms via dibromides.

an important intermediate in producing dimethylamino propyl chloride (Fig. 23, II) for pharmaceuticals.

It is rather surprising that no significant application appears to exist for 1,2-bromochloro ethane despite the ability of a number of producers to provide the intermediate at low cost. The use of dibromo or dichloro ethane for providing an ethylene bridge between different groups requires the first displacement to be performed in a large excess of the dihalo ethane, at some additional cost, to limit disubstitution.

The very different reactivity of the halogens in bromochloro alkanes allows formation of mono-Grignard reagents by bromide displacement alone.[4] Under controlled conditions the intermediate Grignard reagent can be prevented from coupling with the remaining chloride-bearing carbon but can be reacted, for example, with carbonyl compounds (Fig. 24).

The available range of $\alpha$-bromo, dibromo and bromo chloro intermediates is expanding with the introduction of improved technologies for producing terminal olefin and alcohol precursors. Some examples are shown in Fig. 25.

I II

**Fig. 23.** Products from bromo chloro alkanes.

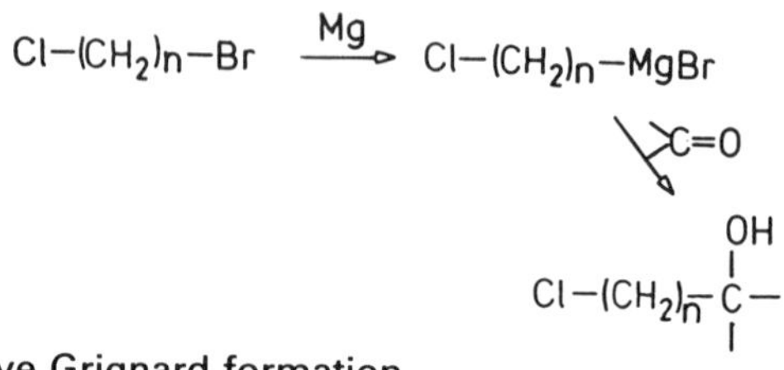

**Fig. 24.** Selective Grignard formation.

# Application—by-products, effluents generated

The milder process conditions normally required for bromide displacement can have substantial benefits in improved yields of desired substitution product, with less organic waste for disposal. Competing reactions of dehydrohalogenation and coupling particularly are more serious in chloride displacements.

In the production of high value specialities, organic waste represents a substantial raw materials loss. Separation of large quantities of by-product adds further cost to processing.

Environmental pressures bearing on the disposal of waste material continue to intensify and the contribution of bromine to providing cleaner processes, particularly for biologically active products, is expected to increase.

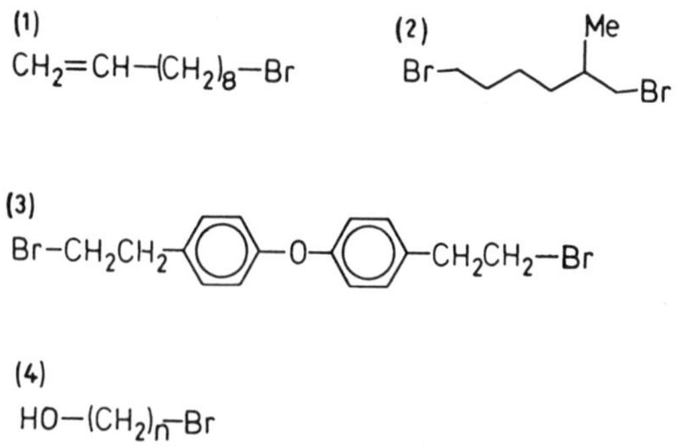

**Fig. 25.** Typical new bromide ranges.

## Conclusions

Organobromine intermediate chemicals are finding increasing application for production of higher value speciality products. The demand for very efficient processes with minimal organic by-product waste is often best satisfied by using the selective incorporation of bromine into molecules and the easy displacement of bromine by other functional groups. The cost penalty incurred in using bromine in place of chlorine for organohalo intermediates is often outweighted by these processing benefits.

Sea water presents an unlimited, although rather expensive, source of elemental bromine. The overall economics of intermediates operations benefit by processing of the organobromine at the site of manufacture, with recycling of displaced bromide.

## References

1. McKinnie, B. G. & Mitchell O. W., Us Patent 4 628 124, 1986 (to Ethyl Corp).
2. Japan Patent 6 1 238 758, (to Ishida).
3. For a review, see Lindley, J., Tetrahedron Report 163, *Tetrahedron,* **40** (1984) 1433.
4. Bailey, W. J. & Fujiwara, E., *JACS,* **77** (1955) 165.

# Some commercially useful organosulphur compounds

**A. G. Badger**

*Robinson Brothers Limited, West Bromwich, UK*

## Introduction

Many thousands of sulphur-containing compounds have been synthesised and characterised and several hundred of these are of commercial significance so that a presentation of this kind has to be severely selective in its coverage. Inevitably, the choice of topics has been influenced to some extent by the nature of the compounds examined over the past 40 years or so by Robinson Brothers. Fortunately we have had cause to consider a wide range of preparative chemistries to provide products for a diverse collection of end-users, although we were able to exploit commercially only some of the opportunities presented to us.

## Types of molecule

Even simple organic compounds can be named in a variety of ways and when the sulphur atom is incorporated into the molecule then the nomenclature possibilities are increased several-fold. Mercapto-, thio-, thia, sulph- all mean that there is sulphur chemically combined into the molecular structure. A detailed exposition of the structure and designation of organosulphur compounds is not appropriate here but a brief examination of the main classes of compound and their relationship to probably more familiar substances may help to clarify the rest of this paper.

## Thiols

The widely encountered aliphatic alcohols, ROH, have as their sulphur counterparts the mercaptans or thiols, RSH, where R represents an appropriate alkyl group. The lower alkyl mercaptans are volatile and highly odoriferous, but mercaptans containing larger groups ($C_{12}$–$C_{18}$) are higher boiling liquids or waxy solids.

## Thiophenols

Aromatic thio compounds are much more difficult to prepare than the corresponding phenols unless particular substituents on the aromatic ring create advantageous reaction conditions. Many have strong and persistent smells, adding to the already severe problems of a potential manufacturer. In contrast to the numerous phenol-based materials produced and consumed, often in large amounts, only a limited range of thiophenols have attained any commercial importance.

## Disulphides

Organic peroxides, RO—OR′, are not noted for their thermal stability but the analogous sulphur compounds, RS—SR′, are much more stable. A great variety of disulphide compounds are known because the S—S link is a strong one due to the presence of additional bonding forces. Mercaptans are readily oxidised to disulphides while the disulphides can be symmetrically cleaved in the presence of a reducing agent:

$$\text{RS-SR'} \underset{[\text{O}]}{\overset{[\text{H}]}{\rightleftharpoons}} \text{RSH} + \text{HSR'}$$

This redox reaction is an important one in respect of both the preparation and use of many mercaptans and disulphides.

## Thioacids

Sulphur can be incorporated into the structure of an organic acid in several ways. For example, oxygen atoms in the carboxyl group can be replaced by sulphur:

| | |
|---|---|
| $CH_3COOH$<br>acetic acid | $CH_3COSH$<br>thioacetic/thiolacetic acid |
| | $CH_3CSSH$<br>dithioacetic acid |
| $NH_2COOH$<br>carbamic acid | $NH_2CSOH$<br>thionocarbamic acid |
| | $NH_2CSSH$<br>dithiocarbamic acid |
| HOCOOH<br>carbonic acid | HOCSSH<br>dithiocarbonic/xanthic acid |

Neither xanthic acid nor the two sulphur-substituted carbamic acids are stable as free acids, but many of the substitution products have good stability and some of these have considerable commercial significance.

In a second mode of incorporation, the sulphur is covalently linked to one of the non-carboxyl carbon atoms. For example:

| | |
|---|---|
| $CH_3COOH$<br>acetic acid | $HSCH_2COOH$<br>mercaptoacetic acid |

Mercaptoacetic acid can also be regarded as a sulphur substitution product of glycollic acid, $HOCH_2COOH$; hence its commonly used alternative name of thioglycollic acid.

Quite different molecular structures are created when sulphur is present in a higher valency state, as typified by the versatile sulphonic acids, $RSO_3H$, and their derivatives, the sulphonates. An important compound with this type of structure is the methylating agent dimethyl sulphate, $CH_3O{-}SO_2{-}OCH_3$.

## Commercial outlets

The industrial utilisation of organosulphur compounds is best illustrated by attempting to group them according to their areas of application. In this presentation four main headings will be considered, namely:

- pharmaceuticals
- polymeric materials
- biocides and agrochemicals
- consumer products

Some of the materials presented find use in more than one area so that an overlap between these somewhat arbitrary groups is unavoidable.

## Pharmaceuticals

A number of new and extremely successful drugs contain sulphur in significant positions in the active ingredient and this theme has been continued in several second and third generation pharmaceutical products.

Thus one area of considerable commercial interest centres around the fairly simple molecule mercaptoethylamine, known also as cysteamine and usually handled as the hydrochloride $HSCH_2CH_2NH_3^+Cl^-$. Cysteamine is used in considerable quantities in the synthesis of two extremely successful anti-ulcer drugs, namely Cimetidine (Tagamet from SmithKline Beecham) and Ranitidine (Zantac from Glaxo):

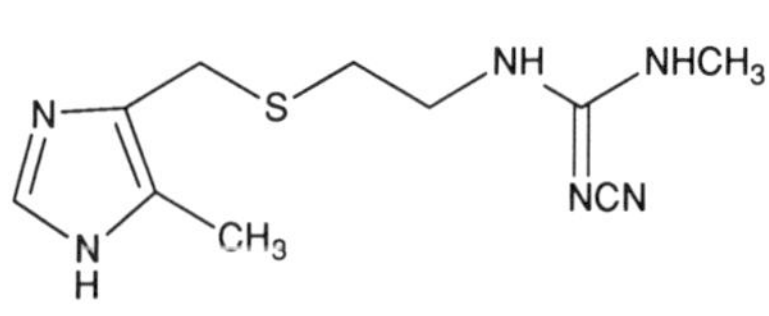


CIMETIDINE

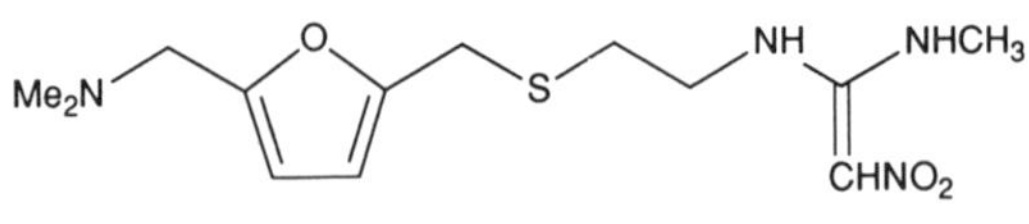


RANITIDINE

Although each molecule has its own distinctive features to maximise pharmaceutical effectiveness and to avoid rival patent claims, both of these compounds retain a central link based on cysteamine.

What is not immediately apparent is that sulphur chemistry is involved also in the synthesis of the cyanoguanidine group of Cimetidine and the nitrovinylamino group of Ranitidine.

In the case of the first named, cyanamide is made into its sodium or potassium salt and this is reacted with carbon disulphide. Both alkali metal atoms are then replaced with methyl groups using dimethylsulphate as the methylating agent:

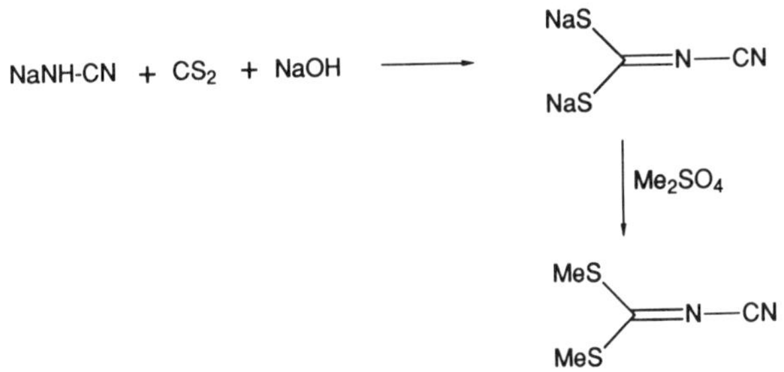

The end-product, dimethyl cyanodithioiminocarbonate, is incorporated into the Cimetidine molecule by reactions involving the sequential elimination of the two methylthio groups as methyl mercaptan.

The analogous precursor to Ranitidine is made using similar chemistry. Nitromethane, $CS_2$ and potassium hydroxide are reacted to form a dipotassium salt which is then methylated to provide the required bis(thiomethyl) nitroethene:

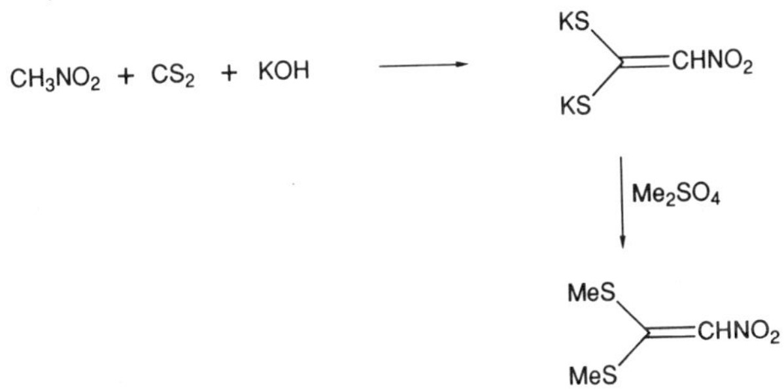

In addition to the precautions needed to handle the starting materials, there is additional hazard in that the dipotassium salt can detonate if allowed to dry out. Both sulphur atoms are lost from the bis(thiomethyl) compound during the subsequent reaction stages to the final Ranitidine molecule.

Encouragingly, a central cysteamine-derived linkage has been preserved in other anti-ulcer candidates such as Elli Lilly's Nizatidine and a product from Bristol-Myers designated BL 6341 A:

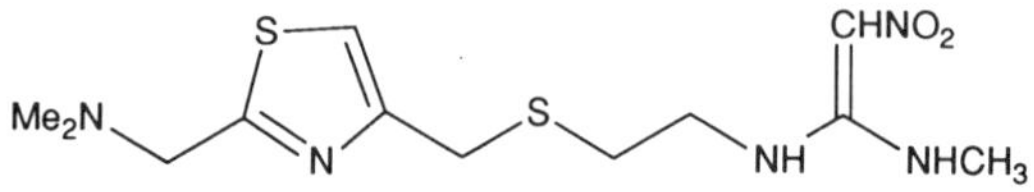


NIZATIDINE

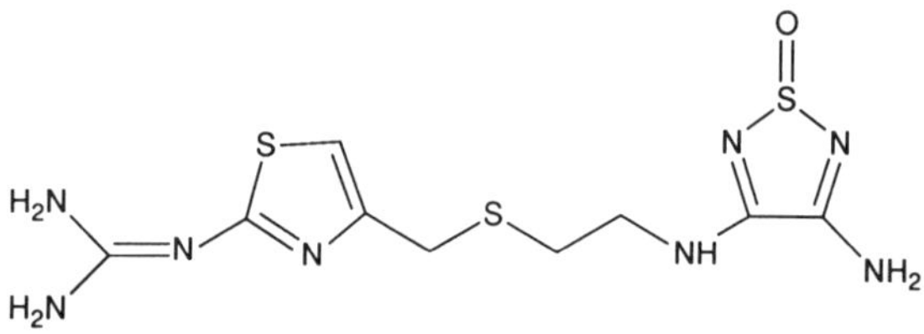


BL 6341 A

The discontinued ICI development product Tiotidine also incorporated a cysteamine linkage:

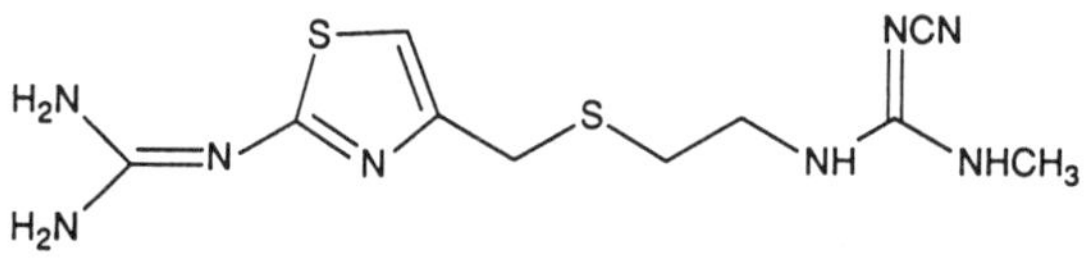


TIOTIDINE

Naturally one looks for further structural similarities between the various histamine H2 receptor molecules. For example, Tiotidine had exactly the same cyanoguanidine group as Cimetidine and therefore would have almost certainly required dimethyl cyanodithioiminocarbonate for its synthesis had it gone ahead. Both Tiotidine and Bristol-Myers' BL 6341 A incorporate the same thiazole ring and this structural unit also appears within the molecule of another recent contender, namely Yamanouchi's Famotidine, made also by Merck Sharp and Dohme under licence.

We quickly established that this feature, a substituted thiazole, was best synthesised from dichloroacetone and amidinothiourea and a viable manufacturing process was developed to make amidinothiourea starting from dicyanidiamide and $H_2S$. The use for amidinothiourea,

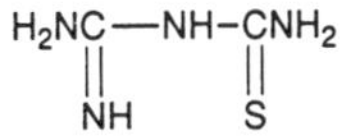

need not be restricted to make intermediates for histamine H2 antagonists. For example, it can be used to build up side chains in penicillin-type antibiotics.

In agriculture, amidinothiourea can be used directly as a source of nitrogen and sulphur, as an inhibitor of nitrification bacteria and as a soil conditioner. The compound is an effective corrosion inhibitor for mild steel pickling in hydrochloric and sulphuric acids and derivatives have been claimed as blowing agents and as flame-proofing agents for textiles and plastics.

Once the pharmaceutical role of cysteamine hydrochloride had been established, further consideration was given to its economic preparation on a commercial scale. Several routes are possible starting from ethanolamine, for example by its reaction with carbon disulphide and the acid hydrolysis of the mercaptothiazoline intermediate that is formed:

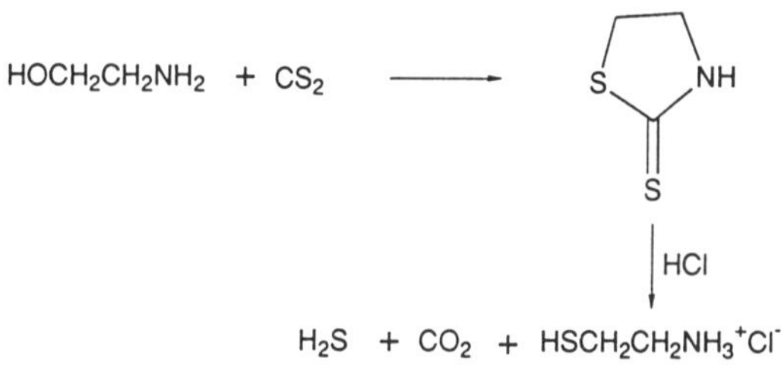

This method is certainly far better environmentally than attempting the direct reaction of ethylene imine and hydrogen sulphide although the use of $H_2S$ as the sulphur donor has obvious attraction in that the sulphur atoms are used more economically. One method of employing $H_2S$ in an acceptable way is to first synthesise an oxazoline, react this with $H_2S$ and then hydrolyse the resultant thioamide. Thus 2-ethyl oxazoline, made from ethanolamine and propionic acid, is reacted with $H_2S$ to form $\beta$-mercaptoethyl propionamide (BMEP). Acid hydrolysis yields mercaptoethylamine and regenerates propionic acid which can be recycled:

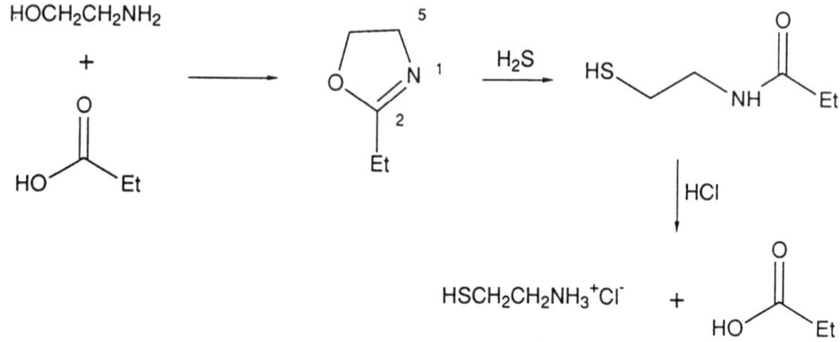

Unfortunately, ring opening of 2-ethyl oxazoline can also occur at the 1,2-position to form an alternative and unwanted product. In addition a second molecule can react with BMEP to form a diamide. Clearly both these reactions need to be minimised in a commercially viable process. It should be noted that BMEP can also be used directly to introduce the —$NHCH_2CH_2S$— group into a molecule, thereby avoiding the hydrolysis stage altogether.

Once mastered, the process for making 2-ethyl oxazoline is capable of providing other 5- and even 6-membered rings as required. This group of compounds is capable of a variety of reactions involving attack at the $\alpha$-carbon atom or ring opening as already mentioned. We hope that improved availability will encourage further commercial exploitation of their potential as chemical reagents, for example to introduce the —$NHCH_2CH_2$— and —$NHCH_2CH_2CH_2$— groups into molecules, or as monomers in polymer formation. Many examples have been published and reviewed.

As we have seen, a mercaptan group can be introduced by direct reaction with hydrogen sulphide but it is often necessary for a number of reasons to provide protection for the SH group. A very useful reagent for introducing a blocked SH group is thioacetic acid, made from acetic anhydride and $H_2S$:

$$CH_3CO{-}O{-}OCCH_3 + H_2S \longrightarrow CH_3COSH + HOOCCH_3$$

The stoichiometric mixture of acetic and thioacetic acids needs to be separated by careful distillation.

As an example, thioacetic acid is used to introduce an acetylthio group into the steroidal structure of Spironolactone marketed by Searle and others as the diuretic drug Aldactone. The acetylthio group provides an ideal balance between long-term stability of the drug during storage and ready availability of the active metabolite once the drug has been administered.

A newer example of the pharmaceutical use of thioacetic acid is the synthesis of the Squibb anti-hypertensive Captopril (Capoten) from L-proline and acetylthio-2-methyl propionyl chloride, the latter being made from methacrylic acid and thioacetic acid followed by chlorination:

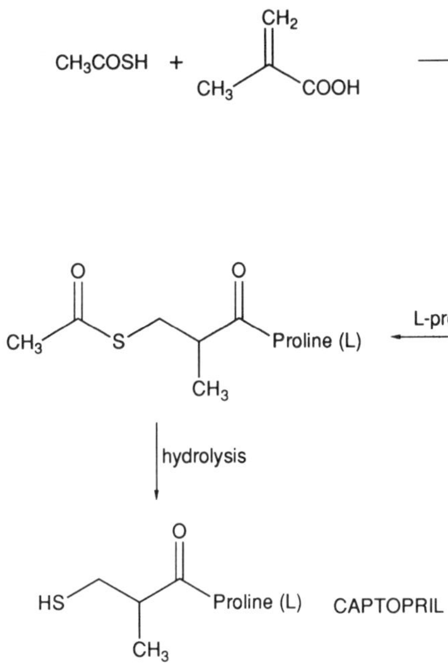


## Polymeric materials

Sulphur-containing materials play an important part in the manufacture and use of some plastics and most rubbers.

Of the several types of polysulphide polymers on the market, the range offered by Morton Thiokol is probably the best established. These liquid polymers are oxidatively cured at ambient temperatures to provide elastomeric products with excellent stress and strain properties over a wide service range. Their high degree of resistance to ageing due to weathering, oxidation or attack by ozone make them particularly useful in the assembly of insulated glazing and as construction sealants.

Chemically the Thiokols are polymers of bis(ethylene oxy) methane containing disulphide linkages and terminated with mercaptan groups. Modulus and elongation properties are controlled by building a certain number of pendant mercaptan groups into the chains and further modification may be possible by substitution on to some of the SH groups.

Sulphur compounds can also play an essential relevant role even when the base polymer itself contains no sulphur. A relevant example of this is the established use of dithiocarbamates and thiuram sulphides as accelerators in rubber vulcanisation processes.

Dithiocarbamates are manufactured by reaction of carbon disulphide with an aliphatic or alicyclic amine in the presence of aqueous alkali, normally caustic soda:

$$RR'NH + CS_2 + NaOH \longrightarrow RR'NCSSNa + H_2O$$

The aliphatic amines most commonly used are methylamine, dimethylamine, diethylamine and dibutylamine. Sodium alkyl dithiocarbamates made in this way do have direct uses in the rubber industry (and elsewhere as will be shown) but of much greater importance are certain water-insoluble salts made from the corresponding sodium dithiocarbamate solution by a process of double decomposition. The dithiocarbamates of zinc, nickel, copper and bismuth are the materials most frequently encountered; a number of permutations between these metallic cations and the starting amines are possible to provide a range of accelerator products under the Robac tradename with precisely the right cure properties. Blends of several active materials are also used extensively.

Further variety and specialisation can be provided. Thus dithiocarbamates can be made using an alicyclic amine such as piperidine and a number of such products have been exploited commercially in specialised areas of rubber processing. Interestingly, amines can also be used as the cation instead of a metal in certain dithiocarbamate molecules and a number of wholly organic accelerators based on piperidine particularly have been developed.

Aqueous dithiocarbamates can also be used as the starting materials for the preparation of thiuram sulphides. Thus sodium dimethyl dithiocarbamate is readily oxidised to tetramethyl thiuram disulphide

(TMT), a cost effective accelerator still used widely in very considerable quantities throughout the rubber manufacturing industries.

Changes in material presentation can also be highly beneficial. For example, the addition of rubber accelerators into mixing equipment as dusty powders tipped from paper bags is no longer acceptable and detailed attention has been directed to this aspect of the rubber processing industry. As a result we now offer our accelerators in a variety of physical presentations. Thus pelletised, dust-free, wax bound and polymer-masterbatched forms are manufactured to provide a range of accelerator products corresponding to precise customer requirements. On occasions we have also made end-poducts incorporating active ingredients not of our own manufacture and we are hoping to extend this activity still further to include active principles used in other areas of industry.

Certain dithiocarbamates have found uses in other areas of polymer technology. Thus zinc dimethyl dithiocabamate and nickel dibutyl dithiocarbamate can act as useful antioxidants in polyolefins. More recently dithiocarbamates of these same two metals but made from a longer chain amine such as diisononylamine have been shown to be useful in polyolefins as antioxidants and UV stabilisers respectively.

## Biocides and agrochemicals

The use of organosulphur compounds in agriculture has already been mentioned briefly in connection with amidinothiourea. Specific dithiocarbamates are also important in this area. For example, sodium ethylene bisdithiocarbamate (Nabam, made from ethylene diamine) is used as a fungicide while sodium *N*-methyl dithiocarbamate (sold as Vapam, Sistan and Metham Sodium) is a well-known soil sterilant. TMT (Thiram) has well-established outlets as an agricultural fungicide, its use by no means entirely superceded by newer but often more expensive products.

The opportunity to commence manufacture of a dithiocarbamate new to our existing range presents itself all too seldom. However, we recently started to produce one such candidate, namely methyl ammonium methyl dithiocarbamate. Commercial interest centres around its use as a precursor to the well-established bacteriocide

Dazomet, made by reacting the aqueous dithiocarbamate with formaldehyde:

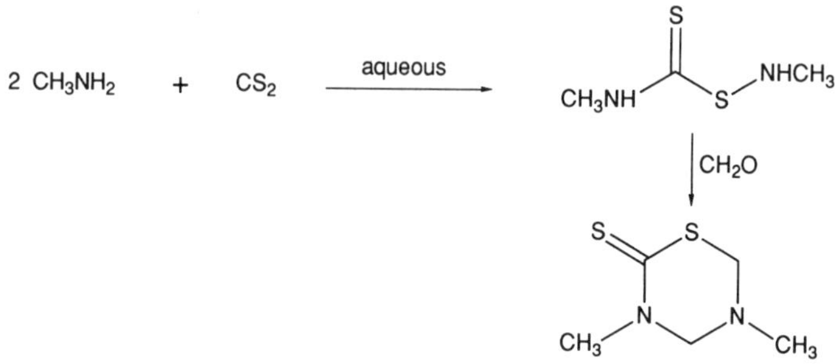

DAZOMET

One of the newer uses of Dazomet is as a biocide in oil well technology including wells drilled under the North Sea. Not surprisingly, dithiocarbamates themselves can also be employed as biocides for this application. Aqueous blends of Nabam and sodium dimethyl dithiocarbamate have proved to be particularly effective and their established outlet into the paper industry was supplemented for a time by their use in the North Sea. There are, however, problems. Although water-soluble dithiocarbamates normally have a low toxicity profile towards mammals they are extremely toxic towards marine life generally so that their use in this area is now discouraged. These compounds are also very destructive towards nitrification bacteria and the grazing worms used in sewage treatment beds. Thus effluent streams generated during the manufacture and use of these materials need to be contained and treated. Several methods for the removal of dithiocarbamate ions from such outflows have been proposed including their precipitation as insoluble metal salts (British Patent 1 446 548, 1976).

## Consumer products

The foregoing sections have exemplified some of the ways in which organosulphur compounds are used within chemistry-based industry. Their presence as discrete molecules in end-products encountered on a day-to-day basis by the public at large is rather less widepread by comparison. However, where sulphur-containing materials are used in this way, their action is again often unique and the end-result difficult or impossible to achieve by other means.

Thus, the longer-chain alkyl mercaptans are relatively involatile materials without excessively prominent odours. The $C_{18}$ mercaptan based on stearyl alcohol is in fact a waxy solid and is used as an anti-tarnish ingredient in silver polish formulations. Silver tarnishes rapidly in the presence of hydrogen sulphide due to the formation on its surface of black silver sulphide. Mercaptans have a similar affinity for silver but adsorption on to its surface of long-chain molecules like stearyl mercaptan provides a hydrophobic coating which protects the metal from further attack.

The volatility and olefactory properties of lower alkyl mercaptans has been referred to already; these characteristics are exploited directly by the use of such compounds as odorants in industrial and domestic fuel gases of natural origin. In the United Kingdom these are mainly derived from off-shore sources of course. Intrinsic odour characteristics vary depending on the source and a universally compatible odorant system is therefore needed to generate an instantly recognisable smell. This is achieved by the inclusion in very low concentration of a blend of $C_{1-4}$ sulphur compounds incorporated into the gas as it leaves the coastal terminals. Specially designed blending, transportation and transfer equipment are required to avoid accidental release of odorant at any point.

A third area in which the public at large may come into contact with thiocompounds—intimately in this case—is the use of organosulphurs as the principal active ingredient in hairwaving formulations.

Most of the established products are based on mercaptoacetic acid (thioglycollic acid) manufactured on a large scale for this and other end-uses. Hairwaving solutions of this type are formulated around a dilute solution of ammonium thioglycollate at a pH value of around 9. In action, the cysteine disulphide links within the hair structure are broken by the reducing action of the thioglycollate radicals and the resultant softening of the hair facilites its reshaping as required. At the end of the curling (or straightening) operation a mild oxidising agent is applied and the disulphide links reformed within the hair. If the concentration of thioglycollate in the applied solution is too great or the pH value too high then damage to the hair will result. This property is turned to positive use in various depilatory formulations, usually based on calcium, sodium and potassium thioglycollates.

The other significant uses for thioglycollic acid are less direct by nature. Thus methyl and ethyl thioglycollates (both are volatile malodorous materials) are used in pharmaceutical and agrochemical syntheses while the $C_8$ alkyl esters are precursors to an important class of organotin stabilisers for PVC. Multifunctional esters of thioglycollic acid have been used as the cross-linking agents in unsaturated polymer systems curable by controlled exposure to ultraviolet light in applications such as the production of printing plates.

Returning briefly to the subject of hairwaving, it is encouraging to note that even in such a well-entrenched market chemical innovation is still possible. Because the operating concentrations and pH values of solutions based on thioglycollic acid are close to the levels at which irreversible hair damage can occur, there have been numerous attempts to develop safer end-products. There is also scope to provide products with better odour characteristics.

One of the potentially more successful innovations in this respect has been the partial or total replacement of thioglycollate with a compound that has been discussed earlier, namely mercaptoethylamine. Suitable formulations are the subject of a number of patent applications, with the French comany L'Oreal having been particularly active in this area. However, the proposed products frequently lack an adequate storage stability and we ourselves considered it worth while to devote some resources to this problem. Work to date has led us to make up development quantities of a stable cysteamine-based concentrate for assessment by potential customers and their detailed comments are awaited with interest.

## Outlook

Organosulphur chemicals have become indispensible components in a wide diversity of commercial processes and products. As has been illustrated, their involvement may be less than obvious.

Once a usage has been established, subsequent displacement by other materials is uncommon because of the specific and often unique properties achieved by the inclusion of chemically bound sulphur as part of a particular system. In fact, novel applications are continually being identified and the list of commercially available organosulphur compounds is steadily increasing both in number and molecular complexity to fulfil new technical and market-place requirements.

Hopefully this presentation will have provided a framework and even a degree of inspiration for further innovation centred around the organic chemistry of sulphur.

# Nitroorganosulphur compounds and their use as pharmaceutical and agrochemical intermediates

**A. Jackson**

*Fine Organics Limited, Middlesbrough, UK*

## Introduction

Cimetidine (I), the first anti-ulcer $H_2$ antagonist drug to be marketed, has the electron withdrawing cyanide group attached to one of the nitrogens of the guanidino moiety. As a variant on this part of the molecule, SKF examined the effect of replacing the cyanamide group with nitromethylene and observed that the compound obtained still retained $H_2$ antagonist activity.

This fact was capitalised on by Glaxo when they introduced the second anti-ulcer drug, Ranitidine (II), on to the market, in which the *N*-cyanoguanidine is replaced by an amidinonitromethylene group.

(I)

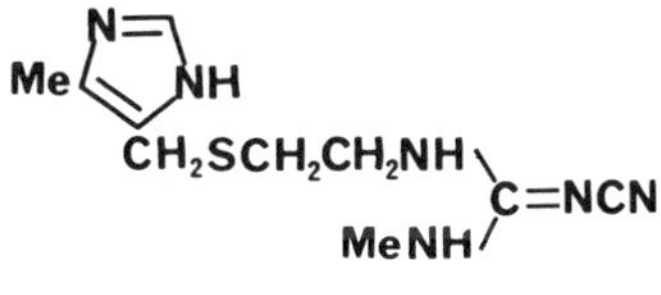

(II)

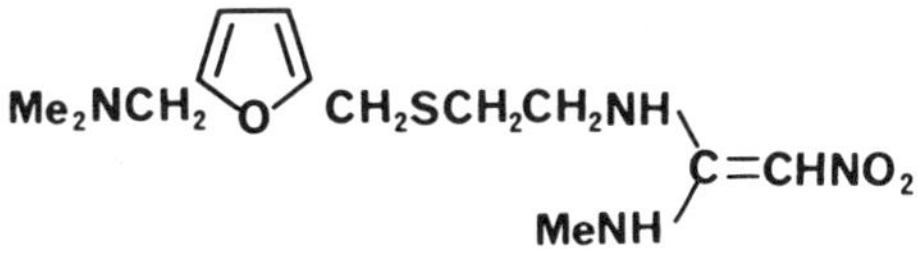

The intermediate required to introduce this function into the molecule is 1,1-bis(methylthio)-2-nitroethene (BMN), a compound that hitherto had not been commercially available. Since this initial application, the potential uses for BMN have expanded, not only in the pharmaceutical field but also in the synthesis of agrochemicals.

## The nitromethylene anti-ulcer drugs

The two anti-ulcer drugs possessing the amidinonitromethylene moiety that have been marketed to date are Nizatidine (III) and Ranitidine.

(III) 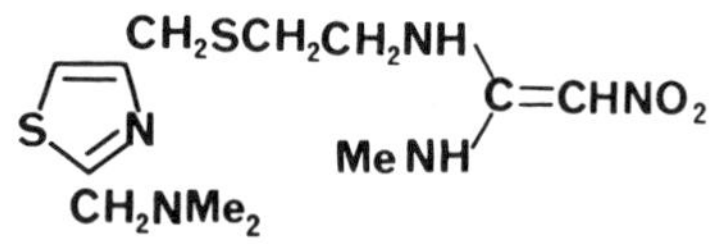

Potentially both of these compounds can be synthesised via two different routes from BMN, depending on which amino group is put in first. However, unlike the Cimetidine intermediate dimethyl cyano-dithioimino carbonate, in which it is very easy to replace one of the methylmercapto groups without further substitution, BMN is not so accommodating and displacement of the second mercaptan occurs very readily.

If, however, methylamine is used as the first amino group for the substitution, then both Nizatidine and Ranitidine can be synthesised via the common intermediate 1-methylamino-1-methylthio-2-nitro-ethene (MMN).

Substitution of the second mercaptan with the relevant thiazole or furan amine leads to either Nizatidine or Ranitidine respectively.

## The synthesis of BMN and MMN

### BMN

The first recorded synthesis of BMN is that of Gompper and Schaefer[1] in 1967 in which the dipotassium salt of nitrodithioacetic acid is methylated in aqueous methanol using dimethyl sulphate as the alkylating agent. The dipotassium salt was prepared following the method of Freund[2] (Fig. 1).

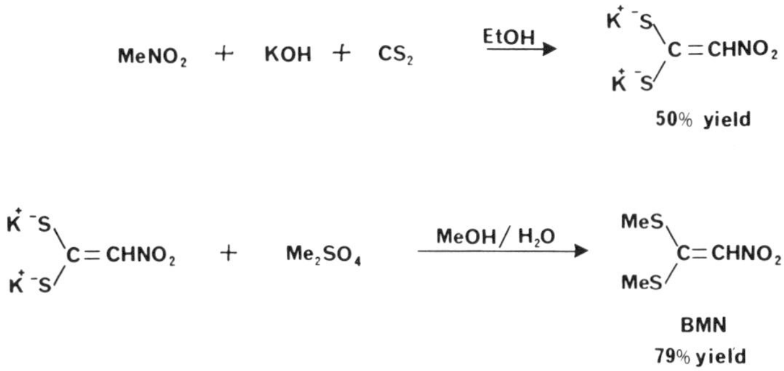

**Fig. 1.** Synthesis of BMN.

Because of the ease of formation of the ethyl xanthate from ethoxide and carbon disulphide, it is surprising that the yield of the dipotassium salt is as high as 50%. Drying of the dipotassium salt is not recommended as it is potentially explosive and even solvent damp material has been known to spontaneously ignite.

Although nothing has been recorded in the literature, considerable efforts have been put into improving this route and procedures have been developed that (i) replace ethanol with an inert solvent that cannot react with carbon disulphide and (ii) avoid the isolation of the hazardous dipotassium salt.

Accelerated rate calorimetry tests have shown that BMN itself can spontaneously decompose at temperatures greater than 80°C and needs to be handled with care as, in addition to the potential decomposition hazard, toxicity tests on the crude product indicate the presence of mutagens.

A more recent synthesis of BMN, patented by Ube,[3] uses 1,1-dichloro-2-nitroethene as the starting material, the two chlorine atoms being replaced by methyl mercaptan. As the starting material is not, as yet, readily available, this route is not commercially viable.

## MMN

Four routes to MMN have been described in the literature, two of which use BMN as the starting compound. The most obvious procedure[4] describes the replacement of one of the mercaptan groups with methylamine which, although attractive as paper chemistry, is not easy to achieve in practice as attempts to fully utilise the BMN leads to formation of considerable amounts of 1,1-bis(methylamino)-2-nitroethene. Although the yield quoted is 68%, separation of the product from the starting material and by-product to achieve the quality required for drug manufacture is difficult.

SKF, realising this difficulty, developed a procedure[5] in which one of the sulphur atoms in BMN is oxidised using peracetic acid to give the sulphoxide derivative. After isolation and reaction with methylamine the sulphoxide group is substituted preferentially to give good quality MMN in 55% yield. Again, however because of decomposition and explosion hazards, the preparation of the sulphoxide has to be handled with great care and large scale preparations should be avoided.

A third procedure uses the dipotassium salt generated in the BMN synthesis[6] as outlined in Fig. 2.

Although the yield quoted in the patent is 63%, the intermediate (IV) is relatively unstable and pH sensitive, and reaction conditions need to be very carefully defined to achieve a significant yield of MMN.

In 1985 Glaxo[7] adapted the procedure of Gompper and Schaefer,[1] which describes the synthesis of 1-anilino-1-methylmercapto-2-nitroethene from methyl isothiocyanate, to produce MMN (Fig. 3).

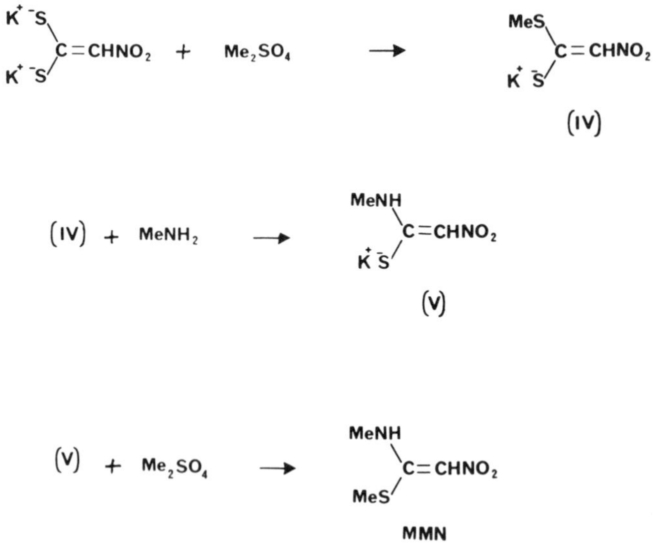


**Fig. 2.** Synthesis of MMN (i).

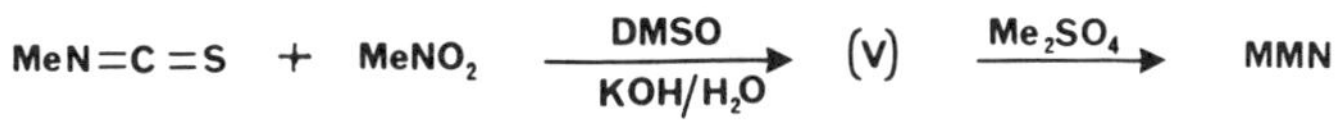

**Fig. 3.** Synthesis of MMN (ii).

The intermediate (V) is the same as for the route described in Fig. 2. The presence of the dimethyl sulphoxide is essential for the reaction to proceed with no product being detected if dimethyl sulphoxide is replaced by dimethylformamide.

The highly substituted dihydropyridine by-product (VI) has been isolated from this reaction, presumably being derived from MMN and a formaldehyde equivalent, and this must be separated from MMN to give material of the required quality for further processing.

(VI)

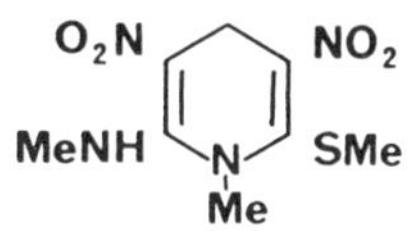

# Potential new uses of BMN

## Pharmaceutical

Bayer have recently patented a series of 2-nitro-1,1-ethenediamines[8] derived from BMN that possess anti-hypertensive properties. An example of the compounds claimed is the 4-aminopyridine (VII), this being related to the newly launched Leo drug, Pinacidil (VIII), in that the nitromethylene group replaces the cyanamide group.

(VII)

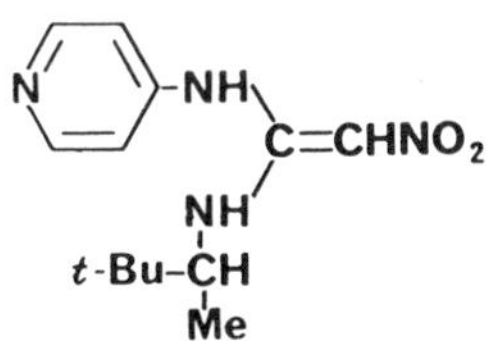

(VIII)

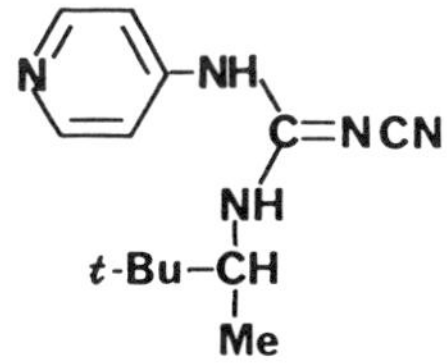

The procedure quoted introduces the pinacolylamine as the final stage of the synthesis, replacement of one of the mercaptan groups with the aromatic amine already having taken place.

Among the potential anti-ulcer drugs based on the nitromethylene group that have been patented is the Bristol Myers propargylamine derivative in which the methylamine group of Ranitidine is replaced by propargylamine. As with Etintidine, the proparglylamine equivalent of Cimetidine, this drug has not been marketed.

## Sweeteners

Further variants on the 2-nitro-1,1-ethenediamine molecule have been claimed as sweeteners, an example being the glycine derivative (IX).

(IX)

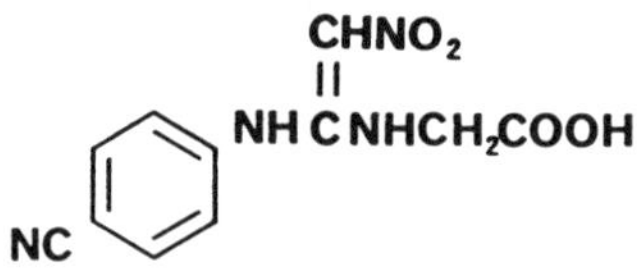


## Insecticides

Among the vast range of potential nitromethylene insecticides investigated by Shell,[10] they have chosen to proceed with 3-formyl-2-nitromethylene tetrahydro-2*H*-1,3-thiazine, this being the compound that combines maximum activity with maximum stability. The intermediate required to synthesise this is the unsubstituted tetrahydro-2*H*-1,3-thiazine (XI), and this in turn can be made from BMN and 3-aminopropanethiol.

However, as the aminopropanethiol is not readily available, Fine Organics Limited have developed a procedure[11] that reacts sodium sulphide with BMN and 3-aminopropylsulphate (APS). Investigation of the reaction has shown that the mechanism is that given in Fig. 4.

The monomethylated intermediate (X) is the same as that (V) proposed for the MMN synthesis outlined in Fig. 2, with the potassium

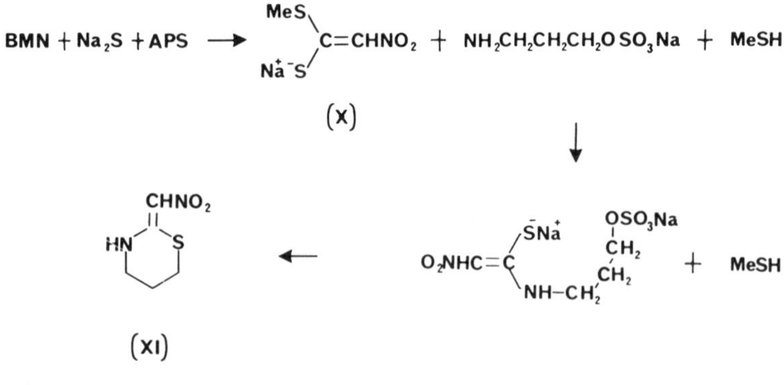


Fig. 4.

anion being replaced by the sodium anion. Reacting (V), prepared as for the MMN synthesis, with aminopropylsulphate does give the expected thiazine but the yield achieved is lower than that obtained when starting with BMN.

A similar range of closely related potential insecticides has been patented by Nikon Tokushu Noyaku Seizo,[12] an example being the *N*-pyridylmethyl derivative of 2-nitromethylene tetrahydropyrimidine (XII).

(XII) 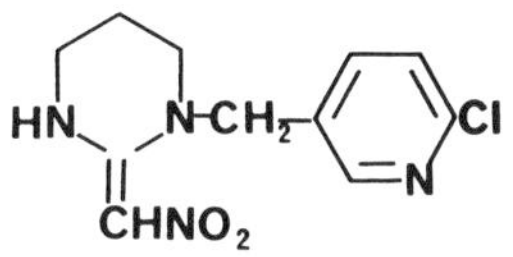

## Conclusion

Following its synthesis in 1967, the chemistry of BMN has grown rapidly, finding commercial application in the manufacture of anti-ulcer drugs, and a large number of potential uses in other areas as outlined above. For a comprehensive review of the chemistry of BMN and derivatives I would recommend that of Tominaga and Matsuda.[13]

## References

1. Gompper, R. & Schaefer, H., *Chem. Ber.*, **100** (1967), 591.
2. Freund, E., *Chem. Ber.*, **52** (1919) 342.
3. Japanese Patent JP60 248 660.
4. *Gass. Chim. It.*, **111** (1981) 217.
5. British Patent GB 1 554 153.
6. Spanish Patent ES 523 448.
7. British Patent GB 2 161 204.
8. Ger Offen DE 3 232 462.
9. European Patent Appl. EP 195 730.
10. Soloway, S. B. *et al.*, Adv. Pestic. Sci. Lect. Symp. Pap. Int. Congr., Vol. 2, 1978, pp. 206–17.
11. European Patent Appl. EP 84 303 421.6.
12. European Patent Appl. EP 192 060.
13. Tominaga, Yoshinori & Matsuda, Yoshiro, *J. Heterocyclic Chem.*, **22** (1985) 937.

# The potential use of nitroparaffins in synthesis

**the Late F. T. Klein & W. Hillenbach**

*ANGUS Chemie GmbH, Ibbenburen, FRG*

## Introduction

The present commercial process to manufacture nitroparaffins began with the work of Prof. Henry Hass *et al.* at Purdue University in 1934.[1] Full industrial production started in 1955 by Commercial Solvents Corporation, predecessor of ANGUS Chemical Company.

This paper is intended to show the industrial chemist an overall practical view of nitroparaffin chemistry. The objective is to point out important nitroparaffin reactions which may be considered by readers in their own development work. The references throughout this paper are generally from open chemical and patent literature.

## Commercial preparation of nitroparaffins and derivatives

Nitroparaffins are produced by a high-temperature, high-pressure, vapour-phase nitration of propane, resulting, after fractionation, in the four commercial-grade nitroparaffins, i.e. nitromethane, nitroethane

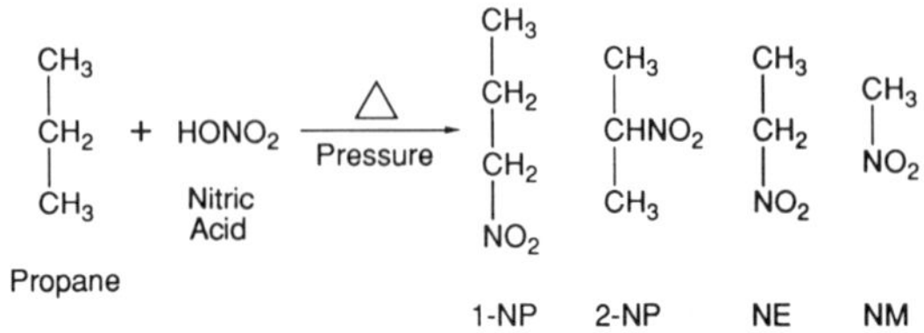

**Fig. 1.** Nitroparaffins synthesis.

and the nitropropanes (Fig. 1). Nitroparaffins are subject to the Henry reaction, whereby they react with aldehydes by addition to form nitroalcohols. The reaction is catalysed by alkalis or KF in alcohol[2] (Fig. 2). Nitroalcohols are reduced to amino alcohols using hydrogen and a suitable metal catalyst, e.g. Raney nickel (Fig. 3). Aromatic nitro compounds may be conveniently reduced at pressures as low as 3·5 bar, but aliphatic nitro compounds are best reduced catalytically at higher pressures to obtain commercially acceptable yields. Chemical reduction of nitroparaffins may be done with Zn/HCl or $Fe/H_2SO_4$ systems.[3] Although chemical reduction is not used commercially, it may be useful for laboratory research purposes.

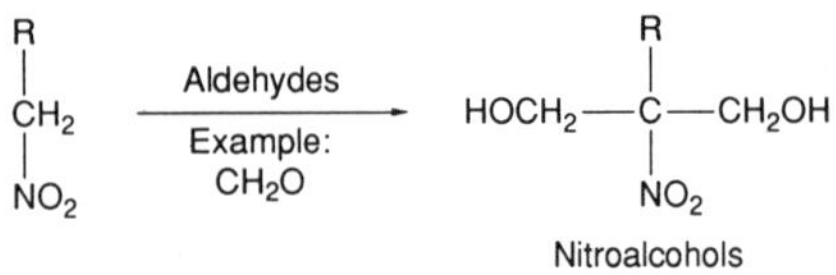

**Fig. 2.**

Commercially available oxazolidines are prepared through the condensation of amino alcohols and aldehydes (Fig. 4). Bicylic oxazolidines can be prepared from amino diols. Although formaldehyde is generally used industrially to give equivalent rings, mixed bicyclics can also be made.[4] The oxazolidines prepared from the

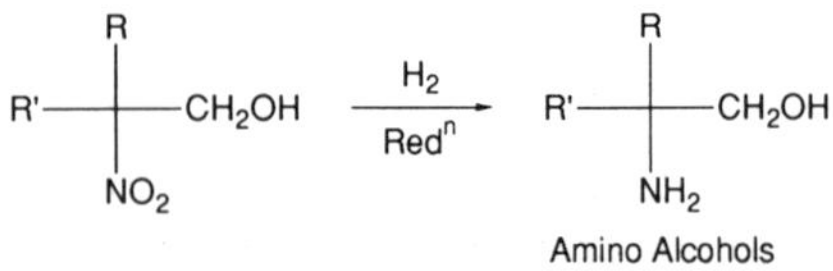

**Fig. 3.**

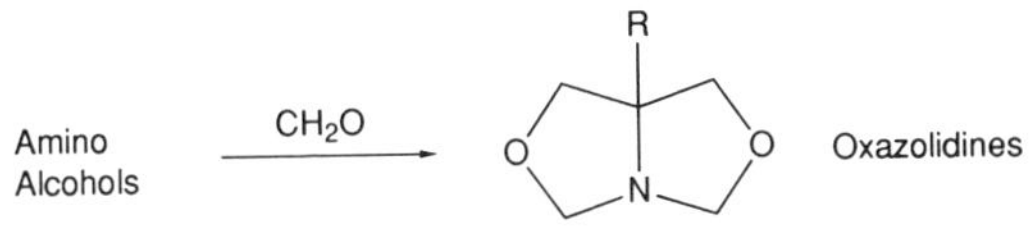


**Fig. 4.**

ANGUS amino alcohols are stable in an alkaline aqueous environment.

Commercially available oxazolines are prepared through esterification of amino alcohols and carboxylic acids (Fig. 5). These oxazolines are soluble in organic solvents, and the ring is stabilized by the presence of the 4,4′-substitution. Hydrolysis of the oxazoline ring will yield the corresponding amide. Oxazolines from $C_1$–$C_3$ carboxylic acids are water soluble.

A comprehensive review of oxazoline chemistry is given by Frump.[5]

# Important reactions of nitroparaffins and derivatives

Nitroparaffins are acidic materials in which the hydrogen atoms on the carbon attached to the nitro group are weakly bound (electron-withdrawing effect of the nitro group). This accounts for their good reactivity. Available nitroparaffins have limited water solubility. When placed in aqueous alkali, however, a water-soluble anion is formed by tautomeric equilibration of the nitroparaffin with its nitronic acid or 'aci' form (Fig. 6). The $\alpha$-carbon of the nitronic acid is a site for entering groups. This enhances the potential carbon atom substitutability of nitroparaffins, yielding nitroalcohols and other derivatives. A large number of reactions of nitroparaffins involve the salt of the parent compound.

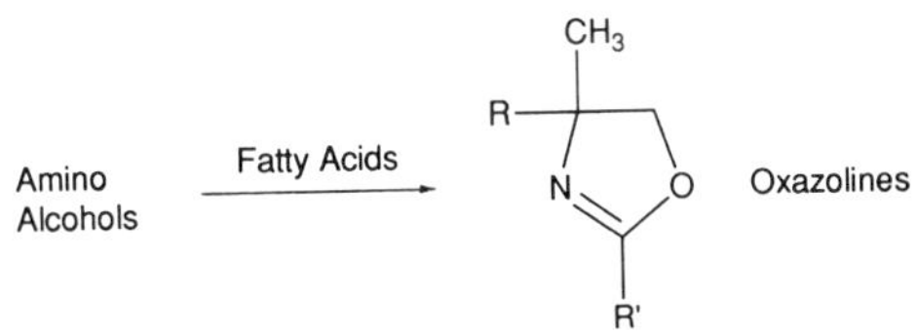


**Fig. 5.**

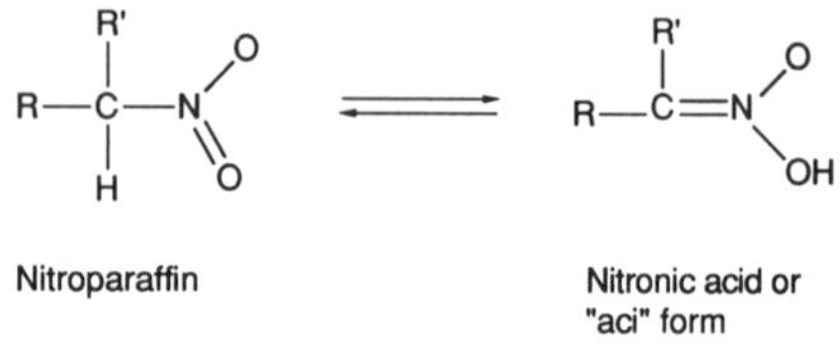


**Fig. 6.**

Nitroparaffins, when properly stored and handled, can be used with safety in industry. A word of caution, however, regarding the handling of alkali salts of nitroparaffins, especially the sodium salt of nitromethane. These intermediates are best used as soon as prepared, and in a wet or slurry form. Isolation and subsequent heating of such salts should be avoided since they are subject to detonation under heat or shock conditions.

Through replacement of the nitro group, the nitroparaffins can be converted to a number of useful derivatives. The Nef reaction, a synthon, converts C—NO to C═O; this can be a useful commercial synthetic tool (Fig. 7).

The classical method is one in which a strong acid is reacted with the alkali nitronate.[6(a)] McMurry describes the titanium trichloride method which is carried out at a more modest pH during the conversion;[6(b)] an electrolysis method has also been reported.[6(c)]

Ono has shown[7] that the nitro group may be replaced with hydrogen. The object in this reaction is to take advantage of the reactivity of the nitroparaffins in preparing the RNO intermediate which is then converted to the target molecule RH. Numerous instances of the use of this reaction have been reported. One paper[8] describes nitroalkanes as alkyl anion equivalents in this reaction, suggesting utility in organic synthesis.

Dehydration of primary nitroparaffins yields nitriles.[9,10] This allows an easy method to prepare substituted alkyl and aryl nitriles, as well as substituted cyanohydrin acetates.

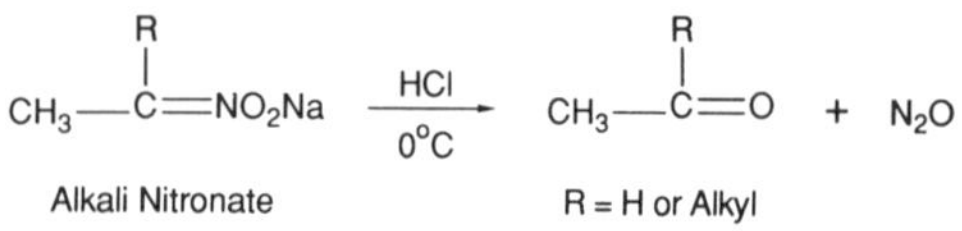


**Fig. 7.** Nef reaction.

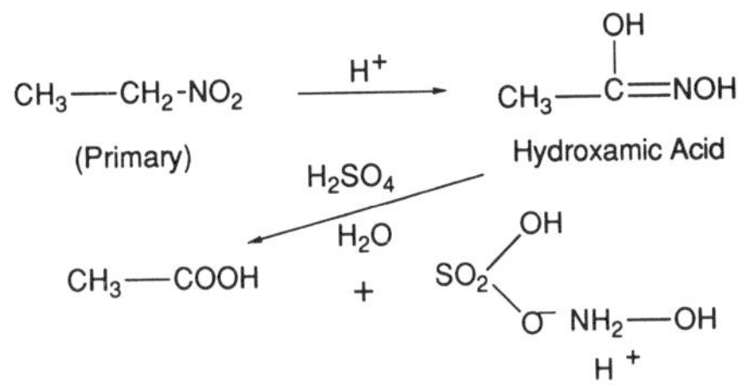


**Fig. 8.**

The action of strong aqueous acid on primary nitroparaffins yields hydroxylamine as the acid salt, along with a carboxylic acid by-product (Fig. 8). The reaction proceeds through a hydroxamic acid intermediate. Hydroxylamine, used in photography as a silver-ion reducing agent, was manufactured for many years by Commercial Solvents using this scheme.

Nitroparaffins are excellent solvents for aluminium chloride in the Friedel–Craft synthesis of compounds such as alkylated aromatics. Nitroparaffins form 1:1 adducts of the Lewis acid catalyst and promote a homogeneous alkylation, which results in a more desirable isomer distribution. Also, polymerizations at low temperature are facilitated. The earlier work was done by Schmerling,[11] but a more recent procedure for chloromethylation of aromatics using nitromethane has also been reported.[12]

Partial reduction of nitroparaffins gives either oximes or alkyl-substituted hydroxylamines[13] (Fig. 9). A tertiary nitro group may be converted to a thiol group.[14]

The Michael addition of nitroparaffins to vinyl compounds, such as acrylates, leads to compounds with both nitro and ester groups. These

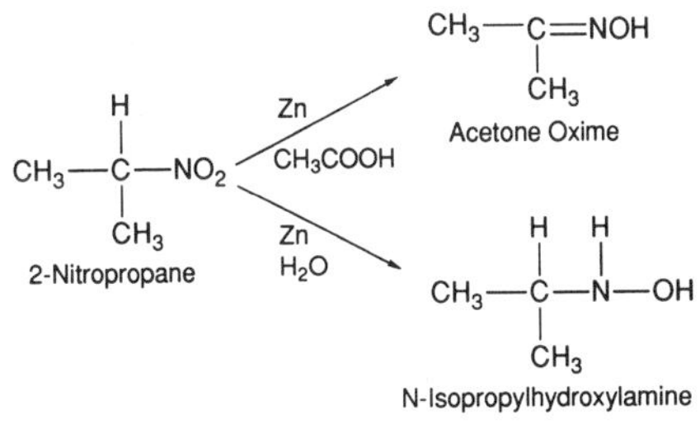


**Fig. 9.** Partial reduction of nitroparaffins.

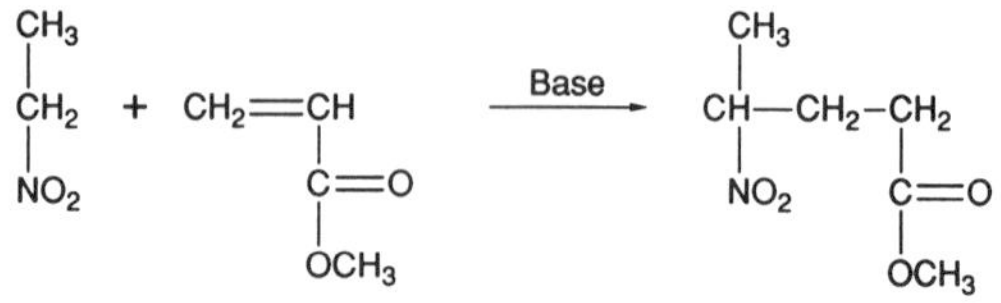

**Fig. 10.** Michael addition of vinyl compounds to nitroparaffin.

adducts may be further reacted to provide useful intermediates[15(a)–(c)] (Fig. 10).

Silyl nitronates, made by reacting a nitroparaffin with trimethyl chlorosilane,[16] show considerable promise for the future (Fig. 11). The versatility of silyl nitronates and the derived 2-isoxazolines as precursors for a number of classes of compounds has been demonstrated. 4-Hydroxy-2-pentenones can be prepared in good yields in few steps from primary nitroparaffins and vinyl ketones. Of special interest is the fact that the cyclization of the hydroxydiones directly gives cyclopentenones suitability functionalized for further elaboration into prostanoids.[17]

Although ketones are generally much less reactive than aldehydes in condensation with nitroparaffins, a publication has detailed the use of high pressure to prepare nitroalcohols from ketones.[18] While Hass' direct nitration scheme does not provide useful yields of nitroketones, oxidation of nitroalcohols gives $\alpha$-nitroketones and has been proven feasible.[19]

The remainder of this paper deals with examples of nitroparaffin chemistry pertaining to specific applications.

# Pharmaceutical applications

—Ethambutol, an anti-tuberculosis drug, is prepared by reacting (*d*)-2-amino-1-butanol with 1,2-dichloroethane.[20] The optically active amino alcohol is prepared from 1-nitropropane (Fig. 12).

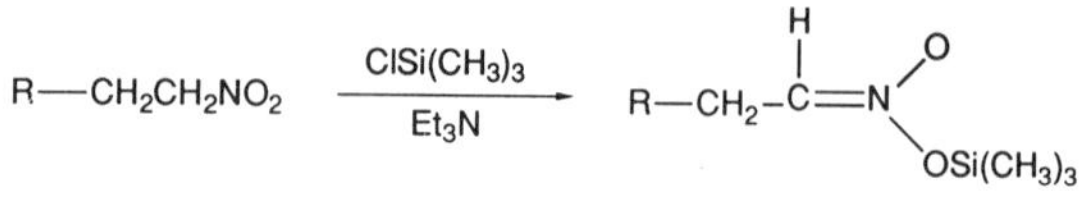

**Fig. 11.**

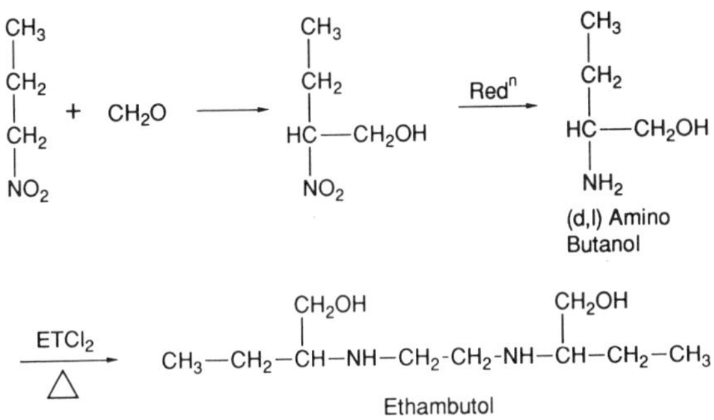


**Fig. 12.**

—α-methyldopa, an anti-hypertensive agent, is manufactured using nitroethane[21] (Fig. 13).
—Ranitidine, an anti-ulcer drug, is synthesized with the aid of a nitromethane derivative.[22(a),(b)]
—Iopamidol, an X-ray contrast agent, is synthesized using 2 moles of the amino diol 2-amino-1,3-propanediol,[23(a),(b)] a nitromethane derivative.
—Prostaglandins, via silyl nitronates, are described above.
—Amino alcohols are used as solubilizers for water-insoluble drugs, e.g. bromotheophylline and 2-amino-2-methylpropanol.[24]
—Amino alcohols are used to prepare physiological buffers, for example from tris(hydroxymethyl) amino methane.
—Bucindolol,[25(a),(b)] a beta blocker, incorporates a tertiary amine moiety. This is conveniently incorporated via the reaction of 2-nitropropane and a tertiary amine.[26]
—An example of the use of nitroparaffins in betalactam chemistry involves nitromethane, Michael addition and the Nef reaction[27] to give an antibacterial precursor.

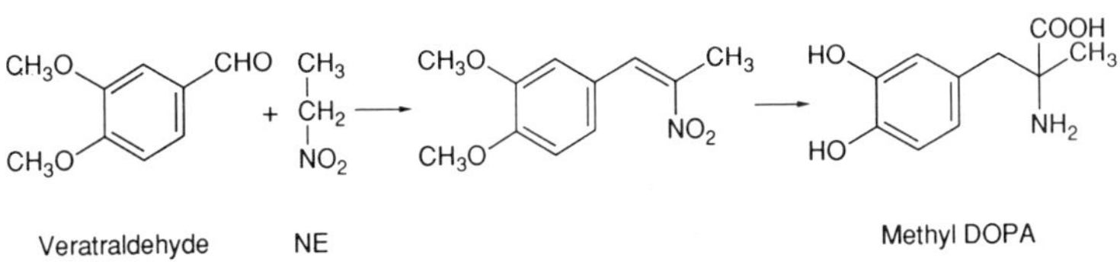


**Fig. 13.**

—It was noted earlier that ketones are not very reactive with nitroparaffins. Steroid chemistry[28(a),(b)] is an exception. It has been shown that nitromethane can be added to a cyclic ketone on a steroid molecule after which the nitroethane moiety is reacted further.
—A new class of anti-tumor agents has been described.[29] These utilize an amino diol made from nitromethane and formylated chrysene to yield the (soluble) hydrochloric salt.

## Agrochemical applications

Some examples on the use of nitroparaffins in known and developmental agricultural–chemical applications, viz.:

—One of the more important products in the area is chloropicrin, derived from nitromethane. This nematocide is easily made by reacting nitromethane with sodium hypochlorite under alkaline conditions (Fig. 14).
—One route to the insecticide Methomyl[30] involves nitroethane, which is reacted with potassium methoxide, methyl mercaptan and methanol to give a hydroxamate ester. This is reduced and methyl isocyanate added to yield the final carbamate.
—The nitromethane derivative BSM (1,1-bis(methylthio)-2-nitroethene) reacts with not only amine groups but with thiols as well. This is used in the preparation of a new class of light- and oxidation-stable insecticides which control earworms and cotton earworms among others.[31]
—In the 1960s Commercial Solvents manufactured the insecticides Prolan and Bulan, based on nitroethane and 1-nitropropane respectively as countertypes to DDT. Although very effective, the products were not a real commercial success because of the 'too short half-life' compared to DDT.

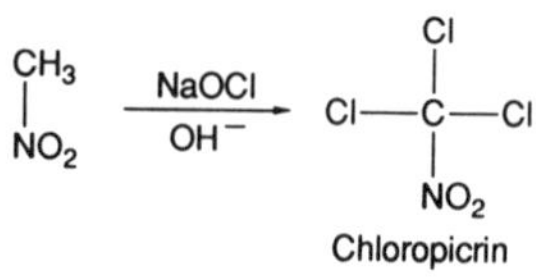


**Fig. 14.**

# Biocide applications

Nitroparaffins find widespread use in the preparation of anti-microbial compounds. Some examples are given below:

—Nitroalcohols are acidic, water-soluble compounds formed via the Henry reaction. Under alkaline conditions this reaction is reversed, yielding formaldehyde. Tris(hydroxy-methyl) nitromethane is a nitroalcohol made from nitromethane and formaldehyde which finds use as a bactericide in metalworking fluids and water treatment.

—The Mannich reaction product of 1-nitropropane, formaldehyde and morpholine, called Bioban P-1487, is also used in metalworking fluids. The Mannich reaction in this case gives a mixture of two products; this mixture is used as the biocide. The oil-soluble product is not a formaldehyde donor, and effectively controls both bacteria and fungi. Formaldehyde releasers are usually not effective against fungi.

—Another Mannich product made from benzaldehyde, nitroethane, and ethylene diamine is an effective slimicide for paper and pulp systems.[32]

—The last Mannich derived product to be cited is Hexetidine. This is made starting with nitroethane, formaldehyde and 2-ethylhexylamine[33(a)–(c)] (Fig. 15). Hexetidine is an anti-bacterial and anti-fungal agent with excellent toxicity properties, used, for example, in mouth-wash formulations.[34(a),(b)] Also, synergistic

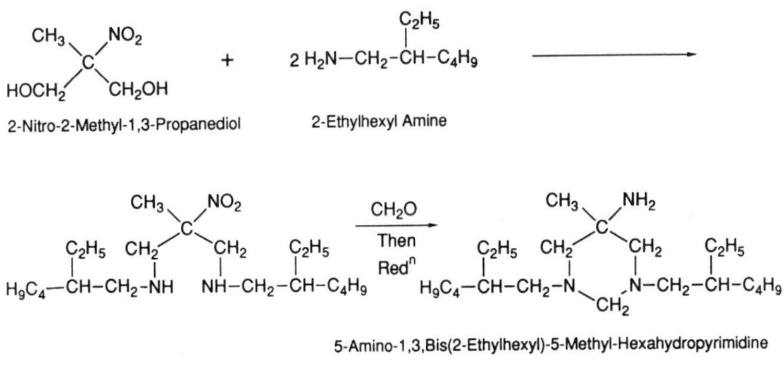


**Fig. 15.**

combinations of Hexetidine and zinc salts have been reported to reduce dental plaque.[35]

—Several types of halo–nitro compounds are effective biocides. BNPD (2-bromo-2-nitro-1,3-propanediol) is prepared from nitromethane. Careful control of reaction conditions in the Henry reaction leads to the sodium salt of the nitronic acid form of the dihydroxy derivative. This is brominated to give the desired product.[36(a)–(e)] Bronidox (Henkel), the formaldehyde acetal synthesized from BNPD, is also used in cosmetic formulations. The acetal form ensures a product with a longer shelf-life.

—Reaction of brominated nitroalcohols with isocyanates leads to bromonitroalkylcarbamates. These products have been shown to be effective against bacteria and fungi, combining these properties, with a favourable toxicity profile.[37]

—Bromonitrosytrene is a powerful anti-bacterial and anti-fungal agent. It has found extensive use in prevention of algae growth in industrial cooling water systems.[37(a),(b)] This halonitro compound is made by first preparing $\beta$-nitrostyrene via the Knoevenagel reaction using nitromethane and benzaldehyde. Under the reaction conditions the methylene group in nitromethane is activated, and the reaction proceeds with water elimination to give the nitroolefin, which is then brominated, and finally dehydrobrominated to yield the algaecide. Bromonitrostyrene has been reported to be effective against molluscs whose growth clogs up water intakes in electrical utility facilities.[39]

—Oxazolidines, the reaction products of aminoalcohols and aldehydes, are used to protect aqueous systems against microbial spoilage. Moreover the hydrogen on the nitrogen atom of the monocyclic oxazolidine can be displaced by halogens. Reaction with halo compounds such as sodium hypochlorite results in derivatives that act as disinfectants, which have been tested and found effective in purifying drinking watcr for farm animals.[40(a),(b)]

## Coatings applications

—Nitroparaffins are used in solvent-based coatings where high performance is required. Nitroparaffins may be the deciding factor to solve problems such as wetting, flow and solvent release.

These properties, combined with the electrical properties, make nitroparaffins choice solvents for the automotive industry. Nitroparaffins are also used as components of stabilizer packages for halogenated solvents which are used in the degreasing of metal parts prior to further treatment or finishing. Many types of aerosol propellant formulations also use this technology. Articles that discuss this subject and point out the value of nitroparaffins have been published.[41(a),(b)] It is noted that nitroparaffins are Lewis bases and may function by passivating metal surfaces by making complexes or salts with the formed metal halides.

—Vinyl oxazolines and esters, prepared from amino alcohols and carboxylic acids, are valuable adhesion promotors for automotive finishes.[42]

—The functionality of the monocyclic oxazolidine includes a secondary amine. This function has been found to be useful as a 'blocking agent' for acid catalysts for high-solids, organic solvent bake finishes.[43] This amine salt permits a stable one-package system which liberates the free acid at a low temperature so as to permit a low bake schedule.

—It is known that some amino alcohols and ketones condense to give hindered amines which are precursors to nitroxide free-radicals.[44] These have value as UV light stabilizers for polyolefin films, referred to as hindered amine light stabilizers (HALS). This chemistry has been advanced to include additional hindered amine structures.[45]

—A British patent claims the use of tertiary nitroso compounds to provide UV resistance for polymer films.[46] This development might utilize the potential value of acrylate monomers made from 2-nitro-2-methyl-1-propanol.[47(a),(b)] Such monomers might be incorporated into polymers and ultimately produce non-extractable UV stabilizers.

## Oil industry applications

Both oxazolines and oxazolidines, based on nitroparaffin-derived amino alcohols, are well known in the oil-related industry. Especially known here is the tris(hydroxymethyl)—amino methane. The derivatives thereof function as cleaners, sludge dispersants, etc., as demonstrated in the following examples, viz.:

—Bicyclic oxazolidines, in transmission fluids, act as copper alloy corrosion inhibitors.[48] Long-chain esters of bicyclic oxazolidines act as friction modifiers, etc.[49(a),(b)] A marine crankcase lubricant is based on an acyl glycine oxazoline.[50] Lactone oxazolines are effective sludge dispersants.[51] Some oxazolines are carburettor detergents.[52(a),(b)] Finally, boric acid derivatives involving amino alcohols are used as detergents and friction reducers.[53(a),(b)]

—Nitroethane has been reported to be a useful dearomatizer and dewaxer of various crude oils.[54(a),(b)]

—Patent literature indicates some amino alcohols, especially 2-amino-2-methyl-1-propanol, may be useful in enhanced oil-recovery operations.[55(a)–(c)]

# Energy applications

—A well-known use for nitromethane is in fuels, where it is used to increase power from the engine. Starkman earlier studied the effect of nitroparaffins on internal combustion engines.[56] Nitromethane has been described as a fuel with a built-in source of oxygen. While the BTU value of nitroparaffins is less than paraffins, their oxygen balance and energy potential provide power when needed.

—Nitromethane is used commercially as a ditching agent and in shaped-charge form to clear underwater obstructions in harbours, etc.

—Liquid plasticizers are used in missile propellant systems to keep the solid component from drying out and caking. These plasticizers must be of the energetic type, i.e. they should not detract from the total energy of the system in use. Nitromethane is effectively used in this area to prepare gem-dinitro compounds which are energetic and which yield a liquid plasticizer.[57] 2,2-Dinitropropanol is the active ingredient made from nitroethane to produce a mixed acetal. Mixtures of the final mixed acetal product and dinitrofluoro alcohols are also known.[58]

—Nitroparaffins are not useful as gasoline octane boosters. However, they are cetane boosters in diesel fuel. Recently it was shown that the nitrate esters of some nitroalcohols may be used as cetane improvers[59] and in alcohol-fuelled diesel engines.

## Acknowledgement

Robert E. Kass, without whose bibliographical research and compilation of data this paper would not have been possible.

In memoriam, F. T. Klein.

## References

1. US Patent 1 967 667.
2. Wollenberg, R. & Miller, S., *Tetrahedron Letters*, **35** (1978) 3219–22.
3. Senkus, M., *Ind. & Engr. Chem.*, **40** (1948) 506–8.
4. Pierce, J. *et al.*, *J. Am. Chem. Soc.*, **73** (1951) 2595–6.
5. Frump, J., *Chemical Reviews*, **71** (1971) 483–505.
6. (a) Noland, W., *Chemical Reviews*, **55** (1955) 137.
   (b) McMurray, J., *J. Org. Chem.*, **38** (1973) 4367.
   (c) Nokami, J. *et al.*, *Synthesis* (1983) 763.
7. Ono, N., *Tetrahedron Letters*, **22** (1981) 1705.
8. Ono, N. *et al.*, *J. Org. Chem.*, **50** (1985) 3692–8.
9. Olah, G., *Synthesis* (January 1979) 36.
10. Wehrli, P., *J. Org. Chem.*, **42** (1977) 3956.
11. Schmerling, L., *Ind. & Engr. Chem.*, **40** (1948) 2072.
12. Long, D. *et al.*, *Tetrahedron Letters*, **24** (1983) 1933.
13. French Patent 1 322 691.
14. Kornblum, N. *et al.*, *J. Am. Soc.*, **100** (1978) 7086.
15. (a) Chasar, D., *Synthesis* (1982) 841.
    (b) Clark, J. *et al.*, *Chem. Letters* (1983) 1145.
    (c) Ono, N. *et al.*, *Synthesis* (1984) 226.
16. Torssell, K., *Acta Chem. Scand.*, **B32** (1978) 118.
17. *Tetrahedron Letters*, **13** (1983).
18. Matsumoto, K., *Angew. Chem.*, **96** (1984) 599.
19. Rosini, G., *Synthesis* (1984) 607.
20. US Patent 3 176 040.
21. US Patent 2 868 818; 3 158 648.
22. (a) Freund, E., *Berichte*, **52** (1919) 542.
    (b) Gompper, R. *et al.*, *Berichte*, **100** (1967) 591.
23. (a) Pitre, D. *et al.*, *Il Farmaco*, **32** (1978) 835.
    (b) US Patent 4 001 323.

24. Bickers, W. *et al.*, *New Eng. J. Med.*, **245** (1951) 453.
25. (a) US Patent 4 234 595.
    (b) German Patent 3 421 252.
26. Synder, H. *et al.*, *J. Am. Soc.*, **69** (1947) 3140.
27. European Patent Appl. 112, 481, November 1983.
28. (a) US Patent 4 490 595.
    (b) Barton, D. *et al.*, *J. Chem. Soc., Chem. Commun.*, **10** (1982).
29. European Patent Appl. 125 701.
30. US Patent 3 576 834.
31. US Patents 3 993 648; 4 483 856.
32. Federal Register 121.2505; EPA Register 618-78-AC.
33. (a) US Patents 2 387 043; 2 415 047.
    (b) Senkus, M., *J. Am. Chem. Soc.*, **68** (1946) 1611.
    (c) US Patent 3 072 529.
34. (a) Saski, W., *J. Pharm. Sci.*, **54** (1965) 277.
    (b) Hexahydropyrimidine Symposium—Northwestern University, Evanston, Ill. 23 October 1958.
35. US Patent 4 522 806.
36. (a) Spanish Patent 453 224.
    (b) UK Patent 1 057 131.
    (c) UK Patent 1 193 954.
    (d) US Patent 3 558 788.
    (e) US Patent 4 488 999.
37. German Patent 1 688 195.
38. (a) US Patent 3 629 465.
    (b) US Patent 3 871 860.
39. US Patent 4 561 983.
40. (a) Worley, S., *Water Resources Bulletin*, **17** (1981) 874.
    (b) US Patent 3 931 213.
41. (a) Gemert, B., *I&EC Prod. Res. & Dev.*, **21** (1982) 296.
    (b) Archer, W., *I&EC Prod. Res. & Dev.*, **18** (1979) 131.
42. US Patent 3 884 852.
43. US Patent 4 200 729.
44. Keana, J. *et al.*, *J. Am. Chem. Soc.*, **89** (1967) 3055.
45. Lai, J., *Synthesis* (1984) 122.
46. UK Patent 2 147 300.
47. (a) Marans, N. *et al.*, *J. Am. Chem. Soc.*, **72** (1950) 2125.
    (b) Rehberg, C. *et al.*, *J. Am. Chem. Soc.*, **72** (1950) 5199.
48. US Patent 4 277 353.
49. (a) US Patent 4 017 406.
    (b) US Patent 4 069 023.
50. US Patent 4 386 001.
51. US Patent 4 174 322.
52. (a) US Patent 3 033 661.
    (b) US Patent 3 033 663.
53. (a) US Patent 4 374 032.
    (b) US Patent 4 016 092.

54. (a) Russian Patent 195 587.
(b) Butovich, N. A., *Refining Liquid Paraffins with Nitroethane,* KAZAN Chemical Engineering Institute.
55. (a) US Patent 4 536 300.
(b) US Patent 4 485 021.
(c) US Patent 4 522 732.
56. Starkman, E. *Ind. Eng. Chem.*, **51** (1959) 1477.
57. Hamel, E. *et al., Ind. & Engr. Prod. Res. & Dev.*, **1** (1962) 108.
58. US Patent 4 453 021.
59. US Patent 4 417 903.

# Phosphine based organophosphorus products and their applications

**N. Weferling**

*Hoechst AG, Hürth-Knapsack, FRG*

## Introduction

The term 'organophosphorus compound' refers here and in the following review to classes of substances that have one to four P—C bonds. Substances that contain one or more P—O—C bonds and are generally also described as 'organophosphorus compounds' are not included here.[1]

Industrially, there are generally two different ways of obtaining these compounds, starting from elemental yellow phosphorus (Fig. 1). The so-called 'Grignard route'[2] requires $PCl_3$ as the basic substance, which is reacted with alkyl magnesium halide, for cost reasons usually the chloride. Trialkylphosphines and phosphinious acid chlorides are formed in various amounts, depending on the nature of the ligand R. If $P(:O)Cl_3$ is used as the starting material, trialkylphosphine oxides and phosphinic and phosphonic acid chlorides are formed. Today, triphenylphosphine, tributylphosphine and trioctylphosphine oxide are being produced on an industrial scale by the Grignard method. For the sake of completeness, mention should also be made here of the production of triphenylphosphine which is obtained by reaction of

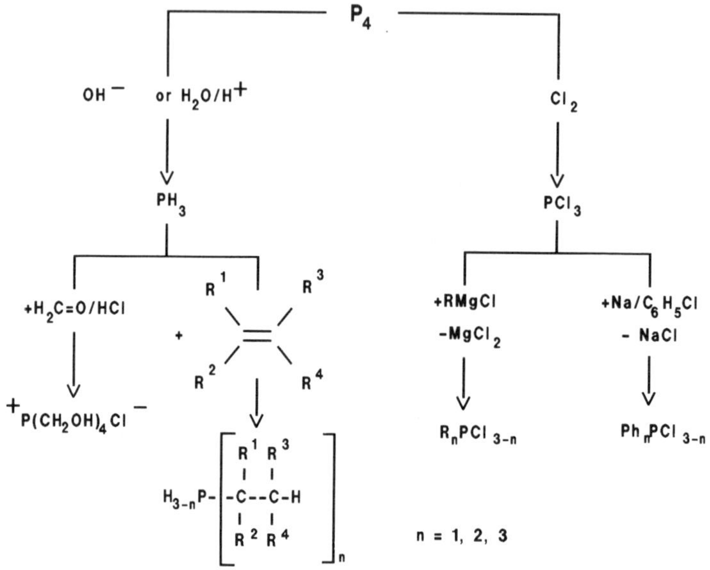


**Fig. 1.** Industrial manufacture of organophosphorus compounds.

$PCl_3$, sodium and chlorobenzene in a synthesis analogous to the Wurtz reaction (BASF, M&T Chemicals[3]).

The alternative method of producing organophosphorus compounds is based on phosphine, which is formed either by acid[4] or alkaline[5] disproportionation of yellow phosphorus. The addition of phosphine to carbon-element double bonds,[6,7] for instance to formaldehyde with formation of tetrakis(hydroxymethyl)-phosphonium compounds, will only be mentioned in passing here, although these phosphonium salts represent an important share, in terms of quantity, of all $PH_3$ derivatives.

A more detailed account is given below of the radically catalysed addition of $PH_3$ to olefins.[7,8] Today, azo initiators, especially the inexpensive azobisisobutyronitrile (ABIN), is used almost exclusively as the radical chain starter for this type of reaction. The reaction takes place along the lines of an anti-Markownikow addition.

In many cases a mixture of primary, secondary and tertiary phosphines is obtained. With regard to a production on a technical scale the process development, which allows access to the desired compound with high selectivity, is essential.

# Industrial $PH_3$ production

The commercial manufacture of products from $PH_3$ is only possible if the gas can be made available in adequate quantities and quality. Of the numerous processes that are described in the literature for $PH_3$ production only two are used on an industrial scale (Fig. 2).

### Acid Process for $PH_3$ Generation

Yellow phosphorus is to a certain degree converted into the red modification at temperatures of more than 300°C and is then converted into $PH_3$ and phosphoric acid with steam at about 280°C.[4] The presence of the acid catalyses the reaction. Under these extreme conditions it is virtually only possible to use special dense graphite as the reactor material. The high investment costs of the plant are offset by an excellent yield of hydrogen phosphide, which is about 60%. Moreover, phosphoric acid as the other relatively valuable reactant product is obtained.

### Alkaline Process for $PH_3$ Generation

The alternative alkaline process[5] is technically less complicated and is less cost intensive, although here, too, serious problems of materials

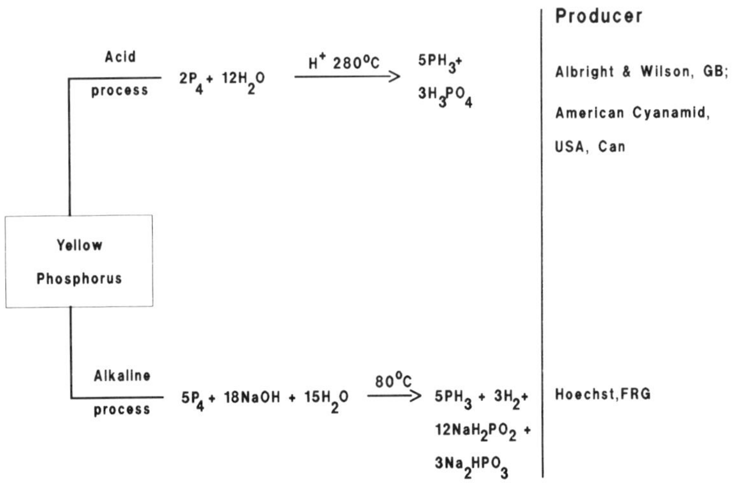


**Fig. 2.** Commercial production of phosphine.

and processing have to be solved beforehand. It is worth noting that the yield obtained by this process, at more than 30% $PH_3$ relative to the yellow phosphorus used, is considerably higher than the 25% expected in theory. Obviously the reaction equation does not accurately reflect the complicated conditions in practice.

Caustic soda in aqueous solution and molten yellow phosphorus are metered into the reactor together with a long chain alcohol which has a dispersing function. By a skilful control of the process the gas used for further reaction is mixed with relatively little hydrogen. The content of $PH_3$ in this gas mixture is more than 95% by volume. The by-product, a solution of phosphite and hypophosphite, can be further processed into condensed phosphates, e.g. by oxidative spray drying.

## Alkyl phosphines from $PH_3$

It has been already mentioned that the selectivity of the $PH_3$/olefin reaction, in other words the amount of primary, secondary and tertiary phosphines formed, is of major importance for the producer. It is clear that the constitution and the configuration of the olefin has a decisive influence on the product distribution and the conversion ratio.

Further variables are the stoichiometric ratio of the reactants, the specific quantity of the initiator and last, but not least, the type of the reactor. Figure 3 gives a summary of those classes of products that are accessible from $PH_3$.

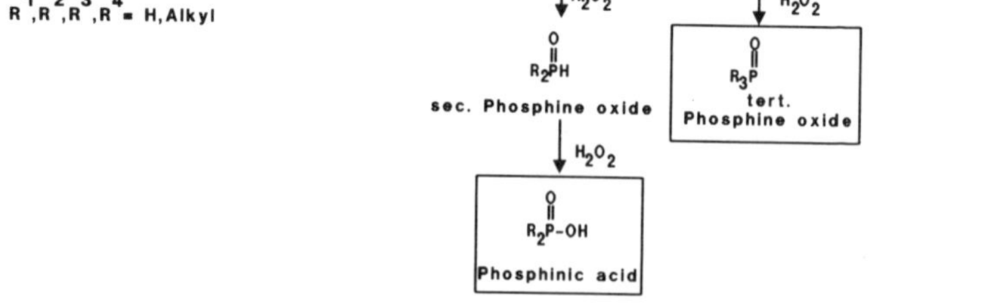


**Fig. 3.** Manufacture of organophosphorus compounds from phosphine.

Secondary phosphines may be converted with oxidising agents, preferably an aqueous solution of hydrogen peroxide, into phosphinic acids. Secondary phosphine oxides are isolable as intermediates. If sulphur is used as oxidising agent, the corresponding thio- and di-thiophosphinic acids, respectively, are formed.

Tertiary phosphines are starting materials either for quaternary phosphonium salts (reaction with alkyl halides) or for tertiary phosphine oxides.

The production of tertiary phosphines, from $PH_3$ and excess linear $\alpha$-olefins, presents the fewest problems[8,9] (Fig. 4). The reaction of, for example, ethylene, butene-1 or octene-1 occurs quantitatively, relative to $PH_3$; little or no secondary and primary phosphines are formed.

Only in a few exceptional cases can cycloolefins, branched $\alpha$-olefins or $\beta$-olefins be converted with $PH_3$ into tertiary phosphines (Fig. 5). In the reaction of $PH_3$ with cyclohexene a mixture of primary and secondary phosphines (molar ratio 2:1) is formed with an acceptable degree of conversion of $PH_3$. This reaction is an ideal case to show the effect of reaction conditions, which can be freely selected, on product selectivity. In fact it is possible, by increasing the molar ratio of olefin/$PH_3$ and the amount of starter, to lead the reaction in a pressure reactor in such a way that the amount of primary phosphines approaches zero in favour of the yield of dicyclohexyl phosphine.

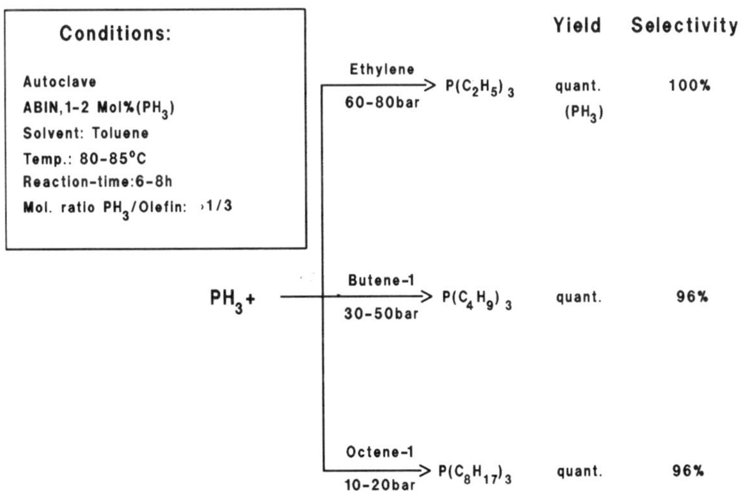


**Fig. 4.** Tertiary phosphines from $PH_3$.

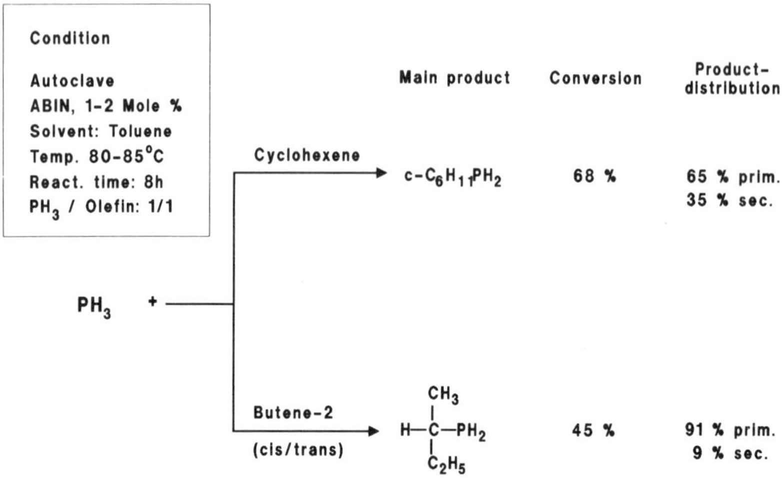


**Fig. 5.** Phosphines with sterically hindered ligands from $PH_3$/Olefin.

Moreover, no tertiary phosphine is formed during this process.[10] Conversely, it is possible to produce monocyclohexyl phosphine almost exclusively.

The reaction of butene-2 with $PH_3$ produces a moderate yield of a product mixture consisting of more than 90 mol% of primary phosphine.[11] With these two examples, the number and kind of those primary phosphines, which are accessible from $PH_3$ and olefins in technically interesting yields and conversion ratios, are nearly complete.

Some olefins react with $PH_3$ in an autoclave mainly to form secondary phosphines (Fig. 6). With a reactant ratio $PH_3$/olefin of 1:2 (in the case of $PH_3$/COD 1:1) and conditions otherwise similar to those described above, with 1,5-cyclooctadiene 73 mol% of an isomeric mixture is formed consisting of two secondary phosphines, 9*H*-9-phosphabicyclononane, small amounts of a bicyclic primary phosphine and not inconsiderable quantities of tertiary phosphines.

The analogous reaction with dimeric cyclopentadiene[12] produces an isomeric mixture of secondary phosphines (more than 10 signals are found in the $^{31}P$ NMR spectrum) in surprisingly good yield and selectivity.

The tendency of isobutene to react with $PH_3$ to form secondary phosphines is less pronounced. The reaction behaviour of isobutene

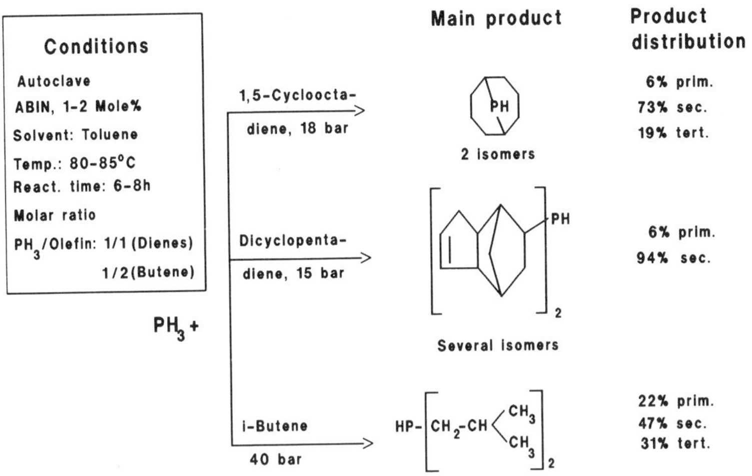


**Fig. 6.** Phosphines with sterically hindered ligands from $PH_3$/Olefin II.

approximates to that of the linear $\alpha$-olefins which would have produced about 60 mol% tertiary, 30 mol% secondary and 10 mol% primary phosphine under the same reaction conditions.

The dependence of the product distribution of the $PH_3$/olefin reaction conditions chosen, especially the reactor type, is shown by Fig. 7. When cyclohexene, butene-2 or 1,5-cyclooctadiene is reacted

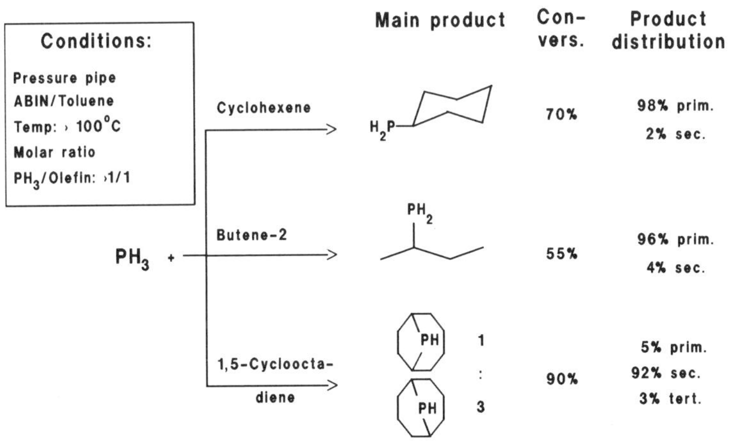


**Fig. 7.** Reaction conditions and selectivity; primary and secondary phosphines from $PH_3$/Olefin.

with excess $PH_3$ at high pressure and temperature in a pressure pipe[12] instead of an autoclave, a narrow product distribution is found in combination with a good-to-excellent conversion ratio; tertiary phosphines are not formed at all, or only in small quantities.

## Industrial production of alkyl phosphines from $PH_3$

The industrial production of alkyl phosphines from $PH_3$ is outlined schematically in Fig. 8. The described plant was constructed at the end of the seventies at our Knapsack site. The process was developed using Hoechst know-how. Strict regulations of the German authorities have been adhered to.

The $PH_3/H_2$ mixture produced in an upstream part of the plant is compressed and fed into the high pressure reactor. At the same time a solution of olefin, solvent and initiator, which was made in a separate vessel, is metered into the reactor by means of a high pressure pump. The $PH_3$ liquefies under the conditions present in the upper part of the pressure pipe so that hydrogen as a non-condensable gas can be removed at the head of the column. The educts react on their way through the column in an exothermic reaction. The crude product is

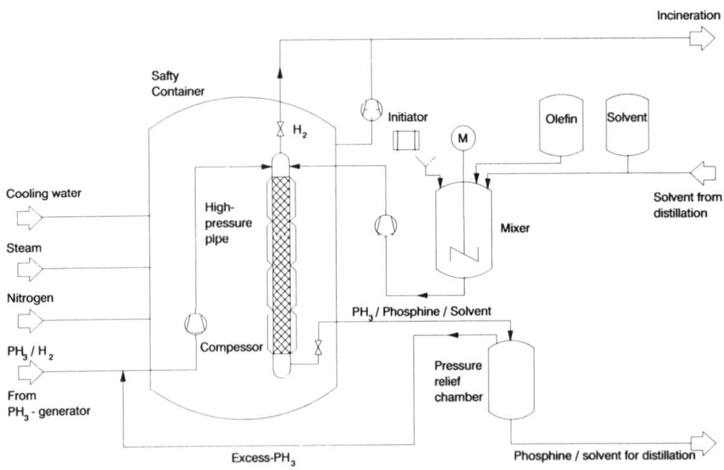


**Fig. 8.** Technical alkylphosphine production.

passed into a pressure releasing vessel. The gaseous phase, excess $PH_3$, is transported back to the compressor. The crude product is distilled and the low ends are returned to the mixing vessel.

The process described is designed for the production of the primary and secondary phosphines mentioned earlier. The continuous production of tertiary phosphines is achieved by exchanging the pressure relief chamber for a pressure reactor of appropriate volume to ensure a sufficient residence time of the reactants.

For safety reasons the parts of the plant in which $PH_3$ is handled under pressure are housed in a safety casing vessel.

## Industrial application of alkyl phosphines

The possible uses of organophosphorus products based on $PH_3$ and described in the literature are too numerous to mention. Their use on a commercial scale, however, is still narrowly restricted. The following examples give a rough insight into the industrial use of phosphororganics so far mentioned.

The isomeric mixture of 9*H*-9-phosphabicyclononanes, products from the reaction of $PH_3$ with cyclooctadiene, is converted with a long chain $\alpha$-olefin into the corresponding tertiary phosphines. The latter are used as ligands in the cobalt-catalysed hydroformylation of $C_{10}/C_{12}$ linear olefins (Fig. 9). The products are distributed by Shell (Dobanol® alcohols) and predominantly used in detergents.

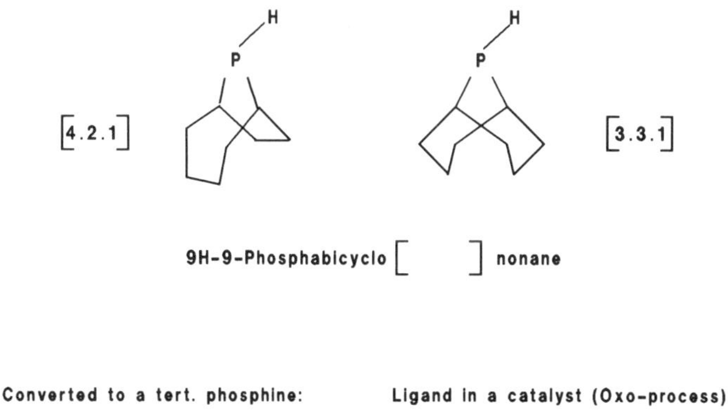


**Fig. 9.** Industrial applications of phosphororganics I.

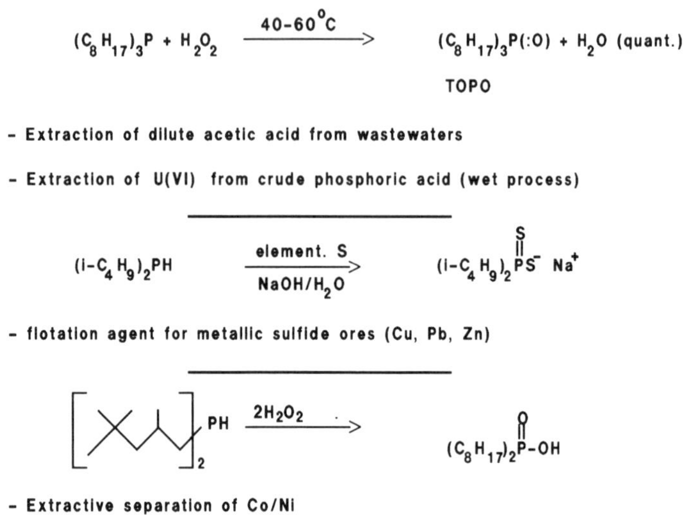


**Fig. 10.** Industrial applications of phosphororganics II.

Trioctyl phosphine is oxidised with $H_2O_2$ solution to form the compound that is commonly known by the acronym TOPO (Fig. 10). The product is used in two different extraction processes. A relatively new process is the extraction of dilute by-product acetic acid[13] from the effluent, which results from cellulose manufacture or the production of acetaldehyde. Thus the pollution of biological purification stages of effluent treatment plants is greatly reduced by the use of TOPO and a saleable substance is recovered. More than a half of the acetic acid requirements for the foodstuff sector in Austria are met by the acid extracted by means of TOPO. The second field of application for TOPO is better known, namely the extraction of uranium (VI) from crude phosphoric acid, which is obtained by decomposing sedimentary phosphates with sulphuric acid.[14]

The sodium salt of diisobutyldithiophosphinic acid is used as a flotation agent[15] in the dressing of copper, lead and zinc ores. Bis(2,4,4-trimethylpentyl)-phosphonic acid, available by oxidation of the secondary phosphine generated by the reaction of $PH_3$ and diisobutylene, is used for extractive Co/Ni separation.[16]

Tributyl phosphine is reacted either at Hoechst or by the customer with alkyl or aryl halides to form the corresponding phosphonium halides (Fig. 11). The reaction conditions vary slightly depending on

$$(C_4H_9)_3P + RHal \xrightarrow[1\text{-}4\ bar]{80\text{-}130\,^\circ C} (C_4H_9)_3\overset{+}{P}R\ \overset{-}{Hal} \quad (95\text{-}100\%\ yield)$$

RHal = n-$C_4H_9$-Br, -Cl; n-$C_{16}H_{33}$Br; $CH_3I$; 2,4-Dichlorobenzylchloride

**Applications**

- Phase-transfer-catalyst
- Promoter, e. g. in the reaction of $P_4S_{10}$ with alcohols
- Biocide (algicide, anti-fouling-paints)
- Solvent/promoter for carbonylation reactions
- Plant-growth-regulator

**Fig. 11.** Industrial applications of phosphororganics III.

the halide. Very good to excellent yields and product qualities can be obtained in every case. The fields of application known at present for phosphonium salts can be described briefly by the following terms: phase-transfer catalyst;[17] promoter in reactions of $P_4S_{10}$ with alcohols;[18] biocides, especially anti-fouling paints;[19] salt-like solvent and promoter in carbonylation reactions;[20] plant growth regulator.[21]

Unlike the compounds reviewed up to now, products that can be manufactured from the above-mentioned *sec*-butylphosphine are, as

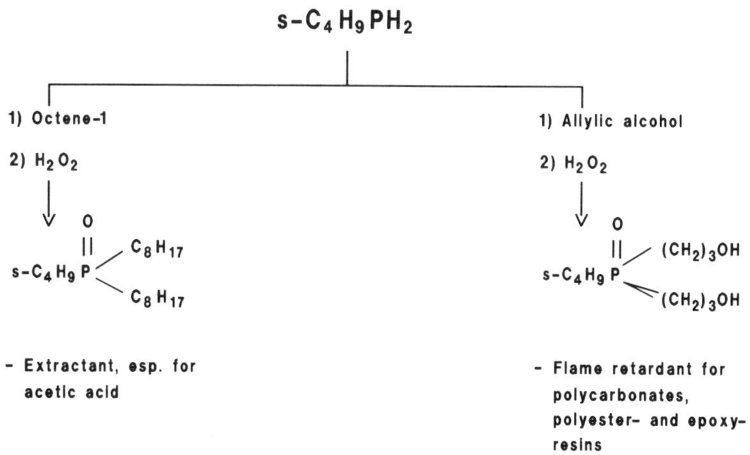


**Fig. 12.** Products based on *sec*-butylphosphine and their applications.

far as we are aware, not yet being produced in industrial-scale quantities. The remarkable properties of some of these products could, however, soon lead to a corresponding demand in the market. We believe this is especially likely in the case of *sec*-butyldioctylphosphine oxide (Fig. 12). Compared with TOPO, this tertiary phosphine oxide possesses decisive application advantages and better properties with regard to the extraction of organic acids, in particular acetic acid, from diluted aqueous solutions.[22]

Secondary-butyl-bis(3-hydroxypropyl)-phosphine oxide conforms to the universally discernible trend away from halogen-containing flame retardant agents. It has proved especially effective in epoxy resins, polyurethanes and polyphenylene ether resins.[23]

## Final remarks

The handling and processing of the highly toxic, spontaneously flammable $PH_3$ on an industrial scale presents a challenge to chemists and engineers. With its high safety standards and qualified staff, Hoechst—as well as other phosphine-manufacturing companies in Great Britian (Albright & Wilson) and USA/Canada (American Cyanamid)—has for years demonstrated that this field of activity is nevertheless controllable. $PH_3$ offers a large number of opportunities for conversion into products of industrial interest, which could only be indicated here in broad outline. It can be considered certain that this field of activity will continue in the future to be characterised by new processes, products and applications.

## References

1. Kosolapoff, G. M. & Maier, L. (eds), *Organic Phosphorus Compounds*, Vols 1–7. Wiley–Interscience, New York, London, Sydney, Toronto, 1972–1976.
2. DE Patent 12 65 746, 1965. Deutsche Advance Produktion, Inventors: H. J. Lorenz, A. R. Zintl & V. Franzen.
3. (a) DE Patent 20 50 095, 1970. BASF, Inventors: H. Müller & A. Stübinger.
   (b) DE Patent 26 38 720, 1975. M&T, Inventor: V. M. Chopdekar.
4. GB Patent 12 942, 1962. Albright & Wilson, Hooker, Inventors: E. J. Lowe & F. A. Ridgway.

5. US Patent 4 258 022, 1981. Hoechst, Inventors: G. Elsner, W. Klose, G. Heymer & C. May.
6. US Patent 4 044 055, 1977. American Cyanamid, Inventor: D. S. Katz.
7. W. Wolfsberger, *Chem. Z.*, **112** (1988) 53.
8. Rauhut, M. M., Currier, H. A., Semsel, A. M. & Wystrach, V. P., *J. Org. Chem.*, **26** (1961) 5138.
9. US Patent 4 324 919, 1980. Hoechst AG, Inventors; G. Elsner, H. Vollmer & E. Reutel.
10. DE Patent 27 03 802, 1977. Hoechst AG, Inventors: G. Elsner, G. Heymer & H.-W. Stephan.
11. Eur. Patent 0 258 693, 1987. Hoechst AG, Inventors: N. Weferling, G. Elsner, H.-W. Stephan & F.-K. Frorath.
12. DE Patent 29 39 588, 1979. Hoechst AG, Inventors: G. Elsner, H. Vollmer & G. Heymer.
13. Eur. Patent 0 038 317, 1981. Vereinigte Edelstahlwerke AG, Inventors: W. Kanzler & J. Schedler.
14. Hurst, F. J., Crouse, D. J. & Brown, K. B., Oak Ridge National Laboratory Report ORNL-TM-2522, 1969.
15. US Patent 3 355 017, 1967. American Cyanamid, Inventors: W. P. Wystrach, M. M. Rauhut & W. L. Frieberger.
16. Danesi, P. R., *Solvent Extr. Ion Exch.*, **3** (1985) 435.
17. Dehmlow, E. V., *Angew. Chem. Int. Ed.*, **13** (1974) 170.
18. DE Patent 30 11 085, 1980. Hoechst AG, Inventors: W. Krause, J. Große & W. Klose.
19. Eur. Patent 0 105 843, 1983. Ciba-Geigy, Inventors: W. Wehner & R. Grade.
20. Eur. Patent 0 240 703, 1987. Hoechst AG, Inventors: H. Erpenbach, K. Gehrmann, W. Lork & P. Prinz.
21. US Patent 3 103 431, 1963. Virginia-Carolina Chemical Corp., Inventor: J. H. Wilson.
22. Eur. Patent 0 259 583, 1987. Hoechst AG, Inventors: N. Weferling, P. Pfüller, W. Kanzler, J. Schedler & H. Thalhammer.
23. Eur. Patent 0 095 453, 1983. FMC Corporation, Inventor: D. P. Braksmayer.

# Synthesis of organic azides and *N,S*-heterocompounds: chemistry and hazards

**Michael Stohlmeier & Klaus Thewalt**

*Dynamit Nobel AG, Leverkusen, FRG*

## Introduction

The first organic azide, phenyl azide, was synthesized by Griess[1] in 1864. Some years later, exactly one century ago, Curtius discovered hydrazoic acid, $HN_3$, and found that acyl azides rearrange to the corresponding isocyanates.[2] The proposal of nitrenes being an unstable nitrogen atom with only six electrons in the outer shell instead of eight was first mentioned by Tiemann[3] in 1891.

A greater interest in azide chemistry began to bloom in the fifties of this century with the reviews of Smith[4] and Boyer and Canter[5] (acyl azides are aryl/alkyl azides, respectively). Up till now more than thirty reviews dealing with azide chemistry have been published.[6]

Industrial interest in azide chemistry began with the use of azido compounds and heterocycles produced thereby, e.g. triazoles and tetrazoles and their increasing use as blowing agents and as antibiotic side chains. Some of them are listed in Table 1.

The group of the above mentioned pharmaceuticals comprises mostly penicillins and cephalosporines. Some of them and their retrosynthetic precursors[8] dealing with this chemistry are shown in Table 2.

**Table 1.** Examples of azido compounds, triazoles, tetrazoles and thiadiazoles used in the chemical industry[7]

| Compound | Use |
|---|---|
| **A. Azides** | |
| $N_3-CH_2-COOCH_2CH_3$ | Antibiotic side chain (Azidamphenicol) |
| $Ph-CH(N_3)-COCl$ | Antibiotic side chain (Azidocillin) |
| $H_3C-C_6H_4-SO_2N_3$ | Photochemical intermediate |
| **B. Triazoles** | |
| 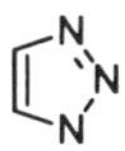 | Antibiotic intermediate |
| 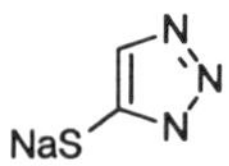 | Antibiotic side chain (Cefatrizin) |
| **C. Tetrazoles** | |
| 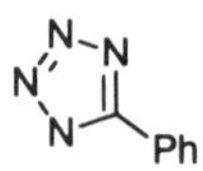 | Blowing agent for polycarbonate, polyacetale and nylon 6 |
| 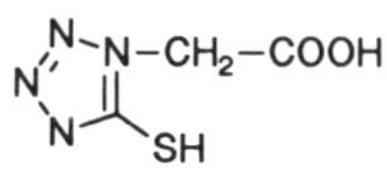 | Antibiotic side chain (Ceforanide) |
| 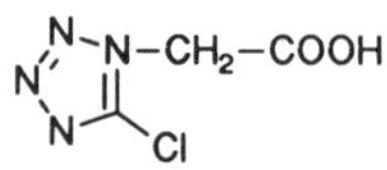 | Pharmaceutical |
| 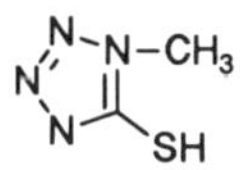 | Antibiotic side chain (Cefamandol, Latamoxef) |
| **D. Thiadiazoles** | |
| 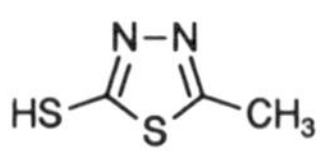 | Antibiotic side chain (Cefazolin, Cefazedon) |
| 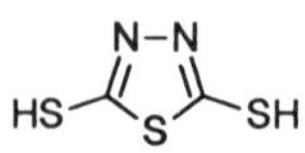 | Corrosion inhibitor |

**Table 2.** Antibiotics and precursors

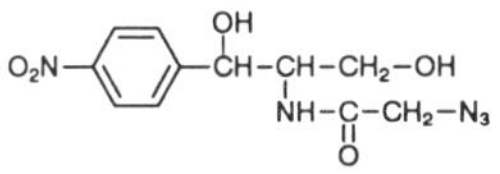

Azidamfenicol

⇒ $H_5C_2OOC-CH_2-N_3$

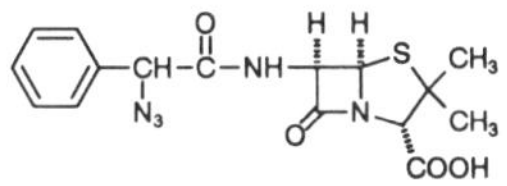

Azidocillin

⇒ 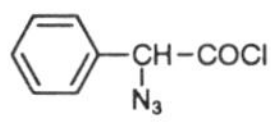

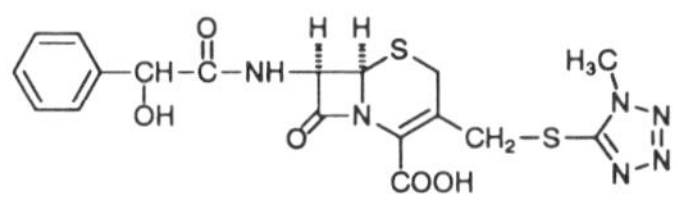

Cefamandol

⇒ 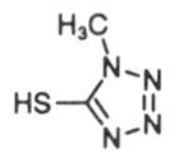

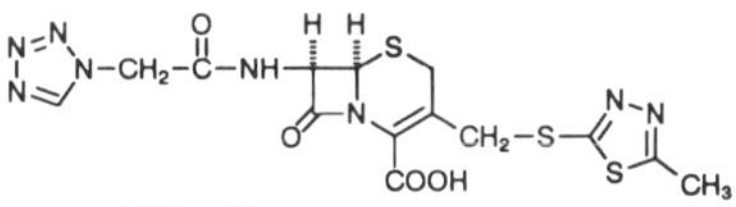

Cefazolin

⇒ 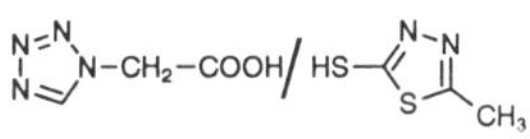

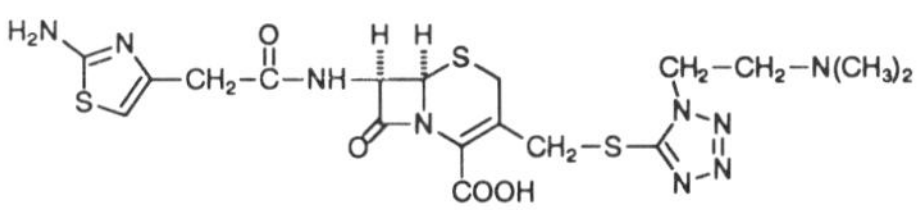

Cefotiam

⇒ 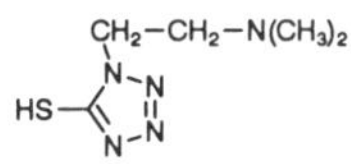

# Chemistry

## Organic Azido Compounds

Organic azido compounds like alkyl and acyl azides are most commonly prepared by displacement reactions of halogenated congeners. Many different leaving groups have been employed, but *p*-toluene sulphonyl, methane sulphonyl and halogens are most frequently used. The reaction takes place by the $S_N2$ mechanism, thus leading to a configurationally inverted species. This affords a stereochemical selectivity which is very valuable in steroid and carbohydrate chemistry. Reaction of the cholestanyl-3$\beta$-*p*-toluene sulphonate with sodium azide in dimethylformamide afforded the 3$\alpha$-azido compound. Reduc-

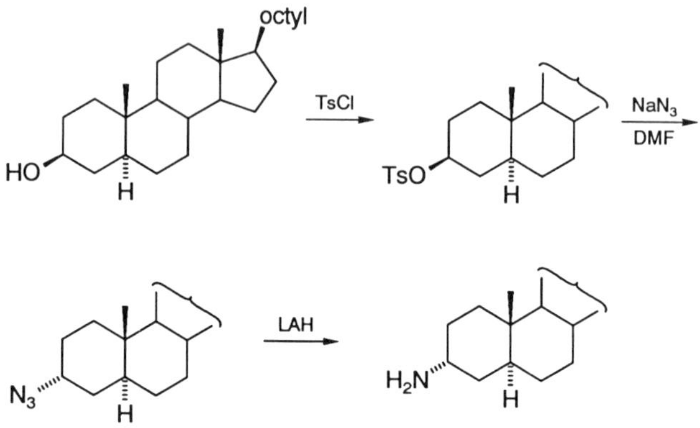


**Fig. 1.** Synthesis of 3-$\alpha$-amino cholestane.

tion with lithium aluminium hydride (LAH) gave 3$\alpha$-amino cholestane in 56% overall yield, including a chromatographic purification.[9] No $\beta$-azido compound has been observed (Fig. 1).

Another example is the synthesis of 3′-azido-2′,3′-dideoxythymidine (AZT), an anti-viral agent against AIDS (*A*quired *I*mmuno *D*eficiency *S*yndrome) and ARC (*A*ids *R*elated *C*omplex). It has been launched in the USA and more than 15 other countries for AIDS and ARC therapy.[10] The synthetic route involves several protection and de-protection steps starting from commercially available 2′-

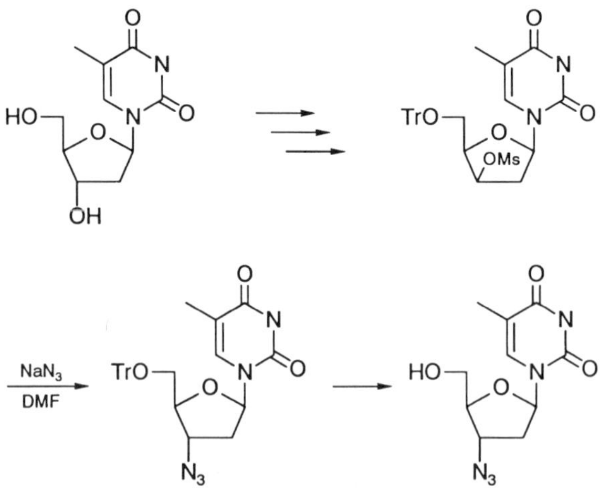


**Fig. 2.** Synthesis of AZT.

deoxythymidine[11] (Fig. 2). We have undertaken similar azidations starting from 2′-deoxyuridine.

Using stereochemically pure mesylated compound the reaction with $NaN_3$ proceeds under total inversion; the stereo isomer derived by retention is not observed at all.

One of the most useful reactions of azides is the rearrangement of azyl azides to isocyanates. This reaction is known as the Curtius rearrangement.[4]

The acyl azides are simply synthesized by reacting acyl halides with $NaN_3$ in a suitable solvent. Thus, alkyl isocyanates are produced as outlined in Fig. 3. The obtained isocyanates are valuable intermediates for a variety of other chemicals like amines, urethanes, carbamates and heterocycles.

In most cases the acyl azides are not isolated. Lower alkyl acid azides undergo Curtius rearrangement at or below ambient temperature; aryl acid azides are more stable. Most of the acyl azides will rapidly lose nitrogen when isolated or moderately heated; some azides explode on isolation (e.g. oxalic acid azide[12]).

## 1,2,3-Triazoles

Addition of azides to activated methylene compounds or alkynes are two well established methods for the synthesis of 1,2,3-triazoles.[13]

The base-catalysed nucleophilic attack of azides to activated methylene compounds leads to substituted triazoles. Thus, diethyl-

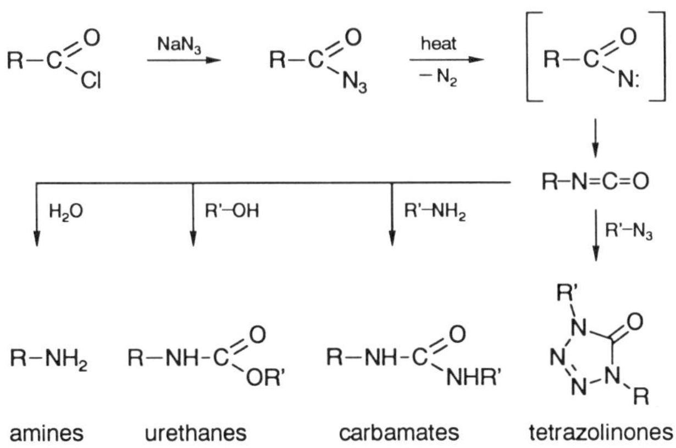


**Fig. 3.** Synthesis and use of isocyanates.

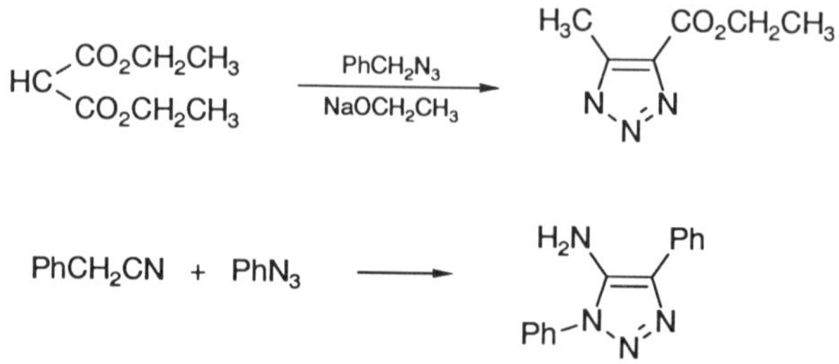

**Fig. 4.** Reaction of activated methylene compounds with organic azides.

malonate reacts with benzyl azide to form the 1-benzyl-4-carbethoxy-5-hydroxy-1,2,3,4-triazole in good to excellent yield.[14] The reactions of methylene groups activated by cyano substituents give 5-amino-1,2,3-triazoles. 1,4-Diphenyl-5-amino-1,2,3-triazole can be obtained from phenylacetonitrile and phenylazide in 99% yield[15] (Fig. 4).

*N*-unsubstituted triazoles need azides R-$N_3$ with a substituent R, being easily removed in the obtained target molecule. Thus, sulphonyl azides and trimethylsilyl azide are frequently used, because they can be removed under solvolytic or hydrolytic conditions,[16] respectively.

The reaction of azides and acetylenes yields 1,2,3-triazoles directly. The parent ring 1,2,3-triazole was obtained by reacting benzyl azide with acetylene under high pressure and consequently hydrogenated catalytically to yield 1,2,3-triazole in 85% yield[17] (Fig. 5).

Acetylene dicarboxylic esters react with trimethylsilyl azide to yield protected triazole diesters, which after deprotection ($H^+$) and saponification can be decarboxylated to 1,2,3-triazole in good overall yield.[18]

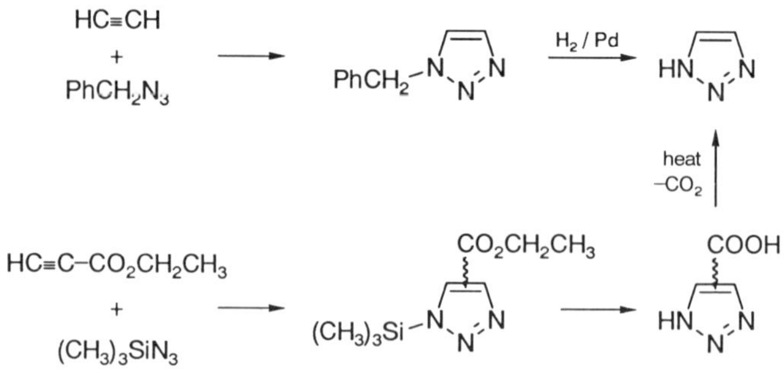

**Fig. 5.** Synthetic routes towards triazole.

It should be pointed out that trimethylsilyl azide is a very safe azide transfer reagent and it allows homogeneous reactions in organic solvents. Nevertheless, it is very moisture sensitive and should be best prepared by reacting sodium azide with chlorotrimethylsilane just before use.

## Tetrazoles

Different interesting ways to obtain tetrazoles in a variety of substitution patterns may be used. They are all starting from sodium azide (Fig. 6).

The mechanism involves the addtion of azide to the C—N triple bond or the substitution by azide ion to form an azavinyl azide with the structural segment N═C—$N_3$. This is not isolable and will stabilize

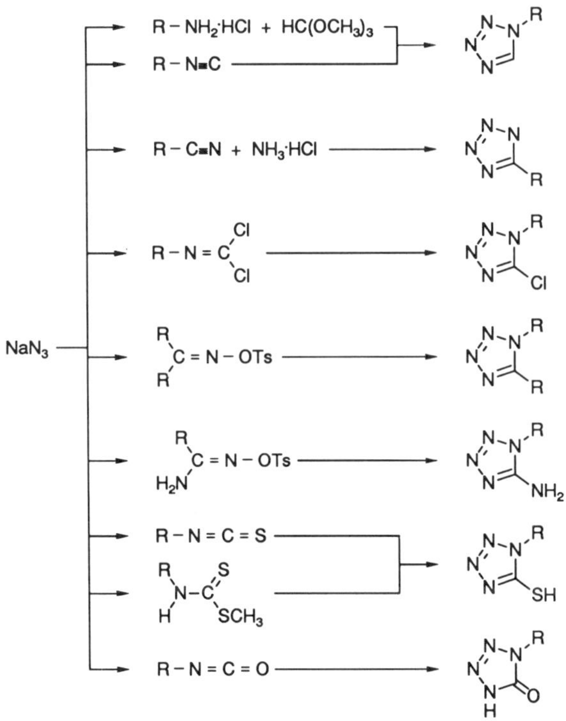

**Fig. 6.** Cyclization reactions with sodium azide.

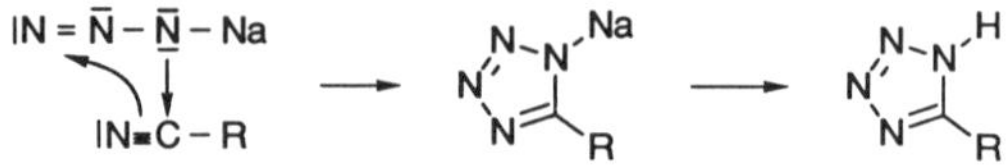


**Fig. 7.** Formation of tetrazoles.

itself by spontaneous intramolecular cyclization to form the tetrazole ring.

In special cases, e.g. amino tetrazoles, an intramolecular migration occurs. Some examples may illustrate the broad variety of this synthetic concept (see Fig. 6).

Most of these reactions are known only for special substituents, but if slightly modified, a wide spectrum of substituted tetrazoles can be obtained in kilogram or ton lots. Thus, 5-substituted tetrazoles are easily obtained by reacting aliphatic or aromatic nitriles with sodium azide in an appropriate solvent like isopropanol or dimethyl formamide to obtain tetrazoles in good to excellent yields. As mentioned, this reaction involves the addition of azide ion to the triple bond and cyclization to the desired product (Fig. 7).

As reported by Huisgen,[19] the dipolar cycloaddition of 1,3-dipoles (such as azides, diazoalkanes, etc.) to a dipolarophil (e.g. nitriles) occurs by a concerted mechanism. As a consequence a stereoselective cis addition is observed.

In the case of mercapto tetrazoles the substituent is an origin of carbon disulphide. The corresponding isothiocyanates and methyl dithiocarbamates are produced by addition of $CS_2$ to the desired amine. The intermediate product may be transformed with chloroformates or dimethyl sulphate, respectively (Fig. 8).

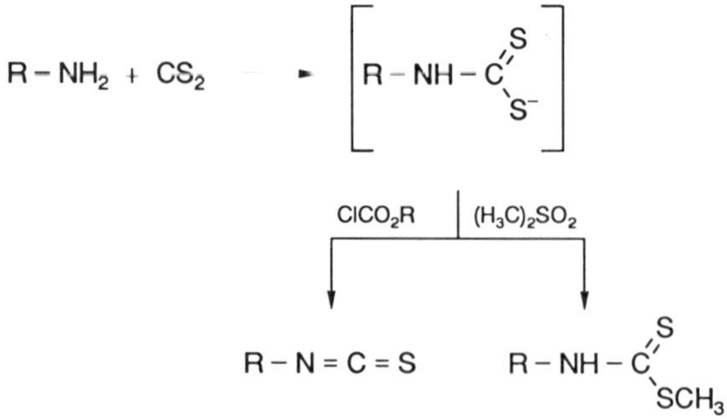


**Fig. 8.** Precursors of mercapto tetrazoles.

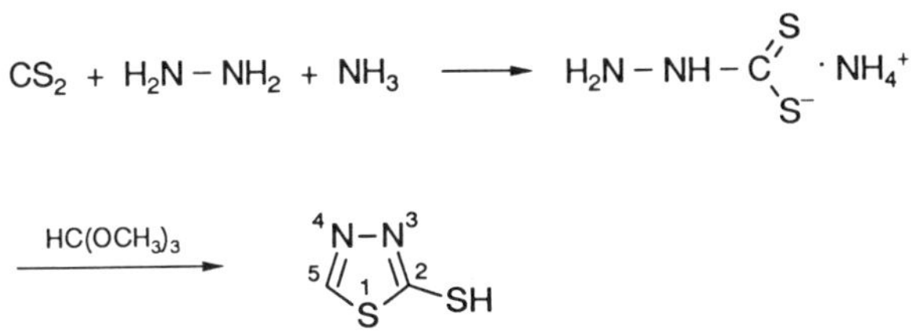

**Fig. 9.** Synthesis of mercapto 1,2,4-thiadiazoles.

### Thiadiazoles

Thiadiazoles are obtained by reaction of hydrazine derivatives with carbon disulphide. Thus, $CS_2$, hydrazine hydrate and ammonia react to form the amino dithiocarbazinate, which on reaction with trimethyl *ortho*-formate cyclizes to the desired product (Fig. 9).

The yields are good to excellent, substituents at $C_5$ depends on the used *ortho*-ester. With *ortho*-formates 5-unsubstituted thiadiazoles are obtained. Substituted hydrazines will lead to *N*-substituted compounds.

## Hazardous potential

The basic materials (sodium azide, hydrazine and carbon disulphide) are readily available worldwide (see Table 3). The production capacities of sodium azide will heavily increase due to the development of the air bag, a safety device for motor cars. Ignited by an accident, $NaN_3$ is catalytically decomposed to produce a great amount of nitrogen which blows up a bag between the driver and the steering wheel. (All new cars in the United States must be equipped with an air bag. Other countries will follow soon.)

However, they are all three very harmful and often dangerous in properties and use. Thus, sodium azide liberates very toxic hydrazoic acid on contact with mineral acids like hydrochloric acid, sulphuric

**Table 3.** World production capacities[20]

| | |
|---|---|
| $NaN_3$ | 1 000 tons/a (estimated) |
| $H_2NNH_2{\cdot}H_2O$ | 60 000 tons/a (1988) |
| $CS_2$ | 1 100 000 tons/a (1984) |

acid, etc., and is liable to explode together with bromine and carbon disulphide.

Carbon disulphide, having one of the lowest flash points of commercially used solvents, has caused many fires and explosions in chemical plants. Together with azides highly explosive azido thioformates are formed.[21,22]

Hydrazine hydrate causes cancer in animal test and hydrazine itself is used as rocket propulsion fuel; carbon disulphide is considered to be teratogen, causing defects in unborn life. These chemicals are therefore very toxic, explosive, cancerogen, teratogen and easily ignited.

Some physical and biological properties are listed in Table 4. Thus, hydrazoic acid is much more toxic than hydrocyanic acid, HCN, causing disturbance in the central nervous system and a lethal sudden hypotonic effect (death).

The explosion limits range from one to a hundred per cent (see Table 4). Condensed hydrazoic acid (boiling point 36°C) is highly shock sensitive and presents a malicious explosive. Carbon disulphide may be ignited on surfaces at 100°C, a temperature commonly found in chemical plant facilities.

These are not all of the dangerous properties of the starting materials. Chemical hazards must be added. The liberated thermal energy when decomposed (e.g. by heavy metal induction) is often

**Table 4.** Physical properties of hydrazoic acid, carbon disulphide, hydrazine and its hydrate, respectively[23]

| | $HN_3$ | $CS_2$ | $H_2N—NH_2$ | $H_2N—NH_2{\cdot}H_2O$ |
|---|---|---|---|---|
| Density ($g/cm^3$) | 1·126 | 1·26 | 1·00 | 1·028 |
| Boiling point (°C) | 35·7 | 46·3 | 113·5 | 118·0 |
| Melting point (°C) | −80 | −112 | +2 | −51 |
| MAK (ppm)[a] | 0·1 | 10 | — | — |
| Vapour pressure (mbar at 20°C) | [b] | 499 | 13 | — |
| $LD_{50}$ (oral, mg/kg) | 20 | — | — | 129 |
| Flash point (°C) | [b] | −30 | 52 | 75 |
| Limits of explosion (vol %) | 20–100 | 1–60 | 5–100 | 5–100 |
| Ignition temperature (°C) | [b] | 102 | — | 310 |

[a] MAK = TLV in air.
[b] Not applicable.

very high. Hydrazine reacts with oxygen to form nitrogen, water and 620 kJ/mol thermal energy. When decomposed without control, one kilogram of hydrazine liberates about 20 MJ of energy in a split second. Hydrazoic acid produces upon decomposition 530 kJ/mol or 12.3 MJ/kg. Furthermore, hydrazoic acid—and in some cases sodium azide too—easily form heavy metal azides, especially with lead, cadmium, mercury and silver. Their possible formation must absolutely be avoided by technical installations. The heavy metal azides are very shock sensitive; they are partly more sensitive to shock than trinitrotoluene (TNT) or nitroglycerine (NGL).

Who is aware that today, used glass connectors in chemical plants are normally fixed with cadmium plated screws because of problems of corrosion? Who bears in mind that in older plants the waste water pipes are still made of lead? This is a potential source of highly explosive lead azide. Who remembers that sodium azide, a very safe and not shock sensitive compound, in contact with mineral acids liberates hydrazoic acid, which in higher concentrations tends to decompose with explosive violence?

In addition, carbon disulphide, a very often used solvent in the chemical industry, is also very harmful. With a flash point of −30°C and an ignition temperature of about 102°C it has to be handled with special care and in special reaction vessels. The vessels must meet the Ex-G5 class. This is the highest class for protecting electrical apparatus against formation of unexpected electrical sparks. They are normally not installed in a chemical plant.

Hydrazine and its hydrate are used as rocket propellants because of their very high decomposition energy. The salts of hydrazine with chlorate, perchlorate, nitrate and azides are readily formed. They are spontanously flammable or explosive when dry or moderately heated. Special care has to be taken to avoid such contact and conditions.

As seen above, the reagents themselves are harmful and very dangerous in unforeseen mutual contact. Therefore special precautions for storage, handling and reacting them in a chemical plant have to be taken. Otherwise accidents like fire, explosions or poisoning may take place.

However, there is another risk not yet discussed. The produced compounds might even be more harmful than the starting materials. In many cases the intermediate products or reaction residues (e.g. distillation residues) do have a hazardous potential, too. Thus, when synthesizing triazole from an azide and propynoic acid we found that the acid tends to polymerize with the liberation of a great amount of heat. The evaluation of the explosive properties of the acid showed a high sensitivity to shock and friction when in the pure state. In solution it can be handled safely.

In other cases the final product itself has explosive properties. One of these compounds is 1,2,3,4-tetrazole, used in many laboratories as an activator in the synthesis of deoxyribonucleic acid sequences. In order to obtain a safe delivery form such compounds have to be phlegmatized with a suitable solvent or other additives or sometimes as alkaline salt.

According to the German explosive law (*Sprengstoffgesetz*) all new materials have to be tested for shock sensitivity, friction sensitivity, steel nozzle test and other tests in order to know their danger potential. Only if all tests show no hazardous potential can they be marketed without restriction.

The broad know-how of Dynamit Nobel as a manufacturer of explosives and sensible compounds for more than one century and its established safety equipment make it possible to determine all necessary parameters to check if a new compound has hazardous properties or not and how to avoid accidents when synthesizing and manufacturing them. Starting from the very first stage in the laboratory over pilot plant to production scale all safety data are determined and collected. Only in this way, together with the great know-how of well-skilled and highly motivated collaborators, can the possibility exist to produce safe products by safe reactions even with harmful substances.

## References


1. (a) Griess, P., On a new series of bodies in which nitrogen is substituted for hydrogen. *Philos. Trans. R. Soc. London,* **13** (1864) 377.
   (b) Griess, P., Notiz über Hyperbromide der Diazosäuren. *Ann. Chem.,* **135** (1865) 131.
2. (a) Curtius, T., Über Stickstoffwasserstoffsäure (Azoimid) $HN_3$. *Ber. Dtsch. Chem. Ges.,* **23** (1890) 3023.
   (b) Curtius T., XX. Hydrazide und Azide organischer Säuren. *J. prakt. Chem.,* **50** (1894) 275.
   (c) Smith, P. A. S., Rearrangements involving migration to an electron deficient nitrogen or oxygen. In *Molecular Rearrangements,* ed. P. de Mayo. Interscience, New York, 1963.

3. Tiemann, F., Über die Einwirkung von Benzolsulfonsäurechlorid auf Amidoxime. *Ber. Dtsch. Chem. Ges.*, **24** (1891) 4162.
4. Smith, P. A. S., The Curtius-rearrangement. *Org. React.*, **3** (1964) 337.
5. Boyer, J. H. & Canter, F. C., Alkyl and aryl azides. *Chem. Rev.*, **54** (1954) 1.
6. Scriven, E. F. V. & Turnbull, K., Azides: their preparation and synthetic uses. *Chem. Rev.*, **88** (1988) 297, Table 1.
7. Taken from the brochure: How to overcome hazardous reactions safely. Dynamit Nobel AG, Special Chemistry, Troisdorf (Germany), 1988.
8. Kleemann, A. & Engel, J., *Pharmazeutische Wirkstoffe*, Band 5. Thieme, Stuttgart, 1982.
9. Bose, A. K., Kistner, J. F. & Farber, L., A convenient synthesis of axial amines. *J. Org. Chem.*, **27** (1962) 2925.
10. (a) International Nonpropriertary Name (INN): Zidovudine, sold as *Retrovir* by Wellcome Foundation Limited, London.
(b) Hardcastle, J. (ed.), *Pharma Projects*. PJB Publication Ltd, Surrey (UK), Update September 1989.
11. Glimski, R. P., Kham, M. S., Kalamar, R. L. & Sporn, M. B., Nucleotide synthesis. IV. Phosphorylated 3′-amino-5′-deoxythymidine and derivatives. *J. Org. Chem.*, **38** (1973) 4299.
12. Roesky, H. & Glemser, O., Über die Darstellung von Oxalsäurediazid. *Chem. Ber.*, **97** (1964) 1710.
13. Gilchrist, T. L. & Gymer, G. E., 1,2,3-Triazoles. In *Advances in Heterocyclic Chemistry*, ed. A. R. Katritzky & A. J. Boulton, Vol. 16. Academic Press, New York, 1974, p. 33.
14. Hoover, J. R. E. & Day, A. R., Metabolite analogs. VI. Preparation of some analogs of 4-amino-5-imidazole-carboxamide. *J. Am. Chem. Soc.*, **78** (1956) 5832.
15. Lieber, E., Rao, C. N. R. & Rajkumar, T. V., Improved procedure for condensation of alkyl azides with penylacetonitril of form *vic*-triazoles. *J. Org. Chem.*, **24** (1959) 134.
16. Harvey, G. R., The reactions of phosphorous compounds. XII. A new synthesis of 1,2,3-triazoles and diazoesters from phosphorous ylides and azides. *J. Org. Chem.*, **31** (1966) 1587.
17. Gold, H., Über *N*-aminoalkyl-1,2,3-triazole. *Liebigs Ann. Chem.*, **688** (1965) 205.
18. Tsuge, O., Kanematsu, S. & Matsuda, K., A silylfunctionalized alkyl azide, trimethyl silylmethyl azide; synthesis and cycloaddition reaction to acetylenic dipolarophiles. *Chem. Lett.* (1983) 1131.
19. Huisgen, R., Kinetik und Mechanismus 1,3-dipolarer Cycloadditionen. *Angew. Chem.*, **75** (1963) 742; *Angew. Chem. Int. Ed. Engl.*, **2** (1963) 633.
20. (a) Eichenhofer, K. W. & Roos, E., Hydrazin. In *Ullman's Encyclopedia of Technical Chemistry* ed. E. Bartholome, E. Bickert, H. Hellman, H. Ley, W. M. Weigert & E. Weise, Vol. A5, 5th revised edn. VCH, Weinheim, New York, 1986.
(b) Grothaus, H., Theis, G. & Freundt, K. J., Schwefelkohlenstoff. In *Ullmann's Encylopedia of Technical Chemistry*, ed. E. Bartholome, E.

Bickert, H. Hellmann, H. Ley, W. M. Weigert & E. Weise, Vol. A13, 5th revised edn. VCH, Weinheim, New York, 1989.

21. Muir, G. D., *Hazards in Chemical Laboratory*. The Chemical Society, London, 1977.
22. Bretherick, L., *Handbook of Reactive Chemical Hazards*. Butterworth, London, 1975.
23. (a) Hollemann, A. F. & Wilberg, E., *Lehrbuch der Anorganischen Chemie*. W. de Gruyter, Berlin, New York, 1976.
    (b) Weast, R. C. & Astle, J. M., *Handbook of Chemistry and Physics*, 63rd edn. CRC Press, Boca Raton, 1982.

# New catalytic pathways to heterocycles

**Wim Klop**

*DSM Research, Geleen, The Netherlands*

## Introduction

In recent years a growing interest is seen in the application of heterogeneous catalysis to the synthesis of fine chemicals. This is, for example, illustrated by the very successful 'International Symposia on Heterogeneous Catalysis and Fine Chemicals' that were organised in Poitiers in 1988 and 1990.[1]

Coming from a long tradition in catalysis research, DSM has already studied the application of heterogeneous catalysis to the synthesis of fine chemicals for a long period of time. One area in which heterogeneous catalysts can be applied with much benefit is the synthesis of heterocyclic compounds.

Heterocyclic ring systems are synthesised classically by liquid phase condensation reactions. When unsaturated starting materials are used these condensations normally give at once the desired aromatic sysems. The starting materials needed are, however, often hard to prepare. This is illustrated in Fig. 1, in which a classical pyrimidine preparation by the so-called 'principal synthesis'[2] is given. The starting materials, (protected) malondialdehyde (derivatives) (I) and alkylamidines (II) are relatively expensive. A second disadvantage of this

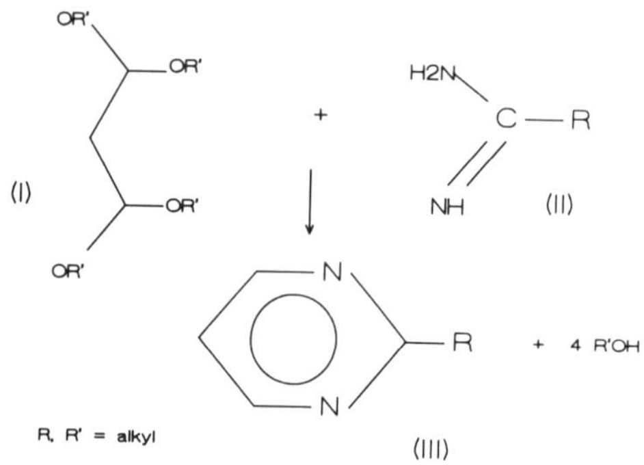

**Fig. 1.** Alkylpyrimidine preparation by the 'Principal Synthesis'.

method are the relatively low yields in the preparation of mono- or disubstituted pyrimidines. The method works best for highly substituted pyrimidines.

At DSM syntheses of heterocyclic compounds have been developed on the basis of a totally different concept. Relatively cheap, saturated starting materials are used. This results in the (*in situ*) formation of (partially) saturated heterocycles as intermediates. By gas phase catalytic dehydrogenation these intermediates are (*in situ*) converted to the desired aromatic systems. As an example of this concept it will be shown in this paper that the alkylpyrimidines of Fig. 1 can also be prepared from simple carboxylic acids and 1,3-diaminopropane (derivatives). By use of very efficient heterogeneous catalysts this can be achieved in a two-step process. These two steps can even be combined into one unit operation by use of a double catalyst bed.

Alkylpyrimidines, for example, are used as intermediates in the preparation of pesticides. Pyrimidines with a *t*-butyl substituent at the 2-position are of special interest in this field. The examples that are shown will be compounds of this type. This does not mean, however, that the techniques that are shown are limited to only these types of compounds. The scope is much wider. Many heterocyclic compounds can be made on the basis of these techniques.

## The overall process

In Fig. 2 the overall chemistry of the process is shown. In the first step an amide is formed from a carboxylic acid and an amine. It is

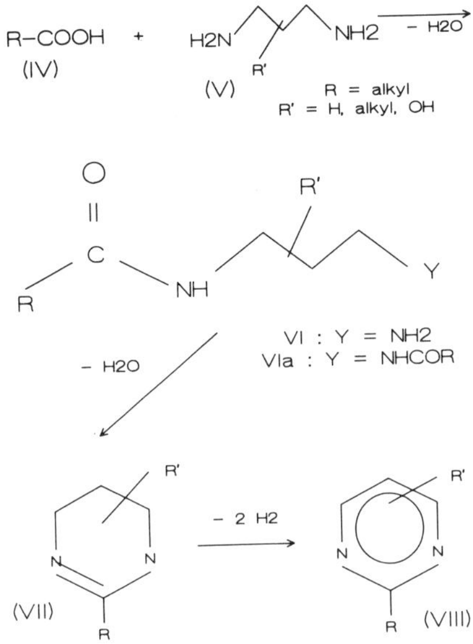


**Fig. 2.** Synthesis of alkylpyrimidines from carboxylic acids and 1,3-diaminopropane (derivatives).

interesting to note that March[3] describes this method for making amides as 'seldom of preparative value'. What is required is an efficient dehydration. Also in the second step a dehydration is needed. The use of zirconia as a catalyst allows these dehydrations to be combined into one step.

The third step is a dehydrogenation. The basic problem in dehydrogenations is formed by the thermodynamics. Favourable thermodynamics can only be reached at a fairly high temperature. Therefore a gas phase process is normally the best choice. Of course special equipment is needed for such a process. DSM has this equipment available from labscale to a scale of several thousand tons a year. The lifetime of the catalyst is always a potential problem in these processes. In this paper a very special way of dealing with this deactivation phenomenon will be discussed.

## Dehydrations over zirconia

The dehydration as shown in Fig. 2 can of course be done in a non-catalytic way, by just heating the compounds in the liquid phase.

For the preparation of the amide (VI) by such a classical dehydration a three hour heating in an autoclave at 240°C is needed. A large excess of 1,3-diaminopropane (V) is needed to prevent the formation of substantial amounts of the diamide (VIa). For example, at a molar ratio 1,3-diaminopropane/pivalic acid of 2, the selectivity towards the corresponding diamide is still 18·6%. Liquid phase dehydration of the amide (VI) by straightforward heating or with an azeotrope forming solvent is very time consuming. It often takes far more than twelve hours.

A gas phase catalytic process is much more efficient than the above mentioned dehydrations. At 340°C, 1,3-diaminopropane (derivatives) and carboxylic acids can be converted catalytically in one step to the tetrahydropyrimidine (VII). Zirconia was found to be a very efficient catalyst for this reaction. Yields are over 90%. Rather surprisingly it was found that only a small excess of 1,3-diaminopropane is needed when using this process. At a molar ratio of 1,3-diaminopropane/pivalic acid of 2 the selectivity towards the diamide (VIa) is only 0·1%. This is a rather dramatic reduction compared to the above mentioned 18·6% for the non-catalytic process. This double dehydration process can be performed conveniently in a continuous fixed bed reactor.

Zirconia would not be the first choice for most people looking for a dehydration catalyst. Alumina is normally the material that is chosen. We found, however, that zirconia is a very efficient dehydration catalyst, despite its very low acidity ($H_0 = 4$). It is clear from this fact that the overall acidity of a material does not determine its dehydration activity.

Tetrahydropyrimidines with non-alkyl substituents can also be prepared using this process. For example, 2-*t*-butyl-5-hydroxy-1,4,5,6-tetrahydropyrimidine can be prepared over zirconia in one step from 2-hydroxy-1,3-diaminopropane and pivalic acid in a 70% yield.

These dehydration processes have been patented by DSM.[4]

## Catalytic gas phase dehydrogenations

Gas phase dehydrogenations are normally performed over noble metal catalysts. Because of the high temperatures (thermodynamics) and the often relatively unstable starting materials and products, catalyst deactivation is always a potential problem.

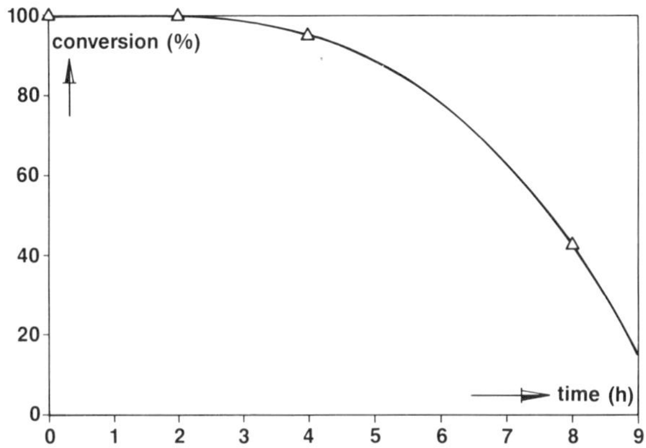


**Fig. 3.** Dehydrogenation of 2-*t*-butyl-1,4,5,6-tetrahydropyrimidine. Standard conditions (Pd catalyst/pyridine solution).

In the case of the dehydrogenation of tetrahydropyrimidines deactivation indeed is a serious problem. When, for example, 2-*t*-butyl-1,4,5,6-tetrahydropyrimidine is dehydrogenated at 340°C over a standard 1% $Pd/Al_2O_3$ catalyst, as a solution in pyridine, the conversion drops very rapidly, as shown in Fig. 3. This is clearly an uneconomical process.

Normally catalyst deactivation is mainly attacked by working on the catalyst. By modifying the catalyst with sodium a major improvement could be achieved. With a 1% $Pd/Al_2O_3$ catalyst, modified with 1% Na, the conversion after 8 hours was still 100%. After 100 hours on stream, however, the conversion falls down again, as shown in Fig. 4. In this case, therefore, catalyst modification only was not enough. A new parameter had to be introduced.

In several bulk catalytic processes a reaction modifier is used. An example of this is the addition of chlorine compounds to the feed in ethylene oxidation.

We found that CO was a useful reaction modifier in the dehydrogenation to tetrahydropyrimidines. This seems somewhat surprising at first sight, because normally CO is considered to be a catalyst poison. The beneficial action of certain poisons in so-called selective poisoning is, however, well known in, for example, liquid phase hydrogenations.[5] Although we do not know the exact course of action of CO in this reaction, we think that CO in this case also acts as a

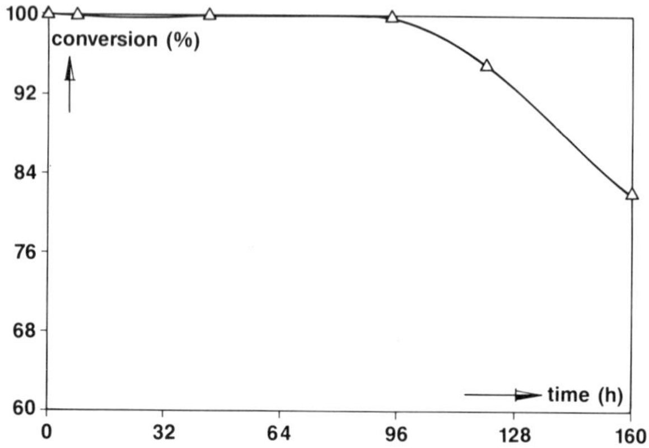


**Fig. 4.** Dehydrogenation of 2-*t*-butyl-1,4,5,6-tetrahydropyrimidine. Effect of sodium promotion (Pyridine solution).

selective poison. It probably deactivates certain sites on the surface that are responsible for the deactivation (polymerisation) phenomena.

A very efficient way of adding CO to the feed in such a reaction is the use of methanol as a solvent. Over noble metal catalysts methanol rapidly decomposes to CO and $H_2$ at temperatures above 300°C.

When methanol was used as a (decomposing) solvent, instead of pyridine, the conversion was still 100% after 270 hours (Fig. 5) and the yield was a stable 89%.

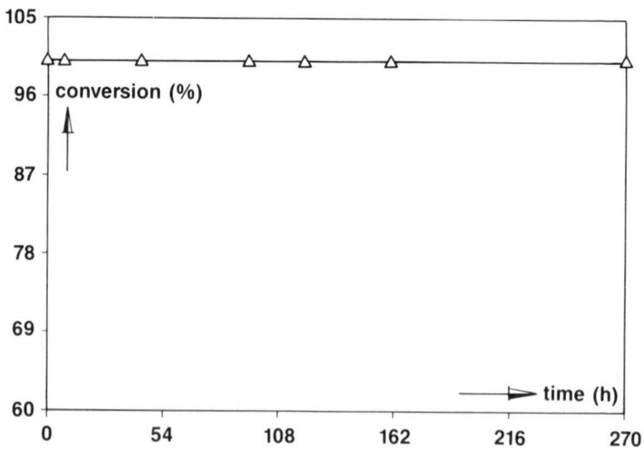


**Fig. 5.** Dehydrogenation of 2-*t*-butyl-1,4,5,6-tetrahydropyrimidine. Methanol as a reaction modifier (Sodium Catalyst).

The use of this very special reaction modifier has been patented by DSM.[6]

## Trifunctional catalysis

The original three-step process has now been improved to a very efficient two-step catalytic process. A further improvement is possible when we use a double catalyst bed in a continuous fixed bed reactor. In the first part of the bed a double dehydration over zirconia takes place and in the second part of the bed the *in situ* formed tetrahydropyrimidine is dehydrogenated over the $Pd/Al_2O_3$ catalyst. With this catalytic system 2-*t*-butylpyrimidine can be prepared in one step from 1,3-diaminopropane and pivalic acid in more than 85% yield.[4]

## Conclusion

Taking the preparation of pyrimidines as an example we have demonstrated that the synthesis of heterocycles can be done much more efficiently than by classical liquid phase, non-catalytical methods, when we take advantage of opportunities that are offered by gas phase heterogeneous catalysis. This is a good example of the benefits that can be obtained from the combination of (gas phase) heterogeneous catalysis and organic synthesis.

## References

1. Guisnet, M. *et al.* (eds), *Heterogeneous Catalysis and Fine Chemicals,* Studies in Surface Science and Catalysis, Vol. 43. Elsevier, Amsterdam, 1988. Proceedings of the Second (1990) Symposium in Poitiers on this subject are to be published by Elsevier in the same series.
2. Brown, D. *et al., The Pyrimidines.* John Wiley & Sons, New York, 1962, and Supplements I (1970) and II (1985).
3. March, J., *Advanced Organic Chemistry,* 3rd edn. John Wiley & Sons, New York, 1985, p. 372.
4. DSM, US-A-4 880 929, EP-A-0 276 503, 1986.
5. Freifelder, M., *Practical Catalytic Hydrogenation,* John Wiley & Sons, New York, 1971, p. 23.
6. DSM, US-A-4 667 034, EP-A-0 192 297, 1985.

# Hydrogen cyanide—a versatile building block

**J. H. Edwards**

*W. R. Grace Limited, Middlesbrough, UK*

## Introduction

Cyanides have been known in nature for thousands of years, for example, in the mandelonitrile glycoside amygdalin (**1**):

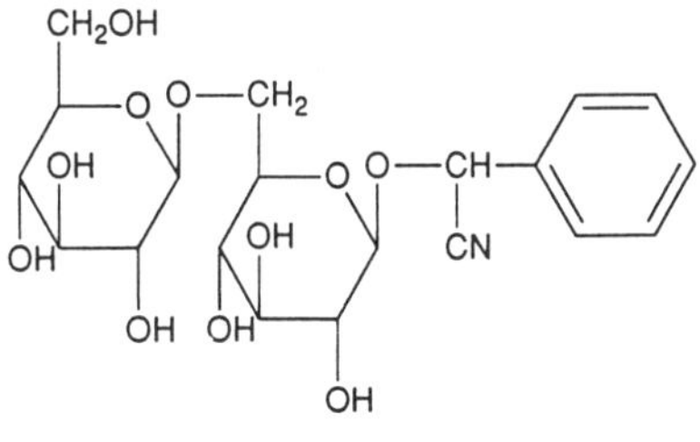

(**1**)

The simplest compound in the series, hydrogen cyanide, was prepared in its pure form as early as 1782 by Scheele who later died whilst using the material.

Cyanides have gained an increasing use in industry over the past fifty years, mainly in ore extraction and the manufacture of a wide range of plastics, dyestuffs, pharmaceuticals and agrochemicals.

The key raw material in cyanide chemistry—hydrogen cyanide—can be produced in a variety of ways ranging from the outdated Carbide

$$2NH_3 + 3O_2 + 2CH_4 \xrightarrow{\text{Pt catalyst}} 2HCN + 6H_2O$$

Andrussow process

$$NH_3 + CH_4 \longrightarrow HCN + 3H_2$$

BMA process

$$HCONH_2 \longrightarrow HCN + H_2O$$

Formamide process

$$2CH_2{=}CHCH_3 + 2NH_3 + 3O_2 \longrightarrow 2CH_2{=}CHCN + 6H_2O$$

Acrylonitrile process

**Fig. 1.** Major processes producing HCN.

process to the electrothermal Shawinigan (Fluohmic) process and the Andrussow, BMA and Formamide processes currently in operation throughout the world. Large quantities of by-product hydrogen cyanide are also produced in the distillation of coal and, more importantly in the Sohio acrylonitrile process (Fig. 1).

W. R. Grace first became involved in hydrogen cyanide chemistry in the early 1960s through its USA based Hampshire Chemicals subsidiary who were interested specifically in the manufacture of a range of amino carboxylate type chelating agents. Since that time we have investigated and produced a wide range of materials based on hydrogen cyanide and developed safe methods of handling this toxic material. The following is an attempt to describe some of the materials produced and handling methods used.

## Use of hydrogen cyanide in organic chemistry

Hydrogen cyanide can be used to add a cyanide group to a molecule; that cyanide group may then be converted to a wide range of alternative chemicals or left unreacted in order to impart a particular property to some agrochemical or pharmaceutical (Fig. 2).

Grace developed its use of hydrogen cyanide with the production of nitrilotriacetic acid (NTA), a detergent builder which can be made from ammonia formaldehyde and hydrogen cyanide in an acid catalysed reaction known as the Singer process (Fig. 3).

By judicious control of the raw material ratios and the pH of the reaction a wide range of other nitriles including glycinonitrile, methylene glycinonitrile and iminodiacetonitrile can be made (Fig. 4).

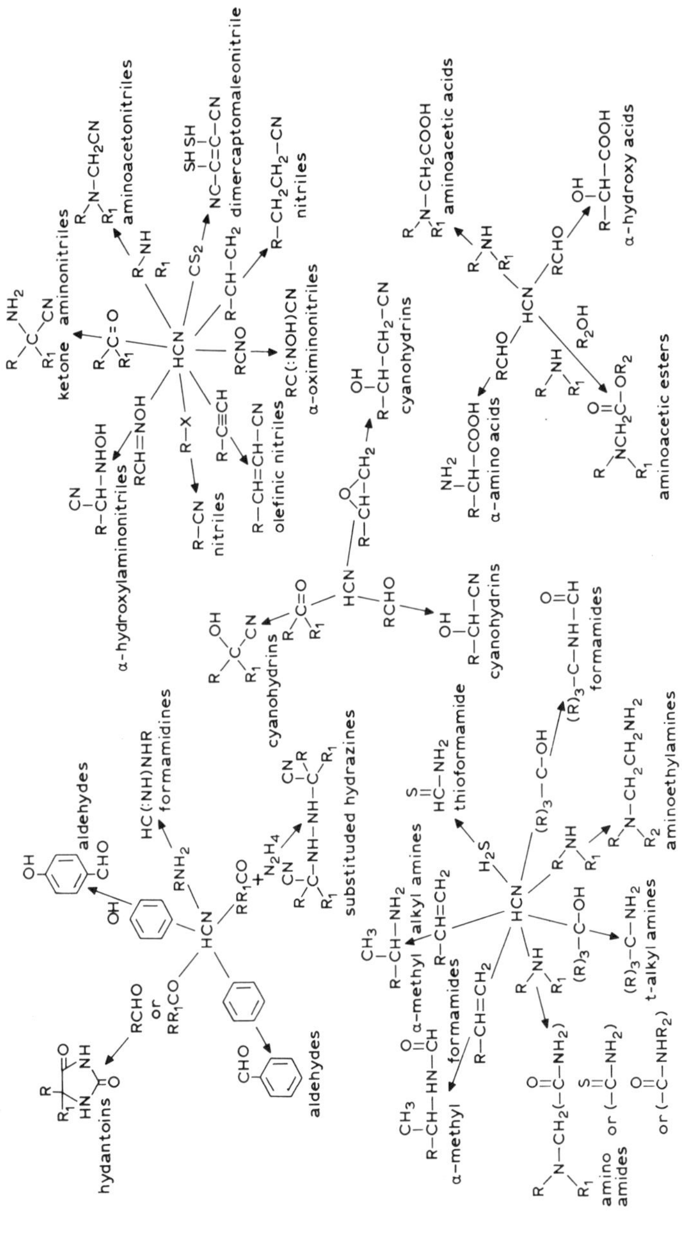


**Fig. 2.** Various reactions of hydrogen cyanide.

$$3HCHO + 3HCN + NH_3 \longrightarrow N(CH_2CN)_3 + 3H_2O$$

$$N(CH_2CN)_3 + 3NaOH + 3H_2O \longrightarrow N(CH_2COONa)_3 + 3NH_3$$

**Fig. 3.** NTA via the Singer process.

Iminodiacetonitrile is produced in large quantities for use in the production of the herbicide glyphosate (**2**):

$$HO\overset{O}{\overset{\|}{C}}CH_2NHCH_2\overset{O}{\overset{\|}{P}}(OH)_2 \qquad (\mathbf{2})$$

whilst glycinonitrile is used (mainly as its sulphate or hydrochloride) in the manufacture of a number of agrochemical and pharmaceutical intermediates including glycine, hydantoin and malononitrile.

A wide range of aminoacetonitriles can be made in a similar manner starting from the appropriate amine, formaldehyde and hydrogen cyanide (Fig. 5).

Polyfunctional amines (such as ethylene diamine and diethylene triamine) can lead ultimately to different chelating agents whereas primary and secondary amines yield aminonitriles which, because of the versatility of the nitrile group, can give rise to very useful intermediates (Fig. 6). Aminonitriles are also intemediates in the Strecker amino acid synthesis and the Bucherer Bergs hydantoin reaction (Fig. 7).

Other hydantoins have been used in the synthesis of keto acids; e.g. hydantoin and benzaldehyde can be used in the production of phenylglyoxylic acid, a precursor of aspartame.

Another group of materials produced by Grace are cyanohydrins and the related aminonitriles. There are many uses for these materials: methylmethacrylate and azobisisobutyronitrile from acetone cyanohydrin, diphenyloxazole from mandelonitrile and the now mature synthetic pyrethroid cypermethrin from meta phenoxybenzaldehyde cyanohydrin (Fig. 8).

Cyanohydrins derived from aldehydes and non-symmetrical ketones contain a chiral centre and it is therefore possible through the use of

$$H_2NCH_2CN \qquad H_2C{=}NCH_2CN \qquad HN(CH_2CN)_2$$

**Fig. 4.** Formaldehyde/ammonia/HCN derivatives.

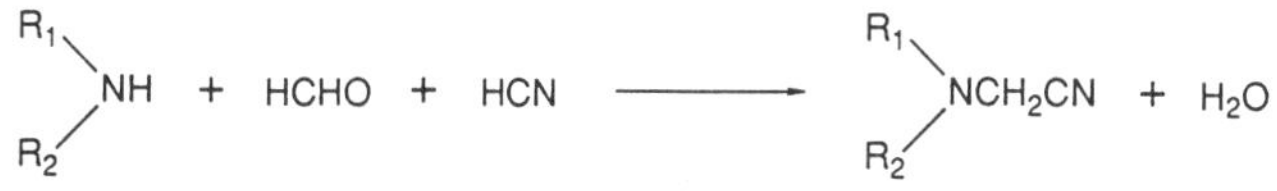

**Fig. 5.** Aminoacetonitriles.

an appropriate catalyst (e.g. oxynitrilase, some peptides and organotitanium esters) to induce a specific stereochemistry in the cyanohydrin often leading to greatly enhanced activity in the end product. Cyanohydrins can also be oxidised to acylcyanides which, due to multiple functionality, are useful intermediates (Fig. 9).

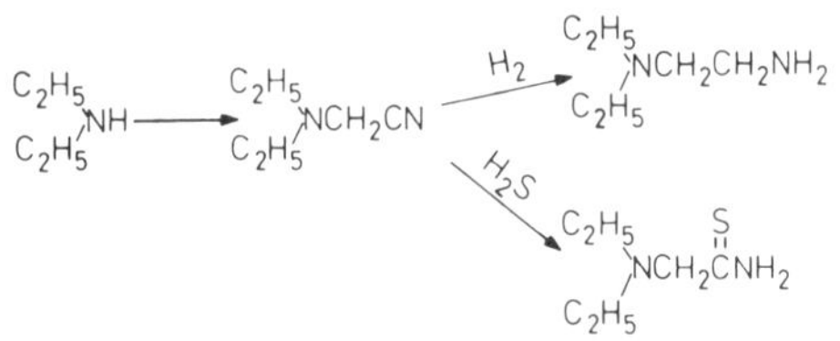

**Fig. 6.** Reactions of diethylaminoacetonitrile.

There are many other methods of incorporating the cyanide group in organic molecules, some of the more interesting being copper catalysed exchange with a diazonium salt (Fig. 10) and the high temperature reaction of cyanogen chloride (itself derived from hydrogen cyanide and chlorine) with a substituted alkyl compound in, for example, the conversion of 3-methyl thiophene to the high valued synthetic penicillin intermediate thiophene-3-acetonitrile.

The above represents a small section of the vast array of cyanide chemistry. There are many reviews covering this interesting area of chemistry.

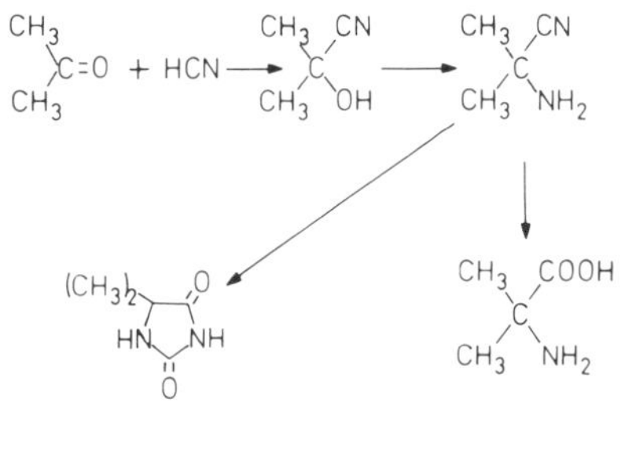

**Fig. 7.** Hydantoins and $\alpha$-amino acids from cyanohydrins.

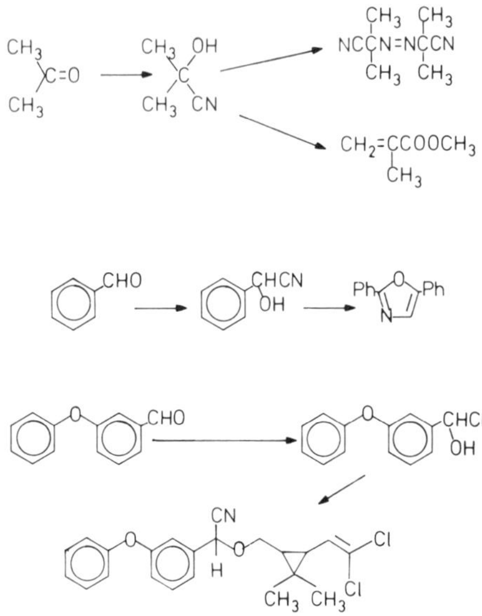

**Fig. 8.** Examples of the industrial use of cyanohydrins.

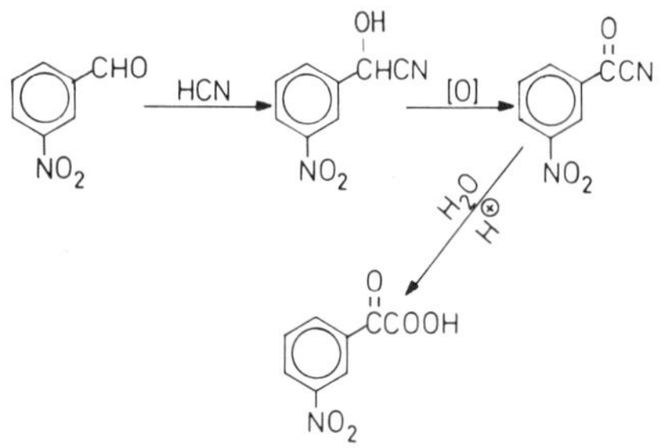

**Fig. 9.** Acyl cyanides from cyanohydrins.

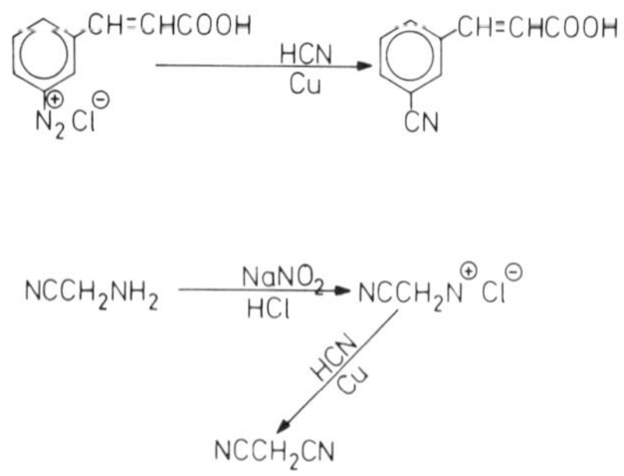

**Fig. 10.** Copper catalysed cyanide exchange.

## Safe handling of hydrogen cyanide

Hydrogen cyanide is a highly toxic material, its $LD_{50}$ and TLV setting it alongside such materials as chlorine, hydrogen sulphide and phosgene. It does, however, have some properties that assist in its safe handling, for instance it is a non-corrosive liquid (b.p. 26°C) which can be transferred in conventional steel pipelines. As far as can be assessed, there are negligible long term effects due to exposure in sub-lethal doses; this is probably because the human body has, over hundreds of thousands of years, evolved an effective cyanide detoxification mechanism as protection against a material that is widespread in nature.

Within W. R. Grace, hydrogen cyanide is handled in the laboratory, pilot plant and production plant on gram, kilo and tonnage scales respectively. In the laboratory the material is received in one litre stainless steel Dewar flasks and transferred to reaction flasks by pump or chilled addition funnels, all reactions being carried out in efficiently vented fume cupboards. In the pilot plant hydrogen cyanide is transferred from storage to a small weigh vessel and from that vessel via fixed pipelines to any one of a range of steel or glass lined reactors. On the full scale plant hydrogen cyanide is transferred from storage via a mass flowmeter and flow control valve set-up which allows very accurate materials charging, so avoiding the necessity of large scale detoxification during product work-up.

## Hazard assessment

Hydrogen cyanide is a reactive and toxic material; therefore great care has to be taken in assessing all processes in order to determine the likelihood of a runaway reaction that could lead to a vapour release.

We have commissioned a series of calorimetric tests to determine the outcome of a runaway reaction involving mixtures of hydrogen cyanide and a number of other materials. One example is the reaction between hydrogen cyanide and an aldehyde or ketone, a process which is exothermic and can lead to secondary reactions resulting in rapid pressure and temperature build-up (Fig. 11).

With the aid of Hazard Evaluation Laboratory Limited's Adiabatic Calorimeter PHI TEC, we have been able to evolve a

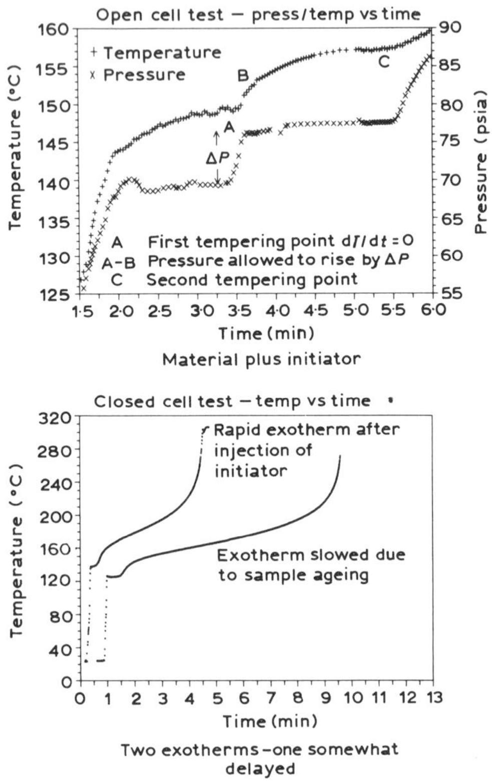


**Fig. 11.** Results from the PHI TEC tests on an NCN/aldehyde mixture.

philosophy for the design of relief systems on a range of batch reactors and will continue to refine the philosophy to encompass different types of reaction.

# The future

Hydrogen cyanide is firmly established as one of the chemical industry's key raw materials. However, transportation legislation in the United States, Europe and other parts of the world is becoming increasingly stringent and it is quite possible that, for instance, this compound, chlorine, ethylene oxide, acetone cyanohydrin and other such toxic materials might not be shippable outside the producer's plant. This will place manufacturers of these materials in the position of having to produce intermediates further downstream from the basic raw materials and could persuade smaller companies to invest in synthetic hydrogen cyanide units.

## Acknowledgements

My thanks go to the many colleagues at W. R. Grace in the United Kingdom, Sweden and the United States who introduced me to cyanide chemistry, to J.R. for opening the door and the numerous organic chemists of the late nineteenth and early twentieth centuries whose work forms the basis for many industrial processes.

# Hydrocyanic acid derivatives

**H. Klenk & K. Huthmacher**

*Degussa AG, Hanau, FRG*

## Introduction

As a carbon-1 basic raw material, hydrocyanic acid justifies its classification as an important synthetic building block in organic chemistry. Figure 1 shows some of the more important derivatives of HCN produced on a large scale.

Very often, the nitrogen atoms in both HCN and ClCN take no active part in most of their reactions. Indeed, from their oxidation states they can be best considered as derivatives of formic and carbonic acids.

The use of HCN and ClCN in production processes is, however, limited due to their low boiling points (HCN 26°C, ClCN 13°C), their tendency to polymerise, and their high toxicity. Additionally, transportation is limited by the strict transport regulations and the attendant safety problems.

Consequently, Degussa has built production units for HCN and ClCN in various countries of the world and has used these to produce bulk derivatives such as DL-methionine and cyanuric chloride. These plants have also been used to produce smaller scale intermediates for their customers. Additionally, Degussa has built a

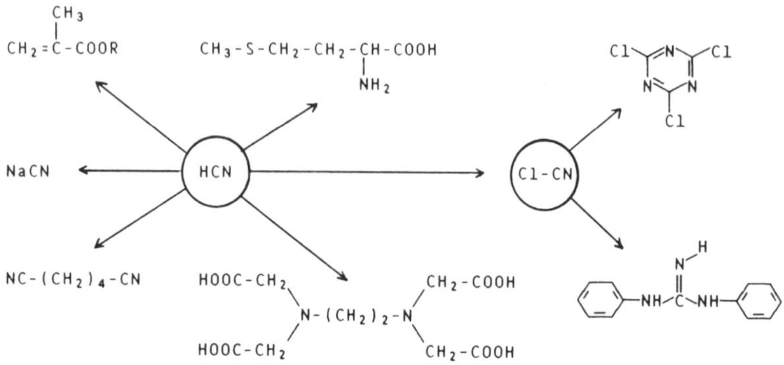


**Fig. 1.** Important derivatives of HCN.

number of multi-purpose plants where both toll production and small scale specialist manufacture are possible to the highest levels of safety.

# Types of compounds

## The Cyanohydrins

Cyanohydrins result from the addition reaction of HCN to carbonyl compounds like aldehydes and ketones (Fig. 2). This is a typical equilibrium reaction and with aldehydes and aliphatic ketones the equilibrium lies normally on the right side. The reaction is catalysed by bases and the mechanism of cyanohydrin formation was the first reaction mechanism in organic chemistry that was recognised as such and studied by Lapworth at the beginning of this century.

Due to the easy elimination of HCN from the adducts, certain cyanohydrins are used as a vehicle for transportation or reaction

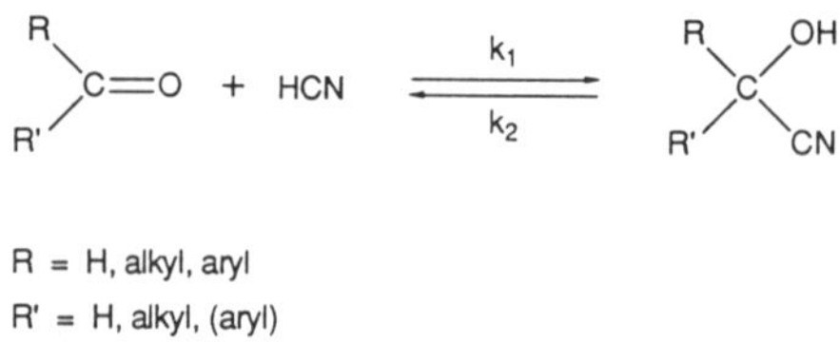


**Fig. 2.** Cyanohydrin formation.

instead of the free acid. If a cyanohydrin should be stored or transported, it has to be stabilised by acids like sulphuric or phosphoric acid.

Commercially important cyanohydrins are, for example, acetone cyanohydrin which is either the starting material for methacrylic acid derivatives or an intermediate for the radical starter azobisisobutyronitrile (Fig. 3).

Mandelonitrile is produced by the reaction of benzaldehyde and HCN but occurs also in nature in a bonded form in the glycoside amygdalin. Saponification of mandelonitrile yields DL-mandelic acid; by reaction with benzene and sulphuric acid the diphenylacetonitrile is produced. This is used as an intermediate for numerous pharmaceuticals.

Depending on the reaction conditions, the addition of HCN to

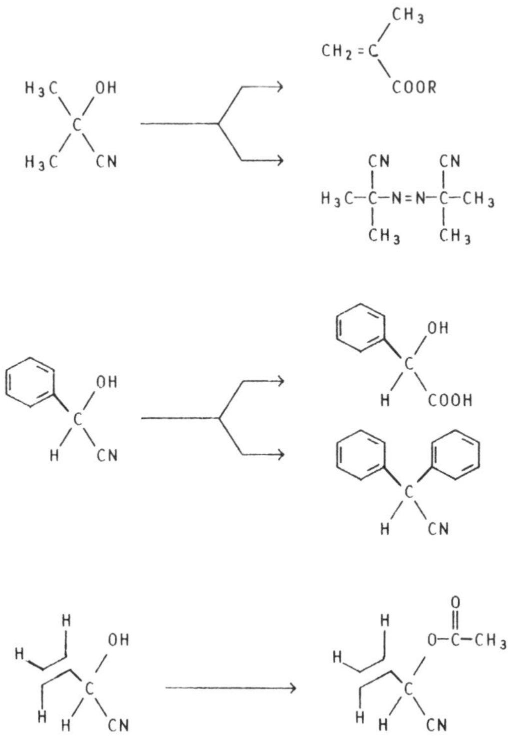

**Fig. 3.**

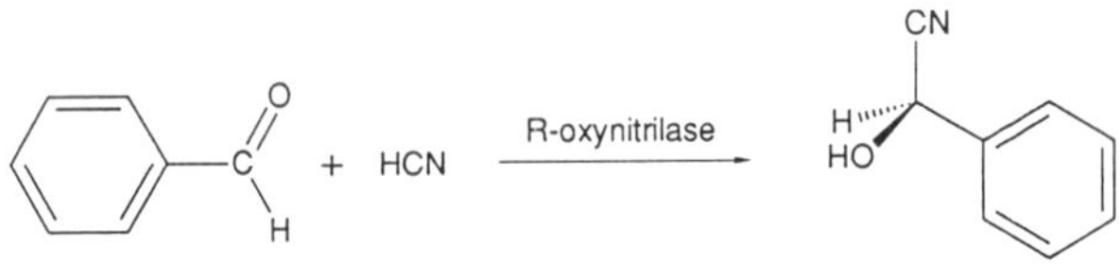


**Fig. 4.**

acrolein yields cyanohydrin or 1,4 addition produces cyanopropionaldehyde (not illustrated). Acrolein cyanohydrin decomposes even after acid stabilisation at low temperatures and therefore its storage is not possible. After acetylation with acetic anhydride, it is transformed in the stable acetate and in this form it is an intermediate in the synthesis of the herbicide 'Basta' from Hoechst AG.

Using an enzyme system such as oxynitrilase, it is possible to add HCN in a stereospecific manner to certain carbonyl compounds (Fig. 4). With such reactions, the difficulty is to demonstrate superiority for the enzyme catalysed reaction compared to the chemically catalysed system. However, a process is available for D-mandelonitrile where high yields and selectivity are obtained using R-oxynitrilase.

## The Aminonitriles

Very closely connected with the cyanohydrins are the basic aminonitriles (Fig. 5).

Starting with formaldehyde as the carbonyl compound, several important aminonitriles are possible. The nature of the reaction products is strongly dependent upon the molecular ratio (Fig. 6) and the applied reaction conditions such as pH. Sometimes preformed mixtures of formaldehyde and ammonia like urotropin are favourably used.

Aminoacetonitrile is only stable as acid salt and it is an intermediate

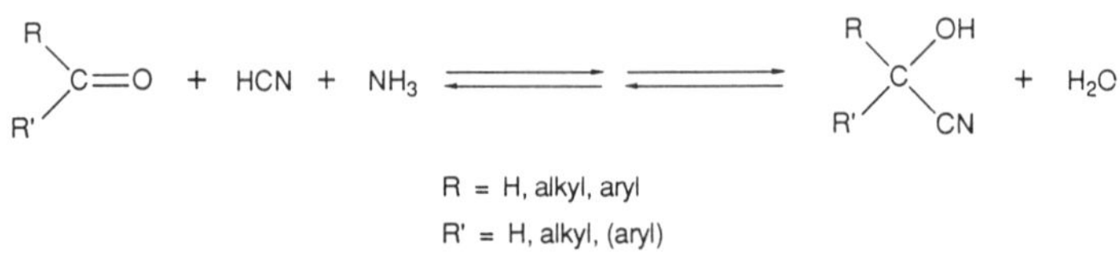


**Fig. 5.** Aminonitriles.

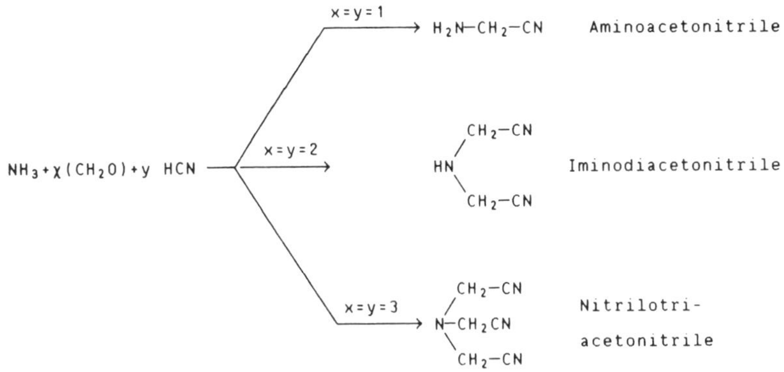


**Fig. 6.**

for the production of glycine, hydantoin, hydantoin acid nitrile and a number of other intermediates and pharmaceutical specialities.

The crystalline and stable iminodiacetonitrile can be used in the synthesis of the worldwide largest selling herbicide—Monsanto's 'Roundup'.

Sodium salt of hydrolysed nitrilotriacetonitrile can be used as a builder in detergents.

Starting from acetone and methylisopropylketone two other aminonitriles are used in large quantities for herbicide production (Fig. 7).

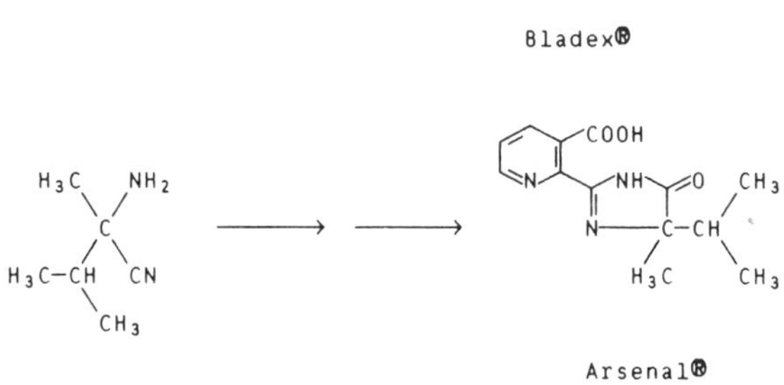


**Fig. 7.** Herbicide production from aminonitriles.

Aminoisobutyronitrile is used for the side chain of the Shell herbicide Bladex® and aminocyanomethylbutane is used in the form of the amino acid amide in the imidazolinone ring system of the new American Cyanamid herbicides like 'Arsenal'.

Aminonitriles are the usual intemediates for the classical amino acid synthesis proposed by Strecker or Bucherer–Bergs (Fig. 8). The Strecker synthesis is normally the acidic hydrolysis of an aminonitrile while a hydantoin is the intermediate state in the process according to Bucherer–Bergs.

## Acyl Cyanides

In spite of the interesting combination of a keto and a cyano group in one molecule, acyl cyanides had found only limited application. This was due to the poor availability and instability of these compounds. This has now changed considerably because the normal hydrolysis products—the $\alpha$-ketocarboxylic acids—have become important intermediates (Fig. 9).

For example, in the mid-1970s a new therapy for the treatment of renal diseases was introduced using the keto analogues of certain essential amino acids like, for example, valine, leucine and isoleucine.

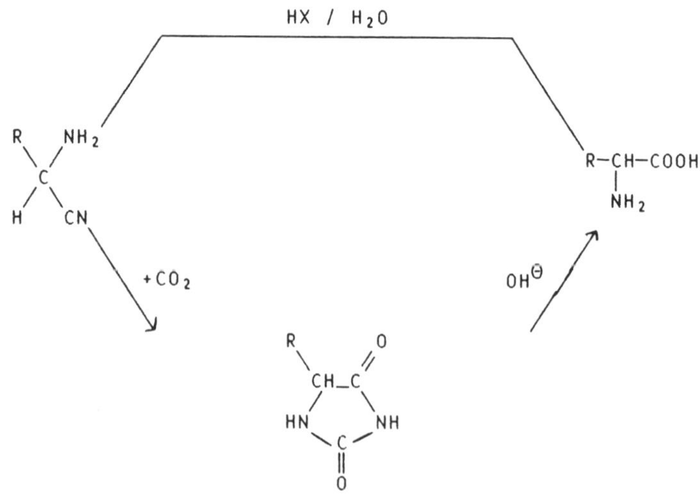


**Fig. 8.** Aminonitriles as intemediates in amino acid synthesis.

**Fig. 9.**

Phenylglyoxylic acid and trimethylpyruvic acid are produced on a large scale because they are intermediates for the important Bayer pesticides 'Goltix' used in sugar beet and 'Sencor' used in soya bean (Fig. 10).

For the production of some of these keto acids new syntheses had to be developed.

Acyl chlorides are the starting materials that yield, in conjunction with sodium cyanide and a selected catalyst/solvent system, the acyl cyanides. These were found to be hydrolysed at very low temperatures under strictly acidic conditions with a minimum quantity of water. However, even under these conditions, the synthesis fails for some acyl cyanides. In these cases, the Ritter reaction can be useful, yielding, for example, a tertiary butylamide (Fig. 11).

## Cyanogen Chloride

Commercially, one of the most interesting derivatives of HCN is cyanogen chloride. This is produced commercially by chlorination of HCN in an aqueous system.

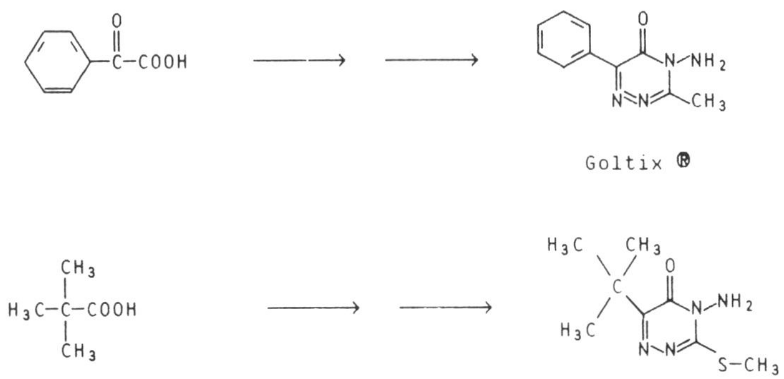


**Fig. 10.**

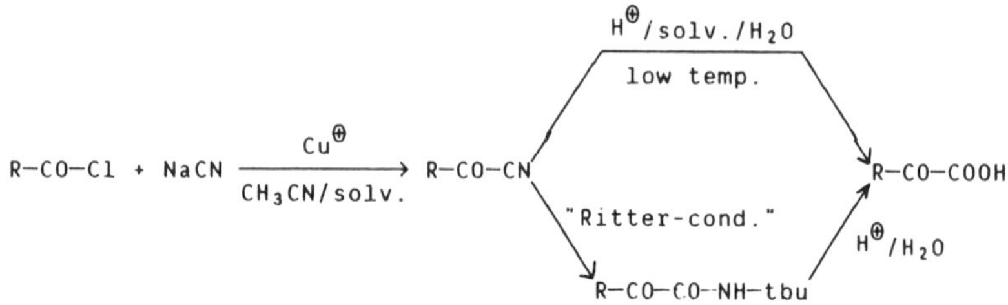


**Fig. 11.**

Cyanogen chloride is one of the few three atomic molecules with three different atoms. Due to its reactive centres, ClCN not only trimerises, yielding cyanuric chloride (the most commercially important product), but additionally, as a disguised acid chloride, can also be substituted at the chlorine atoms through nucleophiles. Supplementary addition reactions at the C≡N triple bond and complexing reaction of Lewis acid at the nitrogen atom are also possible (Fig. 12).

## Cyanamides and Guanidines

Amines can react with cyanogen chloride in a two-stage way. Firstly, a cyanamide is generaed by substitution of the chlorine atom. Then, through addition of another molecule of amine to the triple bond, a guanidine is produced (Fig. 13(a)).

An example of this is with dimethylamine fulfilling the amine role (Fig. 13(b)), and the final tetramethylguanidine produced can be used as a catalyst in polyurethane production or as a base in the synthesis of photochemicals.

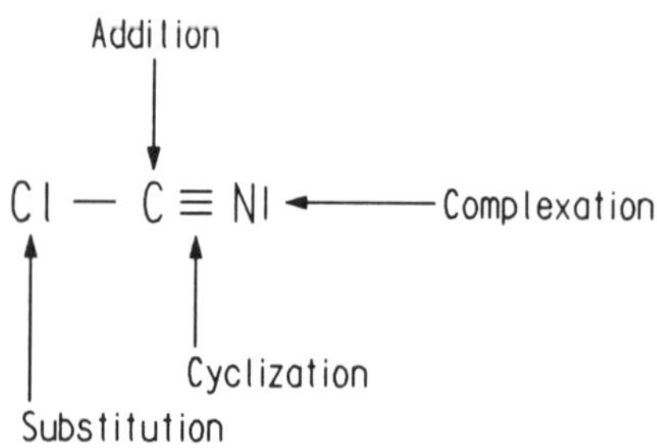


**Fig. 12.**

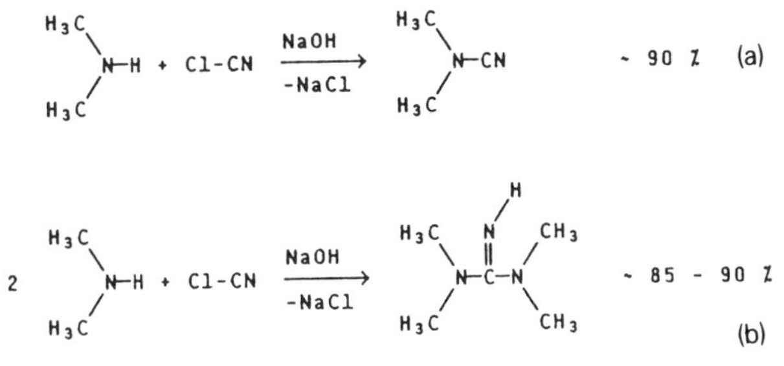

tetramethylguanidine

**Fig. 13.**

Aliphatic cyanamides of primary amines do not react in this way because they trimerise very quickly to the appropriate triamino-*s*-triazines. Cyanamide itself reacts with cyanogen chloride to sodium dicyanamide (Fig. 14(a)).

Sodium dicyanamide is an important intermediate for the synthesis of substituted cyanoguanidines and biguanides, for example in the production of the biocide Chlorohexidine (Fig. 14(b)).

## Acetonitriles

The normal substitution reactions of ClCN proceed in the liquid phase under polar conditions. At high temperatures in the gas phase, radical

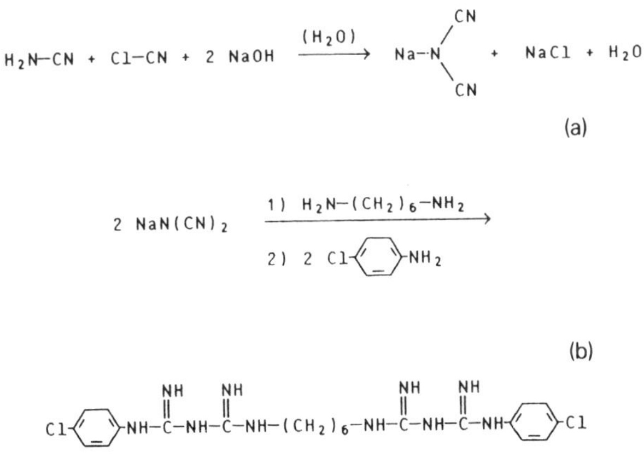

Chlorohexidine

**Fig. 14.** Synthesis of chlorhexidine.

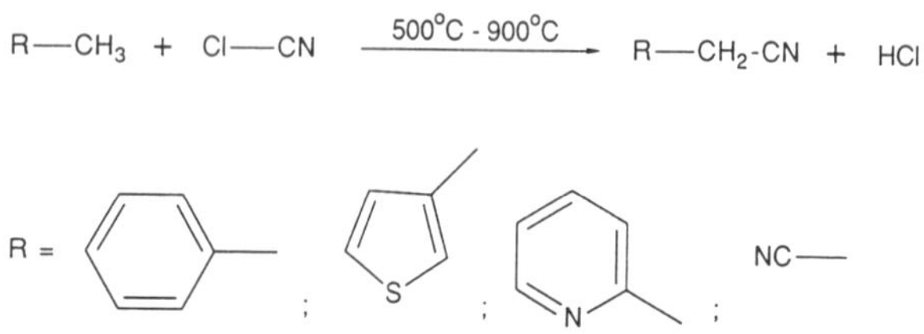


**Fig. 15.**

formation is observed. These yield—starting with methyl compounds as educts—certain acetonitriles (Fig. 15).

Normally these acetonitriles are produced by starting from the alcohols or the methyl compounds, converting them to the chloride and making a halogen-cyanogen exchange reaction. However, with some compounds this sequence is inefficient due to the instability of intermediates or final products under the chosen reaction conditions.

This technique can be used, for example, for the production of thiophene acetonitriles and malonodinitrile, the latter being the only effective example in the aliphatic series.

Malonodinitrile is a fascinating chemical building block and is used in large quantities for the production of vitamin B1. Similarly interesting but not so important in volume terms are the thiophene acetonitriles. Thiophene-3-acetonitrile, for example, could be used for the synthesis of the Beecham antibiotic 'Ticarcillin' (Fig. 16).

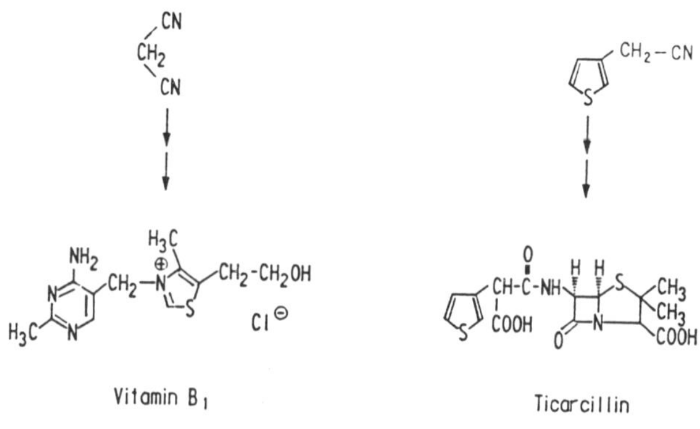


**Fig. 16.**

## Conclusion

This brief review shows that the chemistry of hydrocyanic acid and its derivatives is diverse and versatile offering chemists the opportunity to develop new products or commercially attractive routes to existing compounds.

# Phosgene and related compounds: reagents for the laboratory and industry

**G. Sennyey**

*SNPE, Le Bouchet, France*

Since its discovery by Davy in 1812,[1] the chemistry of phosgene has been thoroughly studied by numerous academic and industrial chemists, thus realizing the prediction of Jean-Baptiste Dumas,[2] a pioneer in the chemistry of phosgene: 'L'acide chloroxicarbonique, . . . , offre une composition si simple et si remarquable que, s'il réalisait toutes les réactions que l'on a droit d'en espérer, on parviendrait à reproduire, à son aide, les combinaisons les plus curieuses de la chimie organique.'

All the investigations led since then have clearly demonstrated the value of phosgene as a building block for the preparation of industrial chemicals and as a reagent for carbonylation, chlorination and dehydration.

It is beyond the scope of this review to describe all the reactions of phosgene and of its derivatives since adequate reviews on the chemistry of phosgene,[3] isocyanates[4] and chloroformates[5] and a remarkable book written by BAYER chemists[6] are available. Phosgene is a very toxic gas (see Table 1) which was used as a chemical weapon during World War I and is mostly prepared for captive uses.

Due to its unique toxicity, captive uses and also to the impact of the Bhopal accident on the public, information related to the production of phosgene is generally kept confidential. However, the total World

**Table 1.** Lethal dose low and threshold limit value for some gases
(Source: RTECS)

| | Human LCLo | Mammal LCLo | TLV |
|---|---|---|---|
| CO | 4000 ppm/30M | 5000 ppm/5M | 50 ppm |
| $Cl_2$ | 873 ppm/30M | 500 ppm/5M | 500 ppm |
| HCN | — | 200 ppm/5M | 10 ppm |
| $NH_3$ | 30 000 ppm/30M | 5000 ppm/5M | 25 ppm |
| $COCl_2$ | — | 50 ppm/5M | 0·1 ppm |
| $CO_2$ | 105 ppm/1M | 9 $10^4$ ppm/5M | 5000 ppm |

capacity amounts to around 2·2 million tons, 85% of which is used for the manufacture of di- and polyisocyanates, TDI and MDI as intermediates for polyurethanes. A further 6% is used for the manufacture of polycarbonates, which leaves only 9% of the total production of phosgene for the manufacture of fine chemicals.

Fine chemical phosgene consumption is divided roughly as 50% for the preparation of peresters and percarbonates used as polymerization initiators, 25% in the agrochemical field and 25% for the manufacture of intermediates for pharmaceuticals and dyestuffs. Although it represents only a small part of the total phosgene production, activity in the field of fine chemicals remains very intensive. It can be roughly divided into two classes of products, whether the structural unit remains in the final product—the intermediates—or not—the reagents.

## Phosgene derived reagents

Phosgene itself remains a unique reagent due to its numerous reactions with a broad spectrum of organic functional groups. However, regulatory requirements in transportation and security know-how in the handling of phosgene have restricted its uses to specialized companies and custom synthesis is becoming a common practice. As the acid chloride of carbonic acid, phosgene reacts with common nucleophiles such as alcohols, phenols, mercaptans and amines to give the corresponding chloroformates, carbonates, carbamyl chlorides, isocyanates and ureas (Scheme 1).

All these derivatives and phosgene itself are also precursors of

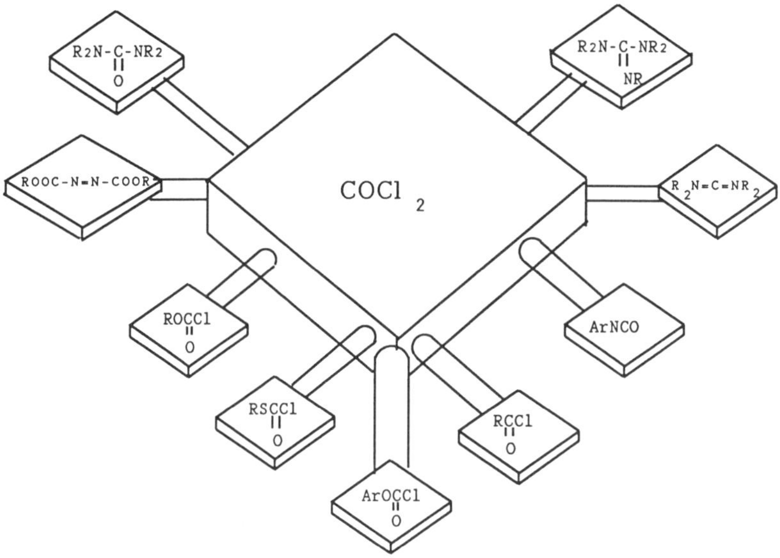


**Scheme 1.**

other reagents such as carbodiimides or bisalkyl azodicarboxylate and are used over a wide range of reactions.

## Dehydrations

Phosgene and its substitutes di- and triphosgene dehydrate formamides in high yields to give the expected isocyanides (Scheme 2) which are used in the Ugi four-components peptide synthesis[7] (Scheme 3) or in the synthesis of alkoxyoxazoles and vitamin $B_6$.[8]

Carbodiimides are also well-known reagents which have been extensively used for the dehydrative coupling of hydroxy compounds such as in the synthesis of anhydrides from acids or in the dehydration of amides and aldoximes to nitriles.

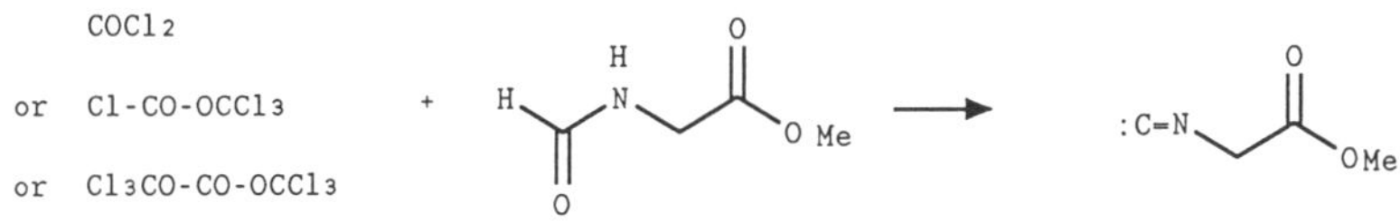


**Scheme 2.**

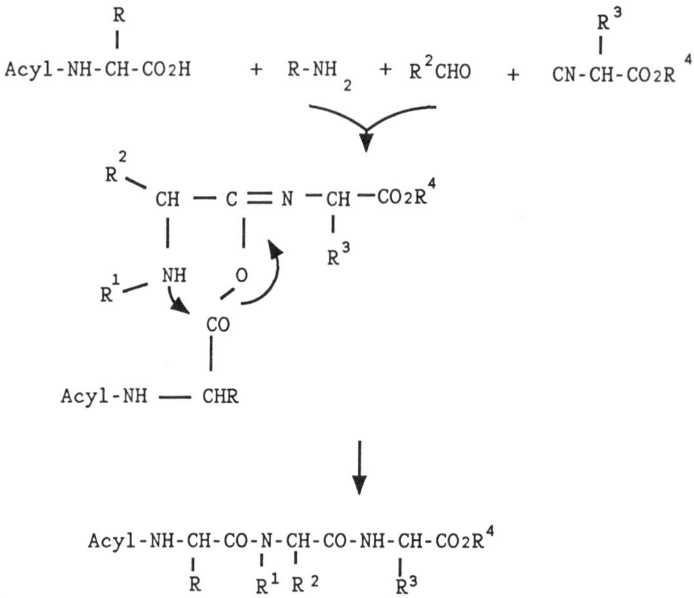

**Scheme 3.** UGI four components synthesis.

## Chlorinations

The most important use of phosgene as a chlorinating reagent is by far the synthesis of acid chlorides which are mainly used as intermediates for polymerization catalysts. The great advantage of phosgene over thionyl and/or phosphoryl chloride is a better purity of the final acid chloride due to the gaseous nature of by-products. This is particularly important for the high boiling fatty acid chlorides which cannot be distilled.

However, since the reaction of phosgene or its congeners di- and triphosgene has to be catalysed, it has been an active field of research to increase the efficiency of these catalysts (Scheme 4 and Table 2).

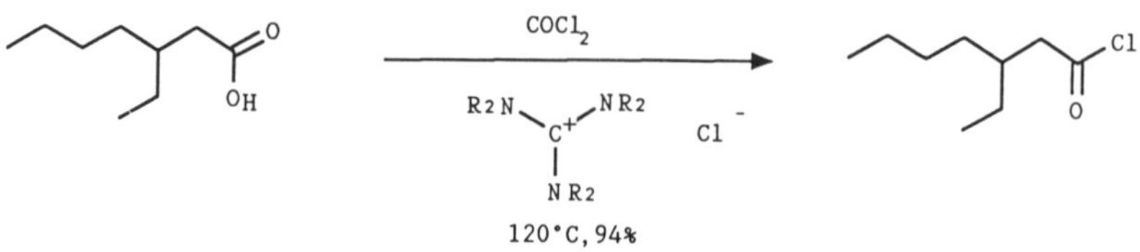

**Scheme 4.**

**Table 2**

| Active catalyst | | k |
|---|---|---|
| 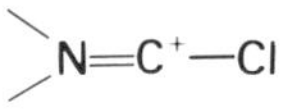 | Cl | — |
| 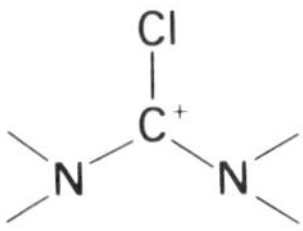 | Cl | 0.023 |
| 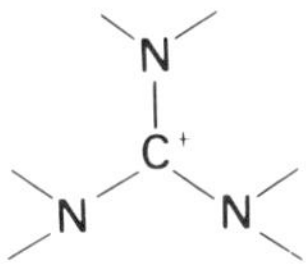 | Cl | 0.035 |

Lactones are converted to beta-chloro acid chlorides when treated with phosgene in the presence of HCl or aluminium chloride (Scheme 5).[9]

Amides are halogenated to give imidoyl or chloroiminium chlorides[10] depending on the nitrogen substitutions. These Vilsmeier salts are effective to transform alcohols into their alkyl chlorides, a reaction that can also be done with a combination of phosgene and dimethylformamide (Scheme 6). These salts can also be used for the preparation of aldehydes, heterocycles and for the esterification of carboxylic acids.

Phosphine oxides are chlorinated to the corresponding phosphine dichlorides which can be easily reduced to the original phosphine with thiols or hydrogen (Scheme 7), thus enabling, for instance, the recovery of triphenylphosphine from Wittig reactions.[11]

Although phosgene is the most used reagent in the chlorination reaction, other derivatives have also been used for this purpose. For instance, *N,N*-dichlorourethane, which is easily prepared from

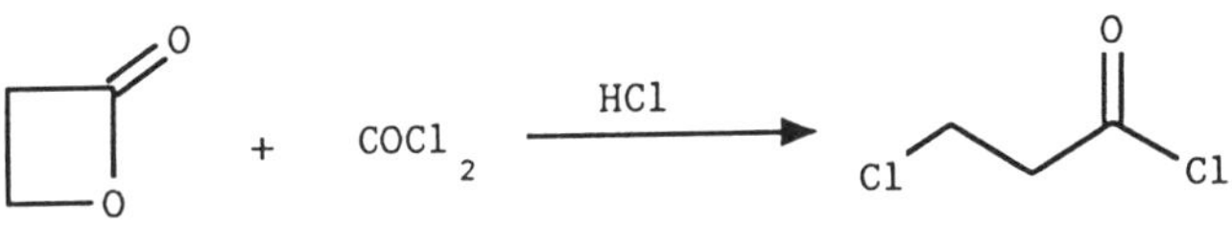


**Scheme 5.**

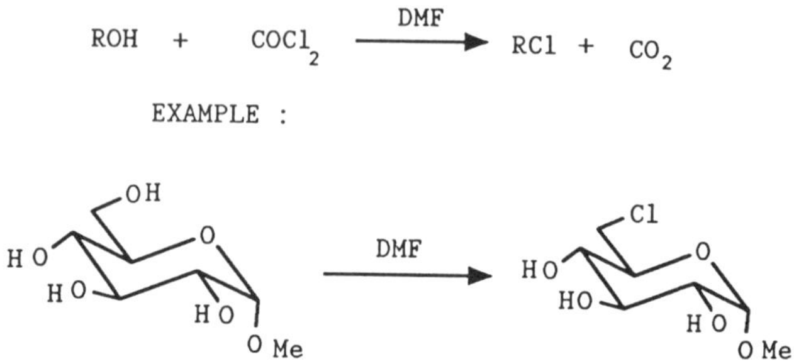


**Scheme 6.**

urethane and chlorine, has been found to effect radical chlorination of carbocycles such as cyclohexane, adamantane or steroids.[12]

## Protection of Functional Groups

The synthesis of molecules of ever-increasing complexity requires the availability of a variety of protective groups to ensure the survival of

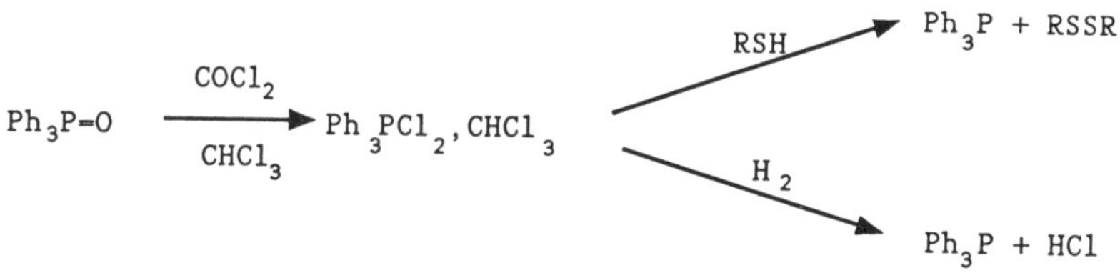


**Scheme 7.**

functional groups such as the hydroxyl and amino functions. Since various carbamates and carbonates are often used to this purpose, phosgene based reagents have gained considerable importance in this field, particularly in complex multistep synthesis and peptide synthesis. Although a great number of protective groups have been proposed, the *tert*-butyloxycarbonyl (BOC) group may now be considered one of

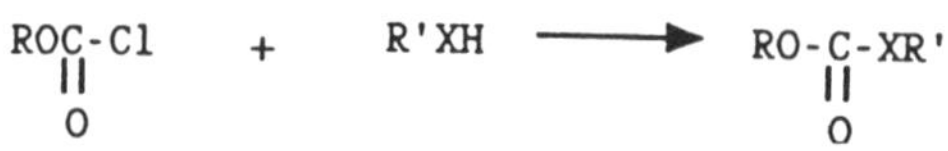


**Scheme 8.**

the 'workhorses' of the field, along with the carbobenzoxy (Z), the fluorenylmethoxycarbonyl (FMOC) and to a lesser extent the allyloxycarbonyl (ALLOC) groups. These protective groups are mostly introduced into a molecule by the reaction of an appropriate chloroformate with the functional group to be protected (Scheme 8 and Table 3).

However, in some cases, chloroformates are not suitable for this purpose because of possible side reactions or more simply because they are unstable under normal conditions. For instance, the instability of *tert*-butylchloroformate led to considerable difficulties in devising suitable techniques for the introduction of the BOC group onto an amino function. As a consequence, a great number of BOCing (Table 4) agents have been developed over the years, mostly carbonates. Di-*tert*-butyldicarbonate—$(BOC)_2O$—is by far the most convenient and widely used reagent for this purpose, although *tert*-butylfluoroformate would be less expensive if both manufacturing and security problems could be overcome.

In recent years, there has been a considerable interest in this field both in academic and industrial laboratories. The increased demand for artificial sweeteners such as aspartame and active research in the peptide based pharmaceuticals as, for instance, in the antihypertensive 'family' has led chemical companies to offer a wide range of blocked amino acids both on a bench or an industrial scale.

## Activation of Functional Groups

In numerous applications of organic synthesis, the manipulation of organic functional groups goes through their prior activation to run the requested reaction. This is often the case for the synthesis of esters, amides, ureas, carbamates and carbonates and some cyclization. Most of these techniques were originally studied for the purpose of peptide synthesis and are used in numerous drugs synthesis, as well as in protein crosslinking and in the activation of functionalized carriers for chromatography. The action of phosgene itself on carboxylic acids has already been mentioned, but other derivatives such as chloroformates, carbonates or imidazolides can be used for this purpose (Table 5). The formation and uses of mixed carboxylic carbonic anhydride has been thoroughly studied by peptide chemists and successfully applied to the synthesis of antibiotics. The formation of active esters through phosgene derivatives has also been very successful.

**Table 3**

| Functional group | Protective group | Stable to | Cleaved by |
|---|---|---|---|
| ROH | $MeO-C(=O)-OR$ | $1 < pH < 8{\cdot}5$ | $K_2CO_3$, MeOH, 25°C |
| | $Cl_3C-CH_2-O-C(=O)-OR$ | $1 < pH < 8{\cdot}5$ | Zn/AcOH<br>−1·65 V/MeOH/ $LiClO_4$ |
| | 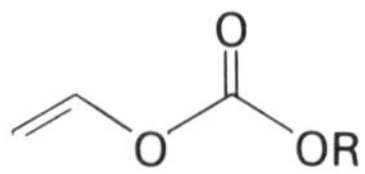 | $1 < pH < 8{\cdot}5$ | $Na_2CO_3/H_2O$ |
| | 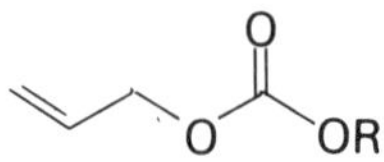 | $1 < pH < 8{\cdot}5$ | $Ni(CO)_4$/TMEDA<br>$Bu_3SnH/Pd(PPh_3)_4$ |
| | 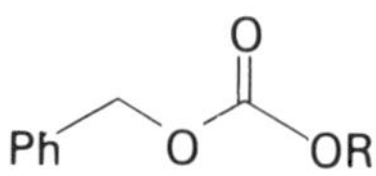 | $1 < pH < 8{\cdot}5$ | $H_2$/Pd/C |
| | 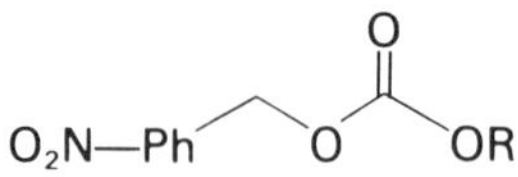 | 1 < ph < 8·5 | Catalyst imidazole, $H_2O$, dioxane |
| $RNH_2$ | 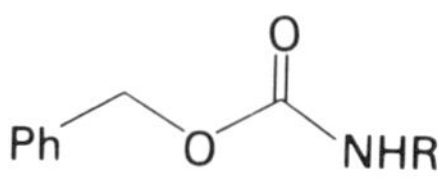<br>(Z) | Dilute AcOH, low alkaline | Anhydrous HF, $H_2$/Pd, HBr/AcOH |
| | 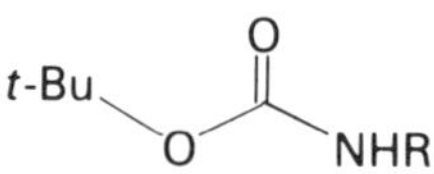<br>(BOC) | $H_2$/Pd, $Na/NH_3$, alkaline hydrolysis | HCl < 30°C, $CF_3CO_2H$, HBr, anhydrous HF |
| | 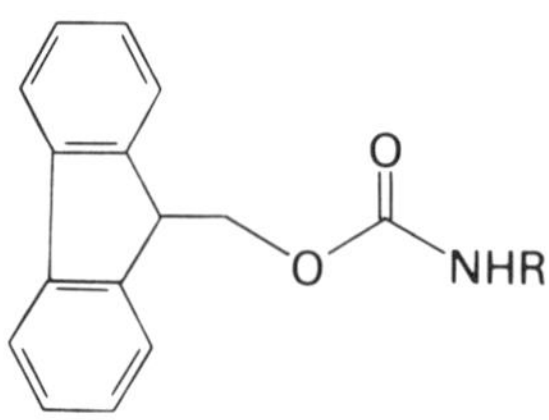<br>(FMOC) | Acid, $H_2$/Pd | Amines, $NH_3$, piperidine |

**Table 3** —*Contd*

| Functional group | Protective group | Stable to | Cleaved by |
|---|---|---|---|
| | 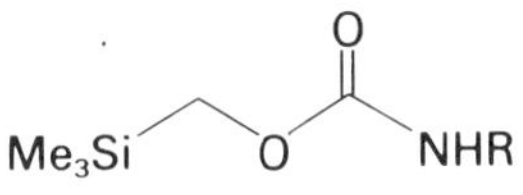 (TEOC) | $H_2$/Pd, low alkaline | $Et_4N^+F$ |
| | 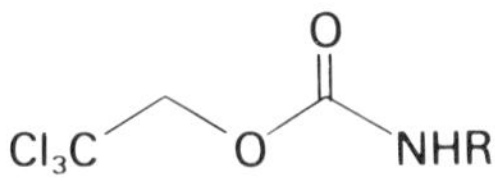 (TROC) | Acids, NaOH | Zn, AcOH |
| | 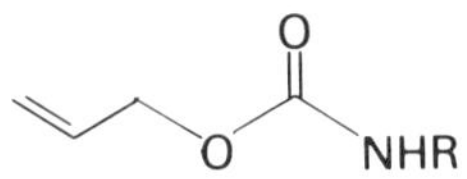 (ALLOC) | Acids, bases | Na/NH, $Bu_3SnH/Pd(PPh_3)_4$ |

Carbodiimides are also very popular reagents for this purpose and have been used extensively both by organic and biochemists. A complete palette of variously substituted carbodiimides is now available and used for the formation of amides, esters and in numerous coupling reactions. When combined with DMAP, carbodiimides (e.g. DCC) are very powerful reagents and even *tert*-butyl esters of hindered acids can be obtained.

Recently, uronium derivatives of benzotriazole, among which TBTU (2-(1*H*-benzotriazol-1-yl)-1,1,3,3-tetramethyluronium tetrafluoroborate) seems to be the most promising, have been proposed by Roche chemists to replace Castro's popular BOP (benzotriazol-1-yl-oxy-tris(dimethyl)-phosphonium hexafluorophosphate) in peptide synthesis.[13]

Hydroxy groups in polysaccharide resins (e.g. Sepharose) have been activated through their hydroxysuccinimide carbonates and coupled to amine containing ligands and proteins to give stable carbamate derivatives useful for protein immobilization and affinity chromatography. Recently, activation of the anomeric position of tetra-*O*-benzyl-D-glucopyranose with isopropenyl chloroformate has also been successfully used in a glycosylation procedure (Scheme 9).[14]

**Table 4**

| Reagent | Abbreviation | Prepared from |
|---|---|---|
| 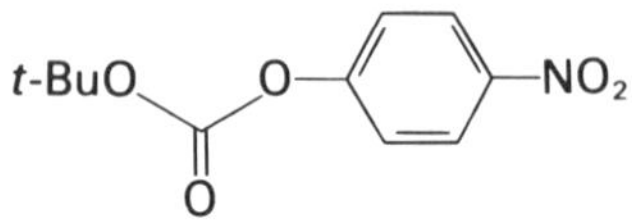 | | 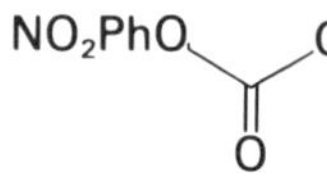 + $t$-BuOH |
| 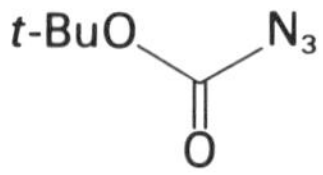 | BOC-$N_3$ | 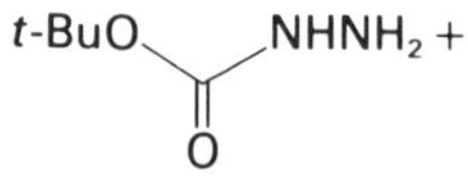 + NaNC |
| 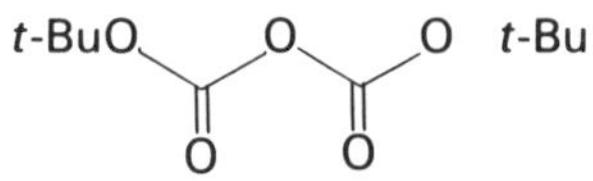 | $(BOC)_2O$ | $t$-BuONa + $CO_2$ + $COCl_2$ |
| 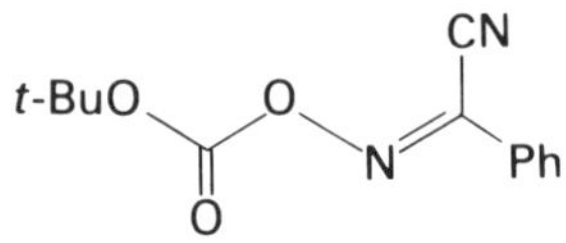 | BOC-ON | $t$-BuO + 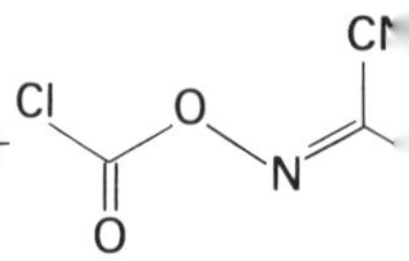 |
| 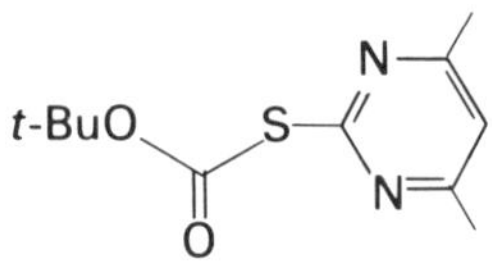 | | $t$-BuOH + 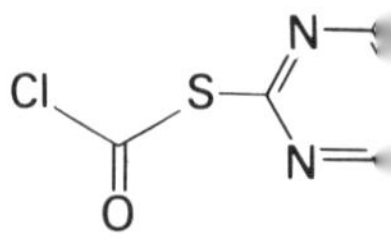 |
| 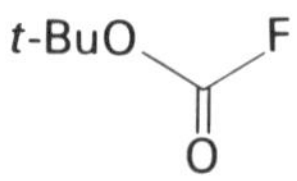 | BOC-F | $t$-BuOH + 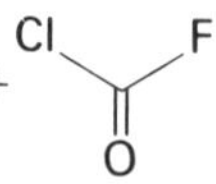 |
| 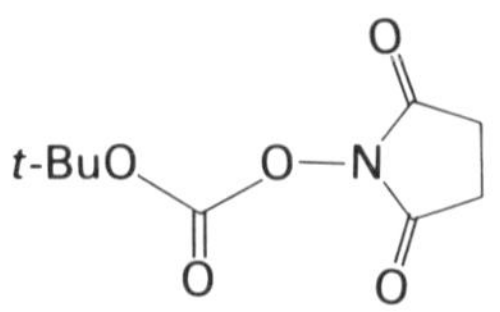 | BOC-OSu | $t$-BuOH + 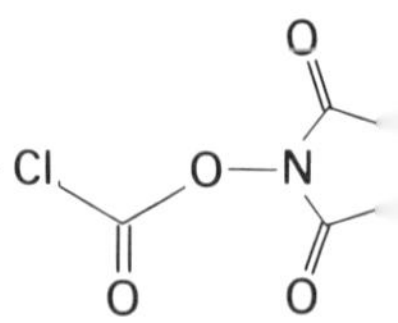 |
| 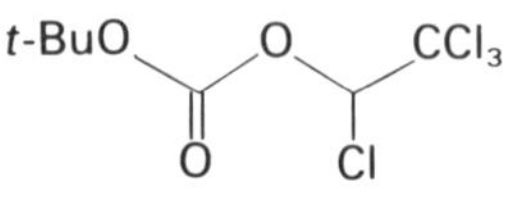 | BOC – OTCE | $t$-BuOH + 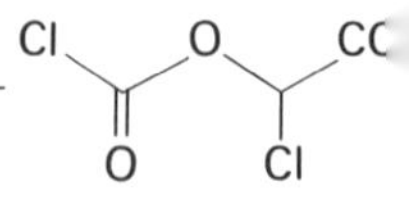 |

**Table 4**—*Contd*

| Reagent | Abbreviation | Prepared from |
|---|---|---|
| 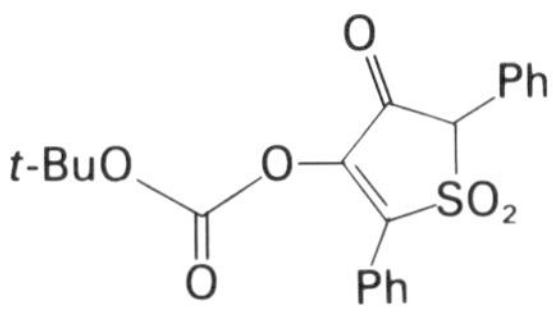 | — | *t*-BuOH + 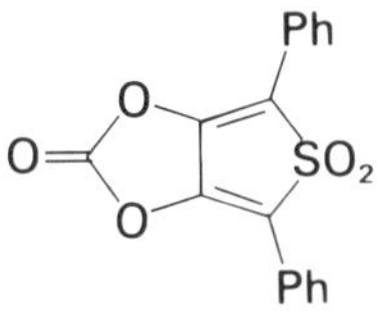 |
| 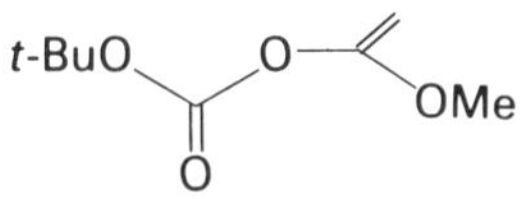 | — | *t*-BuOH + 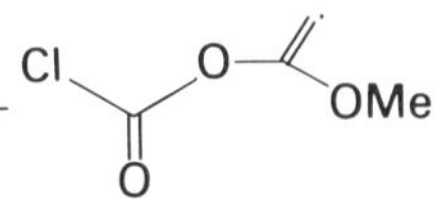 |

Finally, activation of amines or sulphonamides may occur through the corresponding isocyanates or carbamates, a process that is used in the synthesis of the sulphonylurea herbicides (Scheme 10).

## *N*-Dealkylation of Tertiary Amines

The von Braun *N*-dealkylation of tertiary amines has been known since the turn of the century and cyanogen bromide has now been widely replaced by chloroformate reagents which, in addition to their ease of handling, have proven to be more selective and give cleaner reactions. Many chloroformates have been reported in the literature, each one claimed to be an improvement over the last. In 1977 Olofson introduced vinyl chloroformate, which selectivity and ease of removal caused to become the reagent of choice for these transformations.[15] However, the need for mercury salts in the synthesis of this reagent led Olofson and the SNPE team in 1984 to propose the easily available 1-chloroethyl chloroformate as an alternative. Selectivity was in all respects comparable with that of vinyl chloroformate and the cleavage of the intermediate 1-chloroethyl carbamate was done by simple heating in methanol. The strength of this reagent has been demonstrated by comparative *N*-dealkylations of 1-ethylpiperidine (Scheme 11) and also in the synthesis of nalbuphine from oxycodone in five steps with 69% overall yield, the key step *N*-demethylation being almost quantitative. The method is undoubtedly the best that has been published in the literature.[16]

**Table 5**

| Reagent | Intermediate | Product |
|---|---|---|
| $COCl_2$ | RCOCl | RCONHR |
| | 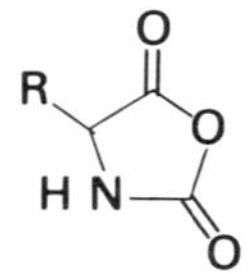 | 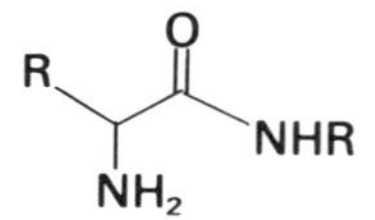 |
| 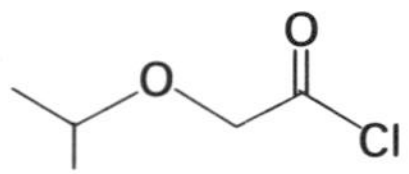 | 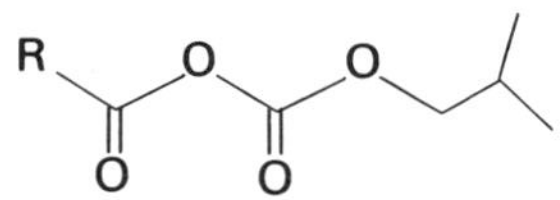 | 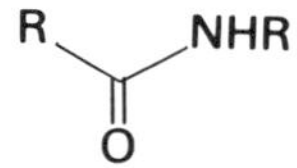 |
| 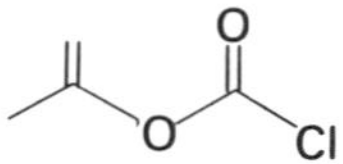 | 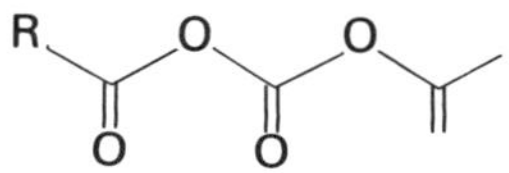 | 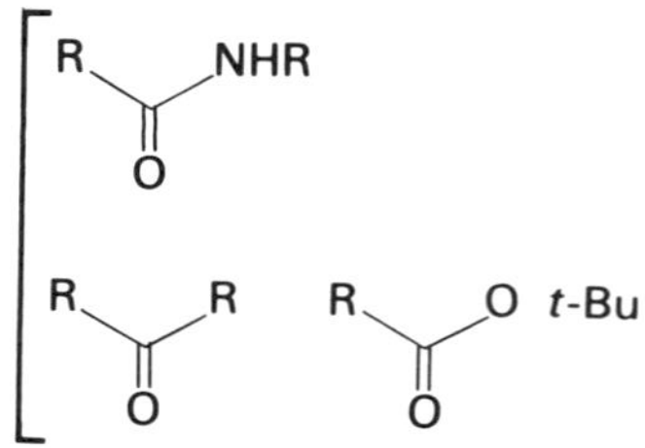 |

| | | |
|---|---|---|
| R′—N=C=N—R′ | 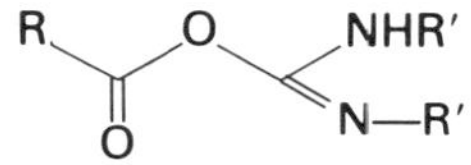 | Amides, esters |
| 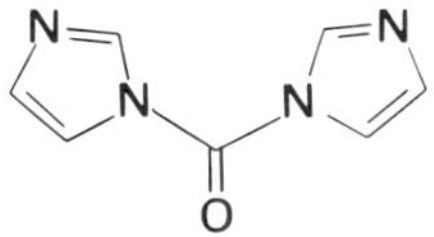 | 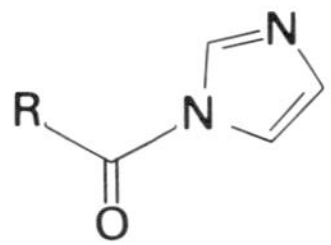 | Acid chlorides, amides, esters, ketones |
| 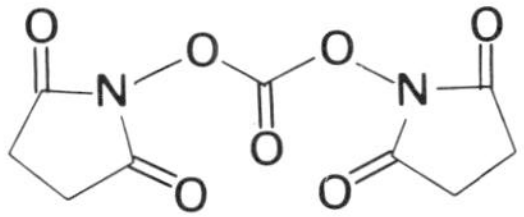 | 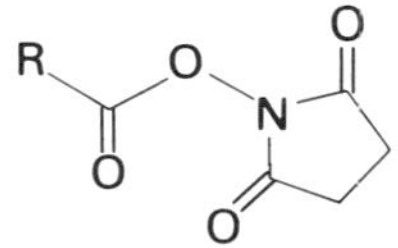 | Amides, peptides |
| 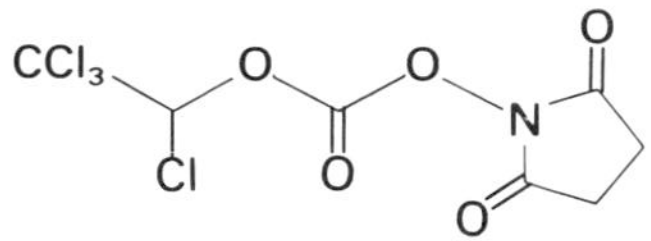 | 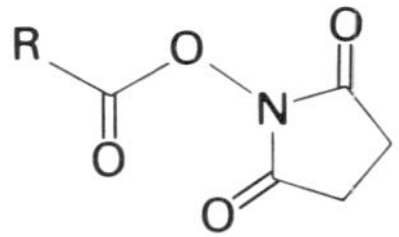 | Amides, peptides |
| 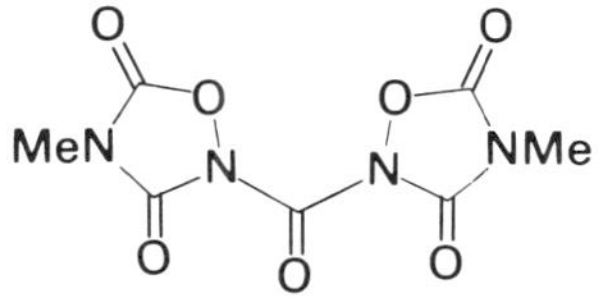 | 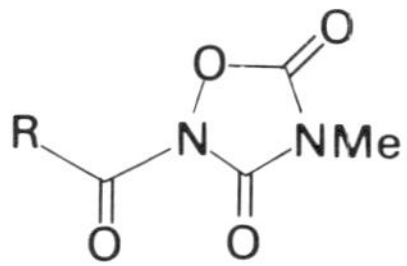 | Amides, peptides, esters |

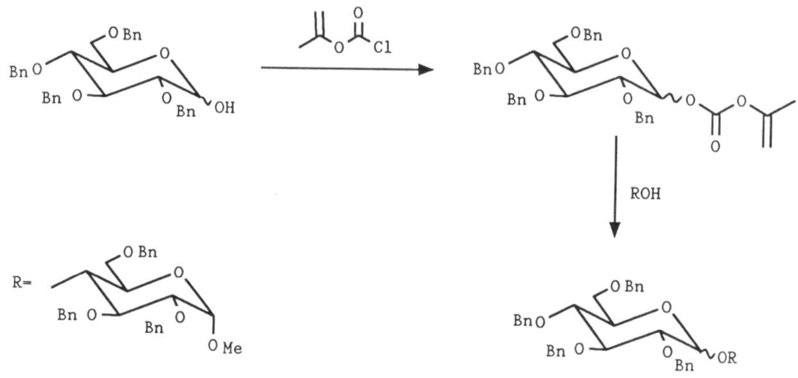

**Scheme 9.**

Although phosgene has been reported to dealkylate triethylamine, it has generally no synthetic value, except for the elegant intramolecular rearrangement which has been recently reported by Olofson and Maryanoff (Scheme 12).[17]

## Cyclizations

The latter reaction could have been considered as an ingenious combination of *N*-dealkylation and cyclization. However, beyond this

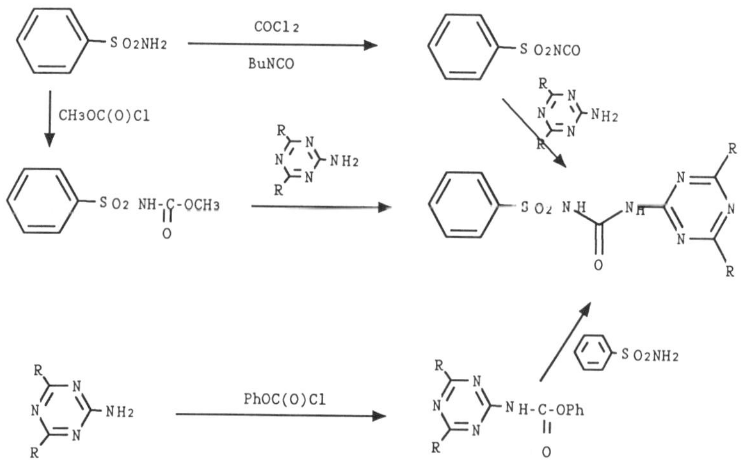

**Scheme 10.**

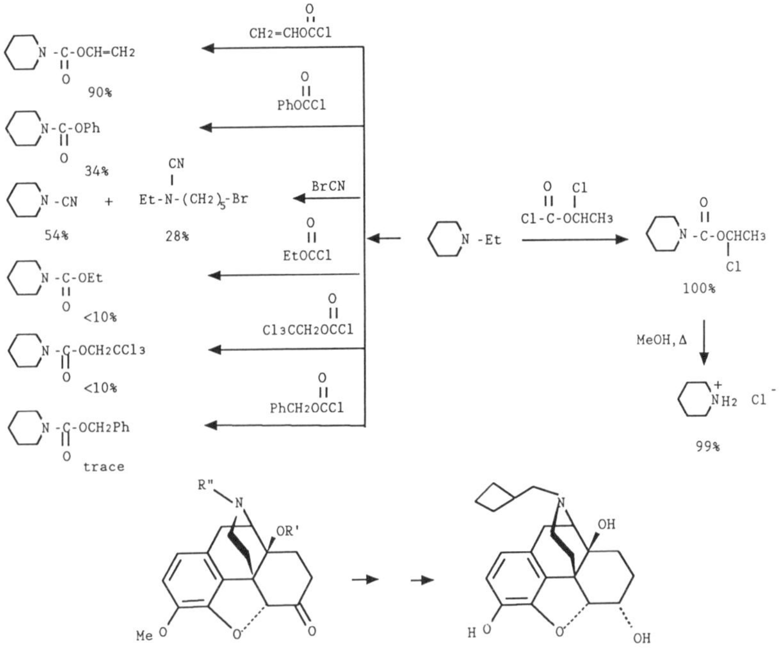

R'=H, R"=Me (oxycodone)

R'=CO-$C_4H_7$-c, R'=Me

R'=CO-$C_4H_7$-c, R'=H,HCl

**Scheme 11.**

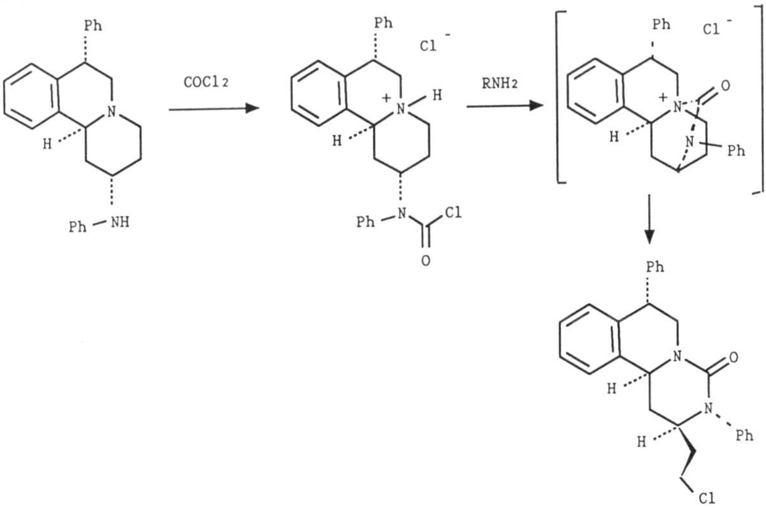

**Scheme 12.**

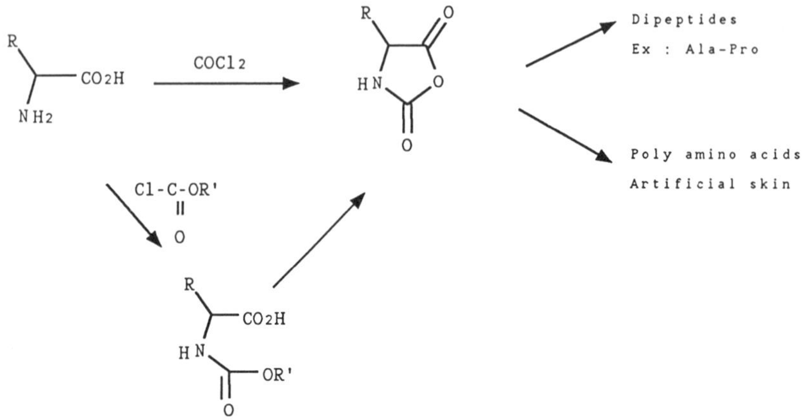


**Scheme 13.**

very particular case, phosgene and derivatives give rise to a large number of cyclizations and so are the source for numerous intermediates in the pharmaceutical, agricultural and polymer fields. These range from simple cyclic carbamates and ureas to more sophisticated heterocycles. Amino acids and related compounds react with phosgene or its derivatives to give *N*-carboxyanhydride (Leuchs' anhydrides) (Scheme 13), useful in the preparation of dipeptides such as Ala-Pro or TFA-Lys-Pro,[18] intermediates in the synthesis of peptidic anti-hypertensive. NCAs are also used in the manufacture of polyamino acid artificial skin.

Hydantoines can also (but not always) be prepared from amino acids and phosgene derivatives. In fact, the use of phosgene chemistry

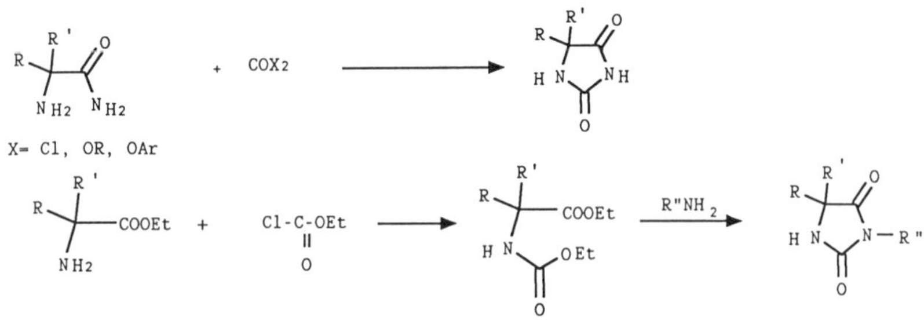


**Scheme 14.**

**Table 6**

| Hydantoine | Use |
|---|---|
| 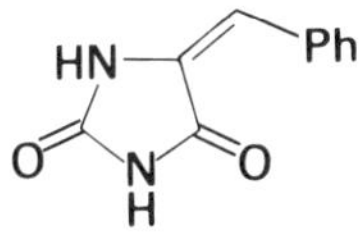 | Phenylalanine synthesis |
| 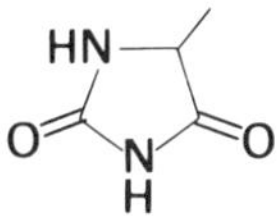 | Alanine synthesis |
| 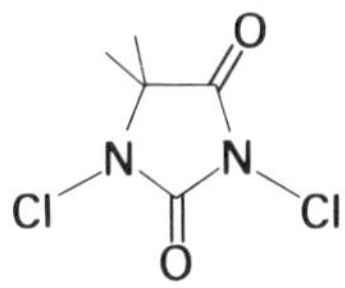 | Bactericide |
| 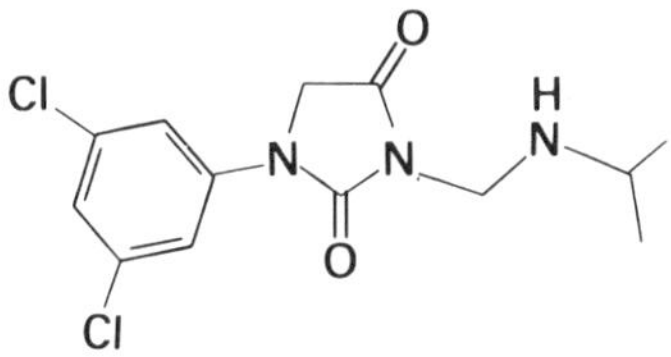 | Fungicide |

is restricted to only a few cases where the amino acid is available (Scheme 14 and Table 6).

*N*-acyl amino acids are also easily cyclized with chloroformates to give substituted azlactones, some of which, like 4,4-dimethyl-2-vinyl-5-oxazolone, are used as monomers for photoresists (Scheme 15).[19]

Hydrazides cyclize to oxadiazolinones which can be further transformed to triazolones. Both structures give rise to important agricultural and pharmaceutical compounds (Scheme 16).

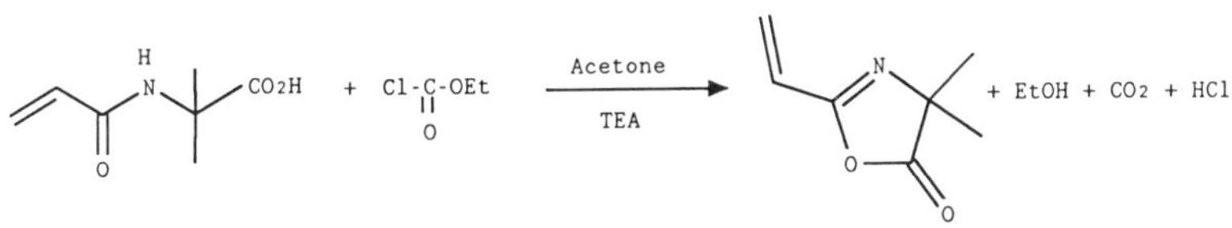


**Scheme 15.**

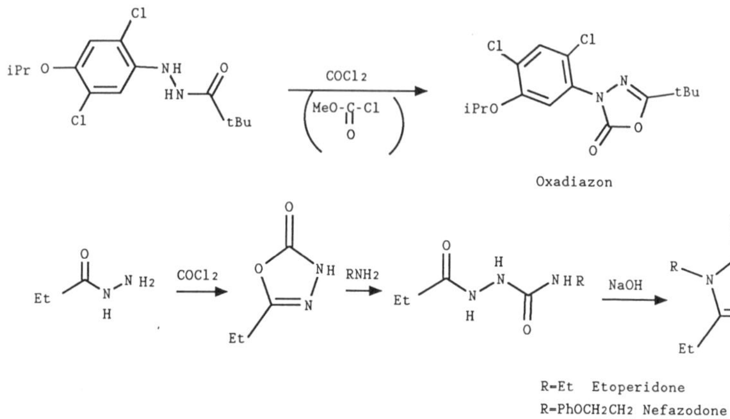


**Scheme 16.**

Alpha hydroxyamides lead to 1,3-oxazolidine-2,4-diones such as the fungicide Vinclozolin (Scheme 17).

Finally, chemists at SNPE[20] have described the new cyclic carbalkoxy aminimides (Scheme 18) which are prepared very simply by heating the 1-chloroalkyl carbazates in THF. These hygroscopic salts are stable indefinitely in the solid form and are transformed to the corresponding cyclic aminimides by treatment with base (e.g. Amberlite IRA-400) in water. These stable and crystalline products are then thermally decomposed at 80–100°C to the transient *N,N*-dimethylamino isocyanate which either dimerizes, reacts with an added nucleophile or can be used *in situ* for the preparation of various

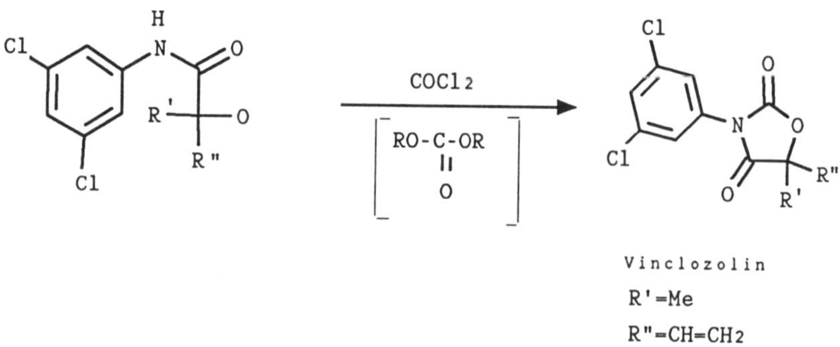


**Scheme 17.**

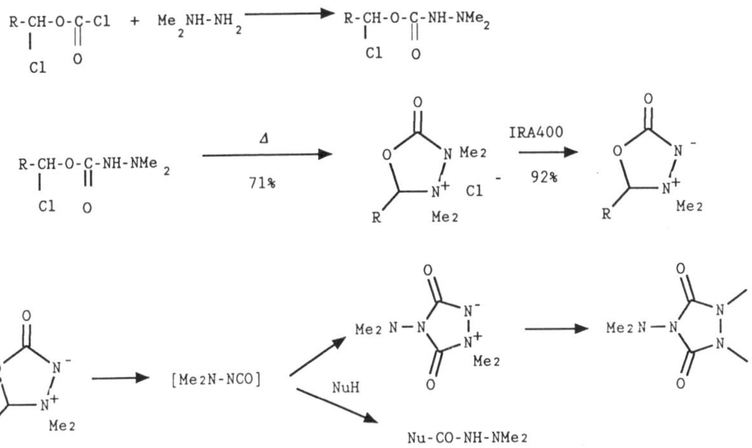

**Scheme 18.**

heterocycles. This method represents a great improvement over already known preparations such as the Curtius rearrangement of carbamoyl azides which is conducted under gas phase vacuum pyrolysis at 300°C.

## Alkylation and Acylation Reactions

Acylation of active methylene compounds and enamines (e.g. carbalkoxylation) with chloroformates and carbonates has now become a trivial reaction. However, acylation of active methylene with mixed carboxylic-carbonic anhydride, imidazolide or similar compounds has been recently described and used in the preparation of various beta-hydroxy amino acids, e.g. the synthesis of BOC statine.[21] This class of reactions is closely related to the activation of carboxylic acids.

The use of phosgene derived reagents for O and C alkylations has been the subject of more intensive work. Dialkyl carbonates have been found to be good alkylating agents of alcohols, thiols and phenols providing the reaction is catalysed with either DMAP or pentaalkyl guanidines (Table 7). Potassium carbonate associated with phase transfer conditions has also been used successfully for this purpose. Dialkylcarbonates and especially dimethyl carbonate are non-toxic and very easy to handle, which is not the case for the classical alkylating agents, and are thus considered as very promising reagents.[22]

**Table 7.** Methylation with dimethyl carbonate

| Substrate | Catalyst | Reaction condition | Yield |
|---|---|---|---|
| PhOH | DMAP | 140°C/4·5 h | 27% |
| $Ar_2NH_2$ | 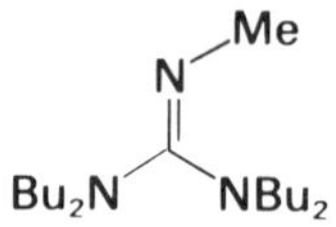 | 140°C/4·5 h<br>160°C/4·5 h | 38%<br>100% |
| | $K_2CO_3$/18-crown-6 | 100°C/8 h | 40% |
| | $K_2CO_3$/PEG-6000 | 180°C/continuous | 99·5% |
| PhSH | NaOEt | 90°C/15 h | 72% |
| 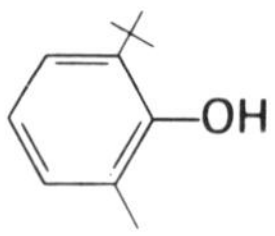 | 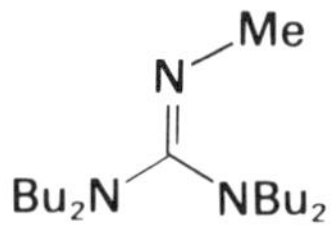 | 180°C/4·5 h | 96% |
| 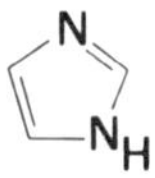 | $K_2CO_3$/18-crown-6 | 100°C/8 h | 91% |
| $Ar_2NMe$ | — | 195°C/4·5 h | 87% |

Chloroformates and carbamyl chlorides have also been found to be valuable reagents to induce regioselectivity in various alkylations. For instance, the reaction of alkynyl or alkenyl–Grignard reagents proceeds in a highly regiospecific manner to afford exclusively 2-substituted 1,2-dihydropyridines (Scheme 19).[23] Regioselective alpha-allylation of various *N*-heteroaromatics has also been obtained by the reaction of allyltin reagents onto *N*-(alkoxycarbonyl)pyridinium salts.

Carbamates prepared from 2-alkenols and *N,N*-diisopropylcarbamoyl chloride (Scheme 20) are easily deprotonated by *n*-butyllithium/tetramethylethylenediamine to form stable alpha-lithio

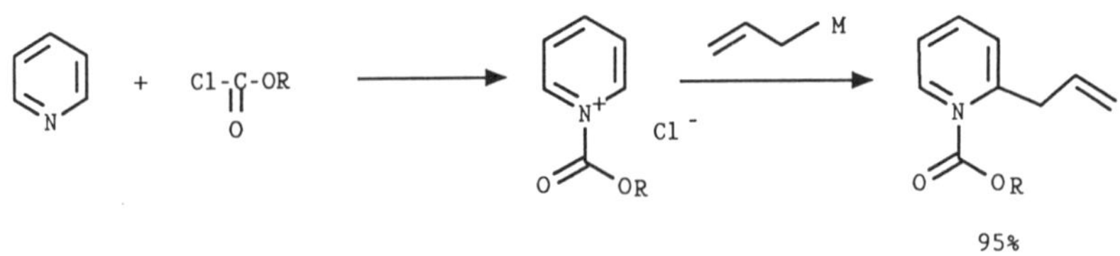


**Scheme 19.**

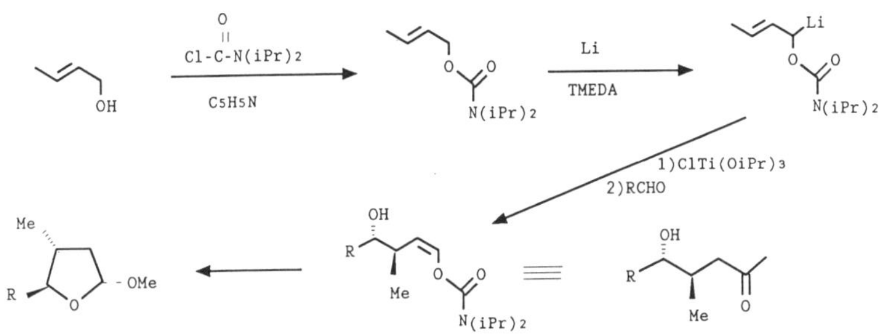


**Scheme 20.**

derivatives. These add to aldehydes or ketones highly gamma-selectively with formation of 4-hydroxy-1-alkenyl carbamates which are equivalent to homoaldols. Moreover, after metal exchange, the reaction also proceeds with high diastereoselectivity depending on which metal is used.[24] Benzyl and aryl *N,N*-diisopropyl carbamates also form stable $\alpha$-lithio derivatives, which are substituted by a wide variety of electrophiles.[25]

Evans' (4*S*)-4-isopropyl-2-oxazolidinone, prepared from valine and ethyl carbonate or benzyl chloroformate, has also seen widespread use as a chiral auxiliary for effecting asymmetric alkylations and aldol condensations.[26]

Finally, sterically hindered guanidines have been demonstrated to be simple inexpensive, hindered organic bases of potential value in organic synthesis (Scheme 21). For instance, alkylation of ethylaceto-acetate with methyl iodide and *tert*-butyl tetramethyl guanidine predominantly gave in 5 minutes at room temperature the 2-methylated compound.[27]

## Analytical Derivatization Reagents

Einarsson has discovered[28] and developed in collaboration with Varian Associates a new method to analyse amino acids. This method

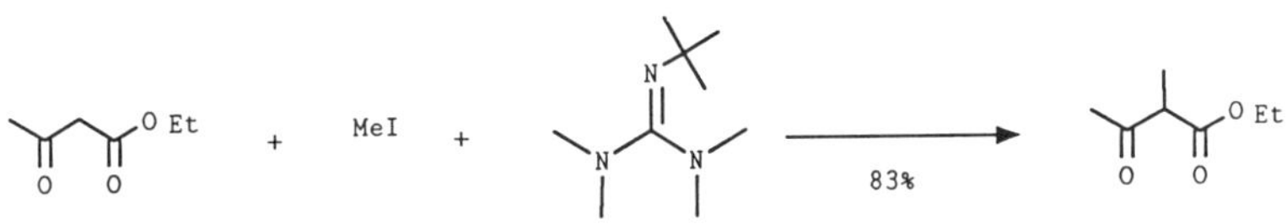


**Scheme 21.**

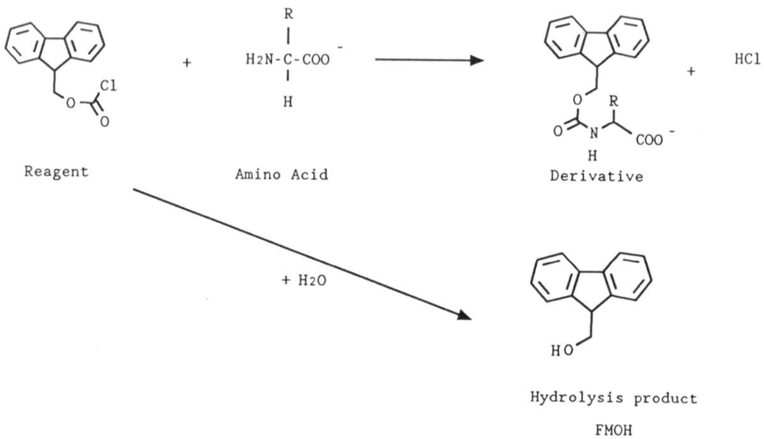


**Scheme 22.**

which has been named AminoTag® uses 1-(9-fluorenyl)methyl chloroformate as a derivitizing agent. The amino acids are converted to their FMOC *N*-protected derivatives which are then analysed by HPLC (Scheme 22). These compounds are stable and well separated on a standard $C_{18}$ reverse phase column. Their high fluorescence allows a very high sensibility. Varian has also developed complete automation to ensure both the derivatization and analysis steps.

The method has then been extended to the chiral separation and determination of amino acids, amines and alcohols using (+) or (−)-1-(9-fluorenyl)ethyl chloroformate.[29] Other reagents such as (−)-menthyl chloroformate or *R*-(−)-1-(1-naphthyl)ethyl isocyanate have also been used for the same purposes.

## Miscellaneous

Many other reagents derived from phosgene or closely related to it are still useful in organic chemistry. For instance, *N*-hydroxyurethane has been found to be a very interesting nitrene precursor;[30] *N,O*-bis(trimethylsilyl)carbamate is extremely suitable for the silylation of alcohols, phenols and carboxylic acids leading only to $CO_2$ and $NH_3$ as by-products;[31] di-*tert*-butyl iminodicarboxylate provides a valuable alternative to the phthaloyl group in the Gabriel synthesis of primary amines;[32] (−)menthyl chloroformate[33] or *R*-(−)-1-(1-naphthyl)ethyl isocyanate[34] are becoming increasingly useful for the resolution of racemic mixtures of alcohols, hydroxy esters and thiols. However, apart from those uses as reactive partners one should not forget the important role of phosgene derivatives as solvents in organic synthesis. Dimethyl carbonate and cyclic ureas such as DMPU ([(dimethyl)-1,3-(tetrahydro)-3,4,5,6]-pyrimidinone-2(1H)) or DMEU ([(dimethyl)-1,3]-imidazolinone-2) have been found to be very useful as polar aprotic solvents. The two cyclic ureas are valuable substitutes for the carcinogen HMPA, their physical and solvating properties being very close to those of HMPA.

## Phosgene derived intermediate and raw materials

As has already been mentioned, the manufacture of polyurethane and polycarbonates represents the largest consumption of phosgene, and TDI and MDI are undoubtedly the leading products among industrial derivatives of phosgene. Apart from these high consuming applications, activity in the fine chemicals area remains very important and represents the greatest part of the accessible market since, because of the toxic hazard, commodities are produced for captive use only. As it is now commonly accepted, phosgene fine chemicals can be classified into three categories: building blocks, custom manufactured products and generics. Only the two first categories are represented here. A large number of chemically different intermediates are derived from phosgene and various markets are concerned, i.e. pharmaceuticals, agrochemicals, dyestuffs, plastics and optical glasses. As previously mentioned, phosgene leads to a large range of chemicals such as chloroformates, carbamates, carbonates, isocyanates, acid chlorides, aldehydes, aromatic ketones or cyclic products.

Carbonates, among which diethyl and dimethyl carbonate are commercially the most important, are used as speciality solvents and in many organic syntheses of pharmaceuticals and agricultural chemicals. Diethylene glycol bis(allyl carbonate) was introduced in 1941 by PPG and polymerizes to give the first commercial polycarbonate used for colourless, optically clear castings. Some recent applications include the use of 1-chloroalkyl carbonates as intermediates for prodrugs (e.g. Bacampicilline) or as UV curing coatings.

**Table 8**

| Use | Intermediates |
|---|---|
| Chlorformates | |
| Methyl | Pharmaceuticals, pesticides |
| Ethyl | Ore flotation agent |
| Isopropyl | Blowing agents |
| Diethylene glycol | Optic polycarbonate |
| 2-Ethyl-hexyl | Percarbonate polymerization initiator |
| Carbonates | |
| Ethyl | Solvent |
| Methyl | Pharmaceutical |
| 1-Chloroethyl | Prodrugs (Bacampicilline) |
| 1-Chloromethyl | Pharmaceutical carbamates |
| 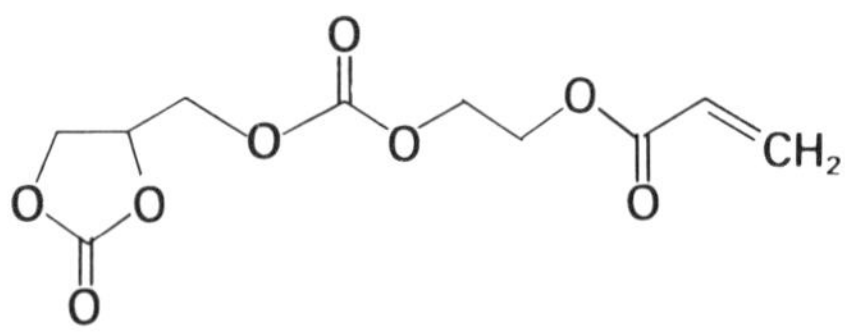 | UV curing coatings |
| Carbamates | Pesticides |
| 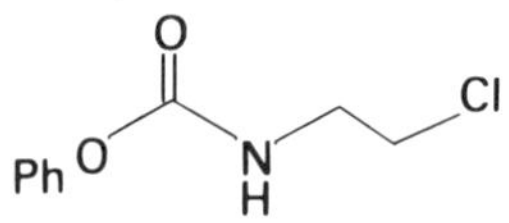 | Anticancer agent |
| 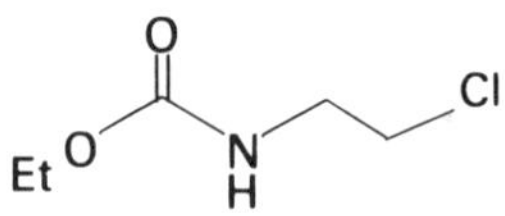 | Insecticide |
| Protected amino acids | Peptides |
| Ketones | |
| 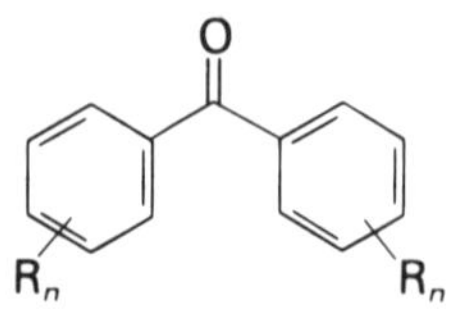 | UV adsorbers, thermostables |
| Isocyanates | |
| Diisocyanates | Polyurethanes, elasomers |
| Acylisocyanates | Pesticides, pharmaceuticals |
| Cyclic products | |
| NCAs | Polyamino acids, antihypertensives |

**Table 8**—(*Contd*)

| Use | Intermediates |
|---|---|
| 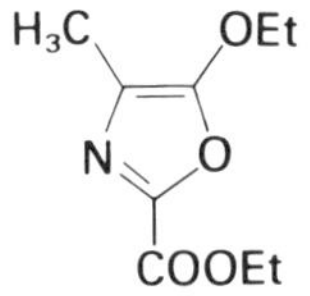 | Vitamin $B_6$ |
| 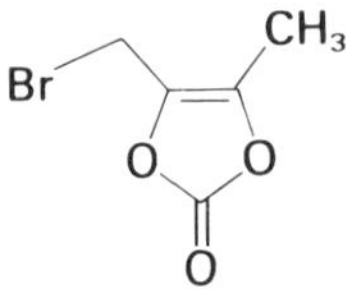 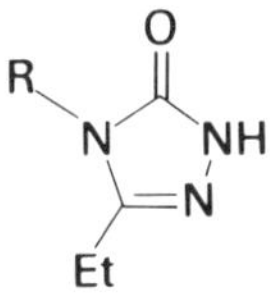 | Antibiotics, antidepressives |

Carbamates and isocyanates are mainly used to manufacture pharmaceuticals, pesticides and insecticides whereas chloroformates are versatile intermediates for various products since they are precursors to the majority of phosgene derivatives. Table 8 lists some of these intermediates together with their corresponding applications.

## Conclusion

There has been a considerable effort devoted to the research of substitutes to phosgene or its more toxic derivatives, isocyanates, but to date the only industrial process that has emerged from this work is the manufacture of dimethyl carbonate from carbon monoxide and methanol and the direct synthesis of aromatic isocyanate from nitro compounds; carbon monoxide, though very promising, still remains on the research level. However, in spite of its acute toxicity, phosgene does remain the leading product of a very rich series of fine chemicals.

## References

1. Davy, H., *Phil. Trans.* (1812) 144.
2. Dumas, J. B., *C. R. Acad. Sci.* (1833).
3. Babad, H. & Zeiler, H. G., *Chem. Rev.*, **73** (1973) 75.
4. Ozaki, S., *Chem. Rev.*, **72** (1972) 457.
5. Matzner, H., Kurkjy, R. P. & Cotter, R. J., *Chem. Rev.*, **64** (1964) 645.

6. Hagemann, H., *Methoden der Organische Chemie* ed. Houben-Weyl, Band E4, Verlag Stuttgart, New York, 1983.
7. Eberle, G. & Ugi, I., *Angew. Chem.*, **88** (1976) 509.
8. Jpn Tokyo Koho, 7920 493, 1979.
9. British Patent 743 557, 1956.
10. Ulrich, H., *The Chemistry of Imidoyl Halides,* Plenum Press, New York, 1968.
11. Eur. Patent 0 005 746, 1979.
12. Mazur, Y. & Cohen, Z., *Angew. Chem. Int. Ed.*, **17** (1978) 281.
13. Knorr, R., Trzeciak, A., Bannwarth, W. & Gillesen, D., *Tetrahedron Lett.*, **30** (1989) 1927.
14. Sinay, P. & Veyriere, A., SCF Meeting, September 1989.
15. Olofson, R. A., Schnur, R. C., Bunes, L. & Pepe, J. P., *Tetrahedron Lett.*, **18** (1977) 1567.
16. Olofson, R. A., Martz, J. T., Senet, J. P., Piteau, M. & Malfroot, T., *J. Org. Chem.*, **49** (1984) 2081.
17. Maryanoff, B. E., Molinari, A. J., Wooden, J. P. & Olofson, R. A., *J. Org. Chem.*, **48** (1983) 5074.
18. Blacklock, T. J., Shuman, R. F., Butcher, J. W., Shearing, W. E., Budavari, J. & Grenda, V. J., *J. Org. Chem.*, **53** (1988) 836.
19. Heilmann, S. M., Jensen, K. M., Krepski, L. R., Moren, D. M., Rasmussen, J. K. & Smith, H. K., *Synth. Comm.*, **17** (1987) 843.
20. Senet, J. P., Vergne, G. & Wooden, G. P., *Tetrahedron Lett.*, **52** (1986) 6319.
21. Join, P. & Castro, B., *J. Chem. Soc. Perkin Trans.*, **I** (1987) 1177.
22. Barcelo, G., Grenouillat, D., Senet, J. P. & Sennyey, G., *Tetrahedron* **46** (1990) 1839.
23. Yamaguchi, R., Hata, E. & Utimoto, K., *Tetrahedron Lett.*, **29** (1988) 1785.
24. Hoppe, D. & Krämer, T., *Angew. Chem.*, **98** (1986) 171.
25. Hoppe, D. & Brönneke, A., *Synthesis* (1982) 1045.
26. Evans, D. A., Ennis, M. D. & Mathre, D. J., *J. Am. Chem. Soc.*, **104** (1982) 1737.
27. Barton, D. H. R., Eliott, J. D. & Géro, S. D., *J. Chem. Soc., Chem. Comm.* (1981) 1136.
28. Einarsson, S., Joseffson, B. & Lagerkvist, S., *J. Chromatogr.*, **282** (1983) 609.
29. Einarsson, S., Joseffson, B. & Möller, P., *Anal. Chem.*, **59** (1987) 1191.
30. Chang, Y. H., Chiu, F. T. & Zon, G., *J. Org. Chem.*, **46** (1981) 342.
31. Birkofer, L. & Sommer, P., *J. Organomet. Chem.*, **99** (1975) C1.
32. Carpino, L. A., *Acc. Chem. Res.*, **6** (1973) 171.
33. Jeyaraj, G. L. & Porter, W. R., *J. Chromatogr.*, **315** (1984) 378.
34. Pirkle, W. H. & Hoekstra, M. S., *J. Org. Chem.*, **39** (1974) 3904.

# *t*-Butyl acetoacetate: a new reagent for the preparation of acetoacetylated materials

**J. S. Witzeman & W. D. Nottingham**

*Eastman Kodak Company, Tennessee, USA*

Acetoacetylated materials (**1**) are of interest as intermediates in the pharmaceutical, agrochemical, chemical and polymer industries.[1,2] Traditionally these materials have been prepared from the reaction of diketene (**2**) with the corresponding nucleophile (Fig. 1).[2] The need for scrupulous exclusion of water, dedicated manufacturing equipment and refrigerated storage facilities makes use of diketene in industrial environments troublesome, particularly for batch processes. These manufacturing concerns have been further exacerbated by concerns regarding the lachrymatory properties of diketene along with strict regulations regarding its shipping. These issues have thus predicted a need for alternative industrial acetoacetylation technologies.

One such alternative described by Clemens, Hyatt and Kato[3] involves heating the nucleophile of interest in the presence of 2,2,6-trimethyl-4*H*-dioxin-4-one (**3**, TKD, the diketene-acetone adduct). While this approach has been shown to have a great deal of synthetic utility, it has yet to gain widespread acceptance in industrial settings. A second diketene-free approach to acetoacetates involves the transesterification of the corresponding nucleophile with an appropriate acetoacetate (transacetoacetylation). While the use of this

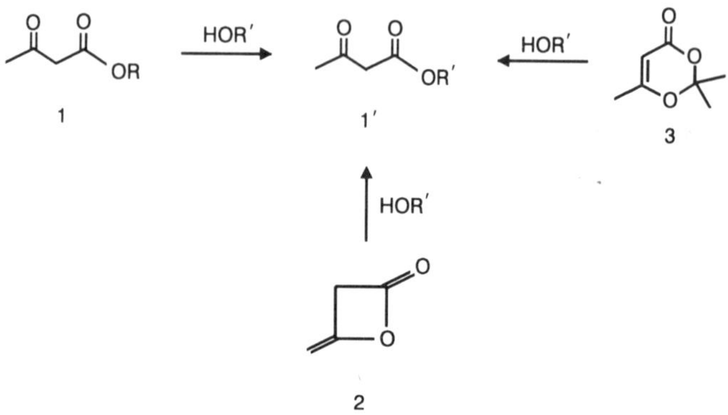


**Fig. 1.** Synthetic approaches to acetoacetylated materials.

reaction has been demonstrated by the work of Bader,[4(a)] Taber[4(c)] and Gilbert[4(d)] the utility of this reaction in large-scale and/or industrial processes has been limited by the long reaction times,[4(a),(c),(d)] excess reagents,[4(a),(c),(d)] dilute solutions[4(c),(d)] and large amounts of 4-dimethylaminopyridine (DMAP) catalyst required.[4(c),(d)] The limitations of these literature transacetoacetylation procedures suggested that the mechanism of this process needed to be understood in better detail in order to optimize the reaction for industrial applications. This work appeared particularly merited since, to our knowledge, there has been no work on the mechanism of this reaction since the pioneering work of Bader and Carroll.[4(a),(b),(e)]

## Kinetic studies on the transacetoacetylation reaction

The rate of reaction of a variety of acetoacetates with *n*-butanol in xylene solvent has been determined under a variety of conditions according to Fig. 2. The results of these studies are summarized in

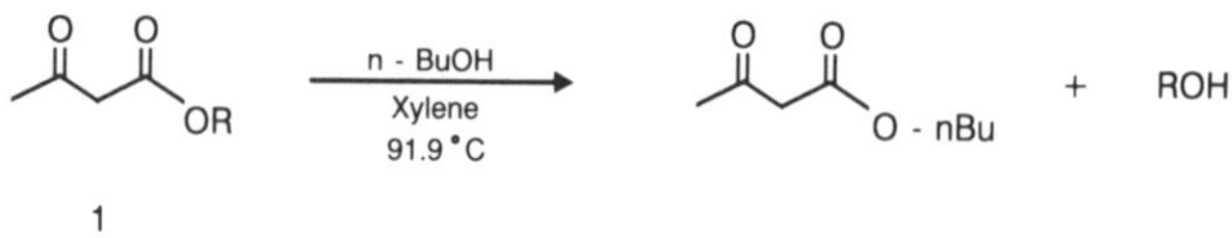


**Fig. 2.** General reaction scheme used for kinetic studies.

**Table 1.** Rate constants for reaction of ROAcAc with *n*-BuOH at 91·85°C

| Entry | Compound | R | $k \times 10^{4a}$(s)[b] | [*n*-BuOH][c] | [ROAcAc][c] |
|---|---|---|---|---|---|
| 1 | **1a** | *t*-Bu | 1·716(0·07) | 0·492 | 0·474 |
| 2 | **1a** | *t*-Bu | 1·662(0·03) | 2·623 | 0·475 |
| 3 | **1a** | *t*-Bu | 1·859(0·03) | 1·027 | 0·472 |
| 4 | **1a** | *t*-Bu | 1·559(0·05) | 0·174 | 0·068 |
| 5 | **1b** | Et | 0·102(0·002) | 0·526 | 0·473 |
| 6 | **1b** | Et | 0·136(0·001) | 4·987 | 0·477 |
| 7 | **1c** | Me | 0·097(0·001) | 0·611 | 0·551 |
| 8 | **1d** | *i*-Bu | 0·138(0·004) | 1·068 | 0·474 |
| 9 | **1e** | *i*-Pr | 0·140(0·007) | 1·070 | 0·537 |
| 10 | **1f** | $HC(i\text{-}Pr)_2$ | 0·083(0·003) | 1·046 | 0·477 |
| 11 | **1g** | *t*-Am | 1·460(0·07) | 1·072 | 0·495 |
| 12 | TKD(**3**)[d] | | 1·07(0·03) | | |

[a] First-order rate constant in $s^{-1}$.
[b] Standard deviation from ANOVA analysis.
[c] Molar concentrations of *n*-BuOH and acetoacetate respectively.
[d] See Ref. 3(c) for a complete listing of kinetic parameters for this compound.

Tables 1, 2 and 3. Examination of the data from these studies indicates that the transacetoacetylation reaction is mechanistically distinct from typical transesterification reactions. Most notably *t*-butyl acetoacetate (*t*-BAA, **1a**) and *t*-amyl acetoacetate (*t*-AmAA, **1g**) react 15–20-fold faster than the more commonly used ethyl or methyl acetoacetates

**Table 2.** Effect of various additives on experimental rate constant for reaction of *t*-butyl acetoacetate with *n*-butanol at 91·85°C

| Additive[a] | $k \times 10^{4b}$(s)[c] | [*n*-BuOH][d] | [*t*-BuOAcAc][d] |
|---|---|---|---|
| DMAP, 0·0237M | 2·08(0·03) | 1·038 | 0·475 |
| TMG, 0·0243M | 2·11(0·05) | 1·061 | 0·489 |
| $Ti(OiPr)_4$, 0·165M | 0·98(0·07) | 1·108 | 0·496 |
| *n*-BuOAc (solvent) | 0·99(0·06) | 0·68 | 0·599 |

[a] Additive and concentration respectively. DMAP = dimethylaminopyridine; TMG = tetramethylguanidine.
[b] First-order rate constant in $s^{-1}$.
[c] Standard deviation from ANOVA analysis.
[d] Molar concentrations of *n*-BuOH and *t*-BuOAcAc respectively.

**Table 3.** Effect of temperature on rate of reaction of ethyl acetoacetate (EAA) and *t*-butyl acetoacetate (*t*-BAA)

| Temperature (°C) | EAA $k(s^{-1}) \times 10^{4a}$ | *t*-BAA $k(s^{-1}) \times 10^{4a}$ |
|---|---|---|
| 91·85 | 0·10 | 1·65 |
| 98·70 | 0·19 | 2·50 |
| 106·00 | 0·37 | 5·16 |

Activation parameters

| Acetoacetate | Log $A$ $(s^{-1})^b$ | $Ea$ (kcal/mol)[b] | $\Delta H^{\neq}$ (kcal/mol)[b] | $\Delta S^{\neq}$ (eu) |
|---|---|---|---|---|
| EAA | 10·22 | 25·43 | 24·69 | −14·2 |
| *t*-BAA | 9·48 | 22·19 | 21·45 | −17·6 |

[a] Average first-order rate constants.
[b] Correlation coefficient (R2) from ANOVA analysis = 0·999 for EAA and 0·978 for *t*-BAA.

(EAA or MAA; **1b**, **1c**) and ca. 50% faster at 92°C than the analogous reaction of dioxinone (**3**). Further evidence that the transacetoacetylation reaction proceeds by a mechanism different from that of typical (second-order) transesterification reactions[5(a)] is apparent from noting the independence of the experimental rate constants on alcohol concentration (entries 1–6 of Table 1). Examination of the data indicates that for both EAA and *t*-BAA the initial first-order rate constants (*vide infra*) are independent of alcohol concentration. Furthermore, unlike quintessential transesterification reactions, this process shows little, if any, rate enhancement if common catalysts are added to the reaction mixture (Table 2).[4(c),(d),5] For example, addition of 5 mol % (relative to *t*-BAA) of either DMAP or tetramethylguanidine results in only a slight (ca. 25%) rate enhancement. Addition of 35 mol % $Ti(OiPr)_4$, a commonly used transesterification catalyst[5(b)] results in an inhibition of the rate by approximately 40%. Changing the solvent from xylene to the more polar butyl acetate also slightly inhibits the rate, while addition of catalytic amounts (0·5–1 wt %) of *p*-toluenesulfonic acid and trichloroacetic acids do not affect the rate of the reaction. The slow rate of reaction observed for diisopropyl carbinyl acetoacetate (**1f**, entry 10, Table 1) indicates that the rate enhancement noted for *t*-BAA and *t*-AmAA may be due to factors other than merely the steric bulk associated with the acetoacetic ester.

The lack of a dependence of the observed rate constants on the nucleophile concentration, the absence of any pronounced rate effect when common transesterification catalysts are used and the surprisingly rapid rate of the more sterically encumbered tertiary esters demonstrates that a mechanism involving the tetrahedral intermediate **4**, such as has been proposed by Bader (path A, Fig. 3),[4(a),(b)] cannot be operative. The reaction must proceed via a rate determining, unimolecular decomposition to a reactive intermediate such as is given in path B of Fig. 3. The most likely reactive intermediate is acetylketene (**5**), based on analogy with dioxinone (**3**)[3(c)] and spectroscopic evidence (*vide infra*).

The rate of a reaction following this mechanism would be dependent on the structure of the acetoacetate and independent of the nucleophile, which is in accord with the experimental results. The activation parameters (Table 3) and concentration dependence (Table 1) for transacetoacetylation of *t*-BAA and EAA suggest that while the rates of the two processes differ, the mechanisms of the two reactions are very similar and thus this proposed mechanism may, in fact, be quite general. The $\Delta S^{\neq}$ for these processes suggest a highly organized transition state, such as might be expected for a cyclic intermediate involving the enol form of the acetoacetate complexing with acetoacetate ester oxygen.

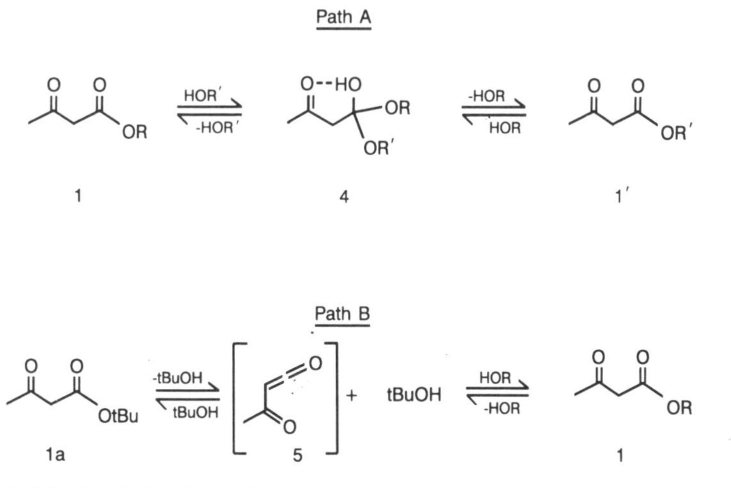


**Fig. 3** Mechanistic alternatives for the transacetoacetylation reaction.

While the formation of the reactive intermediate in path B of Fig. 3 is unimolecular, the concentrations of the nucleophiles can become important in the transacetoacetylation reaction of acetoacetates, particularly during the latter part of the reaction when the concentration of the alcohol co-product becomes significant. Thus, while the initial rate of reaction with butanol is identical for EAA with 1·2 and 6·1 eq of alcohol respectively, the reaction in which 1·2 eq were used slows down considerably in the latter part of the reaction (Fig. 4), reflecting a more complex kinetic scheme in this instance. The *t*-butyl acetoacetate process, on the other hand, displays first-order behavior out to at least 90–95% conversion.

The proposed mechanism explains the results discussed in this work as well as the observations of Bader,[4(a),(b)] Taber[4(c)] and Gilbert.[4(d)] For example, the observation that 2,2-disubstituted acetoacetates do not undergo exchange reactions[4(a),(c),(d)] is readily attributed to the inability of these materials to form the corresponding ketenes. The fact that *t*-butyl acetoacetate (as well as other tertiary beta-ketoesters) cannot be prepared from either ethyl or methyl acetoacetate is presumably due to both the thermodynamic preference (preliminary equilibrium studies on the EAA/*t*-BAA equilibrium at 106°C indi-

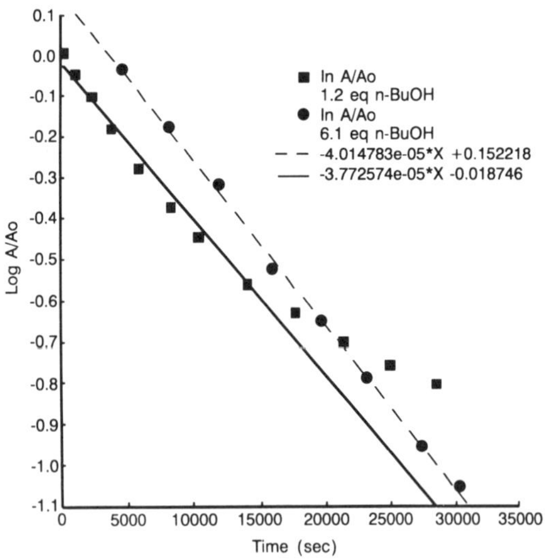


**Fig. 4.** Effect of butanol concentration on reaction rate of ethyl acetoacetate (EAA, **1b**).

cated that this process favors EAA by between 15 and 22 : 1, reflecting a free energy difference ($\Delta G°$) of 2·0–2·3 kcal/mol, which agrees well with the experimental difference in free energy of activation for the two materials) for the less sterically encumbered EAA or MAA as well as the significantly faster back-reaction for decomposition of the tertiary acetoacetate. This hypothesis can be tested by the transacetoacetylation of *t*-BAA with tertiary alcohols. If steric effects play an important role in this process, this reaction should be extremely sluggish. If, on the other hand, this reaction follows a mechanism such as that described in path B, these reactions should proceed readily. The reaction of *t*-BAA with 1-methylcyclohexanol and 2-methyl-3-buten-2-ol produces the corresponding acetoacetates in 83 and 72% yields respectively (*vide infra*), thus supporting the mechanism given in path B.

Finally, the fact that the reactive intermediate involved in this process is acetylketene rather than some other species was demonstrated using matrix isolation-infrared spectroscopy. Thermolysis of either ethyl or *t*-butyl acetoacetate in the transfer line of a GC-IR and trapping of the resulting effluent in an argon matrix at 6 K produced materials with IR spectra identical to that produced from thermolysis of dioxinone (**3**),[3(c)] with characteristic infrared absorbances at 2137 and 1676 $cm^{-1}$.

## Use of the transacetoacetylation reaction in preparative organic chemistry[4(f)]

The preceding work indicated that the mechanism of the transacetoacetylation reaction is distinctly different from that of prototypical transesterification reactions. In particular it was noted that *t*-butyl acetoacetate (*t*-BAA, **1a**) is much more reactive than the more commonly used methyl or ethyl analogs. While the kinetic studies indicated that either *t*-butyl or *t*-amyl acetoacetate would be effective acetoacetylating agents, the lower boiling point of the *t*-butanol by-product and lower cost of *t*-BAA suggested that the *t*-butyl analog is a more industrially viable reagent. Thus, it became incumbent upon us to examine the use of this material in the preparation of various acetoacetic acid derivatives.

1a 8 - 26

**Fig. 5.** General synthetic scheme for preparation of acetoacetylated materials using *t*-BAA.

A variety of acetoacetylated materials have been prepared according to Fig. 5. Table 4 lists the yields obtained from these reactions. The reaction appears quite general, can usually be carried out with equimolar amounts of nucleophile and acetoacetate in concentrated solutions, does not require catalysts and is much more rapid than other transacetoacetylation processes. The major by-product from this reaction is *t*-butanol. The laboratory procedures for preparing these acetoacetic acid derivatives are straightforward and should be readily applicable to industrial scale-up. The preparations essentially involve heating a solution of acetoacetate and nucleophile in solvent either in an Erlenmeyer flask at the boiling point of the solution (method A) or in a flask equipped with a distillation column and still head for distillative removal of *t*-butanol (method B). For more volatile nucleophiles such as allyl alcohol or *t*-amylamine, the reaction can be carried out merely by heating acetoacetate and nucleophile under reflux (method D).

The acetoacetylation of 1-methylcyclohexanol with *t*-BAA illustrates that higher boiling tertiary alcohols can be acetoacetylated with this material. Similarly, 2-methyl-3-butene-2-ol can be acetoacetylated with *t*-BAA in good yield. (This reaction also produced 5–10% of 6-methyl-5-hepten-2-one, presumably produced by the Carroll rearrangement of the acetoacetate.) This material cannot be prepared from **1b** under the transacetoacetylation conditions of Gilbert (EAA/DMAP catalyst/4 Å molecular sieves).[4(d)] This reagent thus offers a decided advantage over other acetoacetates in the acetoacetylation of hindered alcohols. Similarly, the unsymmetrically substituted glycol TMPD (2,2,4-trimethylpentan-1,3-diol) is readily acetoacetylated with **1a** to produce bisacetoacetate (**9**). This material is exceedingly difficult to acetoacetylate with EAA or MAA.

The yield for preparation of phenyl acetoacetate (**13**) under standard reaction conditions was poor, but could be improved significantly when cyclohexane was used to help azeotropically remove *t*-butanol

Table 1. Yield of various acetoacetylated materials using t-butyl acetoacetate

| Compound | Nucleophile (solvent) | Nu (g) | t-BAA (g) | Solvent (mL) | Yield (%) | Experimental method[a] |
|---|---|---|---|---|---|---|
| 8 | TMP[b] (xylene) | 47·5 | 171·00 | 100 | 90[c] | B |
| 9 | TMPD[b] (xylene) | 10·1 | 22·4 | 30 | 87[c] | A |
| 10 | NPG[d] (xylene) | 10·0 | 32·1 | 32 | 87[c] | A |
| 10 | NPG (toluene) | 159·2 | 491·3 | 450 | 81[c] | B |
| 11 | CHDM[e] (xylene) | 9·8 | 22·1 | 28 | 69[c,f] | A |
| 12 | $HNBu_2$ (xylene) | 7·7 | 9·2 | 25 | 96 | A |
| 13 | PhOH (xylene/cyclohexane) | 30·1 | 33·5 | 110/100[g] | 74 | B |
| 14 | $PhNH_2$ (xylene) | 10·0 | 17·2 | 25 | 83 | A |
| 15 | *n*-BuOH (xylene) | 10·0 | 21·4 | 40 | 87 | A |
| 16 | *n*-OctOH (toluene) | 13·0 | 16·6 | 50 | 83 | B |
| 17 | $PhCH_2oH$ (toluene) | 10·8 | 16·6 | 50 | 89 | B |
| 18 | *p*$(NO_2)PhNH_2$ (toluene) | 13·8 | 16·2 | 50 | 86 | B |
| 19 | 1-Me-cyclohexanol (toluene) | 11·4 | 16·2 | 50 | 83 | B |
| 20 | 3-Me-2-buten-1-ol (toluene) | 8·6 | 16·2 | 50 | 97 | B |
| 21 | 2-Me-3-buten-2-ol (toluene) | 8·4 | 9·6 | 33 | 72 | B |
| 22 | 1-diethylamino-2-propanol (toluene) | 26·7 | 32·4 | 40 | 84 | B |
| 23 | *n*-heptylamine (toluene) | 8·2 | 15·9 | 126/50[h] | 54 | C |
| 24 | allyl alcohol (xylene) | 6·3 | 12·9 | 12 | 74 | D |
| 25 | isopinocampheol (toluene) | 7·5 | 7·7 | 13 | 90 | B |
| 26 | *t*-amylamine (toluene) | 12·7 | 29·8 | 250/100[h] | 80 | C |
| 26 | *t*-amylamine (toluene) | 31·33 | 42·9 | 50 | 71 | D |

[a] See experimental section for a description of the method.
[b] TMPD = 2,2,4-trimethyl-1,3-pentandiol. TMP = trimethylolpropane.
[c] Yield of bisacetoacetate for TMPD, NPG and CHDM and trisacetoacetate for TMP.
[d] NPG = neopentyl glycol or 2,2-dimethyl-1,3-propandiol.
[e] CHDM = 1,4-cyclohexanedimethanol (cis/trans).
[f] Yield of crystalline material; some cis isomer lost in recrystallization.
[g] 110 mL xylene, 100 mL cyclohexane.
[h] Milliliter of solvent used with *t*-BAA and amine solutions respectively (see experimental).

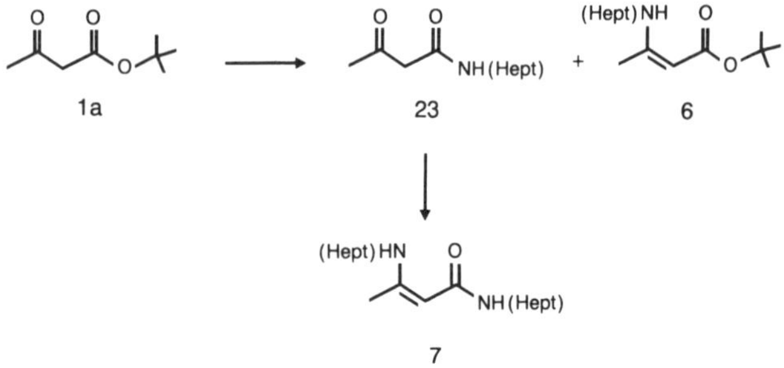

**Fig. 6.** Reaction products from the reaction to *t*-BAA with *n*-heptyl amine.

from the reaction mixture. The sluggishness of the reaction between **1a** and phenol presumably reflects the thermodynamic preference for alkyl acetoacetates over the phenyl analog.[3(c)]

The preparation of acetoacetamides from secondary or aromatic amines and *t*-butyl acetoacetate proceeds easily and in good yield. The corresponding reaction with *n*-heptylamine is complicated, however, by competing formation of the enamines **6** and **7** (Fig. 6). Since the transacetoacetylation process has been shown to be a first-order process while formation of **6** is most likely second-order, it was anticipated that dilution of the reagents would improve the yield of acetoacetamide **23**. Such an improvement in the yield of **23** was indeed noted, along with the concomitant increase in the yield of **7**. By using a slight excess of acetoacetate and slow addition of amine to the hot *t*-BAA solution (method C) it was possible to obtain acetoacetamide (**23**) in fair yield. No such complication was noted with *t*-amylamine, which produced the *t*-amyl acetoacetamide in 70–80% isolated yield.

The preceding clearly demonstrates the synthetic utility of *t*-butyl acetoacetate in the transacetoacetylation reaction. This process is very similar to the acetoacetylation procedures of dioxinone (**3**) and has both industrial as well as laboratory utility.

## Use of the transacetoacetylation reaction in the preparation of acetoacetylated polymers and resins

Polymers and resins that contain pendant acetoacetyl groups have been the subject of a great deal of interest in the coatings and polymer industries.[1] The interest in these materials has been due, in part, to the significant reduction in solution viscosities noted when the hydroxyl end-groups on the polymer or resin are converted to acetoacetyl groups. Such a viscosity reduction enables achievement of the lower volatile organic content (VOC) levels mandated by recent US federal regulations. Another reason for the interest in these materials is due to the reactivity of the acetoacetyl moiety, which enables the use of a variety of novel chemical reactions to crosslink the polymer. In addition, many of the crosslinked coatings prepared from acetoacetylated resins and polymers have superior performance properties relative to traditional systems.[1(b),(c)] (Some of the improvements noted include enhancements in the adhesion, impact strength and corrosion resistance. These properties have been attributed to the ability of the acetoacetyl group to chelate with the metal substrate to which the coating is applied.)

Acetoacetylated polymers and coating resins can be prepared either by use of acetoacetylated monomers[1(a),(b)] in the polymerization process or by acetoacetylation of the material after the fact.[1(c)] The former approach is generally used to prepare acetoacetylated acrylics, where the free radical reaction conditions and availability of the acetoacetyl containing acrylic monomer acetoacetoxyethyl methacrylate (AAEM, **27**, Fig. 7) makes this regime industrially viable. For polyesters, however, the more forcing reaction conditions necessary

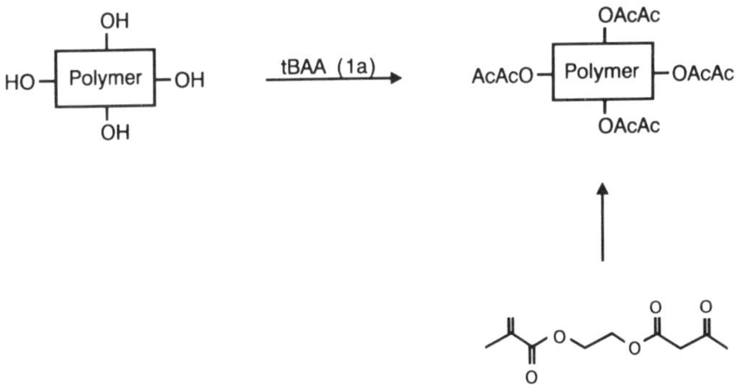


27

+ Other Monomers

**Fig. 7.** Synthetic approaches to acetoacetylated polymers and resins.

**Table 5.** Acetoacetylated polyester resin HS-3-6T[a] polyester resin properties

| % Acetoacetylation | 0 | 11·2 | 18·2 | 25 | 50 | 85 |
|---|---|---|---|---|---|---|
| % solids, determined, 30 min at 325°F | 68·4 | 63·7 | 64·3 | 66·4 | 67·1 | 65·9 |
| APHA color | 120 | 120 | 140 | 140 | 160 | 160 |
| Gardner color | 1 | 1 | 1 | 1 | 1 | 1 |
| Gardner viscosity | H− | B− | B− | C+ | A+ | A− |
| Brookfield viscosity, 23·2°C, centipoise | 264 | 173 | 165 | 144 | 128 | 96 |

[a] HS-3-6T is comprised of 54 wt % 2,2,4-trimethylpentan-1,3-diol, 4·7 wt % trimethylolpropane, 22·0 wt % isophthalic acid and 19·3 wt % adipic acid. See Eastman Publication N-306 for more information.

for condensation polymerization make the use of acetoacetylated monomers unattractive and thus the second approach becomes necessary. In these applications *t*-BAA has been shown to be the reagent of choice, particularly in situations where a high degree of acetoacetylation is desired or where polymer breakdown and/or scrambling can be a problem. It is possible to conduct the acetoacetylation reaction after the polymerization in the same apparatus by allowing the reaction mixture to cool to 150–140°C and subsequently adding *t*-BAA. A second synthetic approach to the preparation of these materials involves heating *t*-BAA and a solution of the resin or polymer at the desired solids level with distillative removal of *t*-butanol. The properties of a typical high-solids coating resin at various extents of acetoacetylation are given in Table 5. It can be seen that the most dramatic reduction in solution viscosity is noted between 10 and 30% acetoacetylation. Improvement in coating performance, however, typically requires between 40 and 60% acetoacetylation. The generality of the *t*-BAA transacetoacetylation reaction enables one to prepare resins with varying degrees of acetoacetylation merely by adjusting the *t*-BAA–resin ratio.

## Extensions of the transacetoacetylation reaction

Consideration of the mechanism proposed in path B of Fig. 3 suggests that other beta-ketoesters that are not 2,2-disubstituted could be

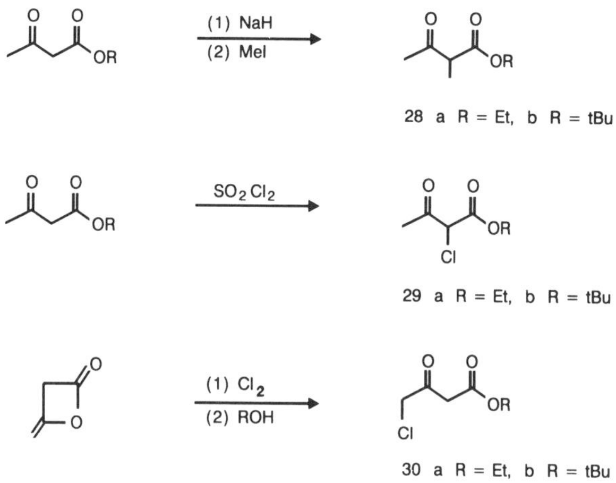


**Fig. 8.** Preparation of 2– and 4–substituted acetoacetates.

**Table 6.** First-order rate constants ($s^{-1}$) for reaction of various beta-ketoesters with *n*-octanol

| Acetoacetate | Temperature (°C) | [AcAc][a] | [*n*-OctOH][a] | $k_1 \times 10^{5b}$(s)[c] |
|---|---|---|---|---|
| **28a** | 114·9 | 0·48 | 2·88 | 0·37(0·02) |
| **28b** | 114·9 | 0·48 | 0·72 | 2·43(0·20) |
| **28b** | 114·9 | 0·48 | 2·88 | 2·97(0·11) |
| **29a** | 114·9 | 0·48 | 2·88 | 0·29(0·01) |
| **29b** | 114·9 | 0·48 | 0·72 | 0·86(0·06) |
| **29b** | 114·9 | 0·48 | 2·88 | 0·86(0·07) |
| **30a** | 91·9 | 0·48 | 0·72 | 2·10(0·20) |
| **30a** | 91·9 | 0·48 | 2·88 | 2·70(0·30) |
| **30b** | 91·9 | 0·48 | 0·72 | 27·80(0·70) |
| **30b** | 91·9 | 0·48 | 2·88 | 30·00(2·00) |
| **1a**[d] | 91·9 | | | 16·50 |
| **1a**[d,e] | 114·9 | | | 109·0 |
| **1b**[d] | 91·9 | | | 1·00 |
| **1b**[d,e] | 114·9 | | | 19·0 |

[a] Concentration (moles/liter) of acetoacetate and *n*-octanol respectively.
[b] First-order rate constant in $s^{-1}$.
[c] Standard deviation from ANOVA analysis.
[d] First-order rate constant for reaction with *n*-butanol.
[e] Rate constants calculated from Arrhenius parameters given in Table 3.

prepared by the transacetoacetylation reaction. To this end the 2-methyl (**28**), 2-chloro (**29**) and 4-chloro derivatives (**30**) have been prepared according to Fig. 8. Preliminary kinetic studies on these materials indicate that the same general trend noted for the unsubstituted acetoacetates; namely $k(t\text{-Bu ester}) \gg k(\text{Et ester})$, is observed for the substituted analogs. The rates of reaction of the 2-substituted analogs are significantly slower than the unsubstituted materials (Table 6), while the rates of the 4-substituted analogs are similar to the unsubstituted materials. The concentration dependence of the rate constants appears to be similar to that noted for the parent compounds and thus a similar mechanism appears to be operative. Further evidence that the mechanism is similar was obtained by matrix isolation studies. Thus, pyrolysis of **28b** in the transfer line of the GC-IR and subsequent trapping of the pyrolysate in an argon matrix at 6 K gave a material with the IR spectrum shown in Fig. 9. The ketene absorbance at $2121\ \text{cm}^{-1}$ is typical of a substituted alkylketene.[6] Similar evidence for the intermediacy of chloroacetylketene in the reactions of **29b** and chloroacetylketene from thermolysis of **30b** has also been obtained.

The results of preliminary synthetic studies on these materials are

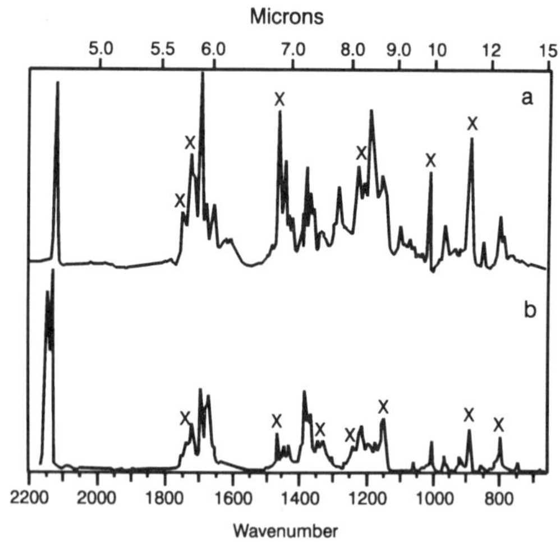


**Fig. 9.** Infrared spectra of acetoacetate pyrolysis products at 6K. (a) is product obtained from **28b**, (b) is product obtained from **1b**. X values indicate starting material.

**Table 7.** Yield for reaction of beta-ketoesters with various nucleophiles by method B

| Compound | Nucelo-phile | Keto-ester (g)[a] | Nucelo-phile (g)[a] | Solvent (mL)[b] | Yield (%) |
|---|---|---|---|---|---|
| **28b** | *n*-OctOH | 3·3 | 7·0 | 60 | 88 |
| **29b** | *n*-OctOH | 5·5 | 3·3 | 50 | 98 |
| **29b** | $PhCH_2OH$ | 2·7 | 4·8 | 50 | 81 |
| **30b** | *n*-OctOH | 1·11 | 0·94 | 10 | 54 |

[a] Grams ketoester and nucleophile respectively.
[b] Toluene solvent (mL).

summarized in Table 7. These preliminary studies demonstrate that variously substituted *t*-butyl acetoacetates are also synthetically useful materials.

## Conclusions

The transacetoacetylation of *t*-butyl acetoacetate and its variously substituted analog is synthetically convenient and can be carried out in concentrated solutions without the use of catalysts. This approach allows industrial preparation of acetoacetic acid derivatives by alternative methods. The mechanism of this reaction is distinctly different from typical transesterification reactions. The results of kinetic and spectroscopic studies suggest the intermediacy of acetylketene (or its substituted derivatives) in the rate-determining step of the transacetoacetylation reaction.

## Experimental

See Refs 1(c), 3(c) and 4(f) for general comments on apparatus and reagents.

### Kinetic Procedure

A solution of acetoacetate, *n*-butanol and *p*-dichlorobenzene internal standard in xylene solvent was placed in a 25 mL round-bottomed flask and immersed in a constant temperature bath maintained at the specified temperature ±0·10C°. Aliquots were periodically withdrawn, the time noted and the sample diluted with cold *p*-xylene. The samples were analysed in duplicate by gas chromatography and the rate constants determined using the protocols described elsewhere.[3(c)] The rate constants given for entries 1–4 and 11 of Table 1 were determined by following the reaction to at least three half-lives, while those given in entries 5–10 were determined by following the reaction to approximately 30% completion.

## Method A

This procedure is analogous to that of Clemens and Hyatt for the preparation of acetoacetylated materials from dioxinone (**2**).[3(a)] The preparation of neopentyl glycol bisacetoacetate is illustrative of the procedure. In an Erlenmeyer flask was placed 10·03 g (0·096 mol) neopentylglycol, 32 mL xylene and 32·1 g (0·203 mol) *t*-butyl acetoacetate. The flask was warmed with stirring on a hot plate. At approximately 120°C the evolution of *t*-butanol was apparent. The reaction was stopped once the solution reached the boiling point of xylene (ca. 10–15 minutes after boiling was first noted). The crude product was concentrated *in vacuo* and distilled to yield 22·83 g of the acetoacetate (87·1%, b.p. 145–148°C/0·05 mm Hg).

## Method B

This procedure consisted of heating the nucleophile, *t*-BAA and solvent in a flask equipped with a distillation column (typically either Vigereaux or Oldershaw), a still head and with thermometers in the base and head of the system. The system was brought to reflux and the *t*-butanol by-product was collected once the temperature in the still head reached the boiling point of *t*-butanol. The distillate was usually 85–99% *t*-butyl alcohol by gas chromatography, with the remainder being the reaction solvent. This method can also be carried out by adding the *t*-BAA to a hot (ca. 90–120°C) solution of nucleophile and solvent.

## Method C

This procedure was developed for the preparation of *N*-(*n*-heptyl)acetoacetamide. In a three-neck, 300 mL flask equipped with reflux column, thermo-couple temperature controller, addition funnel and nitrogen inlet was placed 15·9 g *t*-BAA (0·101 mol) and 126 mL xylene. The solution was warmed to 110°C and a solution of 8·2 g (0·071 mol) *n*-heptylamine in 50 mL xylene was added drop-wise over 45 min during which time the reaction was allowed to warm to 120°C. The solution was heated for 20 additional minutes after addition was complete. The resulting material was concentrated *in vacuo* and recrystallized from petroleum ether/acetone to give 7·54 g (54%) of product which was identical to that prepared by Clemens and Hyatt.[3(a)]

### Method D

This procedure is the method of choice for low boiling nucleophiles and consists of merely heating the nucleophile, *t*-BAA and solvent at reflux with GC monitoring. Reaction times are typically of the order of 30 min to 2 h after reflux had been established, depending on the boiling point of the solution.

## Acknowledgments

We would like to thank Drs Robert Clemens, John Hyatt and John Hubbs for helpful discussions on various aspects of this work, Drs Richard Zimmerman and Candace Sass for conducting the matrix-isolation studies and Dr J. W. Sam Stevenson for helpful comments on this manuscript. The collaboration of Mr F. Del Rector on the work on the acetoacetylated polymers is also acknowledged.

## References

1. (a) Clemens, R. J., Proceedings of the Fifteenth Water-Borne and Higher-Solids Coatings Symposium, 1985, paper 55.
   (b) Rector, F. D., Blount, W. W. & Leonard, D. R., *ibid,* paper 68.
   (c) Witzeman, J. S., Nottingham, W. D. & Rector, F. D., *ibid,* 1989, paper 400.
2. Clemens, R. J., *Chem. Rev.,* **86** (1986) 241.

3. (a) Clemens, R. J., Hyatt, J. A., *J. Org. Chem.*, **50** (1985) 2431
(b) Hyatt, F. A., Feldman, P. L. & Clemens, R. J., *J. Org. Chem.*, **49** (1984) 5105.
(c) Clemens, R. J. & Witzeman, J. S., *J. Am. Chem. Soc.*, **111** (1989) 2186.
(d) Sato, M., Kanuma, N. & Kato, T., *Chem. Pharm. Bull.*, **30** (1982) 1315.
(e) Sato, M., Orgasawara, H., Yohizumi, E. & Kato, T., *Chem. Pharm. Bull.*, **31** (1983) 1902.
(f) Hyatt, J. A., *J. Org. Chem.*, **49** (1984) 5102.
4. (a) Bader, A. R., Cummings, L. O. & Vogel, H. A., *J. Am. Chem. Soc.*, **73** (1951) 4195.
(b) Bader, A. R. & Vogel, H. A., *J. Am. Chem. Soc.*, **74** (1952) 3992.
(c) Taber, D. F., Amedio Jr, J. C. & Patel, Y. K., *J. Org. Chem.*, **50** (1985) 3618.
(d) Gilbert, J. C. & Kelly, T. A., *J. Org. Chem.*, **53** (1988) 449.
(e) Carroll, M. F., Proceedings of XIth International Congress on Pure and Applied Chemistry, Vol 2, 1947, p. 39 (*Chem. Abstr*, **45** (1951) 7015e).
(f) Witzeman, J. S. & Nottingham, W. D., *J. Org. Chem.*, **56** (1991) 1713.
5. (a) See, for example, *The Chemistry of Carboxylic Acids and Esters*, ed, S. Patai. John Wiley, New York, 1969, Chap. 3, p. 117, and references cited therein.
(b) Seebach, D., Hungerbuhler, E., Naef, R., Schurrenberger, P., Weidmann, B. & Zuger, M., *Synthesis*, 1982, p. 138.
6. Gano, J. E. & Jacob, E. J., *Spectrochim. Acta*, **43a** (1987) 1023.

# Substituted ethylene diamines from acrylamide

**M. J. Braithwaite**

*Allied Colloids Limited, Bradford, UK*

## Introduction

### The Background

As part of a continuing programme of organic growth, Allied Colloids has turned its attention towards the fine chemicals market. Traditionally, from its foundation in 1935 to provide chemical auxiliaries to the local textile industry, the business has been regarded as an 'effect chemicals' operation supported by a large technical service capability.

A significant chemical manufacturing operation has been built over the years and this paper attempts to demonstrate the development of a fine chemicals business, initially via a range of unsymmetrically substituted ethylene diamines, from the basic strengths endowed by the position already established in the performance/effect chemicals area.

Allied Colloids' range of substituted ethylene diamines is based on a company strength in being the UK's largest producer of acrylamide from acrylonitrile. Acrylamide is produced by a Raney copper catalysed hydrolysis of acrylonitrile in a continuous process operated

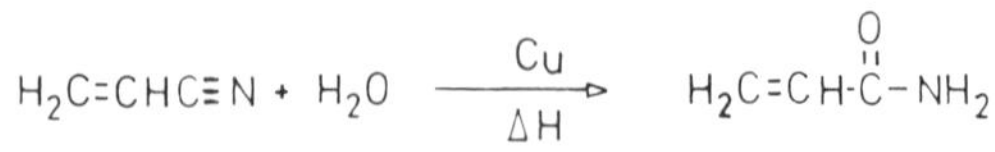

**Fig. 1.** Acrylamide from acrylonitrile.

on the Bradford site (Fig. 1). The major part of the output is a 50% aqueous solution of acrylamide.

## Manufacture of substituted ethylene diamines

Having access to a cheap, readily available source of acrylamide, it was decided to look at classical chemical techniques to derivatise this starting material. Michael addition across the double bond was studied and the addition of unhindered aliphatic secondary amines was shown to proceed readily in virtually quantitative yield to give dialkylamino-propionamides.

When subjected to a Hofmann degradation reaction, a more useful range of diamines can readily be synthesised in good yields (Fig. 2). Where R and R′ are methyl or ethyl, or together formed a morpholine or piperidine ring, the process proceeds well. As the chain length of the alkyl substituent increases, the yield from Hofmann degradation

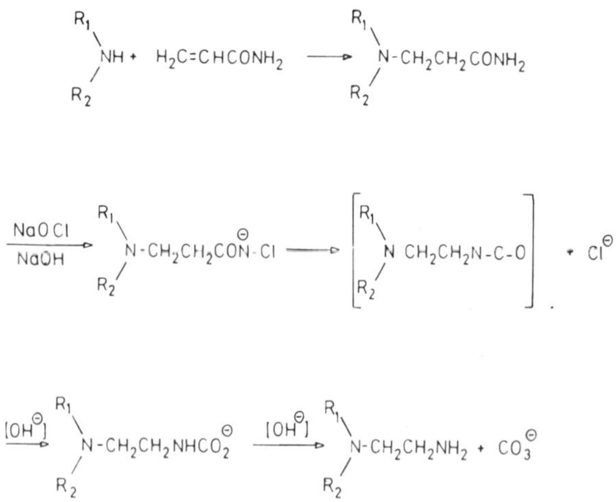

**Fig. 2.** Synthesis of unsymmetrically substituted ethylene diamines.

falls rapidly. However, the dibutyl analogue can be formed by modifying this process and, moving from the aqueous phase to an anhydrous methanolic reaction medium, permits ready synthesis of the higher chain-length derivatives. In the anhydrous medium the formation of alkylacylureas, with subsequent oxidation in the presence of hypohalous acid to nitriles, is minimised.

Addition of a primary amine to acrylamide gives a mixture of mono and bis addition products. Using a molar excess of the primary amine results in an increased proportion of mono addition product which increases with an increase in the molecular weight of the amine. *n*-Butylamine gives virtually 100% mono addition product. By using a 2:1 molar ratio of acrylamide to methylamine, the bis addition product is generated in quantitative yield. Hofmann degradation of this bis addition product proceeds well to give *N*′-methyl diethylene triamine, which is known to have uses as an intermediate in pharmaceutical synthesis and as an epoxy curing agent (Fig. 3). Attempting the Hofmann degradation reaction on the mono addition product does not produce the desired monoalkylated ethylene diamines. It is likely that the secondary amine group is attacked by the hypochlorite leading to complex amine condensation products. Protecting the secondary amine group by acetylation with acetic anhydride prior to the Hofmann degradation reaction leads to reasonable yields of monoalkyl ethylene diamines (Fig. 3).

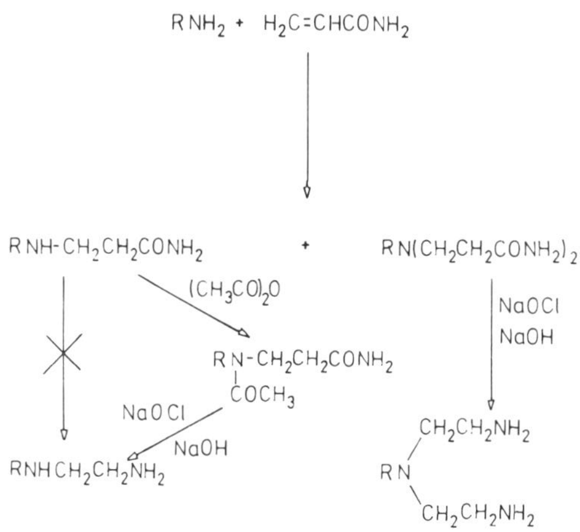


**Fig. 3.** Reactions with monoalkylamines.

## Applications for substituted ethylene diamines as intermediates

There are well established routes to the synthesis of some pharmaceutical activities which utilise members of the group of substituted ethylene diamines.[1]

*N*-ethyl ethylene diamine is reacted with the diethyl ester of oxalic acid to produce *N*-ethylpiperazine-2,3-dione which is an intermediate for the synthesis of the antibiotics Cefbuperazone and Piperacillin (Fig. 4). The reaction between *N*-ethyl ethylene diamine and diethyl oxalate demonstrates two fundamental chemical properties of the ethylene diamines. Primarily the acylation reaction is typically observed with esters, and, secondly, the formation of heterocyclic compounds, through reaction of both primary and secondary amine groups in a monoalkylated species, illustrates a further important feature.

*N,N*-diethyl ethylene diamine appears to have the widest use of the range. Reaction at a 2:1 ratio with diethyl oxalate again demonstrates the reaction with esters to form a precursor for the cholinesterase

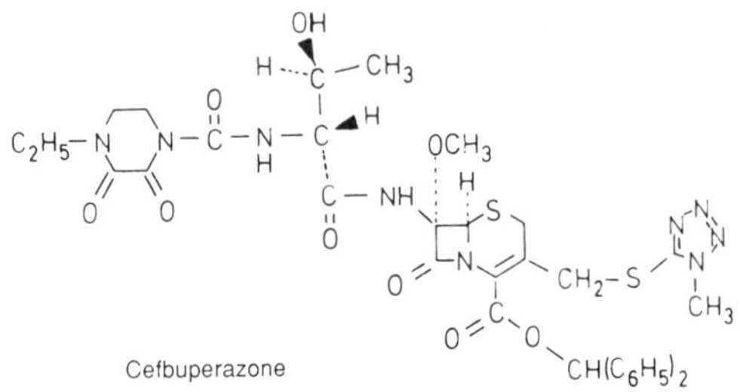

Cefbuperazone

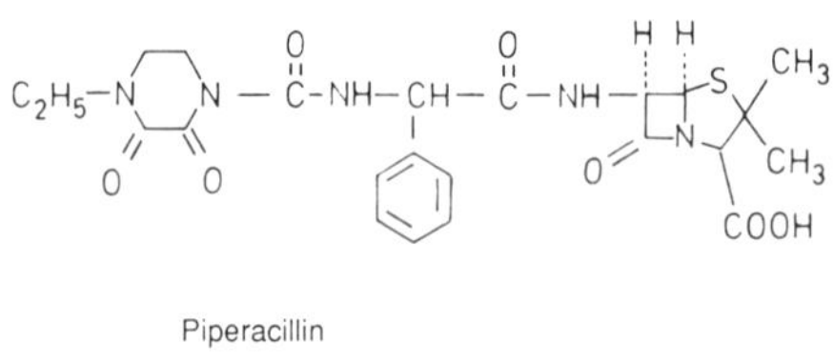

Piperacillin

**Fig. 4.** Pharmaceutical actives derived from substituted ethylene diamines.

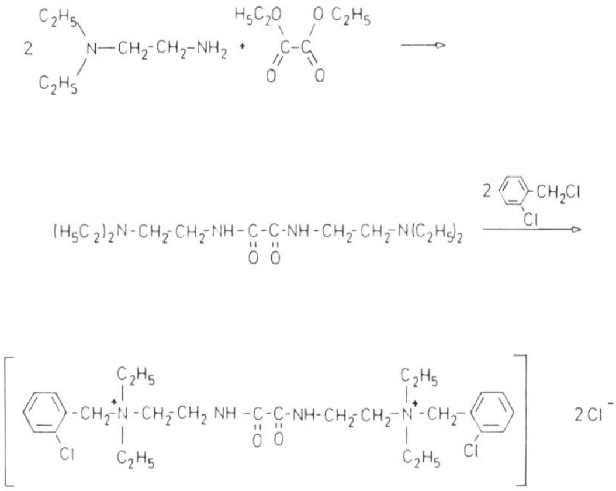

**Fig. 5.** Synthesis of ambenonium chloride.

inhibitor, ambenonium chloride, and subsequent reaction with 2-chlorobenzyl chloride illustrates the quaternisation potential for the tertiary amine (Fig. 5). The *N,N*-diethyl species is also used in the synthesis of Bromopride, Cinchocaine, Clofexamide, Procainamide, Tiapride, Mefexamide and Metoclopramide. This latter product is widely used as an anti-emetic and in the treatment of migraine.

The *N,N*-diethyl analogue is an intermediate for the anti-histaminic, Chloropyramine, but also finds use outside the pharmaceutical industry as an optical brightener intermediate (Fig. 6).

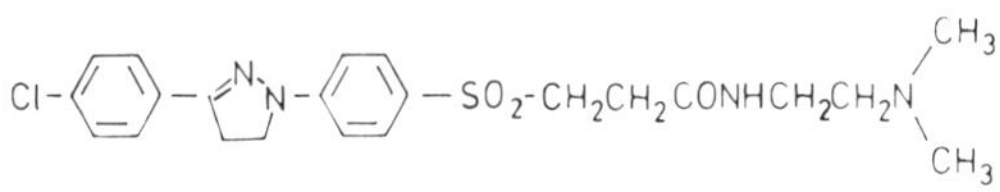

**Fig. 6.** Carboxamide optical brightener.

*N*-(2-aminoethyl)-morpholine is used in the psychotropic, anti-depressant drug, Minaprine (Fig. 7).

Clearly, there are commercial outlets for Allied Colloids' acrylamide-derived, fine chemical intermediates.

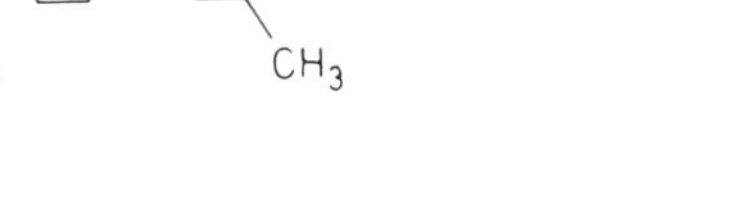

**Fig. 7.** Minaprine.

# Recent developments

In addition to these known uses, an extensive literature survey has revealed that numerous companies and research institutes throughout the world have worked with these products, and the following summary attempts to pick out some of the more interesting applications.

The chemistry reflects the known reactions of amines[2] where the presence of the lone pair on the nitrogen confers the property of basicity. The inductive effect of the alkyl substituent modifies the basicity, and hence the reactivity, particularly of the monoalkylated products.

The tendency for these monoalkylated products, with the 1,2-diamine configuration, to form cyclic products is fully demonstrated in the literature and will be reviewed below.

## Acylation

Primary and secondary amines react with carboxyl compounds to form amides. Acid chlorides react most readily but anhydrides and esters also react and have significant applications. The synthesis of the anti-arrythmic, Procainamide[3] (Fig. 8) demonstrates the reaction between an acid chloride and *N,N*-diethyl ethylene diamine. Also of note is the reported synthesis of an amide via reaction between aminoethyl-piperidine and an acid fluoride[4] (Fig. 9).

Analogous to the carboxylic acid halide reaction, to form carboxamides, is the reaction with sulphonyl chlorides to form sulphonamides. A recent paper[5] describes the synthesis and vasodilatory

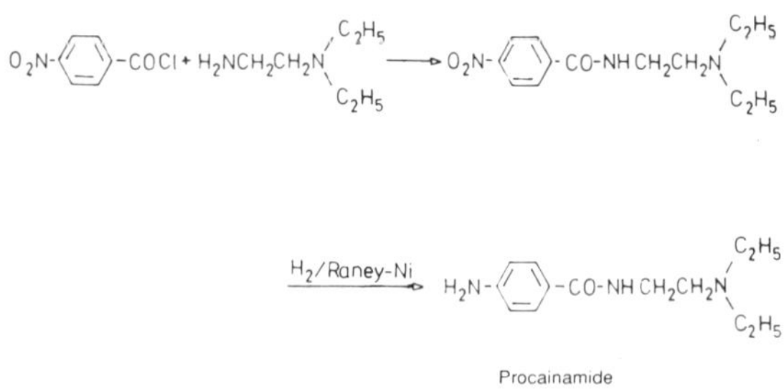


**Fig. 8.** Synthesis of procainamide.

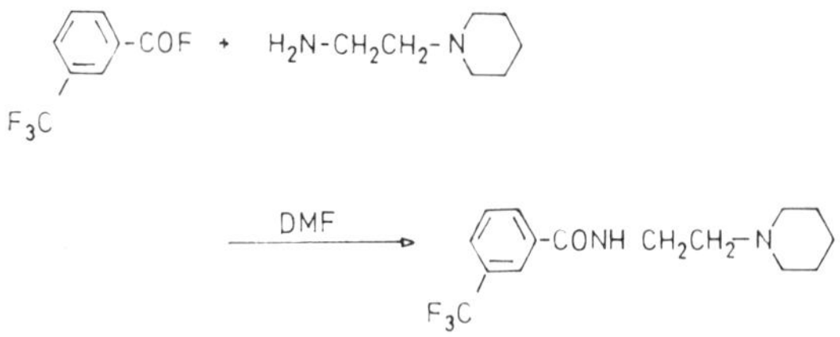

**Fig. 9.** Acid fluoride reaction.

activity of *N*-(2-aminoethyl)-5-isoquinoline sulphonamide derivatives illustrated by the dimethyl analogue shown in Fig. 10.

Reactions with anhydrides are abundant in the literature and are particularly important in the preparation of cyclic compounds. Reaction with esters is well illustrated by the formation of *N*-ethylpiperazine-2,3-dione as described above. However, a more typical example of the reaction between the primary amine group and an ester group giving a pendant diamine chain is described in the patent literature.[6] Anti-tumour activity is claimed for a series of benzophenazine compounds into which various substituted ethylene diamines, as illustrated by the dimethyl analogue, are reacted. The diamine 'tail' is claimed in this, and many other references, to import 'pharmaceutical acceptability' to the molecule (Fig. 11).

The acylation of ethylene diamines by carboxylic acid is also described in the literature.[7] The reaction is catalysed by 1,1-carbonyl di-imidazole and proceeds via an imidazole intermediate which can easily be aminated (Fig. 12).

## Reaction with Carbonyl Groups

Schiff base reactions between carbonyl groups and primary amines are also known for the substituted ethylene diamines.[8] Neither secondary nor teritary amines can form Schiff bases and hence reaction is at the primary amine group only, again giving the 'pharmaceutically accept-

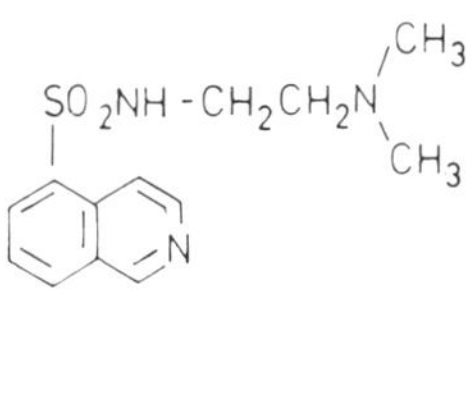

**Fig. 10.** *N*-(*N,N*-dimethylaminoethyl)-5-isoquinoline sulphonamide.

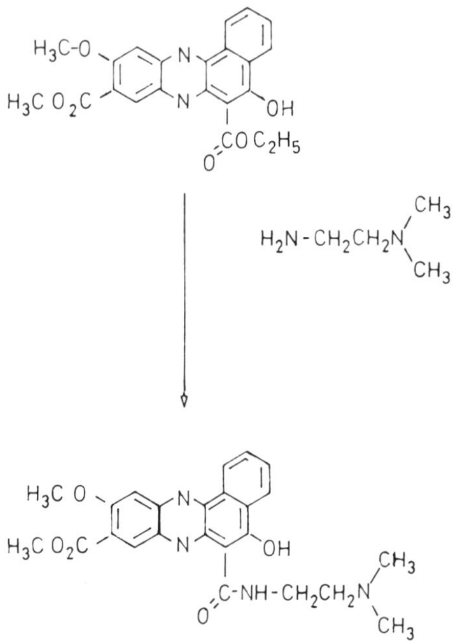


**Fig. 11.** Reaction with ester group.

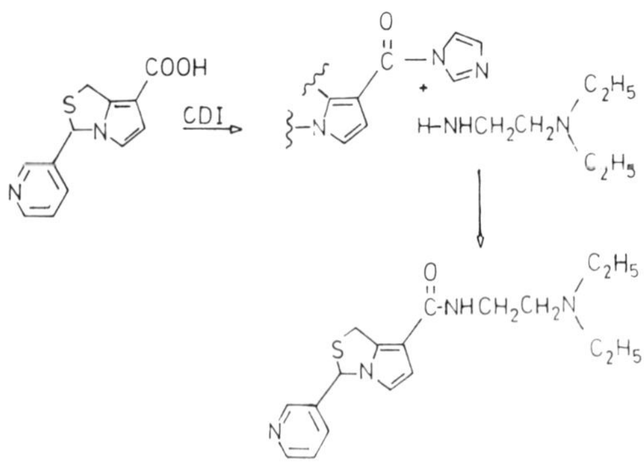


**Fig. 12.** Reaction with carboxylic acid.

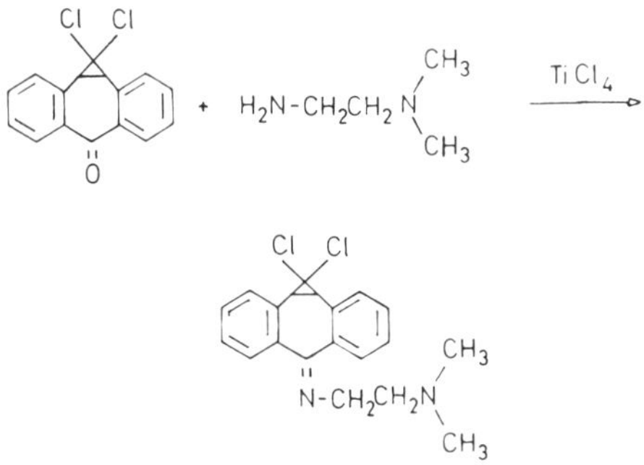

**Fig. 13.** Schiff's base reaction.

able' diamine pendant chain. Titanium tetrachloride is used as a catalyst in the example shown in Fig. 13.

## Alkylation and Arylation

Alkylation and arylation represent the other major category of amine chemistry and again the substituted ethylene diamines are shown to participate in these reactions in the chemical literature. Replacement of halogens, reaction with isocyanates or isothiocyanates and reductive alkylations are amongst the examples given.

### Halide displacement

Both chlorine and bromine are readily displaced by amines, good examples of the reactions of substituted ethylene diamines being illustrated in Fig. 14. The nitropyridine derivative synthesised by the reaction of 2-chloro-5-nitropyridine and 2-aminoethylpiperidine is claimed to sensitise certain cancer cells to X-rays for use in cancer radiotherapy.[9]

Reacting aminoethylpiperidine with 2-bromo-2,5-dinitrothiophene[10] gives the thienylamine derivative as shown in Fig. 14.

In both cases, the presence of the nitro group on the ring activates the molecule to amine attack.

Reaction of amines with isocyanates and isothiocyanates yields ureas and thioureas respectively. By an extension of the Wohler synthesis, *N*-substituted ureas can be synthesised and, in an analogous manner,

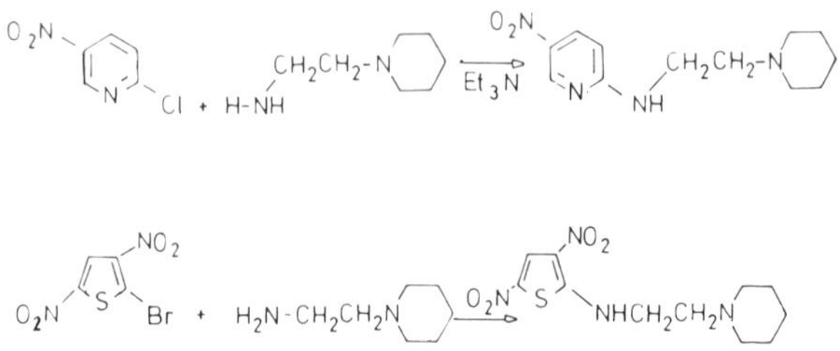

**Fig. 14.** Halogen displacement reactions.

*N*-substituted thioureas are formed. Compounds showing anti-ulcer, anti-secretory and gastric anti-histamine activity were prepared by refluxing an equimolar mixture of methyl isothiocyanate with 1-(2-aminoethyl)-piperidine in a chloroform–DMF mixture[11] (Fig. 15).

## Reductive alkylation

Reaction of the substituted ethylene diamines with aldehydes and ketones followed by reduction achieves alkylation of primary or secondary amine groups. An example[12] would be the family of phenyl amino propane derivatives synthesised via the reductive condensation reaction between trifluoromethyl benzyl methyl ketone and individual members of the substituted ethylene diamines under review. Sodium borohydride is the reducing agent used. The resulting compounds were proposed for use as anorexigenic compounds with low central nervous system activity (Fig. 16). An interesting example of the synthesis of an active pharmaceutical molecule which utilises acylation of *N*-methyl ethylene diamine by an ester, followed by bromine replacement to affect an alkylation at the secondary amine site, followed by a reductive alkylation reaction between a primary amine and an aldehyde, is given in a very recent patent application.[13] This reaction pathway admirably demonstrates the utility of the substituted ethylene diamine series as building blocks for the synthesis of quite complex molecules (Fig. 17).

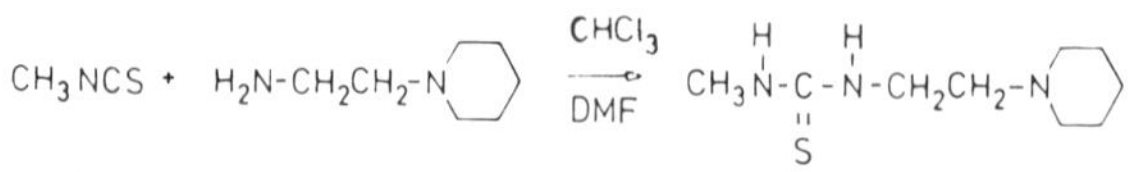

**Fig. 15.** Reaction with isothiocyanate.

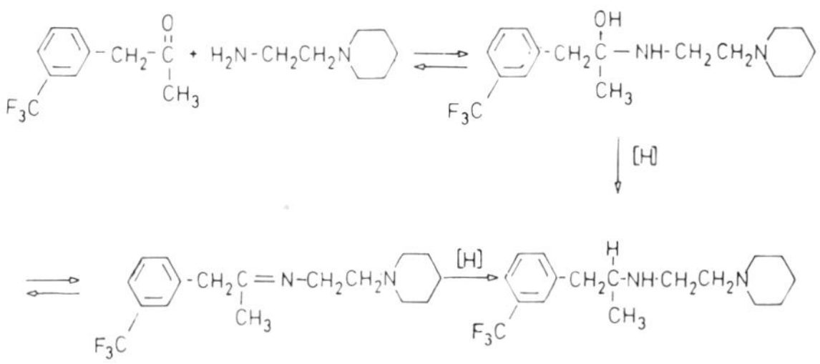

**Fig. 16.** Reductive alkylation.

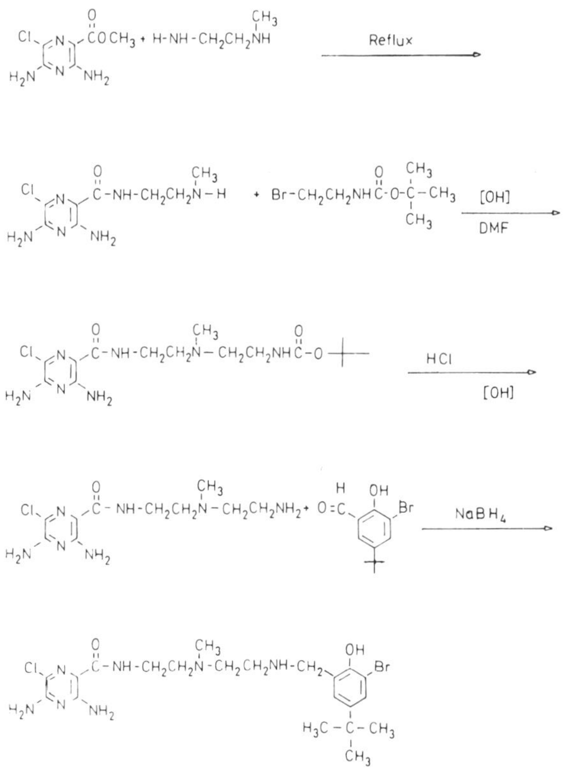

**Fig. 17.** Ester acylation, halogen displacement and reductive alkylation.

# Nitrogen-containing heterocycles

A more useful aspect of the chemical properties of the substituted ethylene diamines is the introduction of one or two nitrogen atoms into a cyclic system to yield nitrogen-containing heterocyclic molecules. The unsymmetrically dialkylated ethylene diamines are reacted at the primary amine site, in similar manner to the reaction examples given above, and have a 'tail' pendant from the heterocyclic ring importing 'pharmaceutical acceptability' to the derivative.

The monoalkylated ethylene diamines also react in this manner but have the added property of secondary amine character at the monoalkylated nitrogen. Both amine sites can therefore participate in the same reaction sequence and both nitrogen atoms can end up in a cyclic system. Imidazoles, piperazines, diazepines and a variety of other nitrogen-containing heterocycles, such as thiazoles, phosphazoles, etc., are produced from the *N*-alkylated ethylene diamines.

## Addition of One Nitrogen Atom via Reaction at the Primary Amine Site

Cyclic amides, pyrroles, piperidines, azepines, quinolines, carbazoles, indoles, oxazoles, thiazoles, etc., are all reported to have been synthesised via reaction of an appropriate substrate with a substituted ethylene diamine. Both mono- and dialkylated species are involved but reaction is at the primary amine site only.

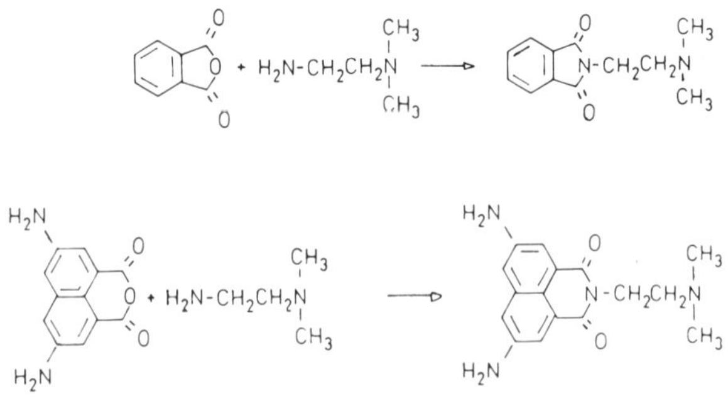

**Fig. 18.** Cyclic imides.

## Examples of Simple Nitrogen Heterocycles

Cyclic imides are readily synthesised by the acylation reaction between a substituted ethylene diamine and a cyclic anhydride.

The synthesis of a series of five and six membered cyclic imides showing prominent anti-leukemic and anti-melanoma activity in both *in vitro* and *in vivo* tumour systems is described in the literature[14] and shown in Fig. 18. Acylation of an ethylene diamine derivative, yielding a carboxamide intermediate,[15] by a cyclic dithiol-one, is an initial step in the synthesis of a series of isothiazolones. Ring closure to form the N—S bond is achieved by oxidation of the amide using iodine in the presence of sodium bicarbonate (Fig. 19).

Alkylation by substitution of a halogen can be used to form azepines, for example by reaction of both the primary amine hydrogens with two halogen atoms.[16] Figure 20 shows two examples. Triazepine derivatives can be formed from 2,3-(bishalo-acetyl)-tetrahydrophthalazines. Similarly thiazocine derivatives can be formed by reacting bishalomethyl sulphide, sulphur monoxide or sulphur dioxide with the primary group of one of the family of substituted ethylene diamines.[17] A family of dibenzothiazocines thus produced is described as having activity on the central nervous system and showing activity as anti-depressants and as treatment for Parkinson's disease with superior activity to existing treatments.

An interesting family of quinazolines derived from substituted ethylene diamines and 2-amino-5-chlorobenzophenone is produced by

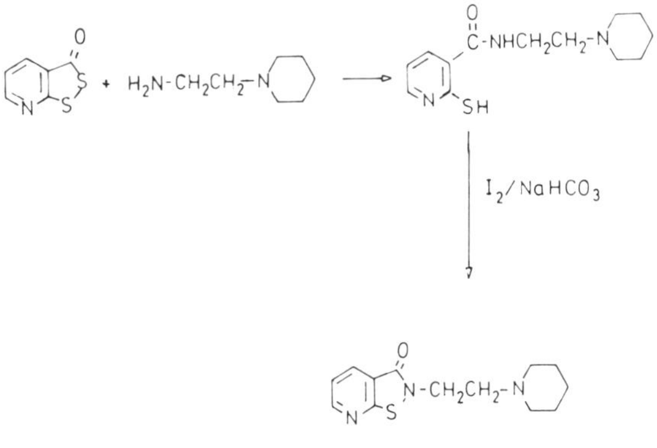


**Fig. 19.** Isothiazolone synthesis.

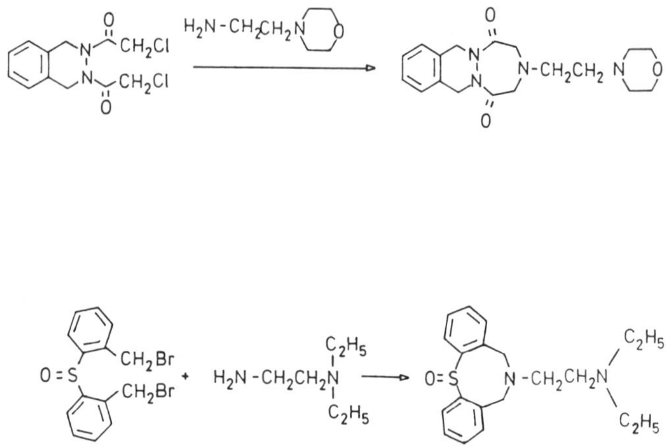


**Fig. 20.** Formation of azepines.

reaction between the primary amine group and the carbonyl group forming the Schiff's base intermediate.[18] Ring closure is achieved by acetylating the aromatic amine group with acetic anhydride, followed by room temperature addition of hydrogen chloride and subsequent hydrolysis. This reaction is shown in Fig. 21.

## Examples of Reactions Involving Both Nitrogens

One of the more important examples of the formation of heterocycles containing two nitrogens has been discussed above. The reaction between *N*-ethyl ethylene diamine and diethyl oxalate to give *N*-ethylpiperazine-2,3-dione shows the reaction of the primary and

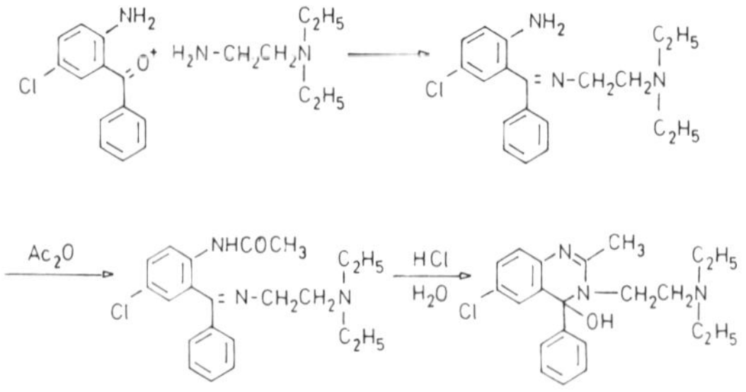


**Fig. 21.** Quinazoline synthesis.

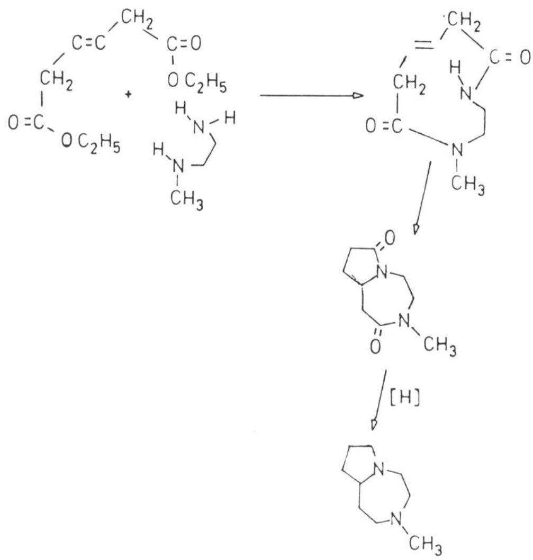


**Fig. 22.** Pyrrolodiazepine synthesis.

secondary amine groups with an ester and the resulting incorporation of both nitrogens into the heterocyclic ring.

A similar acylation reaction by two ester groups with *N*-methyl ethylene diamine, followed by nucleophilic attack by a nitrogen on a double bond carbon site, gives rise to a bicyclic pyrrolodiazepine. Figure 22 shows this reaction. Reaction of both the primary and secondary amine groups with both ester groups of diethyl *trans*-3-hexenedioate leads to an initial cyclisation.[19] However, steric factors lead to addition across the double bond involving a nitrogen and its associated proton, resulting in a further cyclisation to give a pyrrolodiazepine dione. Subsequent reduction of the dione gives the corresponding pyrrolodiazepine in good yield from simple, readily available starting materials.

Similarly, starting with homophthalic acid and reacting with *N*-methyl ethylene diamine in boiling *ortho*-dichlorobenzene, the imidazoisoquinoline derivative shown in Fig. 23 is formed.[20] Steric factors again permit facile ring closure under the reaction conditions in the intermediate proposed.

Amines react with anthraquinones in the presence of certain metal ions to form amino-anthraquinone derivatives which are important as

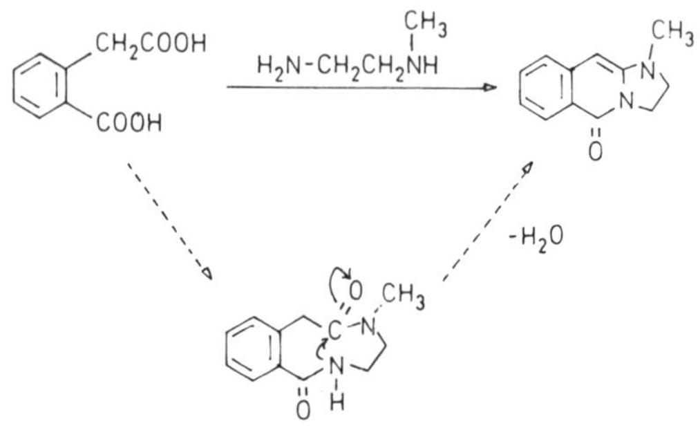


**Fig. 23.** Imidazoisoquinoline synthesis.

dyes or dyestuff intermediates.[21] Figure 24 shows the reaction products when *N*-methyl ethylene diamine is reacted with 1,4-dihydroxyanthraquinone in the presence of copper ions.

A mixture of products, consisting of equal proportions of the 2-aminated product, a ring closure product and unreacted starting

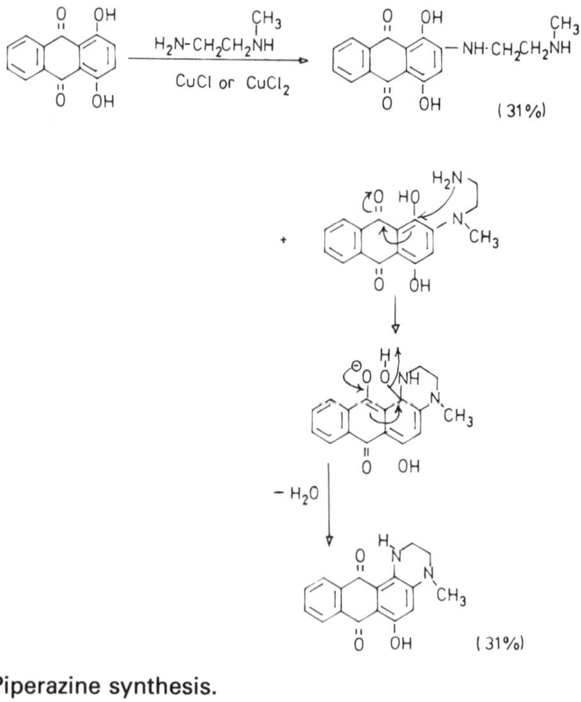


**Fig. 24.** Piperazine synthesis.

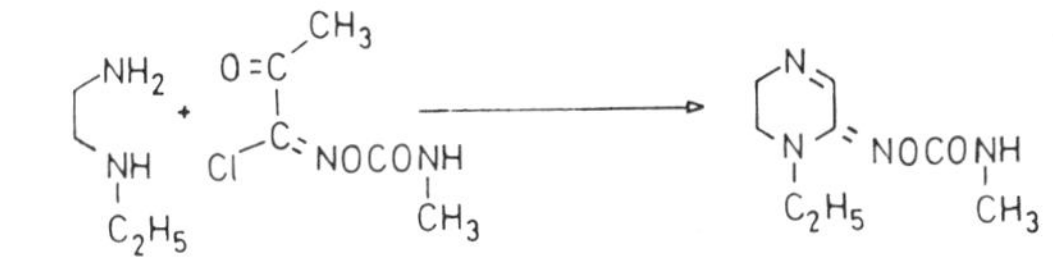


**Fig. 25.** Pyrazine synthesis.

material results. A mechanism is suggested for the ring closure product.

Using ethylene diamine itself gives the ring closure product in 98% yield, whereas *N*-ethylene diamine gives only 4% yield of the ring closure product. The steric effect of the *N*-alkyl substituent is obviously amply demonstrated by these reactions.

Reactions between the substituted ethylene diamines and carbonyl compounds were described above and, in the formation of diazo-heterocycles they again feature. The synthesis of a series of pyrazinone oxime carbamates, with insecticidal and acaricidal properties, demonstrates a reaction between a substituted ethylene diamine and carbonyl group followed by a reaction with the secondary amine group and a chlorine in an alkylation type process.[22] *N*-Ethyl ethylene diamine was reacted with 1-acetyl-1-chloroformaldoxine-*N*-methyl carbamate in the presence of anhydrous potassium carbonate at ambient temperature. The reaction was relatively rapid and proceeded to 80% yield of the pyrazine derivative shown in Fig. 25.

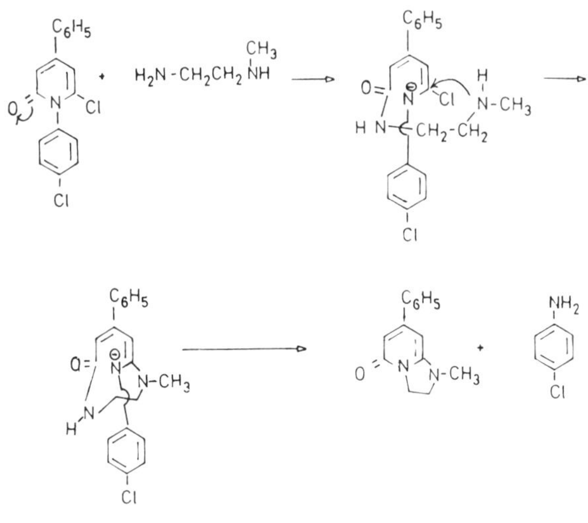


**Fig. 26.** Imidazopyridine synthesis.

Reaction of a substituted ethylene diamine with an *N*-substituted pyridone produces an imidazo-pyridine.[23] This reaction forms the basis of the synthesis of a series of some 90 pyridoheterocycles showing analgesic and anti-inflammatory properties (Fig. 26). It would appear that the primary amine group attacks the carbonyl carbon causing a ring opening. The secondary amine group then substitutes the chlorine forming a new cyclic product. Subsequent hydrogen migration, followed by attack by the secondary nitrogen and the cleaving off of *p*-chloroaniline, results in the formation of the imidazopyridine.

An example of a reductive amination of a carbonyl compound, which eventually leads to a diazepine derivative, is shown in Fig. 27.[24]

*N*-Ethyl ethylene diamine reacts with the aldehyde group and, under reducing conditions, cyclisation occurs, by the mechanism

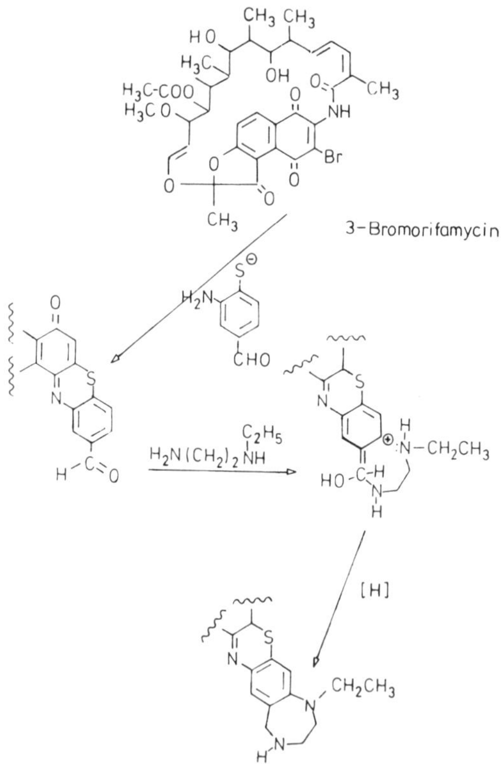


**Fig. 27.** Diazepine synthesis.

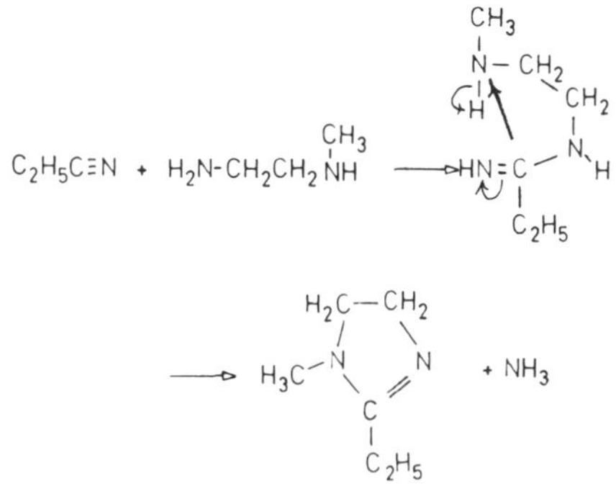


**Fig. 28.** Reaction with nitriles.

proposed in the figure, to give a seven-membered heterocyclic ring. The rifamycins themselves have anti-bacterial activity, but modification of the structure of 3-bromorifamycin in the above manner enhances the antibiotic properties markedly.

Reactions between amines and nitrile compounds give rise to amidines. Examples exist of the formation of nitrogen-containing heterocycles via an amidine intermediate. A simple example[25] of this type of reaction involving an *N*-alkyl ethylene diamine and resulting in a cyclisation to give a substituted imidazoline is shown in Fig. 28. Using rather forcing conditions (high temperatures and pressures with zinc acetate catalyst) this condensation and cyclisation reaction proceeds in high yields giving highly pure product. In an analogous reaction the example in Fig. 29 demonstrates the formation of the amidine derivatives followed by attack from the secondary amine group and the elimination of ammonia to give the imidazoline ring as in the above example. The benzamide derivative shown is described as having 'gastromotive' properties.

An example of the formation of an imidazoline derivative by reaction between a substituted ethylene diamine and a nitrile is provided by the synthesis of a family of imidazobenzothiazines (see Fig. 30).[26] Another typical amine reaction, the formation of a sulphonamide, precedes the reaction of the secondary amine site with the nitrile group. Under basic conditions, cyclisation occurs to give the imidazobenzothiazine derivative. The mechanism was confirmed by a study in which the intermediate (*N*-methylaminoethylaminosulphonyl) benzyl cyanide was isolated and its structure was determined by IR, NMR and microanalysis.

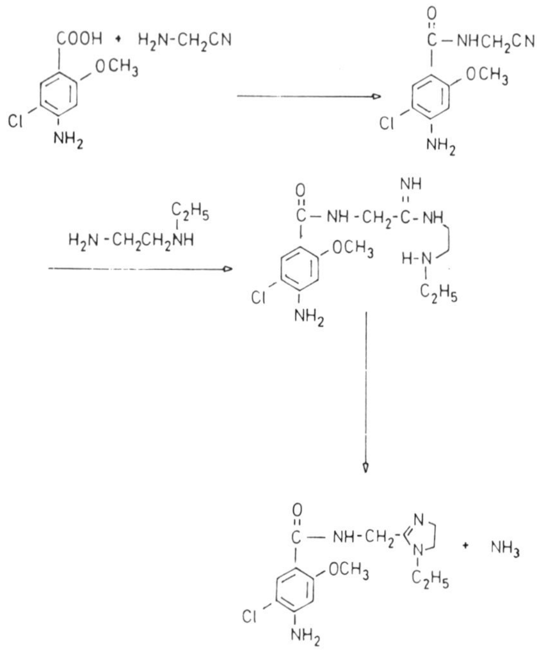


**Fig. 29.** Imidazoline synthesis.

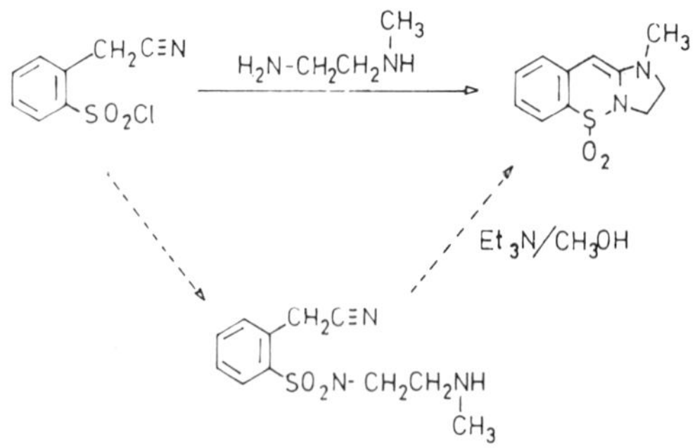


**Fig. 30.** Imidazobenzothiazine synthesis.

# Miscellaneous cyclic compounds containing other heteroatoms

The diversity of reactions reported for the substituted ethylene diamines indicates the scope for using this family of products for the synthesis of novel compounds.

Nitrogen–phosphorus heterocycles can readily be prepared by reacting a monoalkylated ethylene diamine with phosphorodichloridate derivatives.[27] In the presence of triethylamine, condensation occurs to give a phospholane as shown in Fig. 31. Phosgenation of the phospholane derivative produces a carbonyl chloride group which can be reacted with the amine group in ampicillin. Hence, a novel range of phospholane-substituted penicillanic acids has been derived from the substituted ethylene diamines, and the modified properties of the substituted penicillins provide attractive, new commercial possibilities.

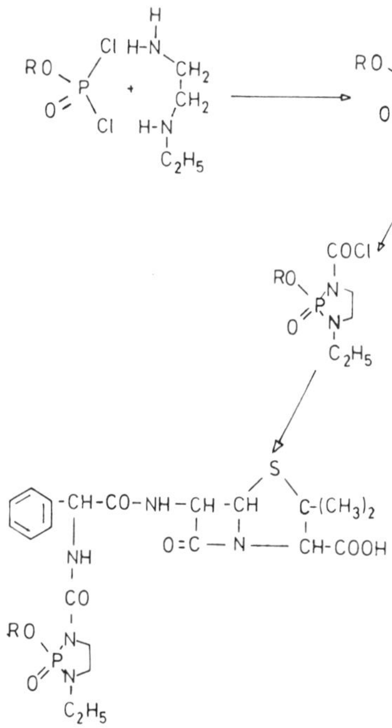

**Fig. 31.** Phospholane-substituted penicillanic acids.

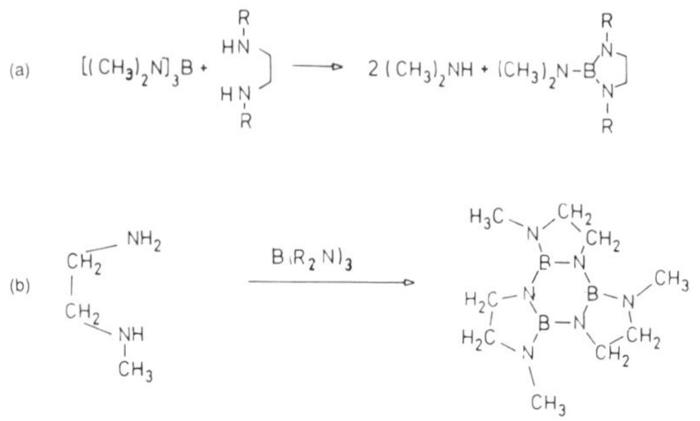

**Fig. 32.** Boron–nitrogen heterocycles.

Boron–nitrogen heterocycles with fungicidal and insecticidal activities have been reported in the literature.[28] By reacting *N,N'*-dialkyl diamines with trisamino boranes an amino diazaboracycloalkane is formed (see Fig. 32(a)). However, if one of the amino groups is primary, as in the mono-substituted ethylene diamines, then further elimination of dimethylamine is observed during the reaction and a polycyclic borazine is produced (Fig. 32(b)).

# Summary

This paper illustrates some of the reported synthetic pathways that utilise the reactive properties of the family of unsymmetrically substituted ethylene diamines. Admittedly, the chemistry reflects that of amine compounds in general but, because of the 1,2-diamine configuration, specific reactions of importance make this series of products useful synthetic tools in the manufacture of nitrogen-containing derivatives and, in particular nitrogen heterocycles.

The 'pharmaceutical acceptability' of the diamines in the side chains of physiologically active molecules is clearly of significance to medicinal chemists, and the great deal of interest shown in the recent literature in utilising the N—C—C—N backbone to form cyclic compounds of pharmacological significance suggests that there will be a growing market for these products in the future. Allied Colloids' strength in being a primary producer of acrylamide for its effect chemicals business provides the starting material for the synthesis of these substituted ethylene diamines—'Fine chemicals from a performance chemicals manufacturer'.

# References


1. Kleemann, A. & Engel, J., *Pharmazeutische Wirkstoffe*. George Thieme Verlag, Stuttgart, 1982, 1987.
2. Patai, Saul (ed.), *The Chemistry of the Amino Group*. Interscience, London, 1968.
3. Baltzy, R., *et al., J. Am. Chem. Soc.*, **64** (1942) 2231.
4. Bruce, W. F., US Patent 3 787 419, 1974.
5. Morikawa, A., Sone, T. & Asano, T., *J. Med. Chem.*, **32** (1989) 46–50.
6. Migita, Y., Eguchi, T., Kumazawa, Y., Nakagami, J., Amano, T., Sota, K. & Sakakibara, J., Eur. Patent Appl. EP 196 910, 1986.
7. Fabre, J. L., Farge, D., James, C. & Lave, D., Eur. Patent Appl. EP 115 979, 1984.
8. Uyeda, R. T., German Offen. 2 348 060, 1974.
9. Kagitani, T., Minagawa, M., Nakahara, Y., Kimura, R., Tsubakimoto, T., Oshiumi, R. & Sakano, K., Jpn Kokai Tokkyo Koho JP61 143 361, 1986.
10. Yan, M., Zhang, Z., Lei, X., Xi, Y. & Xu, Y., *Yaoxue Xuebao,* **20** (1985) 219–23.
11. Curran, A. C. W., British Patent 1 462 487, 1977.
12. Podesva, C., Scott, W. T. & Henson, D. W., British Patent 1 414 223, 1974.
13. Schwartz, J. A., Eur. Patent Appl. EP 293 170, 1988.
14. Zee-Cheng, R. K. Y. & Cheng, C. C., *J. Med. Chem.*, **28** (1985) 1216–22.
15. Baggaley, K. H., Jennings, L. J. A. & Tyrrell, A. W. R., *J. Heterocycl. Chem.*, **19** (1982) 1393–6.
16. Groszkowski, S. & Wesolowska, B., *Arch Pharm* (Weinheim, Germany), **312** (1982) 136–42.
17. Tanaka, S. & Watanabe, H., German Offen 2 246 707, 1973.
18. Bahr, F. & Dietz, G., *Pharmazie,* **35** (1980) 256–63.
19. Lavagnino, E. R. & Ryan, C. W., *Org. Prep. Proced. Int.*, **15** (1983) 251–60.
20. Nagarajan, K., Rao, V. R., Shah, R. K., Shenoy, S. J., Fritz, H., Richter, W. J. & Muller, D., *Helv. Chim. Acta,* **71** (1988) 77–92.
21. Takei, T., Matsuoka, M. & Kitao, T., *Bull. Chem. Soc. Jpn,* **54** (1981) 2735–8.
22. Booth, D. L. & Rodebaugh, R. M., US Patent 3 957 990, 1976.
23. Kubo, K., Ito, N., Souzu, I., Isomura, Y., Homma, H. & Murakami, M., German Offen DE 2 731 982, 1978.
24. Yamane, T., Kondo, H., Hashizume, T., Yamashita, K., Hosoe, K. & Watanabe, K., Jpn Kokai Tokkyo Koho JP 63 35 578, 1988.
25. Franceschini, J., Gardaixlut, R. & Margarit, J., UK Patent Appl. GB 2 178 035, 1987.
26. Kubo, K., Ito, N., Isomura, Y., Sozu, I., Homma, H. & Murakami, M., *Chem. Pharm. Bull.*, **27** (1979) 2372–81.
27. Borrevang, P., Petersen, H. B., Takeda, H. & Okazoe, T., British Patent Appl. 2 036 027, 1980.
28. Cragg, R. H. & Nazery, M., *Inorg. Nucl. Chem. Lett.*, **10** (1974) 481–4.

# Prospects for sodium metal applications

**P. Ashworth & J. Chetland**

*The Associated Octel Company Limited, Ellesmere Port, UK*

## Introduction

The commercial value of most metallic elements lies in 'structural' applications, dependent on physical characteristics of strength and malleability and on good chemical inertness. In contrast, the alkali metals, which make up Group 1A of the periodic table, owe the majority of their applications to high chemical reactivity.

This reactivity arises from the weakly bound single valence electron, which is readily lost in providing a chemical reduction process. The weak electron binding in Group 1A also accounts for characteristic physical properties such as low density, low melting point and high thermal conductivity which are currently exploited in the case of sodium.

Sodium is the second member of the alkali metal group and the most commercially important. The unique combination of physical and chemical properties, together with relatively low cost, provides speciality products essential to a variety of industries. Important physical properties of metallic sodium are shown in Table 1.

**Table 1.** Physical properties of sodium metal

| | |
|---|---|
| Melting point | 97·8°C |
| Boiling point | 883°C |
| Density, solid, 20°C | 968 kg/m$^3$ |
| liquid, 100°C | 928 kg/m$^3$ |
| Vapour pressure, 100°C | $1 \cdot 19 \times 10^{-3}$ mm Hg |
| Viscosity, 100°C | 0·682 Hz |
| Elect resistivity, 20°C | 4·69 $\mu\Omega$ |
| Thermal conductivity, 20°C | 140 W/m °C |
| Ionisation potential | 5·12 V |
| Electrode potential, 25°C | +2·71 V |
| Electronegativity | 0·9 |

# Historical development

Metallic sodium is not found in the free state in nature because of its high chemical reactivity. The metal was first prepared by Humphry Davy in 1807 from electrolysis of caustic soda. Commercial production in the late nineteenth century, by both electrolytic and carbon reduction, was driven by its use in manufacture of aluminium metal from chloride. The Hall electrolytic process providing aluminium from bauxite removed this application for sodium and other outlets were actively sought.

Introduction, in the 1920s, of the Downs electrolytic cell for production of sodium from common salt coincided with the development of a new and major application, as intermediate in the manufacture of lead alkyl motor fuel antiknock. This rapidly became the dominant application for sodium and major Downs production facilities were established at sites of lead alkyl manufacture. While remaining the largest single application for sodium, the lead alkyl business is experiencing severe decline, particularly in the major markets of the Western World.

## Current Production and Supply

All commercial manufacture now uses an improved Downs process employing a ternary salt mixture with calcium and barium chlorides. The latter are added to reduce the melting point of salt from 800°C to approximately 600°C, where the solubility of sodium is sufficiently low to allow high current efficiencies and easy separation of the metal. The

**Table 2.** World production of sodium metal[1]

| Facility | Estimated volumes, 1989 (thousand tonnes) | |
|---|---|---|
| | Capacity | Production |
| USA | 75 | 55 |
| W. Europe (Octel) | 25 | 25 |
| (Other production) | 21 | 10 |
| E. Europe | 5–10 | 5–10 |
| Japan | 6–12 | 6–12 |
| | ~140 | ~100 |

1–2% of calcium metal dissolved at this temperature is precipitated by cooling to 120°C, and filtered. Dry chlorine gas is a valuable by-product of the process.

Currently world production of sodium is estimated at approximately 100 000 t p.a. arising from the major facilities shown in Table 2. Market prices for bulk metal in the United States and Europe are around $2000 per ton. Octel, in the United Kingdom, are the largest European producer of sodium with 25 000 t p.a., the bulk of which is currently consumed internally for lead alkyl manufacture. With the

**Fig. 1.** The Downs Cell Hall at Octel.

steady decline of this captive application, Octel's sodium is increasingly being directed towards other markets.

Figure 1 shows a Downs cell hall within Octel's UK facility, each cell being capable of producing 700 kg of sodium per day.

The metal can be purified to provide grades meeting the specific requirements of each application, and is supplied in a variety of forms and containers. Octel, for example, provides sodium in 45 ton rail tankers. (Fig. 2), in 6 and 10 ton road tankers, in 1·2 ton pots and in 1·6 and 4·8 kg oiled ingots (Fig. 3) packed in 75 kg drums. Ingots can be transferred safely by hand using thick cotton gloves and are obviously more suited to users of small lots.

The development of commercial uses for sodium metal has undoubtedly been retarded by its reputation as a particularly hazardous product. While it is true that sodium is hazardous, due to its high reactivity with oxidising agents and particularly with water, it is no more dangerous than many other chemicals routinely utilised in much larger quantities. Sodium, like many hazardous materials, may be handled and used safely given the correct equipment and procedures for transport, storage, use and disposal.

Octel, along with other major producers of sodium, offer comprehensive safety and advice services to potential users and as support to product sales.

**Fig. 2.** Sodium rail tanker.

**Fig. 3.** Sodium metal ingots.

# Current usage patterns

The world consumption of sodium in 1989 has been estimated at about 100 000 t p.a. As Fig. 4 shows, this is substantially lower than the 1984 total, chosen for comparison, due solely to the decline of the major application in lead alkyl manufacture. Continued decline is expected

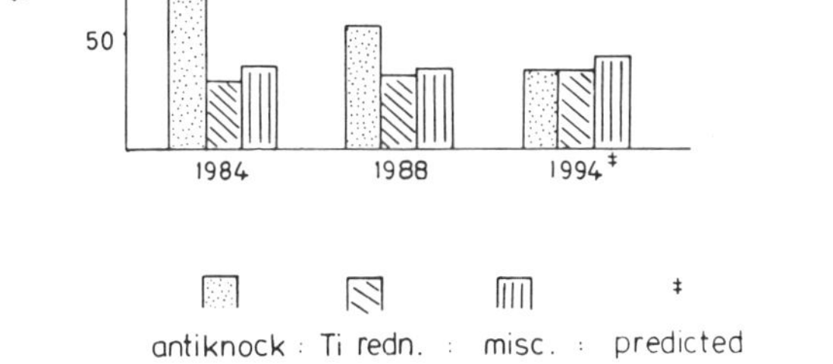


**Fig. 4.** Using patterns for sodium.[2]

as shown in projected figures for 1994. The second largest use, in production of titanium metal, and a variety of miscellaneous applications discussed later show modest growth during the period.

The increasing world over-capacity in sodium metal is obviously a challenge to current producers and is helping to fuel the search for new applications. Positive factors in meeting this challenge are:

- relatively low cost of sodium.
- unique combination of valuable physical properties
- high chemical reactivity as reducing agent

## Current applications and prospects

The capability of sodium metal as a low cost chemical reducing agent currently dominates world demand, chiefly in the production of other metals and organometallics. The reasons for this dominance of sodium use by inorganic rather than organic chemistry merits some discussion.

The major feedstocks for industrial organic chemistry are now obtained from petroleum processing and arise in highly reduced form as hydrocarbons. With the exception of direct hydrogenation of unsaturated hydrocarbons, the demand for reduction processes in derivatising these feedstocks is, therefore, much less than the demand for oxidation.

In contrast, industrial inorganic chemistry uses basic raw materials, i.e. ore deposits, which are generally in a highly oxidised natural state. Most industrial applications necessitate isolation of the pure element form, with a consequent high demand for reduction processing. These reductions (Fig. 5) are either effected electrochemically or chemically with, for example, carbon or highly electropositive metals (e.g. sodium, magnesium, aluminium).

Similarly, in the manufacture of important organometal compounds of weakly electropositive metals, a reducing agent is generally required to provide the metal atom or the organic component in the appropriate oxidation state (Fig. 6). Here, the cheap and highly electropositive

$$\text{Oxide Ore} \xrightarrow[\text{(reductant)}]{+e} \text{Element}$$

**Fig. 5.** Metal oxide reductions.

$$M + R\text{–}X \xrightarrow{+e} M\text{–}R + X^-$$

or

$$M\text{–}X + R \xrightarrow{+e} M\text{–}R + X^-$$

**Fig. 6.** Metal alkyls production.

metals, sodium, magnesium and aluminium, are again employed as reductants, notably in the manufacture of tetraalkyl compounds of silicon, tin and lead.

Despite the dominance of commercial sodium use by these inorganic reduction processes, organics manufacture does provide a diverse range of smaller volume applications for sodium. These organic reductions are both interesting chemically and vital to the production of many speciality chemical products.

A complete survey of important commercial applications of sodium is presented in the remainder of this article, classified according to specific properties and functions of the metal. Some views on future prospects, to meet the challenge of declining antiknock demand, are included.

## Applications Based on Physical Properties of Sodium

Important physical properties were presented in Table 1. Certain unique combinations of these properties are utilised in niche applications where the metal remains chemically unchanged.

### *Heat Transfer Media*

The properties listed below combine to make sodium an excellent heat transfer fluid for high temperature systems:

- low melting point
- wide liquid range
- good thermal conductivity
- low density
- low melt viscosity

The largest current application, as coolant in fast breeder nuclear reactors, employs many thousands of tons of high purity grade sodium. However, since metal is not consumed, the annual turnover is less substantial. Future prospects for this use obviously depend on major decisions in nuclear power generation, but the successful development in this industry of safe systems for the containment of coolant sodium is offering opportunities elsewhere.

Sodium–potassium alloy (NaK) permits much lower operating temperatures but usage has been limited by high cost.

### *Modification of Alloys*

Sodium's wide liquid range finds small volume applications in modification of alloys such as aluminium–silicon. Very small quantities ($<1\%$) can be effective in altering grain size and phase change temperatures to improve casting properties.

This application has prompted some manufacturers to provide small dose packages for use in foundries.

### *Sodium–Sulphur Battery*

Sodium–sulphur is one of the battery systems under intensive study for heavy duty power storage. Although the stored energy derives from chemical reaction, the advantages of the sodium system rely on certain physical properties of the metal.

The low atomic weight of sodium provides, in combination with sulphur, a high energy density system with several times the capacity of an equivalent lead–acid or nickel–cadmium battery. High energy density is of paramount importance to the successful development of battery driven road vehicles and in other areas, e.g. powering of satellites, where weight considerations are critical.[3] The low cost of sodium is an additional virtue in more basic applications, such as storage of excess production in power stations.

Large scale adoption of the sodium battery faces a number of obstacles, not least the problems of containment and disposal of sodium in road vehicle applications. However, the case for providing pollution-free transport systems is accumulating steadily and the potential demand for sodium in this area is clearly very substantial.

### *Other Uses*

The sodium vapour lamp for street lighting is a long established but small volume user of the metal.

The low cost, low density and good electrical conductivity of sodium have stimulated proposals for use in conducting cable. Safety considerations relating to the metal's high reactivity have, not surprisingly, suppressed commercial development.

## Applications Based on Chemical Reduction

Sodium has the electronic configuration

$$[Ne]^3S_1$$

and owes its strong reducing activity to the ease with which it can achieve the stable noble gas configuration by loss of one electron. This activity has been commercialised in the reduction of inorganic and organic compounds, in manufacture of organometallics and in the production of derivatives valued for their strongly basic and nucleophilic properties.

In many applications, particularly involving organic compounds, the bulk metal's activity needs to be modified to provide good process control and optimum yield of desired product. The most effective means of achieving this control (listed below):

- dispersion in inert solvents.
- alloying with other metals
- use in conjunction with co-reagent
- chemical derivatisation

are described in the following review of chemical applications.

### *Reduction of Inorganics*

Titanium metal manufacture, by reduction of the tetrachloride, is the second largest use for sodium (Fig. 7). Direct electrochemical reduction has been investigated, but problems arising from reaction reversibility have not been overcome.

The 'Hunter process', using sodium metal, competes commercially with the 'Kroll' magnesium reduction process, both of which conduct the reduction in a molten mix of metal and metal salts. The sodium

$$TiCl_4 + 4Na \rightarrow Ti + 4NaCl$$

**Fig. 7.** Titanium metal manufacture.

reduction, however, consumes less energy overall and provides a titanium product lower in iron content. Trace metal impurities reduce the hardness of titanium and a high quality grade sodium is required for this application.

Similar processes, consuming smaller quantities of sodium, are used to recover the metals zirconium and hafnium.

Sodium metal has a wide variety of foundry applications in the purification of metals. The strongly reducing properties are used to de-oxidise and desulphurise and to remove trace elements such as arsenic, antimony and cadmium in the smelting of lead and zinc.

In the de-scaling of ferrous metals, the active reductant is sodium hydride produced *in situ* from sodium and hydrogen gas.

Although current usage in these foundry applications is small, the trend towards metal recovery and recycling is expected to increase demand.

### *Production of Organometallics*

Organolead antiknock manufacture remains the largest application for sodium metal, despite the continuing decline. Here, sodium is used in the form of a 50 atom % alloy with lead, the alloy reacting with alkyl (Me, Et) chlorides to provide tetraalkyl lead (Fig. 8).

The sodium serves two distinct functions, as a reducing agent for the alkyl chloride and, in the alloyed state, as an activator for reaction of metallic lead.

The alloy is a brittle material, melting at 370°C, and is produced in an easily handled form as small flakes. Around 200 000 t p.a. are consumed in Octel's lead alkyl manufacture but, despite attracting varied interest, no other significant applications have been found. The material exhibits the properties expected of diluted sodium, oxidising slowly in air and reacting mildly with water. This modified reactivity may offer some prospects for new applications in metallurgic and other reductions.

Organotin compounds have important applications as multi-purpose stabilisers in plastics and as antifoulant biocides. The major commer-

$$4NaPb + RCl \rightarrow PbR_4 + 3Pb + 4NaCl$$

**Fig. 8.** Equation for production of tetraalkyl lead. R = Me, Et.

$$SnCl_4 + 8Na + 4RCl \rightarrow SnR_4 + 8NaCl$$

**Fig. 9.** Tetraalkyl tin manufacture.

cial products are di- and trialkyl (butyl, octyl) tin salts prepared from the tetraalkyl compounds.

The development of manufacturing processes for tetraalkyl tins has seen keen competition between sodium, magnesium and aluminium for the role of reducing agent to provide the alkyl groups in anion form. The manufacture of aluminium alkyls as intermediates for other major products has, however, led to this route becoming dominant for butyl and octyl tins.

The route employing metallic sodium as reductant is nevertheless still used in East Germany. Here, sodium dispersed to 100 $\mu$m particle size is reacted with alkyl chloride and tin tetrachloride (Fig. 9) in a process which is said to be currently competitive with that using aluminium.

The magnesium reduction route, via Grignard reagents, retains advantages in production of speciality aryl and cycloalkyl tins, not readily provided via aluminium or sodium.

In this area of organotins manufacture, the relatively sparse adoption of sodium as reductant owes much to perceived problems in handling and reaction control. In light of this, future prospects for sodium organometals and manufacture may rely on advances in the safe production and use of dispersions.

### *Reduction of Organics*

The strong reducing power of sodium is displayed in reactions involving a wide variety of organic compounds. Where highly polar bonds are present, reactions with bulk metal can often be very violent, leading to major degradation of the organic molecule.

A controlled degradation of this kind forms the basis of processes currently being examined for treatment of toxic wastes, particularly halogenated compounds, dioxins and benzofurans. With growing environmental pressures for safe disposal of such wastes, the prospects for sodium as a cheap degrading de-contaminant appear strong.

In the synthesis of organics, however, the activity of sodium must be well moderated to provide selective reduction of targeted functional groups. While many organic reductions proceed successfully using

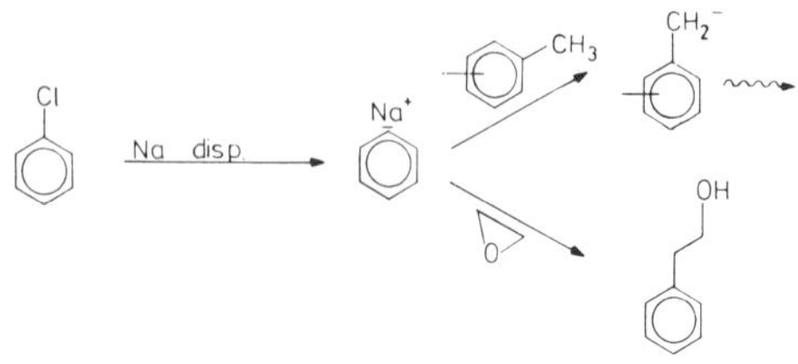

**Fig. 10.** Aryl sodium reactions.

kilogram ingots or smaller sticks of sodium, better control can often be achieved through the use of finely dispersed sodium or a combination of sodium with co-reagents such as ammonia or alcohol.

### *Sodium Dispersions*

Fine dispersions of the metal in an inert hydrocarbon oil medium can be produced by vigorous agitation of a mix at temperatures just above the melting point of sodium. Particle sizes as low as 1–10 $\mu$m in 40% w/w dispersions are readily achieved in Octel's laboratory development of the product.

This form of the metal allows organic reductions to be conducted rapidly, but with excellent process control, by means of gradual addition of the liquid dispersion. Such reductions are in every way typical of sodium, but are accomplished with greater selectivity than afforded by bulk metal.

In the example of Fig. 10, a dispersion in toluene gives smooth metallation of chlorobenzene to provide phenyl sodium. This intermediate may be used either as an extremely strong base for proton abstraction or may be derivatised by reaction with electrophiles.[4] In application as powerful bases, aryl sodiums perform many of the reactions afforded by the more expensive reagent, butyl lithium, and have considerable potential as replacements.

Reduction of aliphatic carbon–halogen bonds by sodium is best performed with dispersed metal. The well-known Würtz and Fittig

$$2RI + 2Na \rightarrow R\text{–}R + 2NaI \qquad \text{Würtz}$$

$$ArBr + RBr + 2Na \rightarrow Ar\text{–}R + 2NaBr \qquad \text{Fittig}$$

**Fig. 11.** Carbon–carbon couplings using sodium.

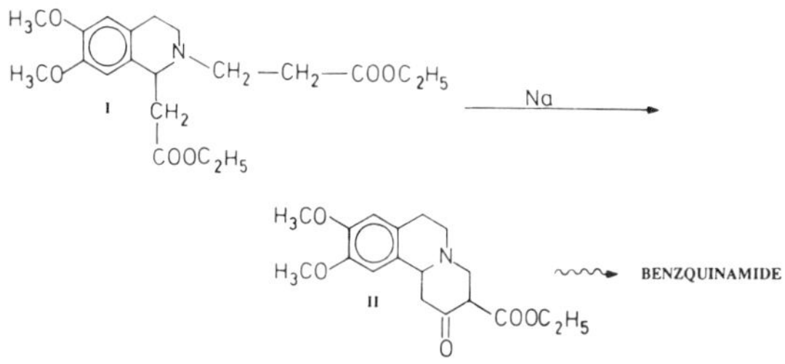


**Fig. 12.** Claisen condensation.

reactions shown in Fig. 11 are, however, of limited commercial value due to the large quantities of by-product generally formed.

Sodium metal readily affords reduction at sites of acidic protons in organic molecules, providing anions for a variety of condensation reactions. Figure 12, for example, shows a Claisen condensation of the isoquinoline derivative (I), providing the benzoquinazoline (II), an intermediate for the tranquilizer 'Benzquinamide'.

These organic anion condensations are often more efficiently catalysed by strong sodium-derived bases, such as hydride, amide or alkoxide, the latter frequently being generated *in situ.*

### *Sodium with Co-reagents*

Alkali metals dissolve in liquid ammonia to form blue solutions which contain solvated electrons. The solutions provided by sodium and lithium are important commercial reagents for the Birch reduction providing partially reduced aromatics, for example dihydrobenzenes. The reduction is used chiefly in the manufacture of steroidal drugs.

Sodium is less strongly reducing than lithium in liquid ammonia and is not effective with electron rich aromatics. However, the much lower cost of sodium (approximately one-tenth of the lithium costs on an equivalent basis) normally assures the application where reaction yields are comparable. The example in Fig. 13 is the reduction of benzamide to the 1,4-dihydro derivative.

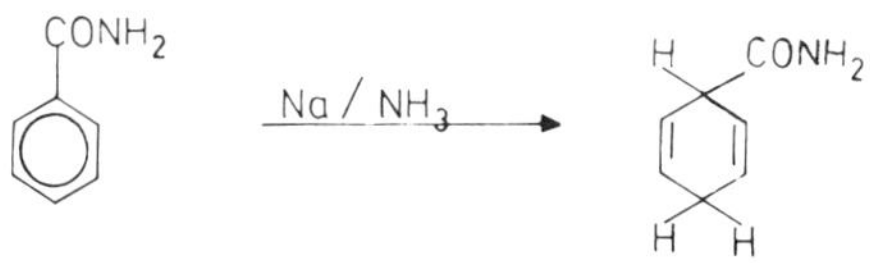


**Fig. 13.** Birch reduction.

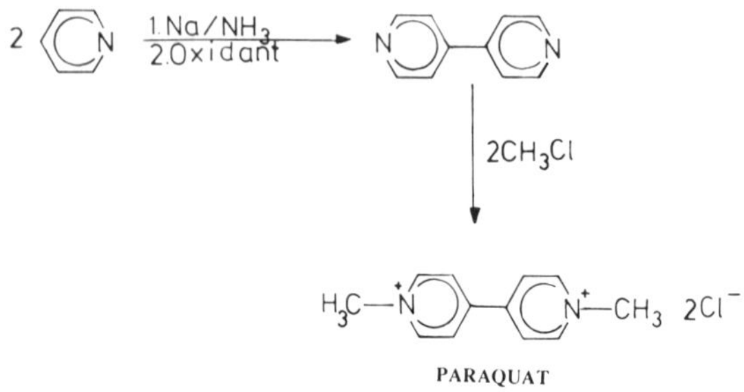


**Fig. 14.** Paraquat manufacture.

Manufacture of the herbicide Paraquat uses sodium in ammonia for the selective reductive coupling of pyridine to 4,4′-tetrahydrobipyridyl, from which Paraquat is obtained by oxidation (Fig. 14).

The ester reduction shown in Fig. 15 is an intermediate stage in manufacture of the high volume antiulcer drug Cimetidine.

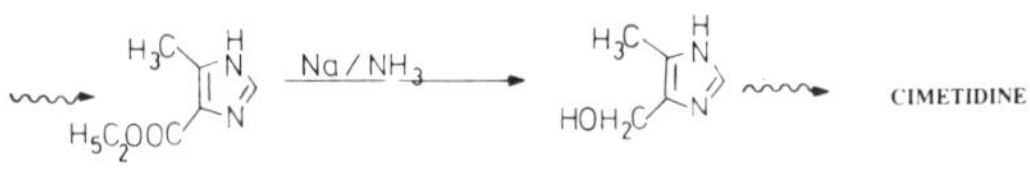


**Fig. 15.** Ester reduction.

A related system, using sodium in reaction with alcohol, is known as the Bouveault–Blanc reduction. This method is effective for reduction of esters and ketones but has been largely replaced in commercial operations by catalytic hydrogenation and metal hydride reductions. Some small volume applications remain, however, in the manufacture of pharmaceuticals, for example Diethylstilbestrol shown in Fig. 16.

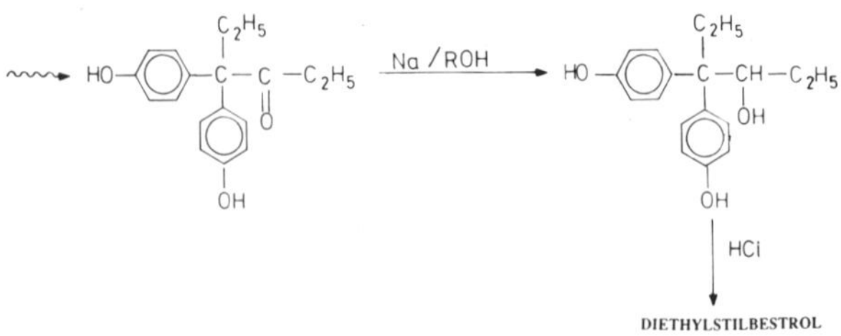


**Fig. 16.** Bouveault–Blanc reduction.

# Production of sodium derivatives

Many simple salts and derivatives of sodium (e.g. halides) are produced by metathesis from caustic soda, this being a cheaper source of sodium ions than the metal. The powerful reducing activity of the metal is, however, required to generate a range of derivatives containing strongly basic and nucleophilic anions, many of which have important commercial applications.

## Sodium Hydride

The hydride is manufactured, normally as a dispersion in oil, by direct reaction of dispersed sodium with hydrogen gas. Several thousand tons per year of hydride are produced as intermediate in borohydride manufacture, while a much smaller volume is recovered and marketed for other applications.

Sodium hydride has little commercial use as an organic reducing agent but is valued as a very powerful base. The base strength is comparable with that of sodium amide and the two compete for application in many organic condensations. While the amide is somewhat cheaper, hydride reductions have the advantage of producing hydrogen gas as by-product, often providing easier product recovery.

Hydride can also offer improved selectivity in certain condensation reactions. In the Michael cyclisation of Fig. 17, only the trans isomer is formed using sodium hydride catalyst.[5] The equivalent process using alkoxide base gives a trans–cis ratio of 3 : 1.

A process with considerable potential is shown in Fig. 18, where hydride is effective in abstracting a proton from the sulphoxonium salt (I). The ylide intermediate (II) formed will add methylene to alkenes and ketones, producing cyclopropanes and epoxides.[6]

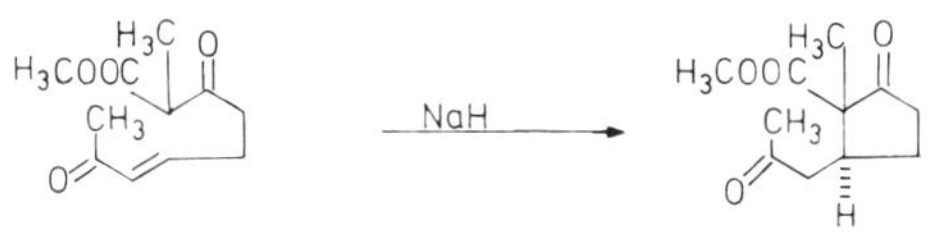


**Fig. 17.** Michael cyclisation.

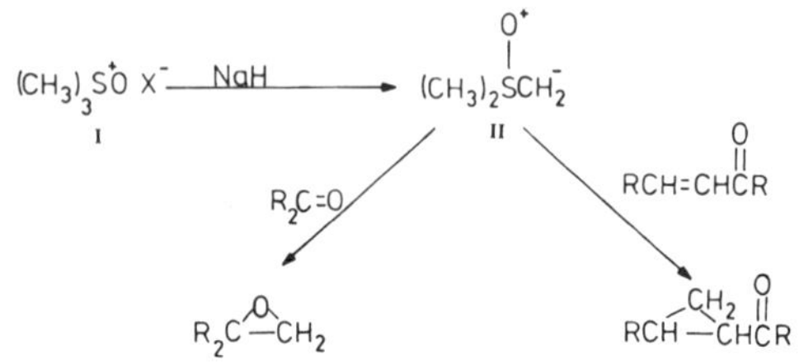


**Fig. 18.** Sulphoxonium salt reactions.

## Sodium Borohydride

The borohydride is manufactured by reaction of hydride-oil dispersions with trimethyl borate. Being a much milder reducing agent than sodium metal, it reacts only very slowly with water and other protic media. As a result, borohydride reductions can proceed successfully in aqueous, alcoholic and amide solutions where lithium aluminium hydride, for example, cannot be employed. Several thousand tons per year of sodium are consumed in producing an aqueous caustic solution of borohydride used in the bleaching of paper pulp.

Much smaller volumes of solid borohydride are consumed as a selective reducing agent for fine organic chemicals.[7] Aldehydes and ketones are reduced efficiently to alcohols, while acids, esters, anhydrides and unsaturated linkages are unaffected. The example shown in Fig. 19 illustrates this selective reduction in one stage of producing the steroidal drug 1-$\alpha$-hydroxyvitamin $D_3$.

The large scale production of borohydride for paper bleaching now affords the fine chemicals reagent at a lower cost per kilogram of available hydrogen than other complex hydrides. Where effective, it has generally become the reagent of choice.

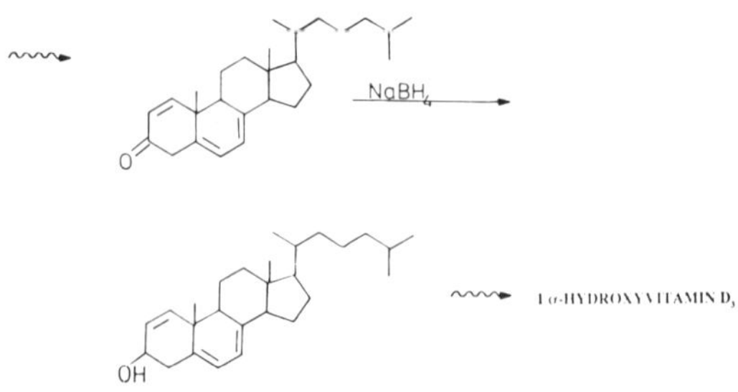


**Fig. 19.** Sodium borohydride reduction.

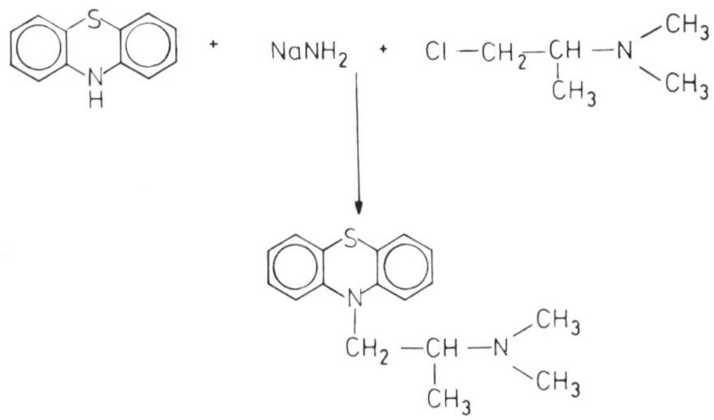

PROMETHAZINE

**Fig. 20.** Sodium amide base catalysis.

## Sodamide

Sodium amide or sodamide is manufactured on a large scale by reaction of ammonia gas with molten sodium at around 350°C.

The amide is a very strong base, comparable with hydride and more powerful than hydroxide and alkoxide. The production of Indigo dye is a major application for the amide as a base catalyst for condensation.

Amide is widely employed as catalyst for C—N condensations, acting by abstraction of hydrogen from functional amide groups. The example in Fig. 20 is the manufacture of Promethazine, one of the group of phenothiazine tranquillizers.

The amide anion is a powerful nucleophile and can introduce the amino group into electron deficient aromatic rings. The well-known Chichibabin reaction of pyridines is illustrated in Fig. 21 and provides intermediates for antihistamine and antibacterial drugs.

## Sodium Alkoxides

Alkoxides are manufactured by controlled reaction of bulk metal with alcohols, and have important applications as bases and as nucleophiles for production of speciality ethers and esters.

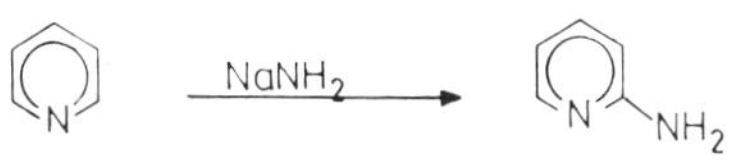

**Fig. 21.** Chichibabin reaction.

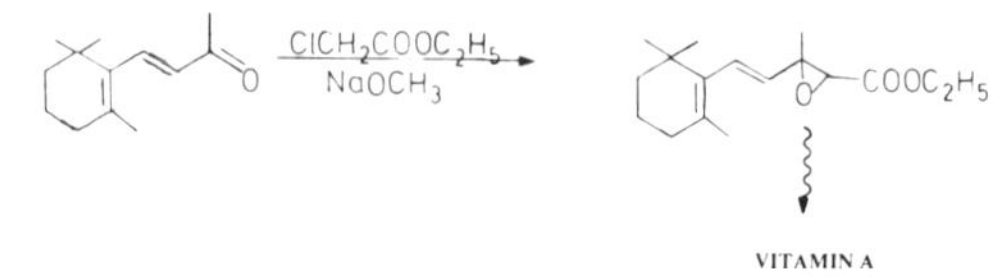


**Fig. 22.** Sodium methoxide base catalysis.

Methoxide is by far the largest commercial product, consuming many thousands of tons of sodium per year. Although not as strong a base as hydride or amide, the low cost assures its use where effective. A major application, shown in Fig. 22, is as base catalyst in production of vitamin A.

As a nucleophilic reagent, the major use of methoxide is in manufacture of trimethyl orthoformate (Fig. 23), an important intermediate for pharmaceutical products.

$$CHCl_3 + 3NaOCH_3 \longrightarrow \underset{\text{TRIMETHYL ORTHOFORMATE}}{CH(OCH_3)_3} + 3\,NaCl$$

**Fig. 23.** Orthoformate manufacture.

Higher alkoxides are manufactured in the same way but have achieved relatively little industrial importance. The higher *n*-alkoxides are quite soluble in parent alcohol and provide basic media with a wide operating temperature range. As nucleophiles for production of ethers and esters, small applications exist in pharmaceuticals production. The alkoxide of dimethyl amino ethanol, for example (Fig. 24), is an intermediate in the manufacture of the antiviral drug Tromantadine.

Somewhat surprisingly, the market for higher alkoxides of sodium, tailor-made to suit specific applications, has received little attention from potential producers.

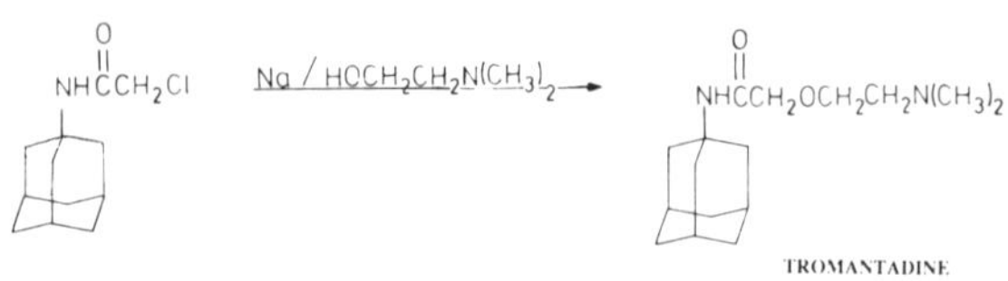


**Fig. 24.** Alkoxide of DMAE.

### Sodium Carbanions

Certain carbanion salts of sodium have achieved commercial importance, but low stabilities generally require that they be made and reacted *in situ*.

Sodium acetylide, produced from acetylene and strong amide base, is generated for some pharmaceutical and terpene products. A range of complex carbanion catalysts (Alfin catalysts) are still used in certain cyclisations and polymerisations of olefins.

## Summary of prospects

The declining use of motor fuel antiknock, which has dominated sodium demand for half a century, is now presenting major sodium producers with the challenge of seeking new markets for the metal. Only modest progress has been made to date but the longer term outlook appears much brighter.

Sodium is by far the cheapest of the alkali metals, is an excellent reducing agent for organics and inorganics and has a valuable mix of physical characteristics. Current applications outside of fuel antiknock are showing steady growth and some new uses under development have excellent prospects.

The sodium battery offers the promise of large demand for clean transport and storage systems and the successful development of sodium coolants for nuclear reactors clears the way to many new heat transfer applications. In the important role of sodium as chemical reductant, new areas of use will continue to emerge, particularly for the speciality derivatives where existing prices can be expected to ease with increasing demand. The waste sodium product from these reductions is sodium ion (generally as sodium chloride), which poses no threat to the environment.

The changing use pattern, from antiknock to more speciality areas, will continue to encourage producers to provide higher purity grades of metal, packaged for safer handling in specific applications and higher levels of technical service to customers.

## References

1. Authors' compilation of data from various sources.
2. Authors' compilation of data from projections from various sources.

3. *Automotive Engineering,* **94** (1986) 46.
4. Nobis, J. F. & Moormeier, L. F., US Patent 2 881 209, 1954.
5. Stork, G., Winkler, J. D. & Saccomano, N. A., *Tetrahedron Letters,* **24** (1983) 465.
6. Corey, E. J. & Chaykovsky, M., *J. Am. Chem. Soc.,* **87** (1965) 1353.
7. For a concise review of hydride reducing agents, see: Stinson, S. C., *Chem. and Eng. News,* 3 November 1980.

# Organolithium compounds and their derivatives in organic synthesis

**Terry L. Rathman**

*FMC Corporation, Gastonia, North Carolina, USA*

This chapter provides a general overview of the properties of commercially available organolithium compounds and their derivatives.† The properties that are described are their structure, reactivity, stability, advantages, and chemical utility. Table 1 contains useful product information for the commercial organometallics available from FMC's Lithium Division.

## General properties

### Structure

Knowledge of the structure of organometallics provides insights into the stability, reactivity, and kinetics of these compounds. Once understood, this information is of considerable value in optimizing their utilization and devising new uses for these products.

† Taken in part from FMC Corporation's borchure 'Organometallics in Organic Synthesis' produced by the Lithium Division, 

**Table 1.** Commercial product information—FMC organometallics

| Commercial product | | *n*-Butyllithium | | | *sec*-Butyllithium | *t*-Butyllithium |
|---|---|---|---|---|---|---|
| Formula (abbreviation) | | $CH_3CH_2CH_2CH_2Li$ (*n*-BuLi) | | | $CH_3CH_2CH(Li)CH_3$ (*sec*-BuLi) | $(CH_3)_3CLi$ (*t*-BuLi) |
| Molecular weight | | 64·06 | | | 64·06 | 64·06 |
| CAS registry number | | 109-72-B | | | 598-30-1 | 594-19-4 |
| Typical analysis | | | | | | |
| Weight % product/other solutes | Butyllithium | 15·0 (1·6M) | 24·0 (2·6M) | 86·0 (10·2M) | 12·0 (1·5M) | 18·0 (1·8M) |
| | Residual Base | 0·3 | 0·3 | 2·0 | 0·3 | 0·3 |
| | % Carbon bound Li | 98 | 98 | 98 | 95–98 | 98 |
| Weight % solvent (alternative solvents) | | Hexane 85 (Heptane) | Hexane 76 (Heptane) | Hexane 14 | Cyclohexane 84 Heptane 4 | Pentane 82 |
| Density at 20°C (alternative solvents) | g/ml | 0·68 (0·72) | 0·69 (0·73) | 0·76 | 0·78 | 0·65 |
| | lb/gal | 5·67 (6·00) | 5·77 (6·11) | 6·34 | 6·50 | 5·42 |
| Stability avg % loss/day | at 20°C | 0·0014 | 0·0015 | 0·012 | 0·05 | 0·0005 |
| | at 35°C | 0·01 | 0·011 | 0·10 | 0·32 | — |
| Pyrophoric | | Yes | Yes | Yes | Yes | Yes |
| Further description | | Least pyrophoric | if the butyllithiums are in the same solvent and concentration → | | | Most pyrophoric |
| Appearance | | All are clear, colorless to light-yellow solutions (86 wt% *n*-BuLi may be cloudy) | | | | |

**Table 1.** *(Continued)*

| Commercial product | | Methyllithium-9307 | Lithium diisopropylamide 9505 | Lithium hexamethyldisilazide |
|---|---|---|---|---|
| Formula (abbreviation) | | $CH_3Li$ (MeLi); $(CH_3)_2Mg$ (DMM) | $LiN(CH(CH_3)_2)_2$ (LDA)<br>$Mg(N(CH(CH_3)_2)_2)_2$ (MDA) | $LiN(Si(CH_3)_3)_2$ (LHS) |
| Molecular weight | | MeLi 21·97; DMM 54·38 | LDA 107·13; MDA 224·69 | 167·33 |
| CAS registry number | | 917-54-4; 2999-74-8 | 4111-54-0; 23293-23-4 | 4039-32-1 |
| Typical analysis | | | | |
| Weight % product/ other solutes | Product | MeLi 2·6 (1M) | LDA 26·5 (2·0M) | LHS 25·3 (1·3M); 21·0 (1M) |
| | Other solutes | DMM 0·5 (0·08M) | MDA 2·75 (0·1M) | |
| | Other | LiCl <0·3 | | |
| Weight % solvent (alternative solvents) | | THF 14<br>Cumene 83 | THF 26·4<br>Heptane 28·5<br>Ethylbenzene 13·1<br>Diisopropylamine 2·75 | THF 66·0; 18<br>Ethylbenzene 8·7; 7<br>Cyclohexane —; 51<br>Heptane —; 3 |
| Density at 20°C (alternative solvents) | g/ml | 0·86 | 0·79 | 0·88; 0·83 |
| | lb/gal | 7·17 | 6·59 | 7·34; 6·92 |
| Stability avg % loss/day | at 0°C | No decomposition | 0·01 (calculated) | No decomposition |
| | at 15°C | 0·008 | 0·05 | No decomposition |
| | at 40°C | 0·09 | 0·45 | Not measured |
| Pyrophoric | | No | No | No |
| Further description | | | Corrosive flammable liquid | |
| Appearance | | Clear, yellow solution | Light-amber solution | Clear, golden-yellow solution |

**Table 1.** *(Continued)*

| Commercial product | Dibutylmagnesium | Phenyllithium[a] | t-Butyldimethylchlorosilane | | |
|---|---|---|---|---|---|
| Formula (abbreviation) | $\begin{pmatrix}(CH_3CH_2CH_2CH_2)_{0\cdot9}\\(CH_3CH_2(CH_3)CH)_{0\cdot9}\\(CH_3(CH_2)_6CH_2)_{0\cdot2}\end{pmatrix}$ Mg (DBM) | $C_6H_5Li$ (PhLi) | $(CH_3)_3CSi(CH_3)_2Cl$ (TBDMCS) | | |
| Molecular weight | 145 (average) | 84·05 | 150·73 | | |
| CAS registry number | 3988-32-8 | 591-51-5 | 18162-48-6 | | |
| Typical analysis | | | | | |
| Weight % product/ other solutes | DBM 20 (1M) | PhLi 20 (2M) | | TBDMCS | 99 |
| | | LiCl 2 (0·05M) | Product impurities | Hydrocarbons | <0·9 |
| | | Biphenyl <0·1 | | Silanol | <0·1 |
| Weight % solvent (alternative solvents) | Heptane 80 | Cyclohexane 55<br>Diethyl ether 23 | | | |
| Density at 20°C (alternative solvents) | 0·72 | 0·83 | Physical properties | m.p. | 86–89°C |
| | 6.0 | 6·92 | | b.p. | 124–126°C |
| Stability avg % loss/day | at 20°C Stable indefinitely | | Stable indefinitely | | |
| Pyrophoric | Yes | Yes | No | | |
| Further description | Less pyrophoric liquid than n-BuLi | | Corrosive, flammable solid (volatile solid) | | |
| Appearance | Clear, colorless to light-yellow solution | Clear, dark-red to brown solution | Clear, crystalline solid | | |

[a] Now available in di-n-butyl ether

Organolithium compounds, such as *n*-butyllithium, are described as polar organometallics in which the lithium atom is bonded directly to carbon. Although the structures of alkyl- and aryllithiums are often represented by RLi, there is considerable evidence to suggest that organolithiums exist as aggregates in the gas, liquid, and solid phases, as well as in solution. The association of the covalently bonded organolithiums is a result of their electron deficient character. Therefore, alkyllithium compounds in the pure state exist as liquids or low-melting-point solids which can be distilled or sublimed, respectively.

*n*-Butyllithium exists mainly as a hexamer, $(n\text{-BuLi})_6$, in both the pure state and hydrocarbon solvents. The degree of aggregation or association of RLi's decreases as the steric bulk about the $\alpha$- and $\beta$-carbons increases. Thus, *t*-butyllithium is a tetramer.

Because of the electron deficient nature of organolithiums, they rapidly undergo coordination with Lewis base solvents. As the coordinating ability of solvent increases, the aggregation number decreases. For example, *n*-butyllithium has been found to exist as a tetramer in tetrahydrofuran (THF) or diethyl ether. However, it can be a monomer or dimer in the presence of strong donors such as tetramethylethylenediamine (TMEDA) or 1,4-diazobicyclo[2.2.2]-octane (DABCO). Contrarily, methyllithium (MeLi) is a tetramer as a pure solid and even in solution in the presence of TMEDA or THF. Temperature and concentration of the organolithium can also cause the aggregation number to vary.[1]

Lithium diisopropylamide (LDA) in the pure form is a solid and is insoluble in hydrocarbon. For solubility, LDA requires the presence of a Lewis base—most commonly THF. At very low temperatures and using NMR, LDA exists as a dimer, 2LDA·2THF in solution.[2] This structure for LDA is in agreement with the results of an X-ray crystal study.[3] Monomeric[4] or ladder-type[5] structures have been reported for solvated lithium organoamides.

The structures of FMC's lithium diisopropylamide, LDA-9505,[6] and methyllithium, Meli-9307,[7] have not been extensively studied, but preliminary high-field NMR information indicates that the additives magnesium diisopropylamide (MDA) and dimethylmagnesium (DMM) are part of the respective aggregates. Interestingly a quick estimation of the 'methyl' concentration of MeLi-9307 can be determined by $^1$H NMR, even at 60 MHz, because all the components of this solution are well separated. (See 'Reactivity', 'Stability and

Storage' and Table 1 for more details concerning LDA-9505 and MeLi-9307.)

Dialkylmagnesium compounds, like dibutylmagnesium (DBM)[8] (see Table 1), are also characterized by bridge-type bonds similar to those described for organolithiums. The cooperative association between carbon and magnesium is often referred to as oligomerization (Fig. 1).

The length of the chain (or value of $n$) is dependent on the mole percent of *normal*-alkyls, i.e. $n$-butyl or $n$-octyl, and the *secondary*-alkyl, i.e. *sec*-butyl. The *normal*-alkyl groups can participate in bridging or be the terminating group of the oligomer. Steric hindrance prevents the *secondary*-alkyl group from participating in bridging and it therefore can only terminate the oligomer. Hydrocarbon solubility and viscosity is controlled by keeping the oligomer short (where $n = 1, 2$) and the presence of a smaller percent of longer chain $n$-alkyl groups ($C_6$ or longer).

More information concerning the structure of organolithiums is amply described and referenced in several sources.[9–12]

## Reactivity

As a general rule, the reactivity of an organolithium increases as the degree of aggregation decreases. The reactivity or basicity varies markedly in the presence of a Lewis base or diamine-type modifier.

In summary, factors that can decrease aggregation and increase reactivity are:

—bulkier organic groups, e.g. *s*-BuLi is more reactive than *n*-BuLi;
—presence of donor solvents, e.g. THF, diethyl ether;

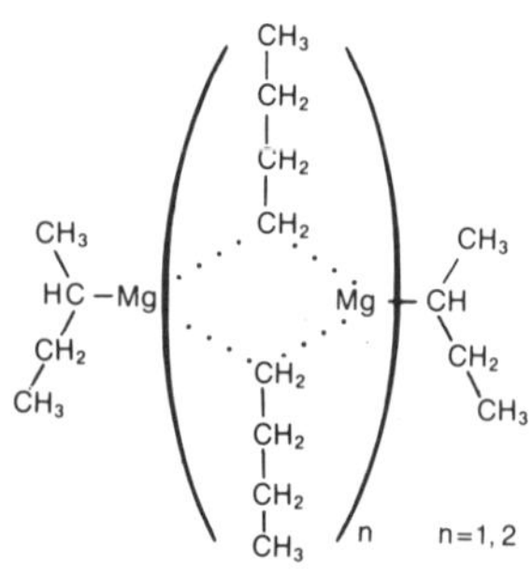


**Fig. 1.** Oligomeric dibutylmagnesium (DBM).

—presence of chelating agents, e.g. TMEDA, DABCO;
—temperature;
—concentration.

Basicity can also be affected by the presence of other modifiers. For example, the addition of potassium *t*-butoxide (*t*-BuOK) produces a more reactive base via formation of butylpotassium. This combination must be prepared at −80°C if THF is present.[13] After the metalation step, the resulting potassium intermediate can be converted to the more selective lithium intermediate by addition of an anhydrous solution of lithium bromide in THF.[14]

*n*-BuLi/TMEDA in hexane (no THF) is less basic and more hydrocarbon soluble than *n*-BuLi/*t*-BuOK, but can be employed at higher reaction temperatures. The *n*-BuLi/TMEDA base system will metalate TMEDA if the latter is in excess.[15] For lithiation of aromatic[16–18] and heteroaromatic[19] substrates that are not susceptible to addition, the metalation temperature with *n*-BuLi/TMEDA/hexane can be elevated to reflux after initially starting at room temperature.

Greater basicity can be achieved with a combination of *n*-BuLi/TMEDA/*t*-BuOK. This mixture is hydrocarbon soluble and normally employed between −30 and −20°C. Brandsma and Verkruijsse published numerous self-tested metalations using *n*-BuLi and various Lewis bases such as THF, diethyl ether, TMEDA, and *t*-BuOK.[20] (See the section on 'Chemical utility—Multiple carbon–carbon bonds' for examples using butyllithium and these modifiers to form vinyl- and allyllithiums.) For LDA, the same modifiers will increase its basicity.[21]

Lithium salts, such as lithium halides, can also be incorporated in the organolithium aggregate in the presence of a Lewis base. The solubility of lithium halides is enhanced by the presence of organolithiums. An increasing number of examples have been reported where lithium[22] or other metal salts (such as $MgBr_2$[23,24] $(i\text{-PrO})_4Ti$[25]) have been shown to significantly affect reactivity. However, these reactivity modifiers are not usually added to alter basicity but rather to improve the stereoselectivity in the subsequent bond formation reactions. These types of reactions are further described in 'Chemical Utility'.

Because LDA-9505 contains 5 mol % of magnesium diisopropylamide, the reactivity effect of the latter has been investigated. Note that

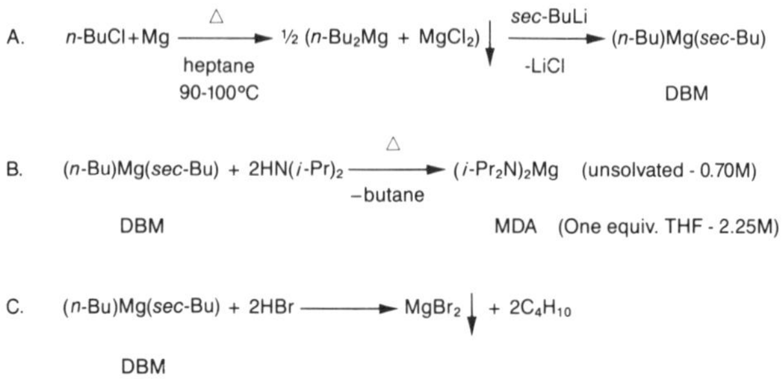


**Fig. 2.** Preparation: (A) and utility (B & C) of DBM.

magnesium diisopropylamide (MDA)[26] is prepared from dibutylmagnesium (DBM) (Fig. 2(B)). The preparation of DBM (Fig. 2(A)) and preparation of active $MgBr_2$ (Fig. 2(C)) are also shown in Fig. 2.

To determine the influence of MDA on the chemical reactivity of LDA, the internal quench technique (Fig. 3) was used as a model system because it produces three different silyl enol ethers.[27] The variation in the amounts of the silyl enol ethers produced should indicate any MDA effect.

As can be seen in entries 1–6 of Table 2, at identical temperatures LDA (made *in situ* from *n*-BuLi) and LDA-9505 gave essentially the same distribution of kinetic and thermodynamic silyl enol ethers. Thus, in this series of reactions, the presence of MDA had essentially no effect on the outcome of the silylation. Also as the mole percent of MDA was increased, the percent of conversion dropped off markedly (entries 5 and 7–9).

Obviously, MDA exhibited poor basicity for this substrate (entries

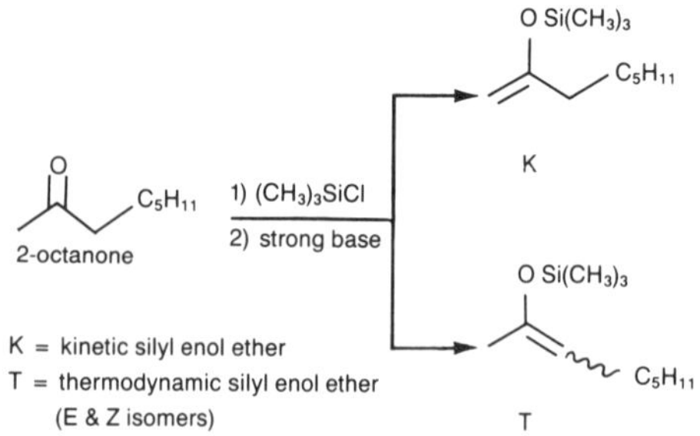


**Fig. 3.** Internal quench of enolates to form silylenol ethers.

**Table 2.** Conversion of 2-octanone to kinetic and thermodynamic silyl enol ethers[27]

| Entry | Base[a] | Temperature (°C) | % conversion[b] | % Silyl enol ethers[c] | | |
|---|---|---|---|---|---|---|
| | | | | Kinetic | Thermodynamic | |
| | | | | | Z | E |
| 1 | LDA | −65 | 96 | 92·0 | 2·4 | 5·6 |
| 2 | LDA | −25 | 99 | 91·7 | 2·1 | 6·2 |
| 3 | LDA | 0 | 99 | 91·8 | 2·4 | 5·8 |
| 4 | LDA-9505 | −65 | 97 | 92·4 | 2·2 | 5·4 |
| 5 | LDA-9505 | −25 | 100 | 92·1 | 1·8 | 6·1 |
| 6 | LDA-9505 | 0 | 99 | 91·4 | 2·4 | 6·2 |
| 7 | LDA-8020 | −25 | 86 | 88·2 | 3·6 | 8·2 |
| 8 | LDA-5050 | −25 | 67 | 77·5 | 5·5 | 17·0 |
| 9 | MDA | −25 | 3·5 | 56·0 | 12·0 | 32·0 |
| 10 | MDA | 0 | 4·2 | 49·0 | 15·0 | 36·0 |
| 11 | MDA | 20 | 6·5 | 30·0 | 40·0 | 30·0 |

[a] LDa is a generic product made from *n*-BuLi and diisopropylamine. LDA-9505 is a proprietary FMC product with the following composition: 95 mol % LDA, 5 mol % MDA (magnesium diisopropylamide).
[b] Reaction time and ratios of 2-octanone/amide/TMCS/THF were held constant.[27]
[c] Determined by GC; TMCS = trimethylchlorosilane.

9–11) and type of reaction. However, MDA does appear to produce more thermodynamic product under kinetic conditions. The magnesium cation of $i$-$Pr_2NMgBr$ has been shown to yield thermodynamic product under kinetic conditions.[28]

For use on a larger scale, higher reaction temperatures and less trimethylchlorosilane (TMCS) were explored. Surprisingly, it was found that:

(1) the silylation may be run at higher temperatures, such as −25°C or even 0°C;
(2) the amount of excess TMCS could be decreased from 5–10 moles to 0.2 mole;
(3) reverse addition (2-octanone to LDA/TMCS) could also be used even at higher temperatures and with less TMCS.

In the silylation scheme described above, it was shown that MDA did not appear to participate in the metalation of 2-octanone. However, MDA does possess sufficient basicity to effect metalation (magnesiation). For instance, in a separate customer evaluation, metalation of acetophenone with MDA, followed by alkylation, gave essentially the same overall yield as LDA. In fact, lithium hexamethyldisilazide (LHS) was the superior base in the customer's evaluation. Also, Eaton and coworkers have shown that both MDA and magnesium bis(2,2,6,6-tetramethylpiperidide), prepared from DBM, regiospecificially magnesiated certain cubane and methylbenzoate derivatives.[29]

Thus, when using LDA-9505 it may be necessary to determine if the contained MDA will metalate your specific substrate.

## Stability and Storage

For specific product information see Table 1.

### *Butyllithium*

Thermal decomposition of butyllithium isomers yields the corresponding olefins and lithium hydride via $\beta$-elimination. In a closed system, there is little increase in pressure because the *in situ* formed olefin is very soluble in the hydrocarbon solvents.

Under identical conditions, *t*-butyllithium is the most stable isomer in hydrocarbons followed by *n*-butyllithium and then by the considerably less stable *sec*-butyllithium, which can be stored at 5°C for three months with less than 1% decomposition.

In a production environment, actual decomposition of butyllithium is a function of the alkoxide concentration ($LiOC_4H_9$), the hydrocarbon medium, and the type and concentration of the butyllithium. The interrelationship between these variables is not clearly understood and comparisons should be made under actual operating conditions.

### *McLi-9703 and LDA-9505*

The commercial methyllithium (MeLi-9307) and lithium diisopropylamide (LDA-9505) contain 7 and 5 mole percent of the respective magnesium analog. These magnesium percentages are the bases for the corresponding 9307 and 9505 nomenclatures. The presence of magnesium in the respective forms of dimethylmagnesium (DMM) and magnesium diisopropylamide (MDA) provides greater thermal stability and a greater solubility over a broader range of temperatures. If long-term storage at 0°C is anticipated for MeLi-9307, provisions for redissolving a MeLi·THF precipitate should be made. Unlike MeLi-9307, LDA-9505 does not precipitate at the recommended lower storage temperature (Table 3).

If recommended storage temperatures are not maintained, the organometallic will become contaminated with decomposition products. Knowledge of the impurities assists the user in deciding if the product can still be used.

The decomposition of MeLi-9307 results from the metalation of cumene at its para/meta (1:2) positions. These metalated species can

**Table 3.** Storage recommendations for LDA-9505 and MeLi-9307

| Product | Storage duration (months) | Recommended storage temperature (°C) |
|---|---|---|
| LDA-9505[a] | 0–1 | 20 |
| | 2–3 | 10 |
| | 4–6 | 0 |
| MeLi-9307[b] | 0–1 | 25 or ambient[c] |
| | 2 | 10 |
| | 3–6 | 0[d] |

[a] If storage recommendations are not followed, solids containing amides and hydrides of Li and Mg form. The solution will darken and the solids could necessitate filtrating before use. Except at temperatures above 25°C, the rate of solids formation is gradual and not catalytic with the LDA losing less than 0·1% per day in assay. *Caution*: when dry and finely divided, these solids are pyrophoric.

[b] MeLi-9307 does not degrade by cleavage of THF, but rather by metalation of the aromatic solvent, forming aryllithiums and methane gas. The recommended storage conditions will significantly limit aryllithium formation. The average decomposition rates are 0·006 and 0·09 wt % per day at 10 and 40°C, respectively. No decomposition was observed at 0°C after nine months.

[c] Slight pressure may be observed.

[d] At 0°C, approximately 20% of the MeLi-9307 precipitates as large clear crystals of a MeLi–THF complex. These crystals are non-pyrophoric and redissolve easily on warming and with agitation. Provisions should be made for the easy redissolution of these crystals at 15°C.

be trapped with TMCS and quantitatively determined by GC. The major route of decomposition of LDA-9505 is $\beta$-elimination to produce lithium hydride and isopropylimine of acetone. Decomposition can also result from metalation of the THF solvent, which further degrades to ethylene and the lithium enolate of acetaldehyde.[30]

In both MeLi-9307 and LDA-9505, the amounts of magnesium and THF have been carefully optimized. The mole percent of magnesium provides the maximum stabilizing effect using the minimum amount of magnesium. This optimization assures that any reaction by-products and workup will be much like those experienced when using lithium reagents alone. Because excess LDA or MeLi will rapidly cleave THF,[31] the ratio of THF to organometallic was adjusted to sufficiently solvate the organometallic while achieving the best solubility and thermal stability.

### *t-Butyldimethylchlorosilane (TBDMCS)*

TBDMCS is a flammable, corrosive, crystalline solid that stores well in a sealed container. After long storage, however, the product may cake, especially if the storage temperature fluctuates significantly. Should this occur, the container can be gently tapped to break up the crystals sufficiently for pouring. The small difference between the melting point (86–89°C) and the boiling point (124–126°C) of TBDMCS makes the solid very volatile. Uncovered TBDMCS will evaporate rapidly in a well-ventilated area.

TBDMCS applications, in organic synthesis are described separately in 'Chemical utility'.

# Product advantages of selected products

## Lithium Diisopropylamide 9505 (LDA-9505)

Several US pharmaceutical companies have chosen to use LDA-9505 on a commercial scale rather than prepare generic LDA from pyrophoric *n*-butyllithium. They have found LDA-9505 offers these process advantages:

* nonpyrophoricity—a very important safety aspect

* convenient and ready-to-use—thus requiring less capital and lowering labor costs
* greater throughput—2.0M concentration is higher than would be created *in situ* from standard 15 wt % *n*-butyllithium
* less hydrocarbon for better solubility of lithiated substrates
* no butane by-product as with LDA prepared from *n*-butyllithium

## Methyllithium-9307 (MeLi-9307)

MeLi-9307 offers major advantages over other methylating reagents on any scale of preparation:

* nonpyrophoricity—even the MeLi·THF crystals which will precipitate from solution at 0°C are nonpyrophoric
* no diethyl ether—a highly flammable solvent
* reduced solvent costs from use of limited THF
* minimum lithium halide—typically less than 2000 ppm LiCl
* greater reactivity and cleaner workups than methyl Grignards
* cosolvent cumene is safe to handle

## *t*-Butyldimethylchlorosilane (TBDMCS)

TBDMCS is an important protecting group. The major user of TBDMCS is the pharmaceutical industry, but references to uses in polymers and other materials are beginning to appear in the literature. When TBDMCS is used, the protected alcohol (or *t*-butyldimethylsilyl ether) has several advantages over other protected groups:

* 10 000 times more stable than trimethylsilyl ether in acidic medium
* no chiral center like that created by tetrahydropyranyl ethers
* stability in the presence of many organometallics such as RLi, RMgX, $LiCuR_2$, $Me_2AlCl$, $Me_3Al$, DIBAL, alkoxides, etc.
* stability against reduction via hydrogenation or from mild reducing agents
* stability against oxidation from reagents like $H_2O_2$, $CrO_3/C_5H_5N$, $NaIO_4/RuO_2$
* stability under Lewis acid conditions caused by the presence of $MgBr_2$, $TiCl_4$, *i*-$PrOTiCl_3$, or $SnCl_4$.

# Chemical utility

## Introduction

Many of the organic reactions cited here are carried out under conditions not yet commonplace to large-scale manufacturing. However, a few low-temperature technologies are being practised and improved. It is also possible that some reactions might run well at temperatures higher than those referenced. For instance, it has been shown that certain reactions involving LDA-9505 and cuprates can be run at higher temperatures than the −78°C previously cited. (See the sections on 'General properties—Reactivity', 'LDA-9505' above, and 'Developmental products and methodologies—Methyl thienylcuprate'.)

Considerable research involving organolithiums has been reported in both patents and chemical journals. There is so much research, in fact, that what follows are only the highlights of the many diverse applications for these organometallics in organic synthesis.

For more background concerning carbanions in organic synthesis, there are several excellent books available.[32–41] For those who are interested in the use of organolithiums in polymer synthesis, several reviews and books are also included in the reference section.[42–50]

## Single Carbon–Carbon Bond Formation

By far the greatest number of useful synthetic methodologies using organolithiums involve single carbon–carbon bond formation.

### *Addition of Organometallics*

These addition reactions are examples in which commercially available organolithiums add to a substrate and become part of the final product.

A Du Pont patent application for preparing rimantadine (Fig. 4(A)) discloses that methyllithium is preferred as a methylating agent over the Grignard analog; this is because methyllithium has greater reactivity toward the sterically hindered adamantyl nitrile.[51] Hydrolysis of the '*N*-lithiated imine' (Fig. 4(B)) should conveniently produce the

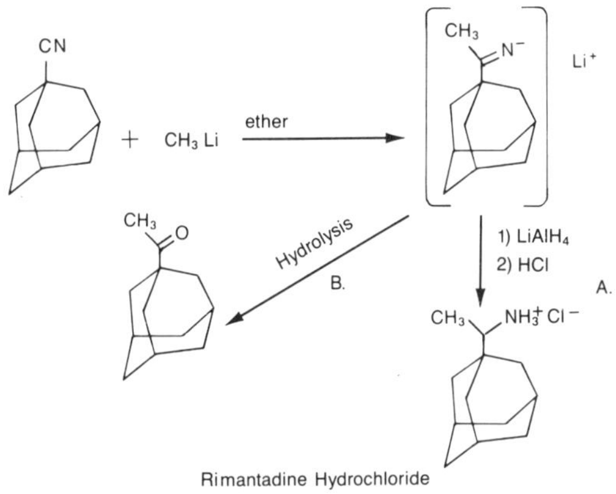


**Fig. 4.** Reduction: (A) and hydrolysis (B) of a methyllithium/nitrile adduct.

corresponding adamantyl methyl ketone, which is otherwise difficult to prepare via methylation of adamantyl carboxylic acid.[52]

On the other hand, cyclohexyl methyl ketone (Fig. 5) is produced in 90% yield by addition of methyllithium to the lithium cyclohexylcarboxylate.[53]

Differences in stereoselective 1,2-addition of methyllithium and methyl Grignards to carbonyls of various heterocycles[54] and of carbohydrates[55,56] have been reported. Addition of *n*-BuLi to carbonyl

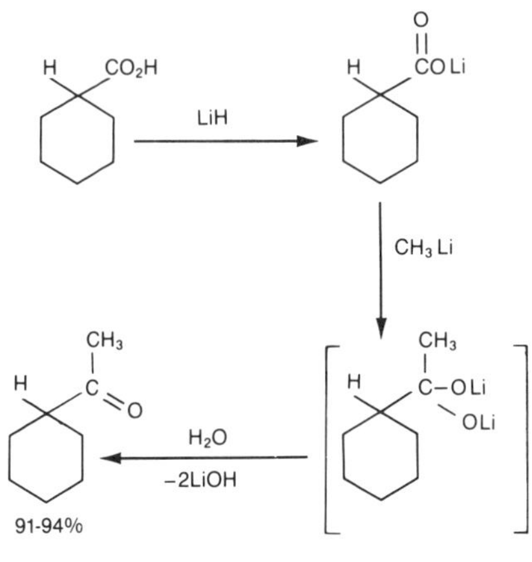


**Fig. 5.** Formation of a methyl ketone from a carboxylic salt.

compounds also gives fewer reduction products than the corresponding Grignards. Asymmetric addition of commercial organolithiums or *in situ* prepared lithium intermediates to various carbonyl compounds has received much attention.[57,58] Successful addition reactions will require an optimization of the solvent system, temperature, chiral auxiliary, and other metals. Other variables such as the order and rate of addition of reactants/reagents may also have to be investigated.

Alkylation of heteroaromatics like quinoline or pyridine via addition of butyllithiums to the C═N bond can be useful. Butylation of quinoline to the 1,2-dihydroquinoline with *n*-, *sec*-, or *t*-butyllithium and subsequent reduction to the tetrahydroquinoline is achieved in excellent overall yields (Fig. 6).[59]

Generally, organolithium compounds are less prone to give reduced products with ketones than the corresponding Grignard reagents. For substrates such as epoxides and $\alpha,\beta$-unsaturated carbonyl compounds, where regioselective addition to these functional groups is possible, organolithiums give $\beta$-substituted ethanols and substituted allylic alcohols (via 1,2-addition), respectively. Grignard reagents most often give $\beta$-substituted carbonyl compounds (via 1,4-addition).

For most substrates such as $\alpha,\beta$-unsaturated carbonyl compounds and epoxides, cuprates derived from organolithiums have greater reactivity than those made from Grignards.[60–63] However, cuprates derived from organolithiums have traditionally required lower reaction temperatures to achieve the desired addition products in good yield. (See the section on 'Developmental products and methodologies—Methyl thienylcuprate'.)

Although enolates are normally formed via deprotonation of $\alpha$-protons of carbonyl-like substrates using bases like LDA (Fig. 7),

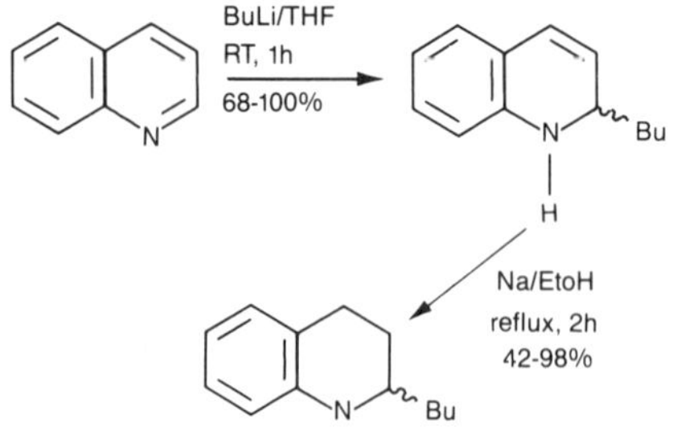


**Fig. 6.** Addition of butyllithium to quinoline.

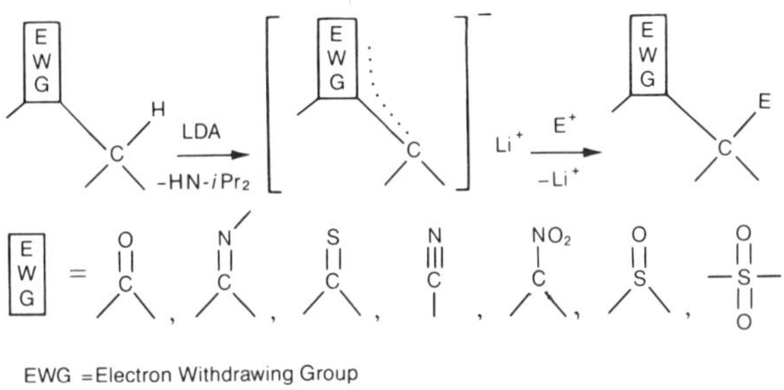


**Fig. 7.** Substrates which are suitable for metalation with LDA.

enolates can also result from:

1. conjugate addition (1,4-addition)[64]
2. addition of methyllithium to enol ethers (Fig. 8(A))[65]
3. cleavage of trimethylsilylenol ethers with MeLi (Fig. 8(B))[66]
4. addition of alkyllithiums to ketenes.[67]

Using these methods, one can generate amine-free enolates and note whether the subsequent reaction with an electrophile is affected by the absence of diisopropylamine. Diisopropylamine is present if the enolate is prepared from LDA and the appropriate activated substrate (Fig. 7).

## *Enolates*

As mentioned above, most enolizations are achieved with sterically hindered and less nucleophilic bases, such as lithium diisopropylamide (LDA) and lithium hexamethyldisilazide (LHS).

Many different substrates possessing the appropriate electron withdrawing group (EWG) and an $\alpha$-proton can be lithiated to provide access to quantitatively formed 'enolates'. The *in situ* formation of 'enolates' is generally achieved by addition of substrate to the

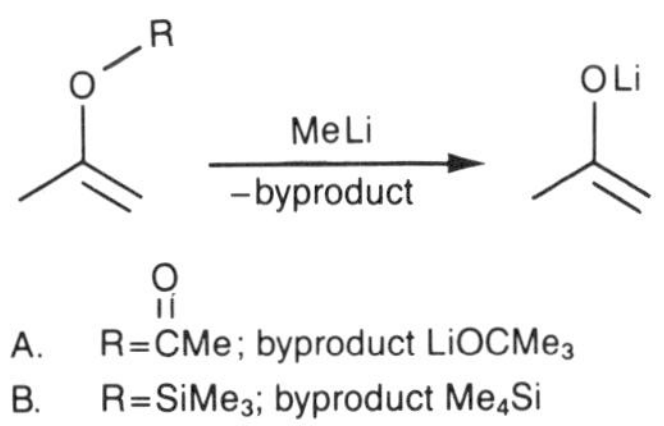


**Fig. 8.** Preparation of amine-free enolates.

hindered base. The resulting 'enolates' are then reacted with various classes of electrophiles (Fig. 7) to form $\alpha$-substituted derivatives. Most commonly, the electrophiles ($E^+$) which react at the carbanionic site of the 'enolates' are organohalides (alkylation) or carbonyl compounds (condensation).

In contrast to electrophilic organohalides and carbonyl compounds, organosilyl chlorides (silylation) normally react at the heteroatom of 'enolates'. Several examples of *O*-silylations can be found throughout the paper.[68] (See the sections on 'General properties,—LDA-9505', 'Heteroatom bond formation', 'Carbon–oxygen bond formation'; and '*t*-Butyldimethylchlorosilane'.)

The understanding of the structure and reactivity of enolates is continually increasing. The following specific examples describe enolates in alkylation and condensation reactions. The compilation of these reactions is not intended to be comprehensive but does generally describe what may be possible in industrial applications. British Petroleum and the research group of Davies have developed a novel chiral iron acetyl which uses *n*-butyllithium in direct routes to many optically active compounds.[69] Extensive reviews of classic work on stereoselective reactions involving enolates can be found elsewhere.[70]

### *Alkylation of Enolates*

Gemfibrozil (Fig. 9) is a good example of an alkylation being performed in industry.[71]

Alkylation of an enolate of a biologically active compound, having the proper protecting groups (Fig. 10), is an excellent approach for developing new pharmaceuticals.[72]

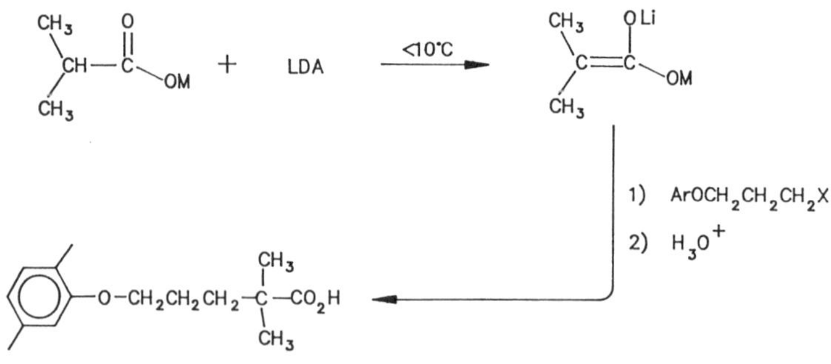


**Fig. 9.** $\alpha$-Alkylation of a carboxylic salt.

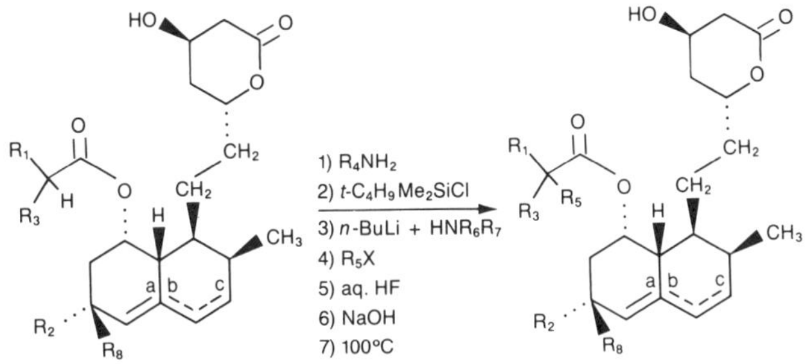


**Fig. 10.** $\alpha$-Alkylation of a natural product.

Improved diasteroselective alkylations result with cyclic and chiral substrates such as one prepared from L-threonine (Fig. 11)[73] and serine.[74]

In an excellent review on enolates, the regiospecifically *C*-methylation of the polyanion of the peptide cyclosporine A is reported using LDA or $n$-BuLi in the presence of lithium halide.[75]

$\alpha$-Alkylation of the cyclohexyldienyl nucleus resulted after lithiation at the $\gamma$-position (Fig. 12). Interestingly, sulfenylation of the enolate with diphenyldisulfide also occurred at the $\alpha$-position but laboratory light promoted migration of the sulfenyl group to form the $\gamma$-sulfenylated product.[76]

Although not possible in this example, problematic disulfenylation of a carbanion having an activated methyl or methylene group can also occur with this electrophile.[77]

It is well known that kinetic enolates can be quantitatively formed at low temperatures. Methylation of the cyclic ketone in Fig. 13 is achieved in greater than 90% yield using the Stork–Danheiser kinetic alkylation procedure.[78]

The enolate of a very similar substrate (Fig. 14) can be used to prepare bicyclo(2.2.2) octanes in an aprotic double Michael addition.[79]

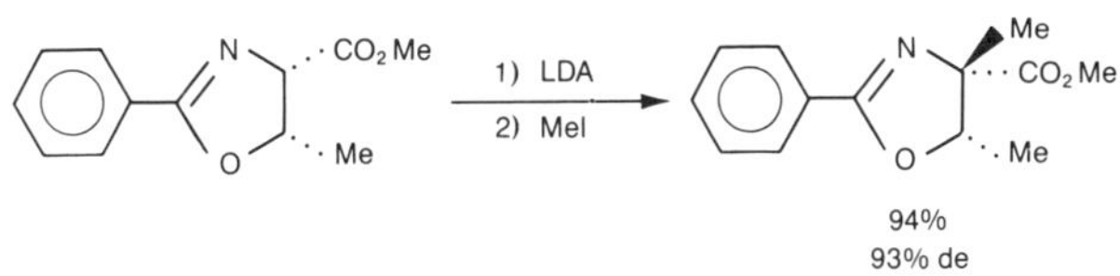


**Fig. 11.** Stereoselective $\alpha$-methylation.

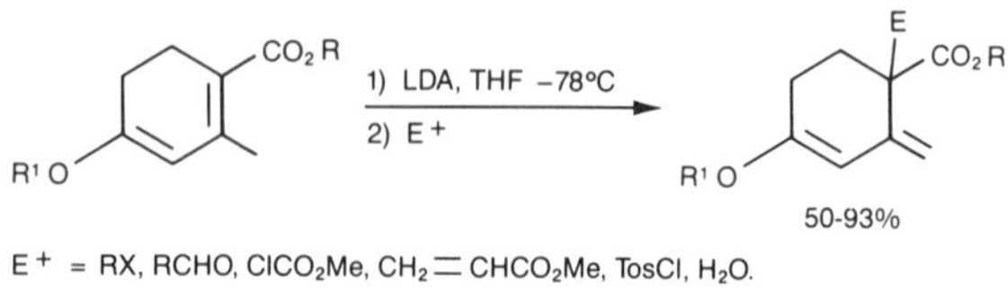


**Fig. 12.** Functionalization of a conjugated enolate.

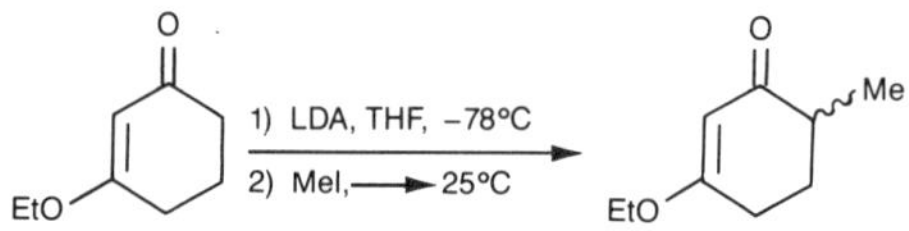


**Fig. 13.** Kinetic $\alpha$-methylation.

For certain unsymmetrical ketones, the kinetic enolate (least substituted) is often slightly different energetically from the thermodynamic enolate (more substituted) which may form if the reaction is allowed to warm up and LDA is not in excess. Alkylations are often slow at low temperatures; yet higher temperatures may cause loss of the kinetic enolate and thus regiocontrol. This problem may be circumvented by using the corresponding dimethylhydrazone derivative (Fig. 15) which favors regioselectivity at higher temperatures, first during lithiation and then during subsequent alkylation.[80]

Enders has pioneered the use of chiral hydrazones based on hydrazines of proline to achieve sterospecific alkylations.[81]

Alkylation of ketimines has also been extensively studied. For acyclic ketimines (Fig. 16), high regioselectivity can be achieved by proper choice of: (1) base, (2) the substituent on the nitrogen of the imine, or (3) the temperatures of deprotonation and alkylation.[82]

Cyclic ketimines such as *N*-cyclohexylimine of 2-methylcyclohexanone required *sec*-butyllithium for achieving *anti*-lithiation at the thermodynamic site (Fig. 17(A)).[83] The mechanism of the lithiation of the corresponding dimethylhydrazone with LDA at the kinetic site using LDA (Fig. 17(B)) was studied in detail using FT-IR.[84]

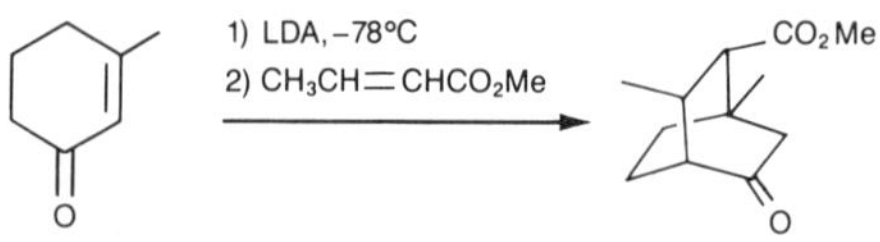


**Fig. 14.** Double 1,4-addition.

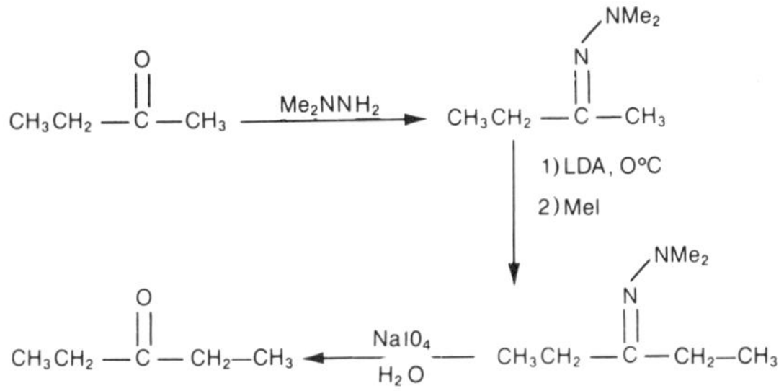


**Fig. 15.** Regioselective methylation of an unsymmetrical ketone.

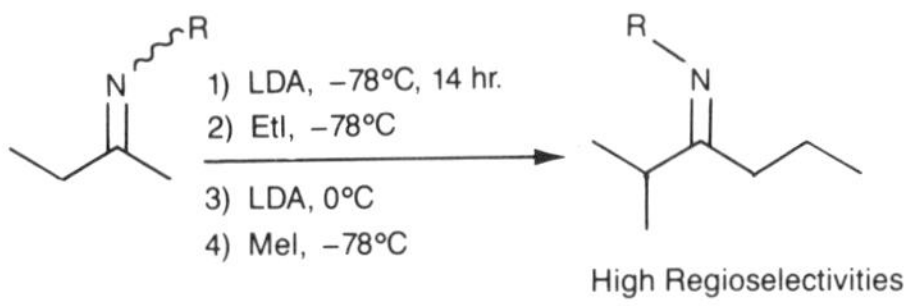


**Fig. 16.** Regioselective alkylation of an imine.

Differences in IR absorbances for the hydrazone, lithiated hydrazone, and mixed aggregate of LDA and lithiated hydrazone were observed.

## *Condensation of Enolates*

Important factors which contribute to the stereochemical outcome of aldol condensations involve the steric and electronic requirements of the base, substrate and electrophile; solvent, temperature, order of addition, rate of mixing, and the cation—whether it be metal or nonmetal. A good example of exploring the cation effect in an aldol condensation can be found in metalation and subsequent reactions of enolates of a chiral $\alpha$-halogenated imide.[85]

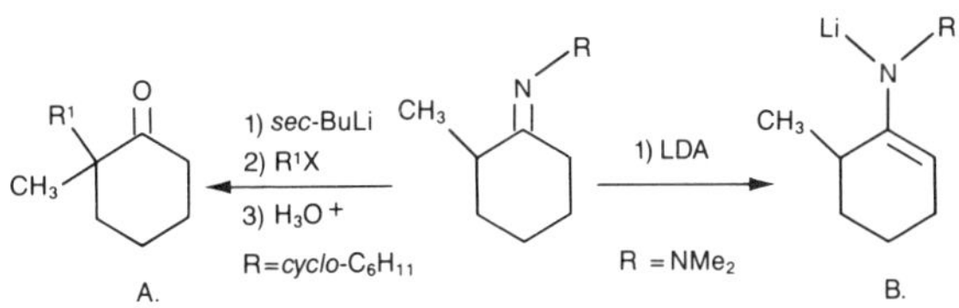


**Fig. 17.** Formation of a thermodynamic (A) or kinetic (B) enolate.

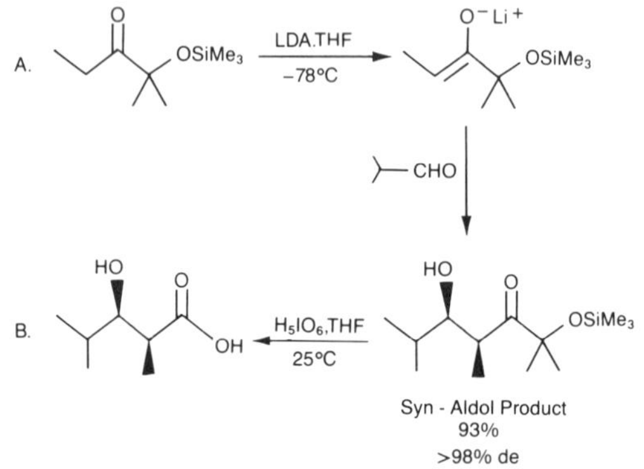


**Fig. 18.** Syn-diastereoselectivity via Z-enolates.

The following examples illustrate the important variables which determine stereochemical selectivity.

In Fig. 18(A), the thermodynamically favored *Z*-enolate yields the *syn*-aldol product with greater than 98% diasteroselectivity. This *syn*-aldol product can be converted to the *syn*-$\beta$-hydroxyacid (Fig. 18(B)) and other *syn*-$\beta$-hydroxy carbonyl compounds.[86]

In contrast to Fig. 18(A) and (B), intermediate enolate of 2,6-dimethylphenyl propionate, when reacted with the same aldehyde, yields the *anti*-$\beta$-hydroxyester (Fig. 19(A)). Saponification of the *anti*-aldol product leads to the *anti*-$\beta$-hydroxyacid (Fig. 19(B)).[87]

An enantiomeric excess of a $\beta$-hydroxyacid (Fig. 20) can similarly be prepared by using diasteroselective methodology in conjunction with a chiral auxiliary which is not racemized during removal and

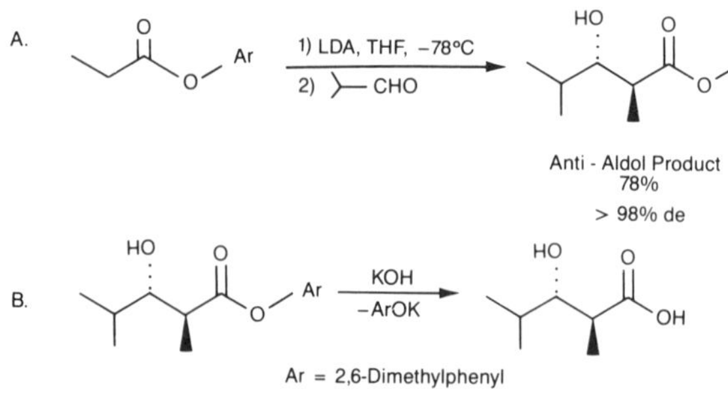


**Fig. 19.** Anti-diastereoselectivity via E-enolates.

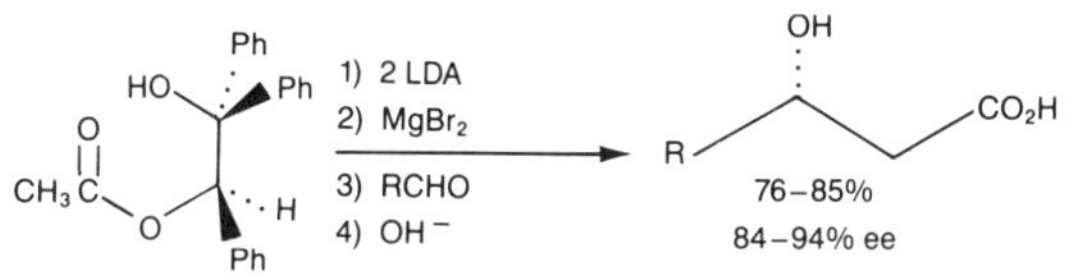

**Fig. 20.** Enantioselectivity via a magnesium enolate.

recovery. The mediated magneisum cation (introduced by transmetalation of the dilithiodianion with magnesium bromide) presumably bridges the oxygens of the alkoxide and enolate to create a preferred diasterofacial attack of the aldehyde.[88]

During the past few years, transmetalation of lithium enolates with transition metals was shown to lead to exceptional diasteroselectivity. The titanium-mediated aldol condensation of a chiral ketone ($R^1$ = cyclohexyl, $R^2$ = —$SiMe_2$—*t*-Bu) produced a *syn*-isomer ratio of 99:1 (Syn I/Syn II) and no *anti*-diastereomer was detected (Fig. 21).[89]

Excess chlorotriisopropoxytitanate or tetraisopropoxytitanate was required to favor equilibration to the more selective titanium enolate.

For a similar ketone as in (Fig. 21) (where $R^1$ = *t*-butyl, $R^2$ = —$SiMe_3$), Heathcock and coworkers have shown that the corresponding lithium, boron, and magnesium enolates gave the relative configurational types of Syn II, Syn I, and Anti I (not shown), respectively.[90]

For several cyclic ketones (Fig. 22), *threo*-selective aldol condensations of the favored lithium *E*-enolates were achieved in the presence of trialkylboranes.[91]

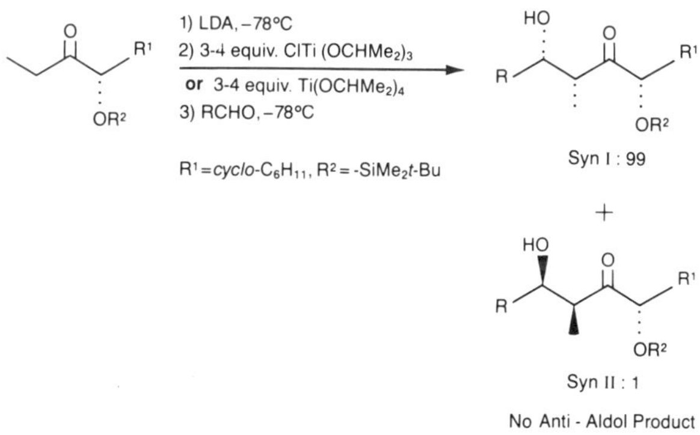

**Fig. 21.** Syn-enantioselectivity via titanium enolates.

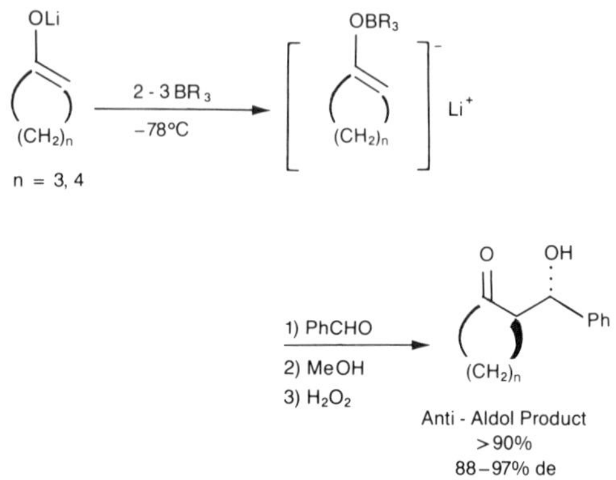


**Fig. 22.** Anti-diastereoselectivity via boron enolates.

Transmetalation of the lithium *E*-enolate of $\gamma$-butyrolactone with one equivalent of zinc chloride gave the unexpected 70:30 (syn/anti) diastereomeric mixture (Fig. 23). The ratio of aldol products is exactly reversed for the corresponding lithium enolate in the absence of the zinc chloride.[92] Although the diasteroselectivity is only modest, it was reported that identical results were observed even when the condensation temperature was raised from −78 to −30°C. Such temperature information is valuable when trying to apply literature methodologies on an industrial scale.

As can be seen from the discussion above, most of current methods give rise to *syn*-aldol products. However, an *anti*-selective aldol process has been indicated using transition metal enolates.[93]

### *Aromatic and Heteroaromatic Systems*

Because of their relative ability to accommodate carbanionic character, metalated aromatic and heteroaromatic ring systems can be readily prepared via metal–halogen (M↔X) exchange or metal–

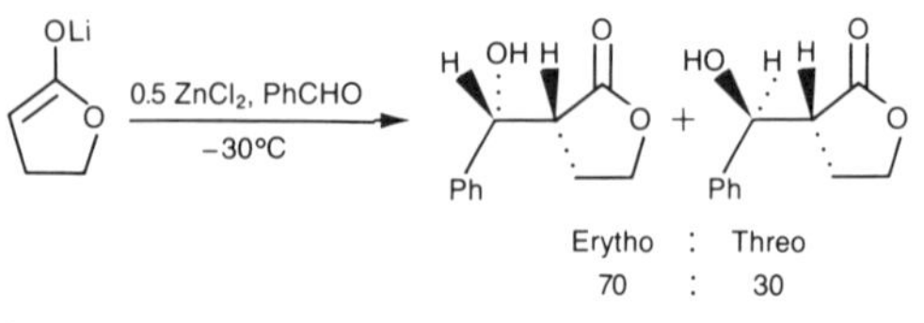


**Fig. 23.** Syn-diastereoselectivity of zinc E-enolates.

hydrogen (M↔H) exchange using alkyllithiums. There are extensive examples of metalations of both types.[94–100]

### *Metalation*

For metal–halogen (M↔X) exchange on a large scale, bromoaromatic compounds are normally employed to achieve rapid and quantitative lithiation using butyllithiums below −70°C (Fig. 24(A)). The facile M↔X exchange has been shown to take place even in the presence of reactive functionalities.[101] However, for certain haloaromatic substrates containing a carboxylic acid group, an alternative mechanism in which an M↔H exchange may in fact take place prior to the M↔X exchange has been proposed.[102]

For M↔X exchanges at higher temperatures, the major complication is subsequent alkylation of the lithiated ring by the butylbromide by-product (Fig. 24(B)). In the laboratory, two equivalents of *t*-BuLi are generally employed with the second equivalent of *t*-BuLi consumed in the decomposition of the butylbromide (Fig. 24(C)). The M↔X exchange is usually performed in ethereal solvents; however, it does occur in hydrocarbons, but at a slower rate.

Regiospecific M↔H of aromatic compounds is activated and directed by an ortho group which contains a heteroatom(s). This ortho-directed metalating group (DMG) exerts its effect by close coordination of the organolithium (RLi) with the heteroatom of the DMG.[103] Coordination of an electron pair of the heteroatom of certain DMGs with Li of the RLi maximizes the electron-withdrawing inductive effect of the DMG. Thus subsequent attack of the $R^-$ of the RLi at the *ortho*-hydrogen is facilitated (Fig. 25(A)). DMGs also direct M↔H at ortho-lateral methyl groups (Fig. 25(B)) which are more acidic than hydrogens on the aromatic ring.

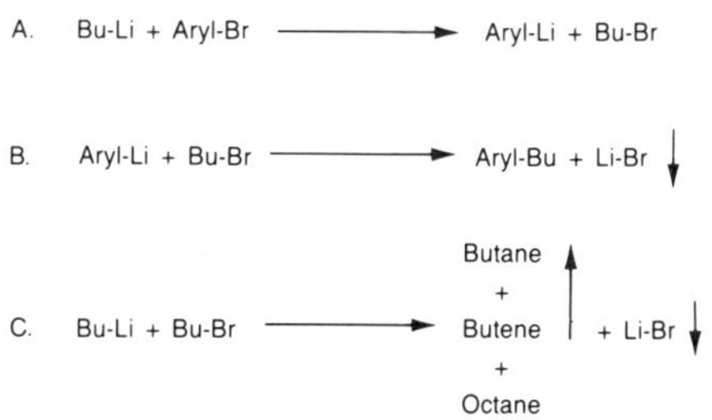


**Fig. 24.** Metal-halogen exchange: (A) and competing reactions (B & C).

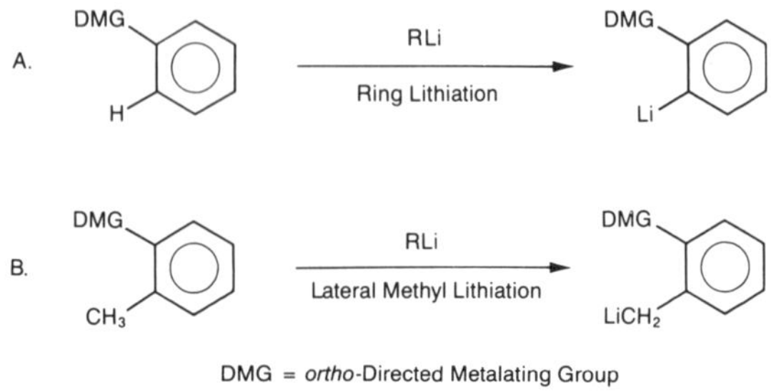


**Fig. 25.** Ortho-ring (A) and ortho-lateral (B) lithiation.

For five-membered heteroaromatics, ring lithiation usually occurs at the carbon position adjacent to the heteroatom; however, when the 2-position is occupied by a strong ortho-DMG, lithiation will occur at the 3-position in preference over the 5-position (Fig. 26).

Ortho-directed metalating groups (DMGs) have varying degrees of effectiveness. In Table 4, three groups of DMGs are listed according to their ability to direct ortho metalation. The order within each group of the strong, medium, and weak DMGs is not a listing according to their relative abilities.

The ability of *ortho*-DMGs to precomplex with butyllithium is further evidenced in the M↔H of six-membered heteroaromatic rings containing C═N bonds,[104] which normally undergo addition reactions. Recall that addition of butyllithium to the C═N was shown earlier to be a useful synthetic method for 2-alkylations of six-membered heterocycles (Fig. 6).

Here are some important points to consider when using LDA in the lithiation of aromatic or heteroaromatic substrates:

—LDA may participate in $\beta$-halogen elimination to give benzyne.

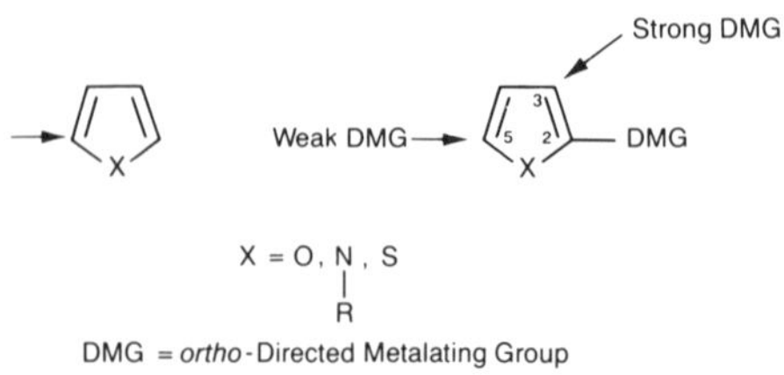


**Fig. 26.** Metalation of aromatic heterocycles.

**Table 4.** Qualitative listing of ortho-directed metalating groups in synthesis[95]

| Strong | | Medium | Weak |
|---|---|---|---|
| $—CONR_2$ | —CONLiR | —OMe | $—CH_2OLi$ |
| —NLiCOR | $—NLiCO_2R$ | $—N{\equiv}C$ | —OLi |
| —CSNLiR | $—OCONR_2$ | $—NR_2$ | |
| 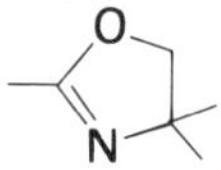 | $—SO_2NLiR$ | —Cl | |
| | $—SO_2NR_2$ | —F | |
| | $—OCH_2OCH_3$ | $—CF_3$ | |

—LDA is sufficiently basic to metalate certain lateral methyl groups of aromatic substrates.
—LDA will only ring-metalate highly activated aryl systems.
—LDA will ring-metalate alpha to the heteroatom of certain five-membered heteroaromatic systems.
—LDA will metalate lateral methyl groups of six-membered heteroaromatics.

### *Metalation and Reaction with Electrophiles*

A few examples of pertinent patents and selected results that illustrate the utility are described below. Comprehensive reviews regarding metalation of aromatic substrates and reaction with electrophiles are available.[96]

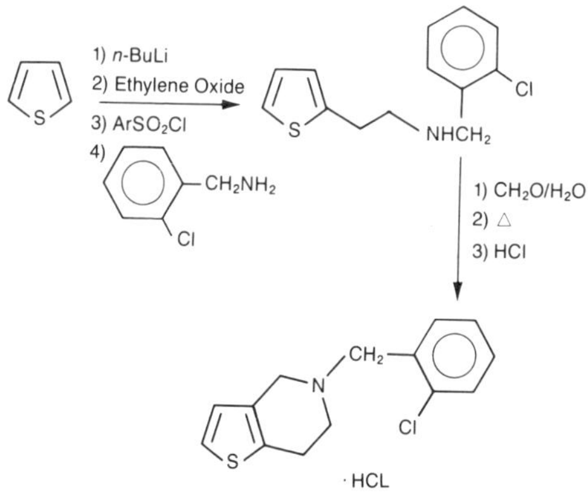


**Fig. 27.** A pharmaceutical from lithiated thiophene.

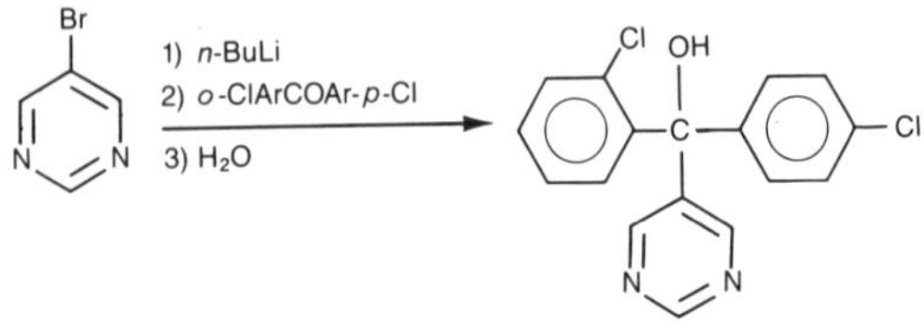

**Fig. 28.** Metal-halogen exchange of 5-bromopyrimidine.

Lithiation of thiophene (Fig. 27) is the initial step in the preparation of ticlopidine.[105]

Facile metal–halogen exchange (Fig. 28) is accomplished at low temperatures with *n*-BuLi in the manufacture of fenarimol.[106]

Triprolidine is also prepared via a metal–halogen exchange reaction (Fig. 29).[107]

Lateral lithiation of 2-methylpyridine with *n*-BuLi (Fig. 30) provides an important intermediate for perhexilene.[108] At FMC, it was found that the lithiation with *n*-BuLi required −30°C, while LDA could be used at 0°C.

Asymmetric lateral alkylation of tetrahydroisoquinolines (Fig. 31) using a chiral auxiliary after lithiation with *t*-BuLi gave high enan-

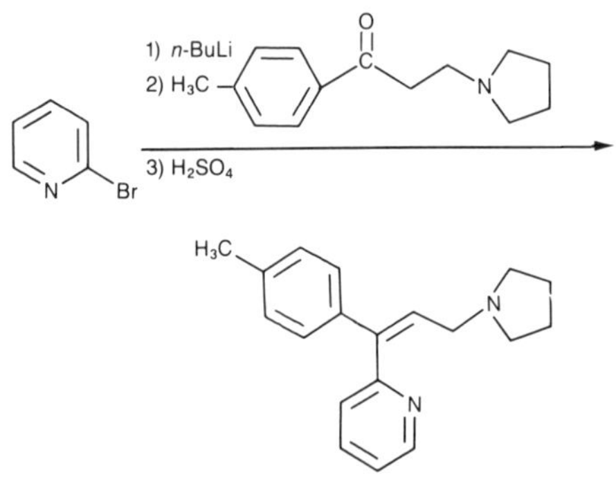

**Fig. 29.** Metal-halogen exchange of 2-bromopyridine.

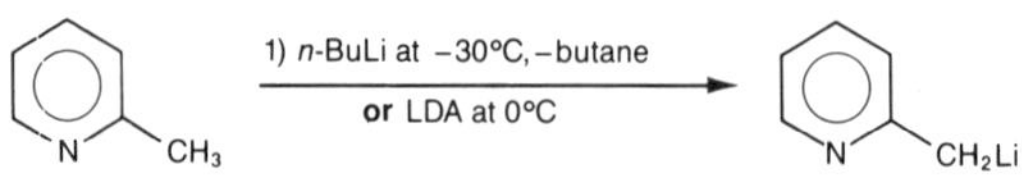

**Fig. 30.** Lateral lithiation of 2-methylpyridine.

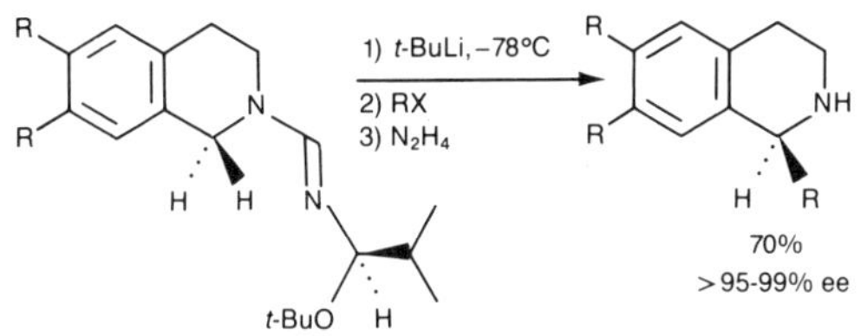

**Fig. 31.** Asymmetric alkylation.

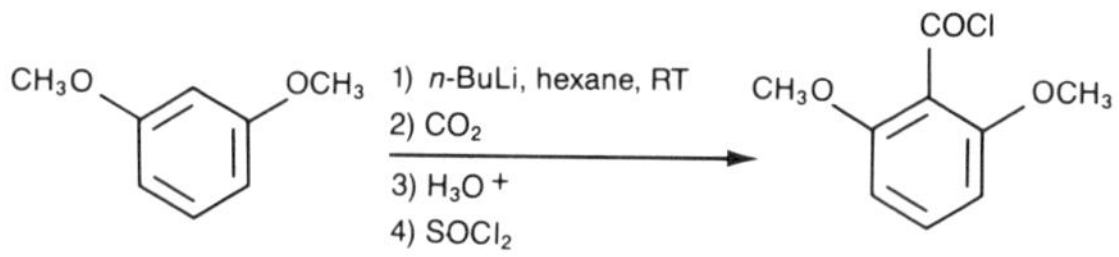

**Fig. 32.** Facile preparation of 2,6-dimethoxybenzoyl chloride (supplied by FMC Lithium Division).

tiomeric excess (>95–99%). Dihydropyrroles and tetrahydropiperidines were prepared in a similar manner.[109]

The intermediate, 2,6-dimethoxybenzoyl chloride (DMBC) can be prepared via a route involving *n*-BuLi (Fig. 32). It was found at FMC that the sparingly soluble 2,6-dimethoxyphenyllithium (DMP) can be prepared in high yield in the absence of THF at room temperature.

Even 1,3-substituted aromatics containing weak and somewhat labile DMGs such as 1,3-dichloro-[110] and 1,3-difluorobenzenes[111] are suitable for preparation of the corresponding 2,6-dihalophenyllithium intermediates.

The relative abilities of two ortho-DMGs, namely —OMe[112] and —$CONEt_2$, are demonstrated (Fig. 33) in this ring lithiation.[113] In Fig. 33, the second lithiation step involving the lateral methyl group can also be accomplished with LDA instead of *sec*-BuLi.[114]

Because amides (—$CONEt_2$) are difficult to hydrolyze and involve an additional reduction step to yield an aldehyde, a novel approach

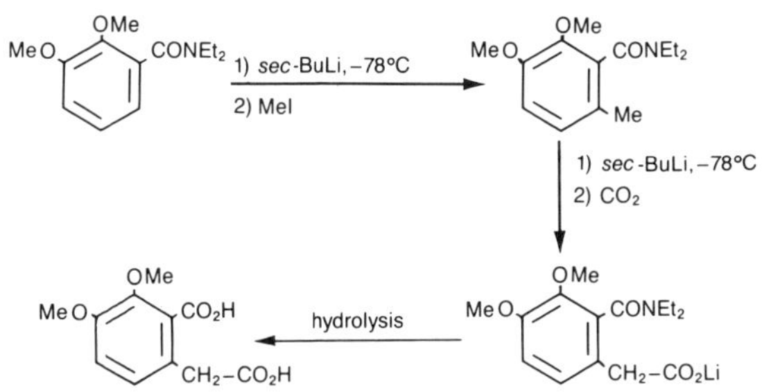

**Fig. 33.** Ortho-directed ring and lateral lithiation.

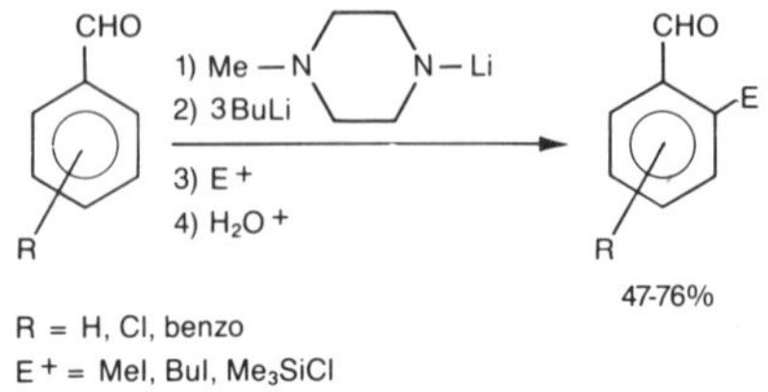


**Fig. 34.** *In situ* protection and ortho-lithiation of an arylaldehyde.

was developed. Here, ortho functionalization to an aldehyde moiety was achieved directly by prior *in situ* protection with lithium *N*-methylpiperidide (Fig. 34). after the addition of the organoamide to the aldehyde, the resulting adduct acts as a DMG for the subsequent lithiation with butyllithium.[115]

Interestingly, attempted ring lithiation of *ortho*-alkoxy aromatic aldehydes and ketones which are similarly protected *in situ* via organoamides can result in *O*-dealkylation using *n*-BuLi.[116]

Using a different approach to *in situ* protection, *C*-lithiation of heteroaromatics which contain an NH group has been developed by reacting the corresponding Li—N monoanions with $CO_2$. This is followed by lithiation with a second equivalent of butyllithium, usually *t*-butyllithium. The resulting dianions of phenothiazines (Fig. 35(A))[117] and indoles (Fig. 35(B))[118] can be reacted with various electrophiles. The nitrogen is then deprotected by simple hydrolysis. Katritzky and coworkers also applied the same methodology in the formation and reaction of dianions of *N*-monoalkylaniline,[119] *N*-monoalkylbenzyl amine, benzyl alcohol,[120] and trialkylsilylmethanol.[121]

### *Cyclization*

In the previous bond formation examples, the lithiated species is subsequently reacted with an electrophile. When this process occurs

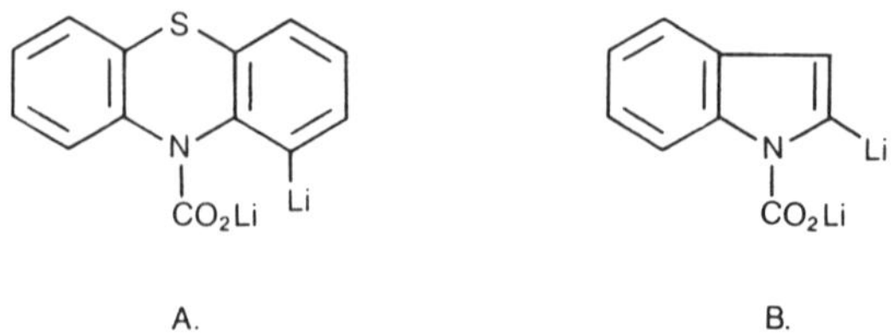


**Fig. 35.** Lithium intermediates having $LiCO_2$-protected ring nitrogens.

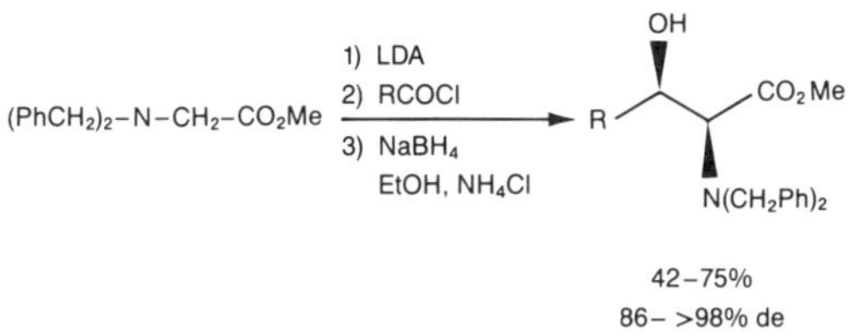


**Fig. 36.** Preparation (1,2) and reduction (3) of a $\beta$-ketoester.

intramolecularly, it is called cyclization. Cyclizations usually produce a ring system containing a heteroatom and at least one $sp^2$ hybridized atom.

## *$\beta$-Lactam Formation*

An important class of compounds, $\beta$-lactams, can be formed using lithium methodologies which involve cyclization. Several examples involving $\beta$-lactam preparations are cited below.

In Fig. 36, the end products can be converted to *threo*-$\alpha$-amino-$\beta$-hydroxyacids (useful starting materials for $\beta$-lactam syntheses) by saponification and hydrogenation.[122] This route possesses greater diasteroselectivity than condensation of the appropriate lithium enolates with aldehydes.

Other recently reported methods have demonstrated the use of LDA, LHS, *n*-BuLi, and TBDMCS to prepare important starting materials for $\beta$-lactam synthesis. Such intermediates were prepared by silylketene acetals of *N*-methylephedrine esters using LDA.[123,124]

In Fig. 37, lithium hexamethyldisilazide (LHS) not only undergoes addition to the nonenolizable aldehyde but provides the nitrogen for the $\beta$-lactam ring which can later be *N*-substituted if desired. The intermediate *N*-trimethylsilyl imines (not shown) resulting from the

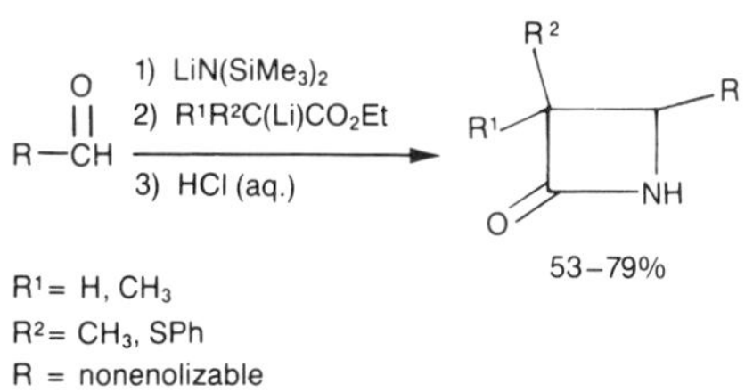


**Fig. 37.** Preparation of N-unsubstituted $\beta$-lactams.

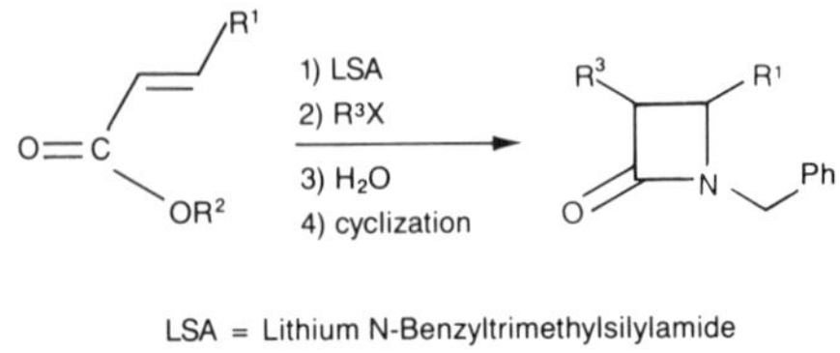


**Fig. 38.** 1,4-Addition of a lithium organoamide.

first step are the electrophiles for lithium enolates to form $\beta$-lactams after cyclization.[125]

In slight contrast, lithium *N*-benzyltrimethylsilylamide (LSA) (prepared from *n*-BuLi and appropriate amine) undergoes 1,4-addition to crotonates. The resulting enolate is alkylated (Fig. 38). As in other methods for producing $\beta$-lactams, the $\beta$-amino alkylated esters are then cyclized.[126]

Carbon–carbon bond cyclization (Fig. 39) has been demonstrated via oxidative coupling of dianions of acyclic amides using Cu(II) carboxylates.[127]

In Fig. 40, two equivalents of LDA produce an intermediate dianion which undergoes addition and cyclization with nonenolizable imines.[128]

## *Labile Cyclic Intermediate*

Intramolecular cyclization can also be used to produce labile intermediates which upon hydrolysis give rise to products that could not be

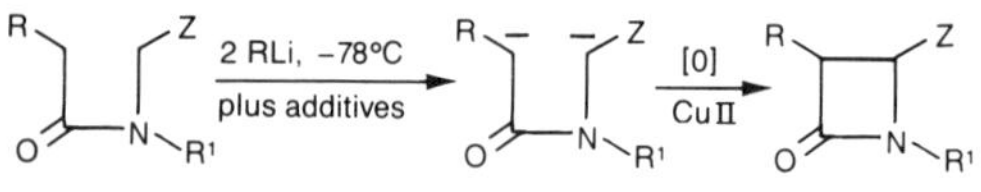


**Fig. 39.** Oxidative-cyclization of a dianion.

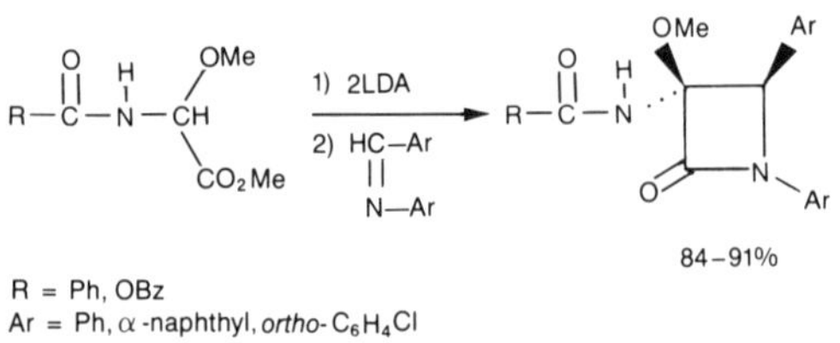


**Fig. 40.** $\beta$-Lactams from an ester enolate and an imine.

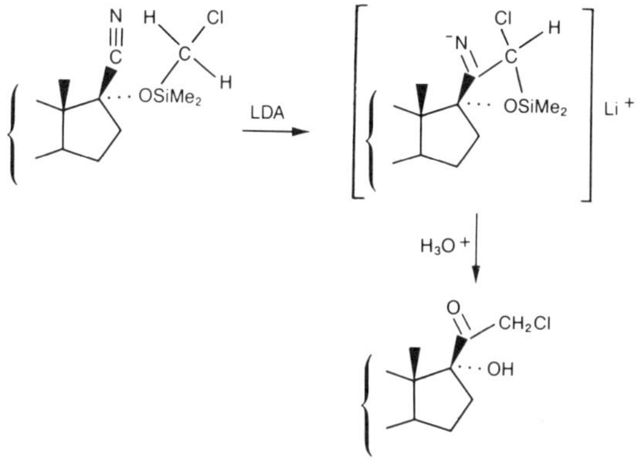


**Fig. 41.** A novel cyclization leading to chloromethylketones.

made via standard intermolecular addition reactions. For instance, a novel silicon mediated synthesis was reported for the homologation of silylated cyanohydrins to various hydroxyacetone steroidal derivatives. Upon lithiation of the chloromethylsilyl group with LDA, the resulting anion cyclizes onto the nitrile. The labile cyclic intermediate is then hydrolyzed to a versatile chloromethyl ketone (Fig. 41).[129]

### *Using Aromatic Systems*

Cyclizations or 'ortho ring closures' are often constructed about an existing aromatic system. The preparation of a 4-aminoquinoline derivative (Fig. 42) was accomplished by ring closure of two ortho groups, an 'enolate' and a nitrile.[130]

Using the selectivity of metal–halogen exchange of aryl versus alkyl, dihydroisoquinoline derivatives (Fig. 43) can be prepared.[131]

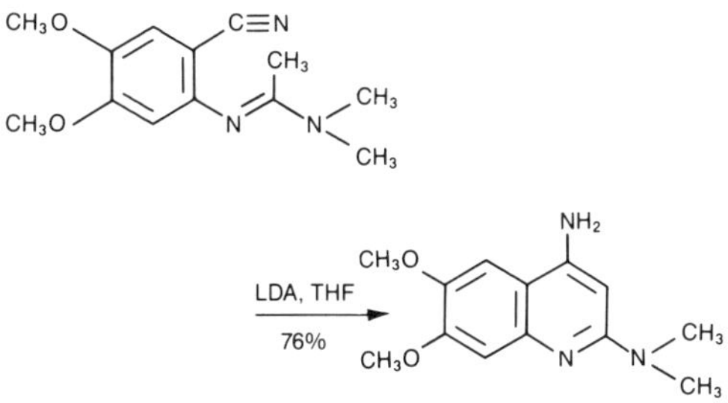


**Fig. 42.** Preparation of a substituted-quinoline.

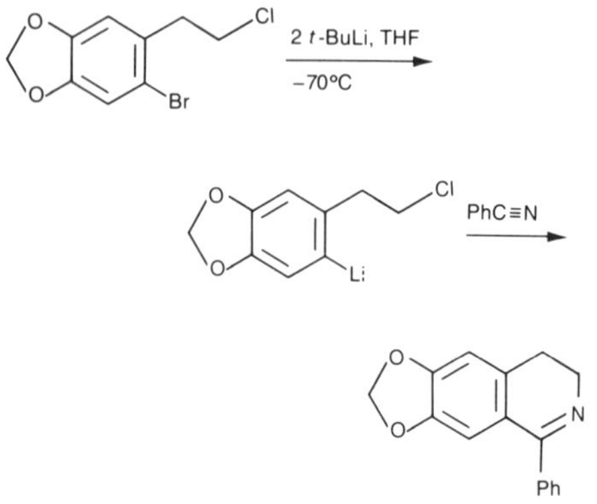


**Fig. 43.** Preparation of 1-aryl-isoquinolines.

Another example of intramolecular alkylation (Fig. 44) was reported for the cyclization step in the preparation of ketorolac in which LDA is one of the bases that can be used to generate the reactive enolate.[132]

Snieckus and coworkers have reported several syntheses of natural products that involve cyclization steps. For preparation of certain key intermediates, it was found that transmetalation of the lithiated intermediate with $MgBr_2$ was required prior to the subsequent condensation step with aliphatic aldehydes (Fig. 45(A)).[133] Transmetalation to the more selective magnesiated anion was also necessary for the electrophile, allyl bromide.[133]

To produce a similar intermediate (Fig. 45(B)), lithiated carbanions of silylated *N,N*-dialkylbenzamides were reacted directly with aromatic aldehydes.[134]

In the literature, most intramolecular cyclizations involving lithium anions occur adjacent to *C*-lithiated ring systems. To a lesser degree, cyclizations can also be mediated through lateral metalated anions (Fig. 42). Intramolecular cyclizations of lithiated lateral methyl groups and anionic cycloadditions have been reported.[135] Intermolecular

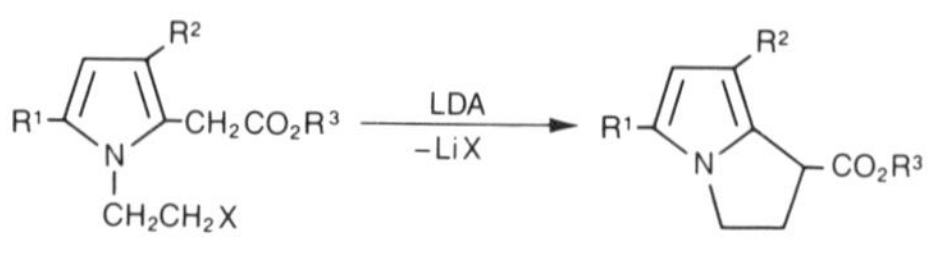


**Fig. 44.** An important intramolecular alkylation.

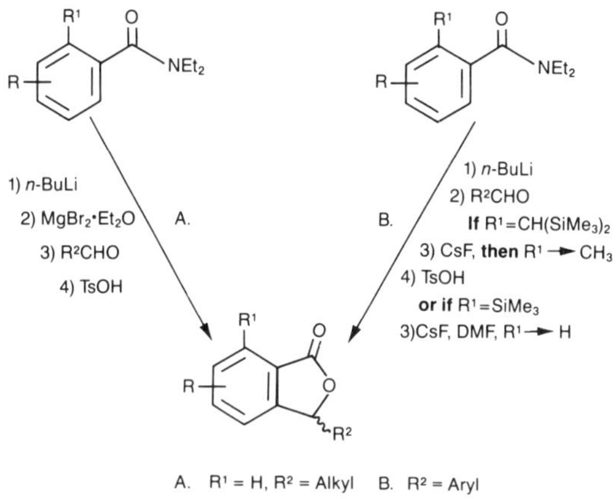


**Fig. 45.** Electrophilic alkyl- and aryl-aldehydes requiring different methodologies.

cylcoaddition[136] and addition[137] reactions also occur involving benzyene intermediates prepared from appropriate haloaromatic substrate and butyllithium or LDA.

## Multiple Carbon–Carbon Bond Formation

### *Dehydration*

In its simplest form, a double bond may arise via a condensation reaction if subsequent dehydration is thermodynamically favored by relief of steric hindrance or by an increase in conjugation. In Fig. 46, only the *trans*-olefin results after the condensation of an ester enolate and a $\gamma$-lactone.[138]

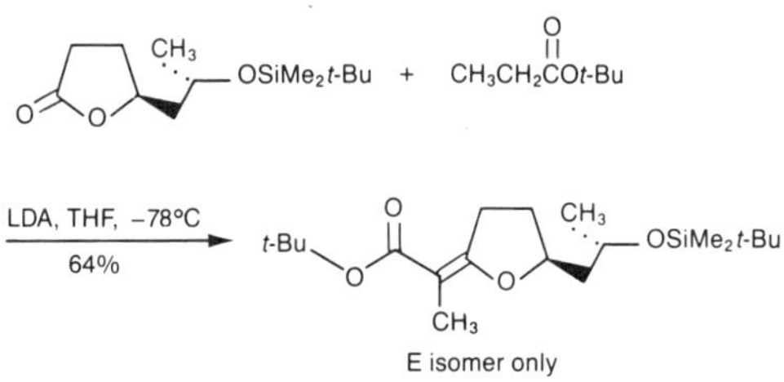


**Fig. 46.** Selective formation of a substituted olefin.

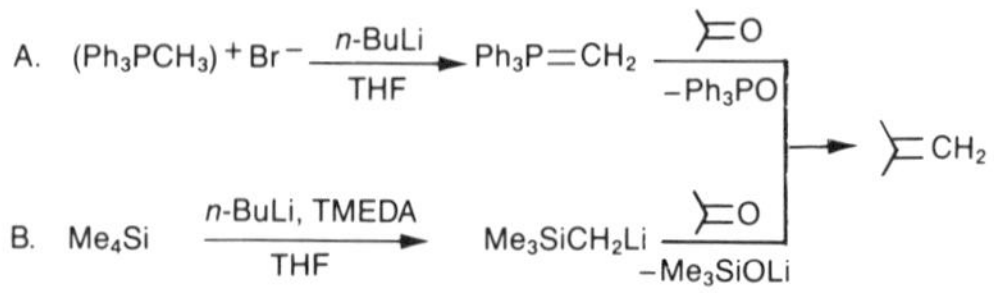

**Fig. 47.** Olefination using Wittig (A) or Peterson (B) intermediates.

### *Olefination*

There has been considerable study of carbon–carbon double bond formation or olefination reactions, especially those mediated by strong bases and involving carbonyl compounds and destabilized Wittig (Fig. 47(A)) or Peterson olefination reagents (Fig. 47(B)). Maryanoff and Reitz recently reviewed the stereochemical and mechanistic aspects of the extensively studied Wittig olefination reaction.[139]

The Peterson reagents are described as being more nucleophilic than the corresponding Wittig reagents.[140,141] Also, the by-product, siloxane or silanoxide, is generally easier to separate from the resulting olefin than the Wittig by-product, triphenylphosphine oxide.

### *Multiple-Bond Nucleophiles*

Although an olefin is not being created, the addition of vinyllithium or allyllithium derivatives to substrates obviously offers a route to >C=C< formation. Vinyllithium derivatives can be formed via metal-hydrogen or metal–halogen exchange.

While preparation of vinyllithiums from tetravinyl tin[142,143] or ketone arenesulfonylhydrazones (Shapiro reaction)[144] is quite viable in the laboratory, large-scale application of these methods is hindered by cost and availability of the raw materials.

Metalation of ethylene is possible with *n*-BuLi/KO*t*-Bu in TMEDΛ/hcxanc at −25°C. The resulting vinyl potassium can be converted to the vinyllithium by metathesis with LiBr (Fig. 48).[145]

For allyllithium, metalation of propene can be achieved with *n*-BuLi and TMEDA (Fig. 49))[146] or KO*t*-Bu (Fig. 49(B)).[147] Schlosser and

**Fig. 48.** Metalation of ethene to give vinyllithium.

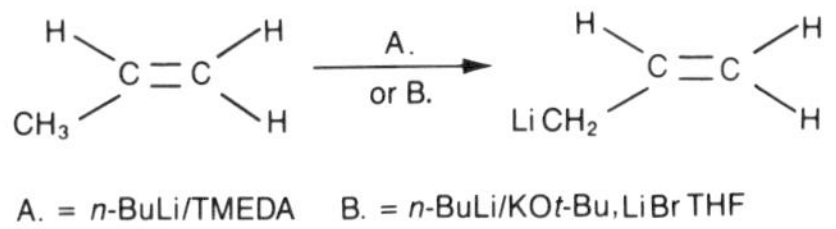


**Fig. 49.** Metalation of propene to give allyllithium.

coworkers have also extensively explored the use of the latter 'superbase' system for the metalation of allyl derivatives.[148]

Although only a small number of lithium methodologies (such as that of Negishi and coworkers[149]) exists that provide a route to forming a terminal alkyne, preparation of triple-bond derivatives via addition reactions using lithium acetylide and monosubstituted acetylides is a common practice. In the laboratory, amine-free lithium acetylide can be prepared from *n*-BuLi and acetylene at −78°C.[150] Lithium acetylide can be conveniently prepared at a much higher temperature from LDA and acetylene (Fig. 50).[151] In contrast to the lithium acetylide ethylenediamine complex, the *in situ* prepared lithium acetylide (from LDA and acetylene) is more reactive in subsequent addition to various steroidal carbonyls.

## *Transition Metal Reagents*

Two versatile schemes in *Organic Syntheses*[152,153] for olefin preparation involve a lithium intermediate and either Cu(I) (Fig. 51) or Pd (Fig. 52). In Fig. 51, the *in situ* formed divinylcuprate can be reacted with a large number of electrophiles to give *cis*-substituted olefins.[152]

In Fig. 52, only the Pd-catalyzed coupling of an *ortho*-aryllithium and a *cis*-alkenylhalide is shown. Other organolithiums and *trans*-

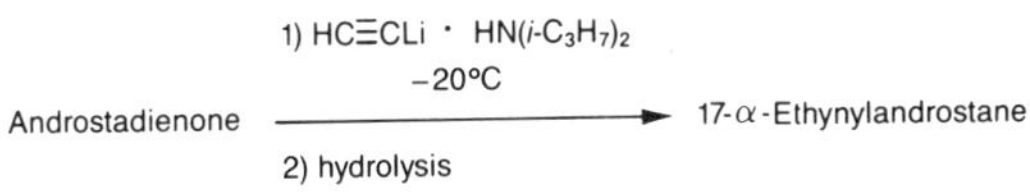


**Fig. 50.** *In situ* preparation of lithium acetylide.

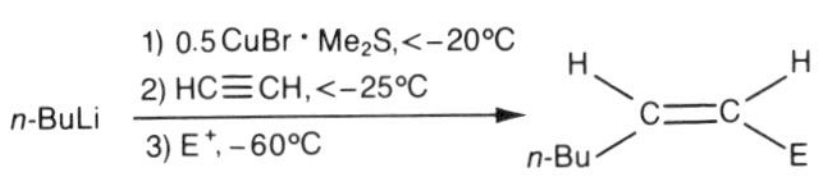


**Fig. 51.** Cis-olefins via cuprate addition to acetylene.

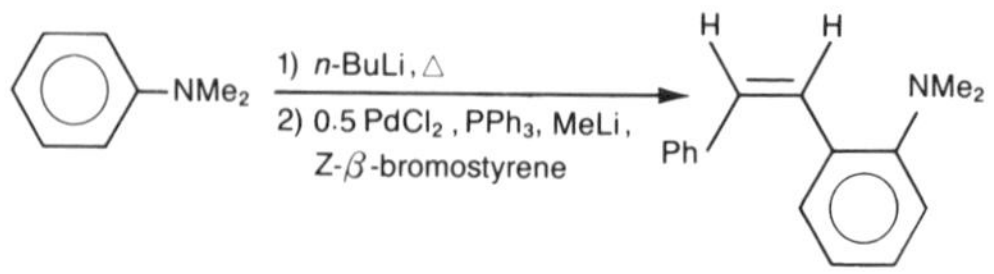

**Fig. 52.** Cis-olefins via Pd coupling.

alkenylhalides or aryliodides could be used in this coupling methodology.[153]

## Heteroatom Bond Formation

### *Carbon–Oxygen Bond*

Preparation of alcohols by the reaction of carbanions with molecular oxygen is only useful for fairly simple aliphatic substrates.[154] Also, no attempt should be made to isolate the solids (possibly lithium organoperoxides) which still may be present from reaction of RLi with oxygen. When aryllithium or benzyllithium intermediates are involved, the resulting oxygenated products are often laced with undesirable coupled products.[155] Hydroxylation of aryllithiums can be accomplished in two steps by treatment with $B(OEt)_3$ and then $H_2O_2$.[156,157]

The existence of lithium alkylperoxides is exemplified by the utility of lithium *t*-butylperoxide as an oxidation reagent. Lithium *t*-butylperoxide can be prepared *in situ* from *t*-butylperoxide and *n*-BuLi in THF and will cause regio- and stereoselective oxidation of $\alpha, \beta$-unsaturated esters and sulfones.[158]

More recently, efforts have been directed toward developing electrophilic reagents which will form a carbon–oxygen bond when reacted with carbanions. However, to date, such reagents have only been demonstrated in the laboratory and may not be practical for large-scale use because of expense and availability. For example, asymmetric oxidation of chiral cyclic enolates (prepared via 1,4-addition) to $\alpha$-hydroxylactones can be accomplished with molybdenum peroxide reagent ($MoO_5$-pyridine–HMPT, MoOPH).[159] Also reported are reactions of MoOPH with enolates prepared with LDA or LHS.[160]

Also, Davis and coworkers have evaluated the chiral oxidative reagent, camphoxylsulphonyl oxazirdine, for the oxidation of acyclic

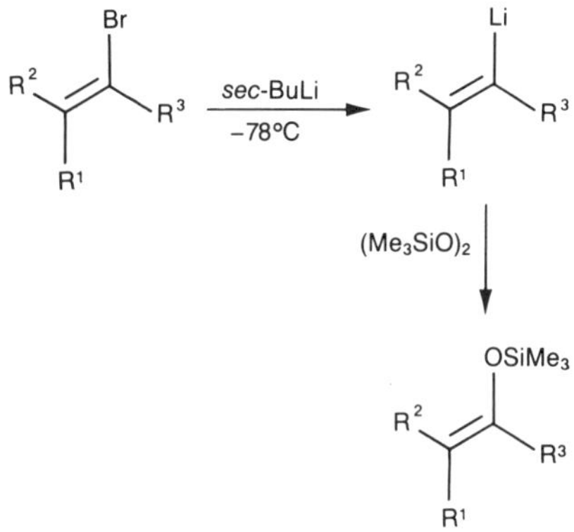


**Fig. 53.** Silylenol ethers via vinyllithium intermediates.

enolates (prepared from silyl enol ethers and MeLi) to α-hydroxyketones. Two achiral oxidative reagents, bis(trimethylsilyl)peroxide (BTSP) and bis(*t*-butyldimethylsilyl)peroxide, can be used to prepare the silyl enol ethers (Fig. 53) by reaction with the appropriate vinyl anions.[161]

Taddei and Ricci have reported that treatment of aryllithiums with BTSP yields phenols.[162] For a broader range of carbanions, reaction with BTSP resulted in products that were explained on the basis of the hard–soft acid base principle.[163] Harder carbanions will react at the silicon and softer carbanions will react at the oxygen.

α-Lithiation of phenylsulfones with *n*-BuLi and subsequent reaction with BTSP gave the expected carbonyl compounds (Fig. 54) via the interesting attack–counterattack mechanisms.[164]

Also using a sulfone, oxidative desulfonylation was achieved with LDA and MoOPH.[165] The oxidation of silyl enol ethers with *m*-chloroperbenzoic acid (Fig. 55) was shown to be a useful route to forming α-hydroxyketones.[166]

α-Hydroxyketones have been prepared by direct oxygenation of enolates.[167] Electrophilic agents which can selectively form C—O

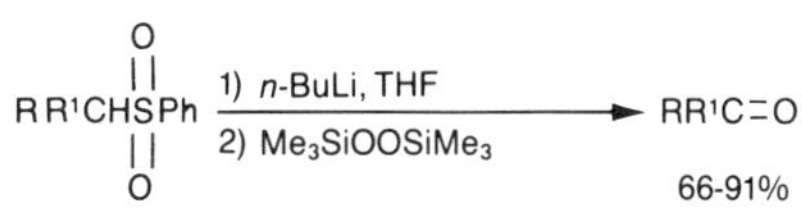


**Fig. 54.** Conversion of sulfones to carbonyl compounds.

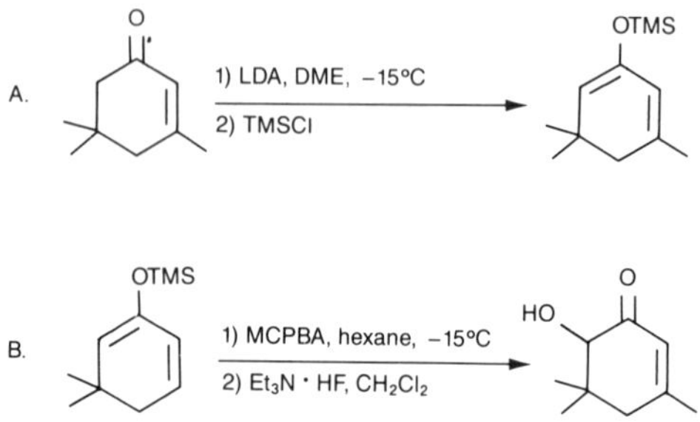


**Fig. 55.** $\alpha$-Hydroxyketones (B) via silylenol ethers (A).

bonds or C—O—Si bonds when reacted with various carbanionic species are still needed. Of course, any such reagent used in production will have to be safe, cost-effective, and versatile.

## *Reduction*

A reduction scheme that involves the use of sequential treatment of an $\alpha,\beta$-unsaturated substrate with organolithiums and lithium metal has been extensively studied (Fig. 56).[168] Depending on the quenching procedure, this process can provide the carbinol or the hydrocarbon. For the reduction step to proceed, the substrate must possess the 'ArCRROLi' functionality. This method of tandem alkylation/reduction was also later extended to esters,[169] 2-acyl pyrroles,[170] nitriles,[171] and heteroaromatic substrates.[172]

LDA-promoted reductions of nonenolizable ketones and aldehydes are known. Majewski recently reported reduction of such aldehydes with LDA and provided a comprehensive reference list of other reductions promoted by LDA.[173]

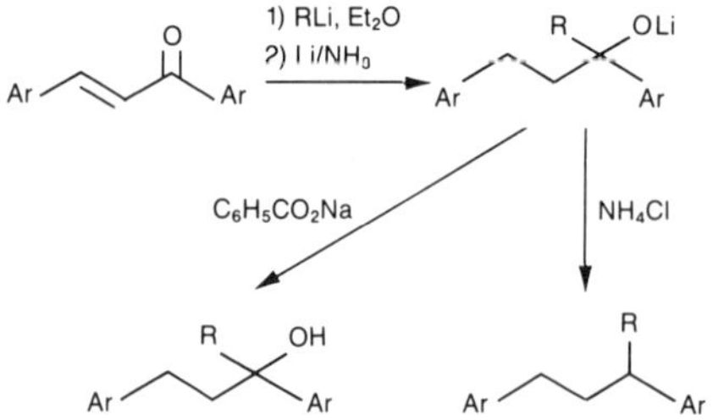


**Fig. 56.** Reductive alkylations of $\alpha,\beta$-unsaturated ketones.

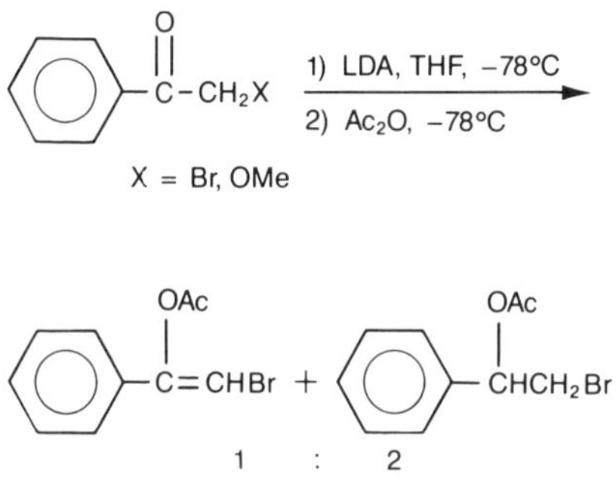


**Fig. 57.** An LDA-promoted reduction.

For certain enolizable, α-substituted ketones, it was found that they undergo Meerwein–Ponndorf–Verley-type reductions (Fig. 57). Two products were obtained in 80% yield with the mixture containing twice as much reduced product over the desired enol acetate. Unwanted reduction was avoided in this reaction by using lithium hexamethyldisilazide (LHS) or lithium 2,2′,6,6′-tetramethylpiperidide.[174]

Because LDA can cause reduction, it was not surprising that magnesium diisopropylamide (MDA) also behaved as a reducing agent. However, in sharp contrast to LDA, MDA is hydrocarbon soluble (0.7M) and thereby offers an organoamide that can be tested in the absence of a Lewis base, such as THF.

Reduction of nonenolizable carbonyl compounds to alcohols (Fig. 58) was achieved in good yields using MDA in heptane at reflux.[175]

In like manner, nitro-, nitroso-, and azoxyaromatic compounds were reduced in a step-wise fashion (Fig. 59) using the unsolvated MDA.[176]

Compounds such as 2-nitrofluorene or nitrocyclohexane were reduced by MDA only to 2-aminofluorene or cyclohexanone hydroxylamine, respectively.

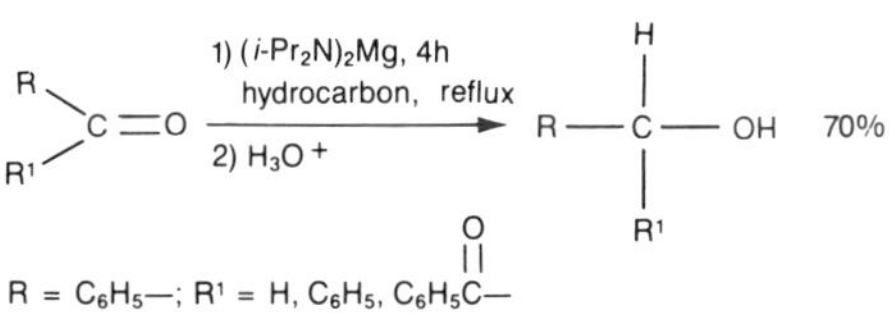


**Fig. 58.** Reduction of nonenolizable ketones with magnesium diisopropylamide.

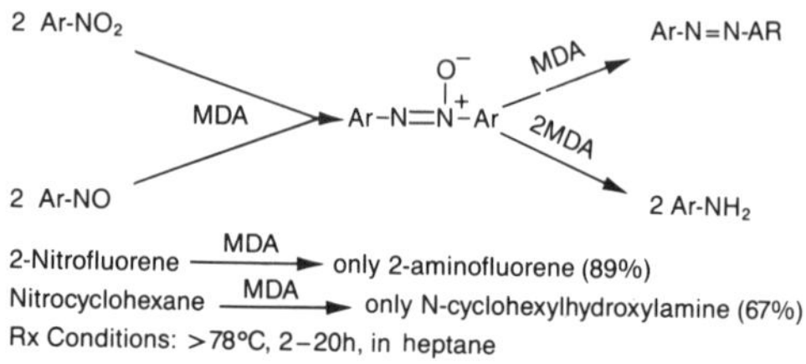


**Fig. 59.** Reduction of nitro-compounds with magneisum diisopropylamide.

## *Protecting Group, t-Butyldimethylchlorosilane (TBDMCS)*

Since the introduction of TBDMCS (or *t*-butyldimethylsilyl chloride) as a protecting reagent for alcohols as silyl ethers, its use has steadily expanded to include protection of other functional groups as well. Lalonde and Chan extensively reviewed the use of TBDMCS as well as many other silyl protecting groups.[177]

### *Protection*

The *t*-butyldimethylsilyl ethers (TBDMSiOR) can be prepared in several ways:

1. By employment of 1·2 and 2·5 equivalents of TBDMCS and imidazole, respectively, in dimethylformamide (DMF) at 35°C for 10 hours.[178] (The occasional problem of product separation associated with DMF may be relieved by using DMF/$CH_2Cl_2$.)
2. Secondary and tertiary alcohols can be easily silylated using *t*-butyldimethylsilyl triflate with 2,6-lutidine in methylene chloride at 25°C.[179]
3. Primary alcohols can be selectively silylated in the presence of secondary alcohols by using catalytic amounts of 4-dimethylaminopyridine (DMAP). (In methylene chloride, triethylamine (TEA) is required to absorb the evolved HCl.)[180]
4. In methylene chloride or acetonitrile, several types of functional groups can be silylated in the presence of 1,8-diazabicyclo-[5.4.0]-undec-7-ene (DBU)(1·1 equiv) at gentle reflux or room temperature.[181,182]
5. Silylation of alcohols in THF can be accomplished via lithium alkoxides generated from LDA. In fact, trapping of lithiated

intermediates resulting from an aldol condensation with TBDMCS (Fig. 60) is a novel step-saving approach.[183]

6. TBDM silylation of N—H bonds of heteroaromatics can be achieved with LDA.[184]
7. In THF, enolates generated with LDA or LHS can be trapped as silyl enol ethers.[185]

### *Stability*

When TBDMCS is used, the protected alcohol (or *t*-butyldimethylsilyl ether) has several advantages over other protective groups:

1. It is $10^4$ times more stable than trimethylsilyl ether in acidic medium.[186]
2. It does not create a chiral center which results from tetrahydropyranyl ethers.
3. It is stable in the presence of many organometallics such as RLi, RMgX, alkoxides, $LiCuR_2$, DIBAL, $Me_2AlCl$, $Me_3Al$.
4. It is stable to reduction via hydrogenation or mild reducing reagents.
5. It is stable in basic medium (aqueous, alcoholic, or Birch reduction conditions).
6. It is stable to oxidative reagents such as $H_2O_2$, $CrO_3/C_5H_5N$, $NaIO_4/RuO_2$.[187]
7. It is stable to Lewis acid conditions such as $TiCl_4$, *i*-$PrOTiCl_3$, or $SnCl_4$.[188]
8. It is stable to $MgBr_2$[189] in ether and $Me_2AlCl$ or $Me_3Al$[190] in methylene chloride when removing the tetrahydropyranyl group without causing silyl ether cleavage.

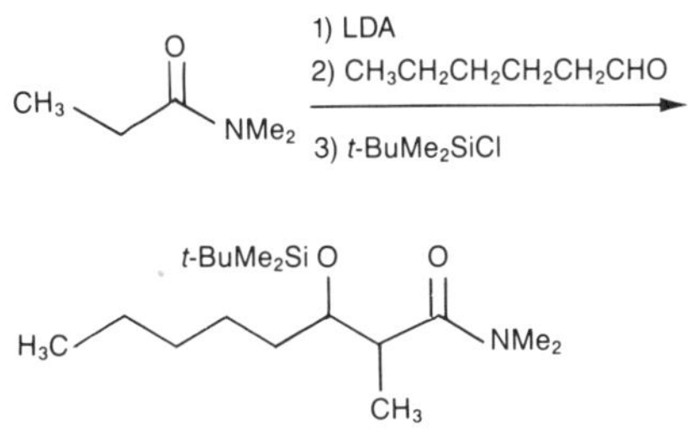


**Fig. 60.** Silylation of an alkoxide resulting from a condensation.

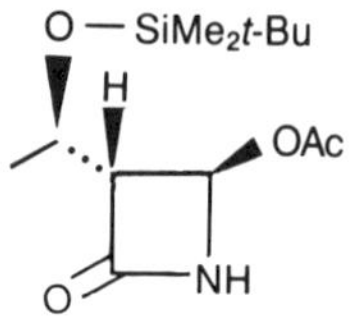


**Fig. 61.** Commercial precursor for antibiotics

As described above, TBDMSilyl ethers are stable to many reaction conditions. However, these same silyl ethers can be selectively cleaved (described below in the section on 'Deprotection'). In addition TBDMSilyl ethers provide derivatives that are crystalline and easy to handle. These properties have made possible the commercial availability of useful intermediates such as (1) TBDMsilylated chiral cyanohydrins,[191] and (2) an intermediate for fourth-generation $\beta$-lactam antibiotics. The latter intermediate, (3*R*, 4*R*)-4-acetoxy-3-[(*R*)-(*t*-butyldimethylsilyloxy)ethyl]-2-azetidinone (Fig. 61), provides facile routes to penems and carbapenems via stereoselective substitution of the acetate group (OAc).[192]

TBDMSilyl enol ethers also possess excellent chemical stability toward organolithiums and provide novel entry into $\beta$-substitution of these silyl enol ethers. For example, a cyclic $\beta$-acylvinyl anion equivalent is formed by transmetalation of a trialkylstannyl derivative with MeLi or *n*-BuLi. The vinyl anion was reacted with alkylhalides, ketones, and benzaldehydes in good isolated yields (Fig. 62).[193]

Using an acyclic substrate (Fig. 63), lithiation was achieved with *t*-BuLi at the $\beta$-carbon.[194]

Unlike the stability of the TBDMSilyl enol ethers in Figs 62 and 63, TBDMSilyl ketene acetals understandably lose their TBDMSilyl group during a base catalyzed Claisen rearrangement. Interestingly, different diastereomers (Fig. 64) resulted from the sequential treatment of the same acyclic chiral ester with (1) LDA or LHS, (2) TBDMCS, and (3) warming and hydrolysis.[195]

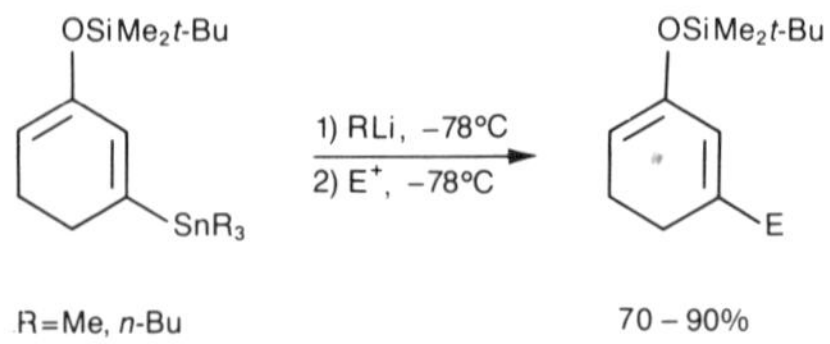


**Fig. 62.** Lithium intermediates via vinyltin derivatives.

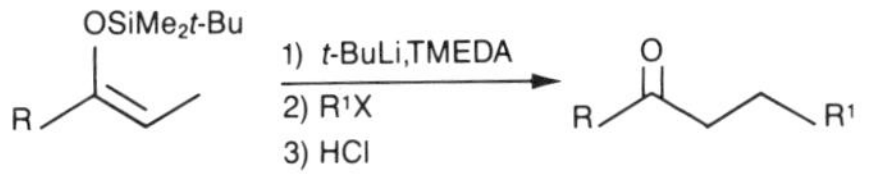


**Fig. 63.** Elaboration of TBDM silylenol ethers at the $\beta$-position.

## *Deprotection*

In the laboratory, cleavage of the *t*-butyldimethylsilyl ether is usually accomplished with a fluoride-containing reagent in various solvent systems that include acetonitrile, THF, pyridine, methanol, dimethylformamide, etc., as shown in Table 5.

One-pot deprotection of the silyl group and reprotection with other protecting groups is also possible. For instance, cleavage of the TBDMSilyl ethers with acetic anhydride/ferric chloride yields the corresponding acetate.[196] Also, desilylation of TBDMSilyl-phenyl ethers to benzyl- or methyl-phenyl ethers can be done with KF/dimethylformamide at 25°C in the presence of benzyl bromide or methyl iodide, respectively.[197]

Under carefully controlled conditions, HF and tetra-*n*-butylammonium fluoride (TAF) will selectively deprotect alkyl- and phenyl-TBDMSilyl ethers, respectively (Fig. 65). However, by having

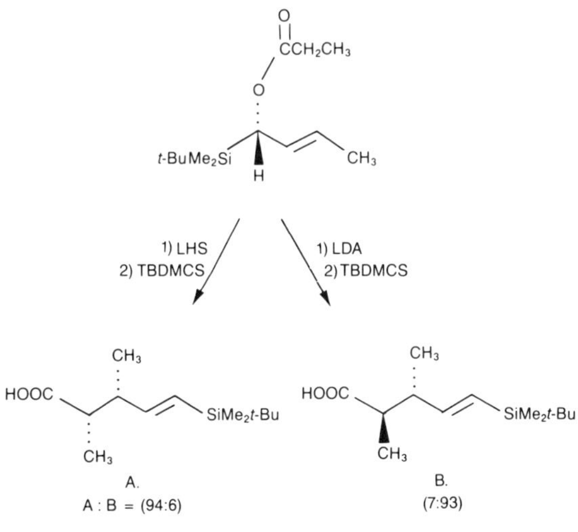


**Fig. 64.** Diastereoselectivity using different lithium bases.

**Table 5.** Deprotection of *t*-butyldimethylsilyl ethers

| Reagent/conditions | Reference |
|---|---|
| Tetra-*n*-butylammonium fluoride/THF at 25°C | 178 |
| Acetic acid/$H_2O$/THF (3 : 1 : 1) at 25°C | 178 |
| $Ac_2O$/$FeCl_3$ at 0°C, 15 min | 196 |
| KF/dimethylformamide | 197 |
| $LiBF_4$/$CH_3CN$/$CH_2Cl_2$ | 198 |
| Sulfonic acid, ROH | 199 |
| KF/crown ether | 200 |
| $KO_2$ 18-crown-6 DMSO | 201 |
| Tetra-*n*-butylammonium chloride/KF·$2H_2O$ $CH_3CN$ | 202 |
| Aq. HF in $CH_3CN$ | 203 |
| Dowex® 50W-X8/methanol | 204 |
| $BF_3$·$Et_2O$ | 205 |
| *N*-bromosuccinimide | 206 |
| Organotin phosphate condensates | 207 |

Dowex is a trademark of The Dow Chemical Company.

two different types of TBDMSilyl ethers, migration of the TBDMSilyl group can result during desilylation. For instance, if the TBDMSilyl ethers are in close proximity (ortho) to each other, the TBDMSilyl group may migrate from the phenyl ether to the free *ortho*-alcohol which was initially desilylated.[208]

Lithium tetrafluoroborate is a viable alternative to either of these reagents and acid hydrolysis.[198]

Although TAF is generally used in the laboratory, hydrogen fluoride (HF) in pyridine has been attempted on the pilot scale. The cost of TAF or the handling equipment for HF will likely make both of these options uneconomical. The use of sulfonic acids in methanol for desilylation on an industrial scale may be a more practical option.[199]

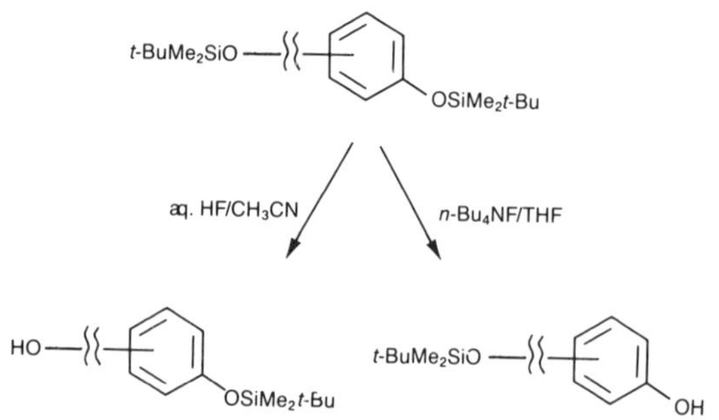


**Fig. 65.** Selective cleavage of phenoxy- or alkoxy- silyl ethers.

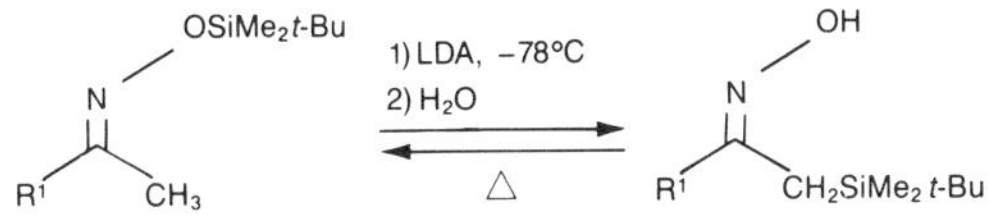


**Fig. 66.** Migration of the t-butyldimethylsilyl group.

### *Migration*

As mentioned above, migration is possible during desilylation. The 1,4-oxygen-to-oxygen migration has also been observed for vicinal diols of ribonucleosides[209] and of polyols.[210] Although less likely for the TBDMSilyl group, migration can also be facilitated by lithiation of a neighboring group. For instance, migration of the TBDMSilyl group has been observed for $\alpha$-lithiated oximes (Fig. 66).[211]

Intramolecular migration (oxygen-to-carbon) via lithiation can be incorporated into a synthetic strategy. In Fig. 67, a facile oxygen-to-carbon silyl migration produces an lithium alkoxide. In this example, the resulting alkoxide acts as an ortho-DMG and assists in the subsequent lithiation at the 4-position. Once formed, the dianion can then be reacted with various electrophiles.[212]

## Developmental products and methodologies

The commercial availability of the products listed in Table 1 are a direct result of responding to an identified need in the marketplace. These products, which are produced in bulk quantities, have become stepping stones to new products and methodologies. The following sections describe some of these new developments which are in the early stages.

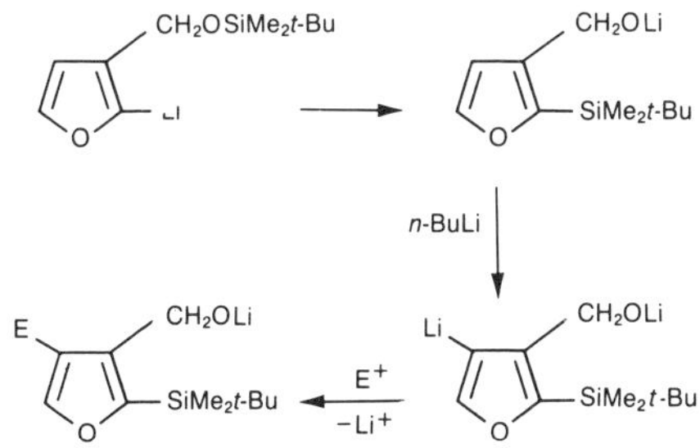


**Fig. 67.** Useful migration of the t-butyldimethylsilyl group.

**Table 6.** Comparison of the reactivities of Me(2-thienyl)CuLi and $Me_2CuLi$. 1,4-Addition to three $\alpha,\beta$-unsaturated ketones.

| No. | Substrate | Reagent | % 1,4 adduct | % 1,2 adduct | % substrate remaining | % Excess $Me^-$ used | Conditions |
|---|---|---|---|---|---|---|---|
| I | 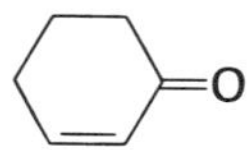 | Me(2-thienyl)CuLi | 97 | ~0·0 | ~0·0 | 20 | −78°C, 30 min |
| | | $Me_2CuLi$ | 98 | ~0·0 | ~0·0 | 110 | −78°C, 30 min |
| II | 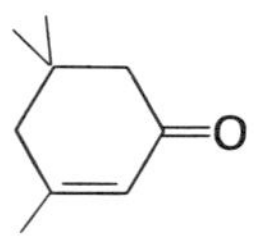 | Me(2-thienyl)CuLi | 89 | 1·4 | ~0·0 | 40 | 0°C, 1·5 h |
| | | $Me_2CuLi$ | 94 | <1% | 4·4 | 140 | −40°C, 1 h |
| | | Me(2-thienyl)CuLi | 86 | 2·0 | ~0·0 | 20 | Room temperature, 1·5 h |
| III | 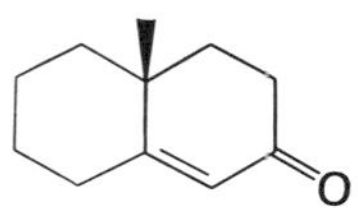 | Me(2-thienyl)CuLi | 60 | 15 | 25 | 10 | 0°C, 30 min |
| | | $Me_2CuLi$[a] | 80–90 | 10 | ~0·0 | 110 | 0°C, 30 min |

[a] Smith, Robin A. J., unpublished results. In diethyl ether.

## Methyl Thienylcuprate

Using MeLi-9307, a lithium methyl thienylcuprate composition in a THF/aromatic solvent has been developed.[213] In utility studies involving 1,4-addition of the methyl group to cyclohexenone, the reactivity of the methyl thienylcuprate product was compared with lithium dimethylcuprate. Two other $\alpha,\beta$-unsaturated ketones which are less reactive because of steric hindrance were also tested. The results of these studies are shown in Table 6. Methyl thienylcuprate showed identical reactivity in adding to cyclohexenone (substrate I) and very comparable reactivity in adding to the hindered substrates II and III.

It was also found that low temperatures are not needed for this proprietary methyl thienylcuprate.[213] The current composition (Fig. 68) is stable at room temperature. Tests show no loss of active methyl concentration after ten weeks, and NMR spectra obtained before and after the ten-week storage were identical. These results were in agreement with copper, lithium (by inductively coupled plasma analysis), and bromine (by titration), as well as by 1,4-addition to cyclohexenone.

Greater stability means that this methyl thienylcuprate can also be used at higher temperatures than lithium dimethylcuprate. Higher reaction temperatures are possible if the adduct, or enolate intermediate, is relatively stable at higher temperatures as well. As shown in Table 6, very high yields for 1,4-addition to substrate II (isophorone) were obtained at room temperature.

## Organoamides

Still another utility of dibutylmagnesium (DBM) in organic synthesis has been reported. It was found that carboxylic acids can be amidated with magnesium bis(diethylamide) and bis(cylohexylamide), prepared from DBM, at reflux conditions in the absence of ethereal solvent (Fig. 69).[214]

0.7 M in THF/cumene
0.1 M LiBr

**Fig. 68.** Stable methyl thienylcuprate.

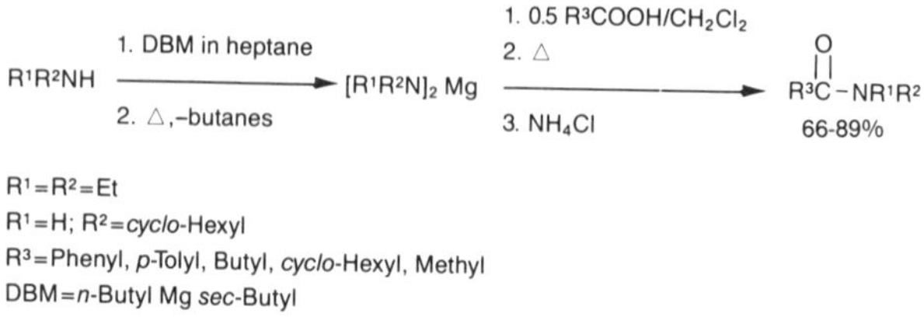

Fig. 69. Direct amidation of carboxylic acids.

This amidation does require 2 moles of the magnesium organoamide; however, compared to the other available methods, this route eliminates the activation of the acid group to an acid chloride or ester prior to addition of the amine.

FMC's DBM will provide new opportunities for the preparation of other useful magnesium organoamides (See 'General properties—Reactivity' and 'Chemical utility—Reduction'.) Eisch and Sanchez,[215] as well as Lappert and his coworkers,[216] have been the pioneers in the use of DBM in organic and organometallic syntheses.

## Future Products

The use of commercial organolithiums will only increase in the development of new products and methodologies because of their diverse utility in organic synthesis. Products already worthy of mention are chiral lithium bases[217] for asymmetric synthesis and *t*-butylsilanes such as di-*t*-butylsilane for use in the electronic industry.[218]

# References

1. Fraenkel, G., Halasa, A. F., Mochel, V., Stumpe, R. & Tate, D., *J. Org. Chem.*, **50** (1985) 4563.
2. Collum, D. B. & Galiano-Roth, A. S., *J. Am. Chem. Soc.*, **111** (1989) 6772.
3. Williard, P. G. & Salvino, J. M., *J. Am. Chem. Soc.*, **112** (in press); for X-ray crystal of lithium hexamethyldisilazide and lithium tetramethylpiperidide, see: Lappert, M. F., Slade, M. J., Singh, A., Atwood, J. L., Rogers, R. D. & Shakir, F., *J. Am. Chem. Soc.*, **105** (1983) 302; Rogers, R. D., Atwood, J. L. & Gruning, R., *J. Organomet. Chem.*, **157** (1978) 229; Mootz, D., Zinnius, A. & Bottcher, B., *Angew. Chem. Int. Ed. Engl.*, **8** (1969) 378.

4. Seebach, D. & Bauer, W., *Helv. Chim. Acta.*, **67** (1984) 1972.
5. Armstrong, D. R., Barr, D., Clegg, W., Mulvey, R. E., Reed, D., Snaith, R. & Wade, K., *J. Chem. Soc., Chem. Commun.* (1986) 869; Barr, D., Clegg, W., Mulvey, R. E., Snaith, R. & Wright, D. S., *J. Chem. Soc., Chem. Commun.* (1987) 716.
6. Morrison, R. C., Mehta, V. C., Hall, R. W. & Rathman, T. L. (FMC Lithium Division), Patent Appl. USSN 160 371, 1988; Morrison, R. C., Mehta, V. C., Hall, R. W., Rathman, T. L. & Kamienski, C. W. (FMC Lithium Division), Patent Appl. CIP USSN 334 667, 1989; Morrison, R. C., Hall, R. W. & Rathman, T. L. (FMC Lithium Division), US Patent 4 595 779, 1986.
7. Morrison, R. C. & Rathman, T. L. (FMC Lithium Division), Patent Appl. USSN 160 388, 1988; Mehta, V. C., Morrison, R. C. & Kamienski, C. W. (FMC Lithium Division), Patent Appl. USSN 488 901, 1990.
8. Kamienski, C. W., McElroy, B. J. & Bach, R. O. (FMC Lithium Division), US Patent 4 069 267, 1978.
9. Wakefield, B. J., *The Chemistry of Organolithium Compounds.* Part I. Pergamon Press, New York, 1974 (currently out of print).
10. Bates, R. B. & Ogle, C. A., *Carbanion Chemistry.* Springer-Verlag, New York, 1983, Chapter II; Buckingham, J. (Executive ed.), *Dictionary of Organometallic Compounds.* Chapman and Hall, London, 1984.
11. Buncel, E., *Carbanions: Mechanistic and Isotopic Aspects, Reaction Mechanisms in Organic Chemistry,* Monograph 9. Elsevier Science Publishing Company, Inc., New York, 1975.
12. Li, W.-K., Nobes, R. H., Poppinger, D. & Radom, L., Molecular orbital theory of carbanions, Chapter 1; Fox, M. A., Electrochemistry of carbanions, Chapter 2; Grovenstein, E. Jr, Structures of organodiaalkali metal compounds ('dianions'), Chapter 3; Tolbert, L. M., The photochemistry of resonance-stabilized anions, Chapter 4; Chambers, R. D. & Bryce, M. R., Fluoro-carbanions, Chapter 5; Koch, H. F., Reactions of hydrogen-bonded carbanion intermediates, Chapter 6. In *Comprehensive Carbanion Chemistry,* Part C, *Ground and Excited State Reactivity,* ed. E. Buncel & T. Durst. Elsevier Science Publishing Company, Inc., New York, 1987.
13. Schlosser, M., *Polare Organometalle.* Springer-Verlag, Heidelberg, 1973; Schlosser, M. & Strunk, S., *Tetrahedron Lett.*, **25** (1984) 741 and references therein.
14. Bradsma, L., Verkruijsse, H. D., Schade, C. & Schleyer, P. von R., *J. Chem. Soc., Chem. Commun.* (1986) 260; Schlosser, M. & Hartmann, J., *Angew. Chem.*, **85** (1973) 544; *Int. Ed. Engl.* (1973) 439.
15. Peterson, D. J., *J. Org. Chem.*, **9** (1967) 373.
16. Rausch, M. D. & Ciapenelli, D. J., *J. Org. Chem.*, **10** (1967) 127.
17. Slocum, D. W., Book, G. & Jennings, C. A., *Tetrahedron Lett.* (1970) 3443.
18. Meyer, N. & Seebach, D., *Angew. Chem.*, **90** (1978) 553; *Int. Ed. Engl.*, **17** (1978) 521.
19. Brandsma, L. & Verkruijsse, H., *Preparative Polar Organometallic Chemistry,* Vol. 1. Springer-Verlag, New York, 1987, pp. 136, 146, 167.

20. Brandsma, L. & Verkruijsse, H., *Preparative Polar Organometallic Chemistry,* Vol. 1. Springer-Verlag, New York, 1987.
21. Lochmann, L. & Trekoval, J., *J. Organomet. Chem.,* **179** (1979) 123.
22. Seebach, D., *Angew Chem. Int. Ed. Engl.,* **27** (1988) 1624 and references therein.
23. Sibi, M. P., Miah, M. A. J. & Snieckus, V., *J. Org. Chem.,* **49** (1984) 737.
24. Braun, M. & Devant, R., *Tetrahedron Lett.,* **25** (1984) 5031.
25. Siegal, C. & Thornton, E. R., *Tetrahedron Lett.,* **27** (1986); Siegal, C. & Thornton, E. R., *J. Am. Chem. Soc.,* **111** (1989) 5722.
26. Mehta, V. C., Morrison, R. C., Rathman, T. L. & Sanchez, R. (FMC Lithium Division), Patent Appl. USSN 160 229, 1988.
27. Corey, E. J. & Gross, A. W., *Tetrahedron Lett.,* **25** (1984) 495.
28. Krafft, M. E. & Holton, R. A., *Tetrahedron Lett.,* **24** (1983) 1345.
29. Eaton, P. E., Lee, C.-H. & Xiong, Y., *J. Am. Chem. Soc.,* **111** (1989) 8016.
30. Bates, R. B., Kroposki, L. M. & Potter, D. E., *J. Org. Chem.,* **37** (1972) 560.
31. Gilman, H. & Gaj, B. J., *J. Org. Chem.,* **22** (1957) 1165.
32. Hase, T. A., Introduction: classical and umpoled synthons, Chapter 1; Ager, D. J., Formyl and acyl anions $^{-}$CH=O and $^{-}$CR=O, Chapter 2; Kolb, M., Hydroxycarbonyl anions and related synthons, Chapter 3; Saavedra, J. E., Heteroatom-substituted $sp^3$ carbanionic synthons, Chapter 4; Hase, T. A. & Koskimies, J. K., Carbonyl alpha-cations; $^{+}$C—C=O, Chapter 5; Werstiuk, N. H., Homoenolate anion equivalents: $^{-}$C—C—CO, Chapter 6. In *Umpoled Synthons, A Survey of Sources and Uses in Synthesis,* ed. T. A. Hase. John Wiley & Sons, New York, 1987.
33. Negishi, E., *Organometallics in Organic Synthesis.* John Wiley & Sons, New York, 1980.
34. Fieser, M., *Fieser and Fieser's Reagents for Organic Synthesis,* Vols 1–14. John Wiley & Sons, New York, 1967–1989.
35. Seebach, D. & Geiss, K.-H., Organolithium compounds in organic synthesis, recent developments. In *New Applications of Organometallic Reagents in Organic Synthesis,* ed. D. Seyferth. Elsevier Science Publishing Company, New York, 1976, Chapter 1.
36. Schollkopf, U., Lithiumorganische Verbindungen. In *Houben/Weyl Methoden der organische Chemie.* Bc XIII/1. Georg Thieme-Verlag, Stuttgart, 1970.
37. Stowell, J. C., *Carbanions in Organic Synthesis.* John Wiley & Sons, New York, 1979.
38. Cram, D. J., *Fundamentals of Carbanion Chemistry.* Academic Press, New York, 1965.
39. Wardell, J. L., Preparation and use in organic synthesis of organolithiums and Group 1A organometallics. In *The Chemistry of the Metal–Carbon Bond,* Vol. 4, *The Use of Organometallic Compounds in Organic Synthesis,* ed. F. R. Hartley. John Wiley & Sons, New York, 1987, Part 1, Section 1.

40. Wakefield, B. J., Compounds of the alkali and alkaline earth metals in organic synthesis. In *Comprehensive Organometallic Chemistry,* vol. 7, ed. G. Wilkinson, F. G. A. Stone & E. W. Abel. Pergamon Press, New York, 1982, Chapter 44.
41. Wakefield, B. J., *Organolithium Methods, Best Synthetic Methods.* Academic Press, New York, 1988.
42. Hogen-Esch, T. & Smid, J. (eds), *Recent Advances in Anionic Polymerization.* Elsevier, 1987.
43. Allen, G. & Bevington, J. C. (eds), Anionic Polymerization, 14 chapters in *Comprehensive Polymer Science,* Vol. 3, "*Chain Polymerization I.* Pergamon Press, New York, 1989.
44. Van Beylen, M., Bywater, S., Smets, G., Szwarc, M. & Worsfold, D. J., Developments in anionic polymerization, a critical review. *Adv. Polym. Sci.,* **86** (1988) 87.
45. Rempp, P., Franta, E. & Herz, J. E., Macromolecular engineering by anionic methods. *Adv. Polym. Sci.,* **86** (1988) 145.
46. Quirk, R. P., Kinning, D. J. & Fetters, L. J. The preparation and morphological features of block copolymers having well-defined structures. In *Comprehensive Polymer Science,* Vol. 7, ed. G. Allen & J. C. Bevington. Pergamon Press, New York, 1989.
47. Morton, M., *Anionic Polymerization: Principles and Practice.* Academic Press, New York, 1982.
48. Halasa, A. F., Recent advances in anionic polymerization. *Rubber Rev., Rubber Chem. Tech.,* **54** (1981) 627.
49. Bywater, S., Anionic polymerization. In *Encyclopedia of Polymer Science and Engineering* Vol. 2, 2nd edn. Wiley–Interscience, New York, 1985, pp. 1.
50. Quirk, R. P., Recent advances in anionic polymerization. In *Advances in Polymer Synthesis,* ed. B. M. Culbertson & J. E. McGrath. Plenum, 1985, p. 321.
51. Liu, J. J., (Dupont), US Patent Appl. 663 663, 1984.
52. Van Leusen, A. M. & Van Leusen, D., *Synth. Commun.,* **8** (1978) 397.
53. Bare, T. M. & House, H. O. *Organic Syntheses,* Vol. V. John Wiley & Sons, New York, 1973, p. 775.
54. Wade, P. A., Price, D. T., McCauley, J. P. & Carroll, P. J., *J. Org. Chem.,* **50** (1985) 2804.
55. Miljkovic, M., Gligoijevic, M., Satoh, T. & Miljkovic, D., *J. Org. Chem.,* **39** (1974) 1379.
56. Mukaiyama, T., Soai, K., Sato, T., Shimizu, H. & Suzuki, K., *J. Am. Chem. Soc.,* **101** (1979) 1455.
57. Evans, D. A., *Science,* **240** (1988) 420.
58. Noyori, R., Suga, S., Kawai, K., Okada, S. & Kitamura, M., *Pure and Appl. Chem.,* **60** (1988) 1597.
59. Goldstein, S. W. & Dambek, P. J., *Synth. Commun.* (1989) 221.
60. House, H. O., Respess, W. L. & Whitesides, G. M., *J. Org. Chem.,* **31** (1966) 3128; House, H. O. & Fischer, W. F., *J. Org. Chem.,* **33** (1968) 949.

61. Erdik, E., *Tetrahedron,* **40** (1984) 641; Herr, R. N. & Johnson, C. R., *J. Am. Chem. Soc.,* **92** (1970) 4978.
62. Ouannes, C., Dressaire, G. & Langlois, Y., *Tetrahedron Lett.* (1977) 815; Hallnemo, G. & Ullenius, C., *Tetrahedron Lett.,* **27** (1986) 395.
63. Lipshutz, B. H. & Sengupta, S., *Organic Reactions* (in press); Posner, G. H., *An Introduction to Synthesis Using Organocopper Reagents.* Wiley, New York, 1980.
64. Posner, G. H., *Organic Reactions,* (1972) **19** 1; Posner, G. H., *Organic Reactions,* **22** (1975) 253; Taylor, R. J. K., *Synthesis* (1985) 364.
65. House, H. O., Phillips, W. Y. & Vanderveer, D., *J. Org. Chem.,* **44** (1979) 2400.
66. Bertz, S. H., Jelinski, L. W. & Dabbagh, G., *J. Chem. Soc., Chem. Commun.* (1983) 388; Stork, G. & Hudrlik, P. F., *J. Am. Chem. Soc.,* **90** (1968) 4462, 4464.
67. Baigrie, L. M., Seiklay, H. R. & Tidwell, T., *J. Am. Chem. Soc.,* **107** (1985) 5391; see also Haner, R., Laube, T. & Seebach, D. *J. Am. Chem. Soc.,* **107** (1985) 5396; Baigrie, L. M., Lenoir, D., Seiklay, H. R. & Tidwell, T., *J. Org. Chem.,* **20** (1985) 2105.
68. Rathman, T. L., The synergisym of lithium and silicon chemistry. Chemspec 1990 USA, Cherry Hill, NJ, and references therein for examples of *C*-silylation.
69. Davies, S. G., Dordor-Hedgecock, I. M., Easton, R. J. C., Preston, S. C., Sutton, K. H. & Walker, J. C., *Bull. Soc. Chim. Fr.,* No. 4 (1987) 608: A Review; Davies, S. G. & Wills, M., *J. Organomet. Chem.,* **328** (1987) C29; BP Chemicals Limited, New Specialities Business, London.
70. Evans, D. A., Stereoselective alkylation reaction of chiral metal enolates. Chapter 1; Heathcock, C. H., The aldol addition reaction, Chapter 2; Lutomski, K. A. & Meyers, A. I., Asymmetric synthesis via chiral oxazolines. Chapter 3; Enders, D., Alkylation of chiral hydrazones, Chapter 4. In *Asymmetric Synthesis,* Vol. 3, *Stereo-differentiating Addition Reactions,* Part B, ed. J. D. Morrison, Academic Press, Orlando, 1984; Evans, D. A., Nelson, J. V. & Taber, T. R., Stereoselective aldol condensations. In *Topics in Stereochemistry,* Vol. 13, ed. N. L. Allinger, E. L. Eliel & S. H. Wilen. John Wiley & Sons, New York, 1982, Chapter 1.
71. Creger, P. L. (Parke-Davis & Co.), US Patent 3 674 836, 1972; Creger, P. L., *J. Am. Chem. Soc.,* **89** (1967) 2500.
72. Verhoeven, T. R. & Askia, D. (Merck & Co, Inc.), US Patent 4 820 850, 1989.
73. Seebach, D. & Aebi, J. D., *Tetrahedron Lett.,* **24** (1983) 3311.
74. Seebach, D. & Aebi, J. D., *Tetrahedron Lett.,* **25** (1984) 2545.
75. Seebach, D., *Angew. Chem. Int. Ed. Engl.,* **27** (1988) 1624 and references therein.
76. Baker, M. V., Ghitgas, C., Hayes, R. K., Hilliker, A. E., Lynch, G. J., Sherwood, G. V. & Yeo, H.-L., *Tetrahedron Lett.,* **25** (1984) 1625.
77. Rathman, T. L., Sleevi, M. C., Krafft, M. E. & Wolfe, J. F., *J. Org. Chem.,* **45** (1980) 2169.
78. Kende, A. S. & Fludzinski, P., *Organic Syntheses,* **64** (1985) 68.

79. Spitzner, D. & Engler, A., *Organic Syntheses,* **66** (1987) 37.
80. Corey, E. J. & Enders, D., *Tetrahedron Lett.* (1976) 3.
81. Enders, D., Kipphardt, H., Gerdes, P., Brena-Valle, L. J. & Bhushan, V., *Bull. Soc. Chim. Belg.,* **97** (1988) 691.
82. Newcomb, M., Bergbreiter, D. E., Meyers, A. I., Smith, K. J. & Williams, D. R., *Tetrahedron Lett.,* **24** (1983) 3559; Smith, K. J., Bergbreiter, D. E. & Newcomb, M., *J. Am. Chem. Soc.,* **105** (1983) 4396; Honek, J. F., Mancini, M. L. & Belleau, B., *Tetrahedron Lett.,* **24** (1983) 257.
83. Hosomi, A., Araki, Y. & Sakurai, H., *J. Am. Chem. Soc.,* **104** (1982) 2081.
84. Collum, D. B. & Galiano-Roth, A. S., *J. Am. Chem. Soc.,* **111** (1989) 6772; see also Collum, D. B. & Galiano-Roth, A. S., *J. Am. Chem. Soc.,* **110** (1988) 3456; Wanat, R. A. & Collum, D. B., *J. Am. Chem. Soc.,* **106** (1984) 4865; **108** (1986) 3415.
85. Abdel-Magid, A., Pridgen, L N., Eggleston, D. S. & Lantos, I., *J. Am. Chem. Soc.,* **108** (1986) 4595
86. Bal, B., Buse, C. T., Smith, K. & Heathcock, C. H., *Organic Syntheses,* **63** (1984) 89.
87. Montgomery, S. H., Pirrung, M. C. & Heathcock, C. H., *Organic Syntheses,* **63** (1984) 99.
88. Braun, M. & Devant, R., *Tetrahedron Lett.,* **25** (1984) 5031.
89. Siegal, C. & Thornton, E. R., *Tetrahedron Lett.,* **27** (1986) 457; Siegal, C. & Thornton, E. R., *J. Am. Chem. Soc.,* **111** (1989) 5722.
90. Heathcock, C. H. & Arseniyadis, S., *Tetrahedron Lett.,* **26** (1985) 6009.
91. Yamamoto, Y., Yatagai, H. & Maruyama, K., *Tetrahedron Lett.,* **23** (1982) 2387.
92. Widdowson, D. A., Wilbecke, G. H. & Williams, D. J., *Tetrahedron Lett.,* **23** (1982) 4285.
93. Murphy, P. J., Procter, G. & Russell, A. T., *Tetrahedron Lett.,* **28** (1987) 2037.
94. Wardell, J. L., Preparation and use in organic synthesis of organolithiums and Group 1A organometallics. In *The Chemistry of the Metal–Carbon Bond,* Vol. 4, *The Use of Organometallic Compounds in Organic Synthesis,* ed. F. R. Hartley. John Wiley & Sons, New York, 1987, Part 1, Section 1; Wakefield, B. J., *Organolithium Methods, Best Synthetic Methods.* Academic Press, New York, 1988, Chapters 3, 6, 9, 12; Wakefield, B. J., *The Chemistry of Organolithium Compounds.* Pergamon Press, New York, 1974, Part I, Chapters 3, 4, 7, 8, 9, 10, 17 (currently out of print).
95. Snieckus, V. Short Course Notes, *The Aromatic Directed Metalation in Organic Synthesis,* 3rd edn. University of Waterloo, 1989.
96. Snieckus, V., *Chemical Reviews,* in 1990.
97. Beak, P. & Snieckus, V., *Acc. Chem. Res.,* **15** (1982) 306.
98. Beak, P., Zajdel, W. J. & Reitz, D. B., *Chemical Reviews,* **84** (1984) 471.
99. Gschwend, H. W. & Rodriguez, H. R., *Organic Reactions,* **26** (1976) 1.
100. Parham, W. E. & Bradsher, C. K., *Acc. Chem. Res.,* **15** (1982) 300.

101. Wakefield, B. J., *Organolithium Methods, Best Synthetic Methods.* Academic Press, New York, 1988, Chapter 3, pp. 27–32.
102. Beak, P., Musick, T. J. & Chen, C.-W., *J. Am. Chem. Soc.,* **110** (1988) 3538 and references cited therein.
103. Bauer, W. & Schleyer, P. von R., *J. Am. Chem. Soc.,* **111** (1989) 7191; for a theoretical discussion of lithiation of aliphatic amides, see Bachrach, S. M. & Ritchie, J. P., *J. Am. Chem. Soc.,* **111** (1989) 3134.
104. Turner, J. A., *J. Org. Chem.,* **48** (1983) 3401; Marsais, F. & Queguiner, G., *Tetrahedron,* **39** (1983) 2009.
105. Braye, E. (Parcor), US Patent 4 127 580, 1978.
106. Davenport, J. D., Hackler, R. E. & Taylor, H. M. (Lilly, USA), FR Patent 1 569 940, 1969.
107. Adamson, D. W. (Burroughs Wellcome & Co.), US Patent 2 712 023, 1955.
108. Richardson-Merrell, Inc., UK Patent 1 025 578, 1964; Screttas, C. G., Eastham, J. F. & Kamienski, C. W., *Chimia,* **24** (1970) 109.
109. Meyers, A. I., Boes, M. & Dickman, D. A., *Organic Syntheses,* **67** (1988) 60.
110. Kress, T. H. & Leanna, M. R., *Synthesis* (1988) 803.
111. Roe, A. M., Burton, R. A., Wiley, G. L., Baines, M. W. & Rasmussen, A. C., *J. Med. Chem.,* **11** (1968) 814.
112. Slocum, D. W. & Jennings, C. A., *J. org. Chem.,* **41** (1976) 3653.
113. DeSilva, S. O., Ahmad, I. & Snieckus, V., *Can. J. Chem.,* **57** (1979) 1598.
114. Watanabe, M., Sahara, M., Kubo, M., Furakawa, S., Billedeau, R. J. & Snieckus, V., *J. Org. Chem.,* **49** (1984) 742.
115. Comins, D. L., Brown, J. D. & Mantlo, N. B., *Tetrahedron Lett.,* **23** (1982) 3979; Comins, D. L. & Brown, J. D., *J. Org. Chem.,* **54** (1989) 3730.
116. Gillies, I. & Loft, M. S., *Synth. Commun.,* **18** (1988) 191.
117. Katritzky, A. R., Vazquez de Miguel, L. M. & Rewcastle, G. W., *Synthesis* (1988) 215.
118. Katritzky, A. R., Akutagawa, K. & Jones, R. A., *Synth. Commun.,* **18** (1988) 1151.
119. Katritzky, A. K., Fan, W.-Q. & Akutagawa, K., *Tetrahedron Lett.,* **42** (1986) 4027.
120. Katritzky, A. R., Fan, W.-Q. & Akutagawa, K. *Synthesis* (1987) 415.
121. Katritzky, A. R. & Sengupta, S., *Tetrahedron Lett.,* **28** (1987) 1847.
122. Guanti, G., Banfi, L., Narisano, E. & Scolastico, C., *Tetrahedron Lett.,* **25** (1984) 4693.
123. Colvin, E. W., McGarry, D. & Nugent, M. J., *Tetrahedron,* **44** (1988) 4157.
124. Gennari, C., Schimperna, G. & Venturini, I., *Tetrahedron,* **44** (1988) 4221.
125. Hart, D. J., Kanai, K., Thomas, D. G. & Yang, T.-K., *J. Org. Chem.,* **48** (1983) 289.
126. Asao, N., Uyehara, T. & Yamamota, Y., *Tetrahedron,* **44** (1988) 4173.
127. Kawabata, T., Sumi, K. & Hiyama, T., *J. Am. Chem. Soc.,* **111** (1989) 6843.

128. Manhas, M. S., Bose, A. K. & Khavjavi, M. S., *Synthesis* (1982) 407.
129. Livingston, D. A., Pearlman, B. A. & Denmark, S. (Upjohn), EPO PCT WO Patent Appl. 8 803 534, 1988; Livingston, D. A., Petre, J. E., Bergh, C. L. & Perrault, W. R., 1989 International Chemical Congress of Pacific Basin Societies, Honolulu, Hawaii, Organic Div., Paper 33, 17–22 December 1989.
130. Campbell, S. F., Hardstone, J. D. & Palmer, M. J., *Tetrahedron Lett.*, **25** (1984) 4813.
131. Jakiela, D. J., Helquist, P. & Jones, L. D., *Organic Syntheses*, **62** (1984) 74.
132. Khatri, H. N., Fleming, M. P. & Schloemer, G. C., (Syntex (USA), Inc.), EP 275 092 Al, 1988.
133. Sibi, M. P., Miah, M. A. J. & Snieckus, V., *J. Org. Chem.*, **49** (1984) 737.
134. Mills, R. J., Taylor, N. J. & Snieckus, V., *J. Org. Chem.*, **54** (1989) 4327; Snieckus, V. & Mills, R.J., *J. Org. Chem.*, **54** (1989) 4386.
135. Wada, A., Yamamoto, H. & Kanatomo, S., *Heterocycles*, **27** (1988) 1345.
136. Gribble, G. W., Keavy, D. J., Branz, S. E., Kelly, W. J. & Pals, M. A. *Tetrahedron Lett.*, **29** (1988) 6227.
137. Pansegrau, P. D., Rieker, W. F. & Meyers, A. I., *J. Am. Chem. Soc.*, **110** (1988) 7178.
138. Barrett, A. G. M. & Seth, H. G., *J. Org. Chem.*, **48** (1983) 5017.
139. Maryanoff, B. E. & Reitz, A. B. *Chemical Reviews*, **89** (1989) 863: see Refs 11 and 12 therein for more synthetic applications; Gosney, I. & Rowley, A. G. In *Organophosphorus Reagents in Organic Synthesis*, ed. J. I. G. Cadogan. Academic Press, New York, 1979, p. 107; Bestmann, H. J. & Vostrowski, O., *Top. Curr. Chem.*, **109** (1983) 85.
140. Magnus, P., *Aldrichimica Acta*, **13,** No. 13 (1980).
141. Ager, D. N., *Snythesis* (1984) 384: review contains 162 references.
142. Fraenkel, G., Hsu, H. & Su, B. M., The structure and dynamic behavior of organolithium compounds in solution, 13C, $^{6}$Li, and $^{7}$Li NMR. In *Lithium: Current Applications in Science, Technology and Medicine*, ed. R. O. Bach. Wiley–Interscience, New York, 1985, p. 273.
143. Seyferth, D. & Weiner, M. A., *J. Am. Chem. Soc.*, **83** (1961) 3585.
144. Chamberlin, A. R., Liotta, E. L. & Bond, F. T., *Organic Syntheses*, **61** (1982) 141.
145. Brandsma, L., Verkruijsse, H. D., Schade, C. & Schleyer, P. von R., *J. Chem. Soc., Chem. Commun.* (1986) 260; Schlosser, M. & Hartmann, J., *Agnew. Chem.*, **85** (1973) 544; *Int. Ed. Engl.* (1973) 439.
146. Akiyamama, S. & Hooz, J., *Tetrahedron Lett.* (1973) 4115.
147. Heus-Kloos, Y. A., deLong, R. L. P., Yerkruijsse, H. D., Brandsma, L. & Julia, S., *Synthesis* (1985) 958.
148. Schlosser, M., *Pure & Appl. Chem.*, **60** (1988) 1627.
149. Negishi, E. I., King, A. O. & Tour, J. M., *Organic Syntheses*, **64** (1985) 44.
150. Midland, M. M., *J. Org. Chem.*, **40** (1975) 2250.
151. Timko, J. M., VanRheenan, V. H. & Cha, D. Y. (Upjohn Co.), US Patent Appl. 84 308 826, 1984.

152. Alexakis, A., Cahiez, G. & Normant, J. F., *Organic Syntheses,* **62** (1984) 1.
153. Murahashi, S. I., Naota, T. & Tanigawa, Y., *Organic Syntheses,* **62** (1984) 39.
154. Longone, D. T. & Wright, W. D., *Tetrahedron Lett.* (1969) 2859; Seyferth, D. & Cohen, H. M., *J. Organoet. Chem.,* **1** (1963) 15.
155. For aromatic carbanions, see: Reed, J. N., PhD Dissertation, University of Waterloo, 1985; Parker, K. A. & Koziski, K. A., *J. Org. Chem.,* **52** (1987) 674; for benzylic carbanions, see: Parker, K. A. & Kallmerten, J., *J. Org. Chem.,* **45** (1980) 2620; Parker, K. A. & Kallmerten, J., *J. Am. Chem. Soc.,* **102** (1980) 5881.
156. Ladd, D. L. & Weinstock, J., *J. Org. Chem.,* **46** (1981) 203; Royen, J. J., *J. Heterocyclic Chem.,* **22** (1985) 39.
157. Sinhababu, A. K., Kawase, M. & Borchardt, R. T., *Tetrehedron Lett.,* **28** (1987) 4139.
158. Meth-Cohn, O., *J. Chem. Soc., Chem. Commun.* (1988) 1378.
159. Hanessian, S., *Aldrichimica Acta,* **22,** No. 1 (1989) 8.
160. Vedejs, E. & Larsen, S., *Organic Syntheses,* **64** (1985) 127.
161. Davis, F. A., Sankarlal, G. & Wei, J., *Tetrahedron Lett.,* **29** (1988) 4269.
162. Taddei, M. & Ricci, A., *Synthesis,* (1986) 633.
163. Camici, L., Dembech, P., Ricci, A., Seconi, G. & Taddei, M., *Tetrahedon,* **44** (1988) 4197.
164. Hwu, J. R., *J. Org. Chem.,* **48** (1983) 4432.
165. Okamura, W. H., Peter, R. & Reischl, W., *J. Am. Chem. Soc.,* **107** (1985) 1034.
166. Rubottom, G. M., Gruber, J. M., Juve, H. D. Jr & Charleson, D. A., *Organic Syntheses,* **64** (1985) 118.
167. Paquette, L. A., DeRussy, D. T., Pegg, N. A., Taylor, R. T. & Zydowsky, T. M., *J. Org. Chem.,* **54** (1989) 4576.
168. Hall, S. S. & Lipsky, S. D., *J. Org. Chem.,* **38** (1973) 1735.
169. Hall, S. S. & Srisethnil, S. T., *J. Org. Chem.,* **42** (1977) 4266.
170. Hall, S. S. & Schumacher, D. P., *J. Org. Chem.,* **46** (1981) 5060.
171. Hall, S. S. & Weiberth, F. J., *J. Org. Chem.,* **57** (1986) 5838.
172. Hall, S. S. & Farahat, S. E., *J. Heterocycl. Chem.,* **24** (1987) 1205.
173. Majewski, M., *Tetrahedron Lett.,* **29** (1988) 4057.
174. Kowalski, C., Creary, X., Rollin, A. J. & Burke, M. C., *J. Org. Chem.,* **43** (1978) 2601.
175. Sanchez, R. & Scott, W., *Tetrhaedron Lett.,* **29** (1988) 139.
176. Sanchez, R., Vest, G., Scott, W. & Engel, P. S., *J. Org. Chem.,* **54** (1989) 4026.
177. Lalonde, M. & Chan, T. H., *Synthesis,* **9** (1985) 817.
178. Corey, E. J. & Venkateswarlu, A., *J. Am. Chem. Soc.,* **94** (1972) 6190.
179. Corey, E. J., Cho, H., Rucker, C. & Hua, D. H., *Tetrahedron Lett.,* **22** (1981) 3455.
180. Chaudhary, S. K. & Hernandez, O., *Tetrahedron Lett.* (1979) 99.
181. Kim, S. & Chang, H., *Bull. Chem. Soc. Japan,* **58** (1985) 3669.
182. Aizpurua, J. M. & Palomo, C., *Tetrahedron Lett.,* **26** (1985) 475.

183. Mandai, T., Mori, K., Hasegawa, K., Kawada, M. & Otera, J., *Tetrahedron Lett.*, **25** (1984) 5225.
184. Sauer, G., Heindl, J. & Wachtel, H., *Tetrahedron Lett.*, **29** (1988) 2425; Sauer, G., Schroter, B. & Kunger, H., *Tetrahedron Lett.*, **29** (1988) 6429.
185. Piers, E., Burmeister, M. S. & Reissig, H., *Can. J. Chem.*, **64** (1986) 180.
186. Sommer, L. H., *Stereochemistry, Mechanisms and Silicon.* McGraw-Hill, New York, 1965, pp. 132, 138.
187. Zibuck, R. & Seebach, D., *Helv. Chim. Acta,* **71** (1988) 237.
188. Shirai, F. & Nakai, T., *Tetrahedron Lett.*, **29** (1988) 6461.
189. Kim, S. & Park, J. H., *Tetrahedron Lett.*, **28** (1986) 439.
190. Ogawa, Y. & Shibasaki, M., *Tetrahedron Lett.*, **25** (1984) 663.
191. Contact Peboc Limited, Llangefni, Gwynedd LL77 7YQ, Great Britain; Brussee, J., Loos, W. T., Kruse, C. G. & Van Der Gen, A., *Tetrahedron,* **46** (1990) 979.
192. See references in Azetidon-Kaneka borchure by Kaneka America Corp., Pharmaceutical Div., New York; Kaneka Deutschland GmbH., Pharmaceutical Div., Wiesbaden, Germany; Kangafuchi Chemical Industry Co. Ltd., Pharmaceutical Div., Osaka, Japan.
193. Piers, E. & Morton, H. E., *J. Org. Chem.*, **44** (1979) 3437.
194. Trimitsis, G., Beers, S., Ridella, J., Carlon, M., Cullin, D., High, J. & Brutts, D., *J. Chem. Soc., Chem. Commun.* (1984) 1088; Kuwajima, I., Kato, M., Mori, A., Oshimo, H., Enda, J. & Kobayashi, K., *J. Am. Chem. Soc.*, **106** (1984) 1773.
195. Ireland, R. E. & Varney, M. D., *J. Am. Chem. Soc.*, **106** (1984) 3668. See earlier paper of Ireland, R. E. & Daub, J. P., *J. Org. Chem.*, **46** (1981) 479.
106. Ganem, B. & Small, V. R. Jr, *J. Org. Chem.*, **39** (1974) 3728.
197. Sinhababu, A. K., Kawase, M. & Borchardt, R. T., *Tetrahedron Lett.*, **28** (1987) 4139; Sinhababu, A. K., Kawase, M. & Borchardt, R. T., *Synthesis* (1988) 710.
198. Metcalf, B. W., Burkhart, J. P. & Jund, K., *Tetrahedron Lett.*, **21** (1980) 35.
199. Baldwin, J. E. & Li, C. S., *J. Chem. Soc., Chem. Commun.* (1988) 261.
200. Stork, G. & Hudrlik, P. F., *J. Am. Chem. Soc.*, **90** (1968) 4462, 4464; Liotta, C. L. & Harris, H. P., *J. Am. Chem. Soc.*, **96** (1974) 2250.
201. Torisawa, Y., Shibasaki, M. & Ikegami, S., *Chem. Pharm. Bull.*, **31** (1983) 2607; Torisawa, Y., Shibasaki, M. & Ikegami, S., *Tetrahedron Lett.* (1979) 1865.
202. Carpino, L. A. & Sau, A. C., *J. Chem. Soc., Chem. Commun.* (1979) 514.
203. Newton, R. F., Reynolds, D. P., Finch, M. A. W., Kelly, D. R. & Roberts, S. M., *Tetrahedron Lett.* (1979) 3981.
204. Corey, E. J., Ponder, J. W. & Ulrich, P., *Tetrahedron Lett.*, **21** (1980) 137.
205. Kelly, D. R. & Roberts, S. M., *Synth. Commun.*, **9** (1979) 295.
206. Batten, R. J., Dixon, A. J. & Taylor, R. J. K., *Synthesis* (1980) 234.

206. Batten, R. J., Dixon, A. J. & Taylor, R. J. K., *Synthesis* (1980) 234.
207. Otera, J., Niibo, Y., Chikada, S. & Nozaki, H., *Synthesis* (1988) 328.
208. Collington, E. W., Finch, H. & Smith, I. J., *Tetrahedron Lett.*, **26** (1985) 681.
209. Jones, S. S. & Reese, C. B., *J. Chem. Soc., Perkin Trans.*, **1** (1979) 2762.
210. Dodd, G. H., Golding, T. & Ioannou, P. V., *J. Chem. Soc., Chem. Commun.* (1975) 249.
211. Mora, J. & Costa, A., *Tetrahedron Lett.*, **21** (1984) 3493.
212. Bures, E. J. & Keay, B. A., *Tetrahedron Lett.*, **28** (1987) 5965; Bures, E. J. & Keay, B. A., *Tetrahedron Lett.*, **29** (1988) 1247; for carbon to oxygen silyl migration, see Spinazz, P. G. & Keay, B. A., *Tetrahedron Lett.*, **30** (1989) 1765.
213. Wedinger, R. S. & Hatch, H. B. (FMC Lithium Division), Patent Appl. USSN 344 416, 1989; Wedinger, R. S. & Hatch, H. H. (FMC Lithium Division), Patent Appl. CIP USSN 404 820, 1989.
214. Sanchez, R., Vest, G. & Despres, L., *Synth. Commun.*, **19** (1989) 2909.
215. Eisch, J. J. & Sanchez, R., *J. Organomet. Chem.*, **296** (1985) C27.
216. Duff, A. W., Hitchcock, P. B., Lappert, M. F., Taylor, R. G. & Segal, J. A., *J. organomet. Chem.*, **293** (1985) 271.
217. Simpkins, N., *Chemistry and Industry* (in London), 20 June 1988, p. 387.
218. Boeglin, H. J., Kenney, L. D. & Rhodes, P. W., Systems approach to liquid-source deposition and diffusion technology. In *Microelectronic Manufacturing and Testing*. April 1990. For more information contact Olin Hunt Speciality Products Inc., West Paterson, New Jersey, USA.

# Industrial uses of cesium compounds*

**H. Prinz**

*Chemetall GmbH, Frankfurt, FRG*

## The origin of cesium

A long time ago, when the earth's surface was much hotter than today, molten silicates solidified to form huge granite bodies. Some chemical elements, including cesium, were concentrated in the last fractions to freeze and crystallized as separate minerals. One of these is pollucite, a cesium bearing ore, which is mined in Canada, Zimbabwe, and at other locations throughout the world, from where it is shipped to Chemetall's factory at Langelsheim, West Germany. Here it is processed into a full range of cesium containing products. Increasingly, a new source for cesium is also used: cesium containing process waste from customers is recycled for cesium recovery.

## The properties of cesium

In elemental form, cesium is a soft, shiny metal, which melts slightly above room temperature. As an alkali metal, cesium will react readily

* A part of this text (from 'The origin of cesium' to 'Organic synthesis') is taken with slight modifications from Bick, M., 'Cesium chemicals from the world's leading producer', Chemetall, 1989.

with most elements, to form, for example, alloys with other metals or salts when reacted with halogens. Vigorous reactions will occur upon contact with air, leading to superoxide formation, and with water, producing hydroxide and hydrogen.

Normally, cesium products come in the form of salt , with properties similar to those of the corresponding potassium salts, although they may be much more pronounced. The high atomic number, extremely low charge density and large radius of the monovalent cesium ion have important consequences, which are directly related to current and future applications.

## Products available

At our Langelsheim factory, we produce a wide range of cesium salts:

| | |
|---|---|
| cesium acetate | cesium hydroxide |
| cesium alcoholates | cesium iodide |
| cesium bicarbonate | cesium metal |
| cesium bromide | cesium nitrate |
| cesium carbonate | cesium oxalate |
| cesium chloride | cesium phosphate |
| cesium fluoride | cesium sulphate |
| cesium formate | cesium trifluoroacetate |
| | and others |

These products can be supplied from 99·5 to 99·999% purity, most of them in tonnage quantities.

## The uses

Since cesium products have become commercially available, several industries have started to make use of their unique properties. Some applications are briefly described here.

### Biotechnology

DNA and RNA contain the genetic information needed for the reproduction of every organism. High purity cesium salts are used in the recovery and purification of DNA and RNA by ultracentrifugation, for application for example in the development of pest resistant agricultural crops and the like. A typical procedure ('Isolation of high molecular weight chromosomal DNA from eukariotic cells') is given in Ref. 1.

## Catalysis

Because of its low ionization potential, cesium effectively promotes the performance of many metal oxide catalysts for heterogeneous processes. This effect is often enhanced by cesium's ability to stabilise high oxidation states of transition metal oxo anions.

By formation and stabilisation of complex salts with transition metal halide catalysts, cesium will also promote heterogeneous halogen transfer processes. In any case, cesium is equivalent to potassium, yet more powerful, as an alkaline promoter. Partial exchange of K with Cs in a catalyst may already demonstrate this.

In homogeneous catalysis, reaction rates are sometimes limited by the solubility of an anionic compound in an organic solvent, in which case the use of cesium as cation may solve the problem.

Catalysts—most of them used on a commercial scale—for production of the following chemicals are activated by addition of cesium:

- Styrene monomer
  —from ethyl benzene at metal oxide contacts
  —from toluene and methanol at Cs-exchanged zeolites
- ethylene oxide
- ammonoxidation, (meth-) acrylonitrile
- (meth-) acrolein
- (meth-) acrylic acid, methyl methacrylate monomer
- fuel methanol, pure methanol
- phthalic anhydride
- anthraquinone
- olefins (metathesis, oligomerisation, from methanol, butadiene, conversion to aromatics, isomerisation)
- low pressure ammonia synthesis
- chlorinations (dimethyl dichloro silane, etc.)
- conversion of $SO_2$ to $SO_3$

## Glass Manufacture

Various forms of cesium, especially the nitrate, are used as glass components to achieve various objectives.[2] The refractive index of optical glass—in the bulk or on the surface only—can be modified by addition of cesium salts. Through surface ion exchange with cesium salt melts or solutions, glass surfaces can be made resistant to corrosion or breakage.

Recently, the manufacture of planar waveguides by ion exchange has been published.[3]

## Radiation Monitoring

Cesium iodide and fluoride have the ability to absorb X-rays, gamma and particle radiation, emitting visible light. This so-called scintillation effect is used in medical diagnostics, exploration for natural resources and in nuclear physics research.[4]

For this purpose, crystals are grown from the melts of highest purity cesium iodide and cesium fluoride, or deposited by evaporation.

## Organic Synthesis

The solubility of some cesium salts in polar solvents allows the very strong basicity of cesium bases to be used in organic reactions.

In cases where the sodium or potassium counterparts fail to give good results due to the strong influence of the cation on the reaction mechanism, cesium may be the cation of choice.

It will not decompose as many organic compounds do. Thus formation of by-products due to interference of the base component with the desired route of reaction is not possible.

Recovery of the used cesium component and recycling to form the desired cesium base again are possible.

Below we will focus on some typical reactions with industrial potential.

## Esterifications

Since cesium fluoride is very soluble in carboxylic acids[5] the reactant acid can be used as solvent in ester forming reactions. Specific

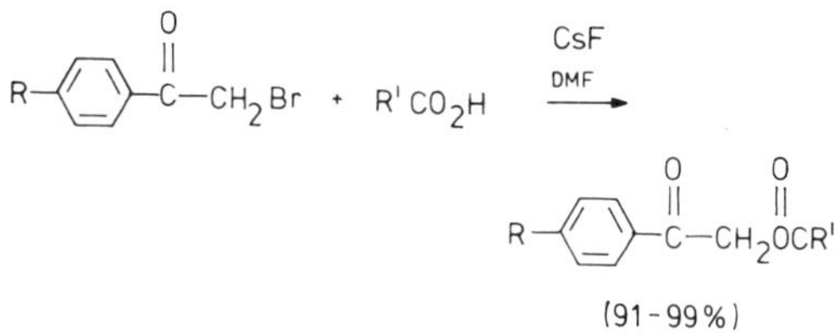

R = H, Br, or Ph; R' = Me, Et, $Me_3C$, Ph, 2-$MeOC_6H_4$, 4-t-$BuC_6H_4$, or 2,4,6-$(Me)_3C_6H_2$

**Scheme 1.**

problems in ester synthesis can be solved; for example phenacyl esters[6] can be produced at high reaction rates[7] using an aprotic solvent (Scheme 1).

A potential application is for synthesis of Merifield resin esters of *N*-protected amino acids (Scheme 2).[8]

## Transesterifications

Ogilvie and Beaucage[9] report complete conversion of triphenyl phosphate into trialkyl phosphates (Scheme 3), which are used as plasticisers.

The general reaction (Scheme 4)[10] may be of importance for the synthesis of insoluble fatty acid esters and polyesters.[11]

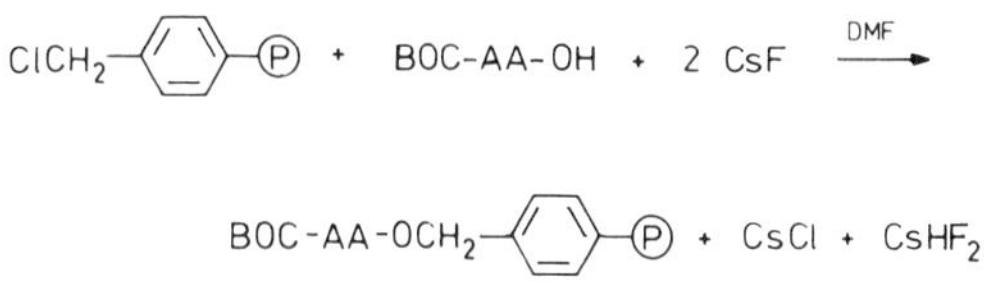

P = polystyrene backbone; BOC-AA-OH = tert-butoxycarbonylamino acid

**Scheme 2.**

$$(PhO)_3PO \xrightarrow[\text{8hr, Room temperature}]{\text{CsF, ROH}} (RO)_3PO + 3PhOH$$

R = Me, Et, *n*-Pr, *i*-Pr, *n*-Bu and *n*-$C_5H_{11}$

**Scheme 3.**

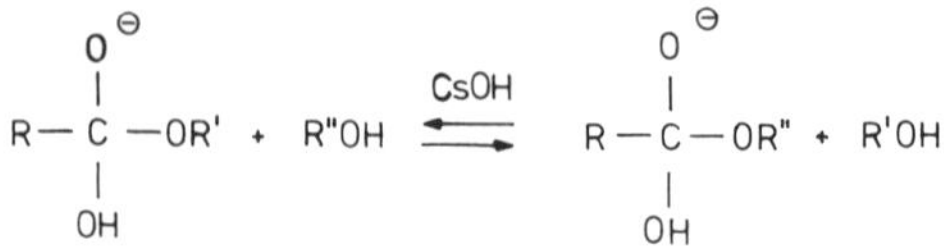

**Scheme 4.**

## Polymerizations

The anionic oligomerization of tetrafluoroethylene yields a mixture of oligomers:[12]

$$n\mathrm{F_2C=CF_2} \xrightarrow{\text{CsF/diglyme}} \mathrm{C_8F_{16}} + \mathrm{C_{20}F_{20}} + \mathrm{C_{12}F_{24}} \qquad (\mathrm{C_{10}F_{20}} = 80\%)$$

Oxiranes may react with fluorinated ketones forming either cyclic dioxolanes or higher polymers (Scheme 5), depending on reaction conditions.[13,14]

Synthesis of a bi-functional fluorinated polymer of formula described in Ref. 15 involves:

$$\mathrm{FOC\underset{\displaystyle CF_3}{\underset{|}{C}}F}\!-\!\left(\mathrm{OCF_2\underset{\displaystyle CF_3}{\underset{|}{C}}F}\right)_m\!-\!\mathrm{OCF_2CF_2O}\left(\mathrm{\underset{\displaystyle CF_3}{\underset{|}{C}}FCF_2O}\right)_p\!-\!\mathrm{\underset{\displaystyle CF_3}{\underset{|}{C}}FCOF}$$

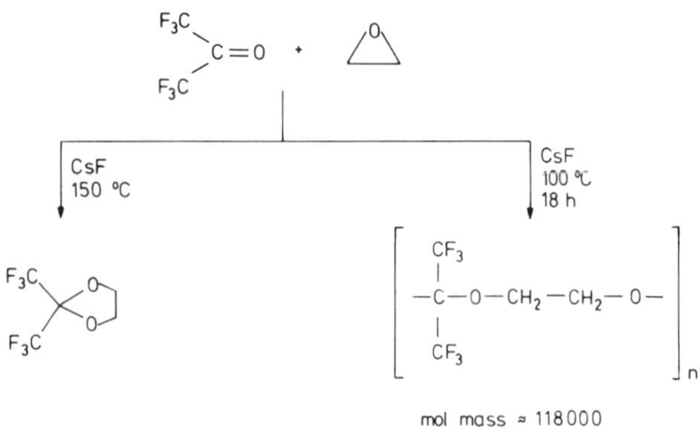

**Scheme 5.**

1. Preparation of an initiator from cesium fluoride and oxalyl-fluoride in tetraethyleneglycol dimethylether.
2. Preparation of the catalyst from the initiator and hexafluoro propylene oxide (HFPO) forming:

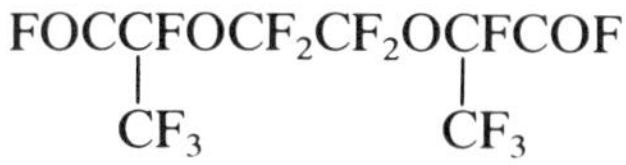

which is then reacted with an excess of cesium fluoride (10%) to form the catalyst

3. Polymerization of HFPO using the catalyst to form a Cs—O—$CF_2$ group containing polymer. The functional groups are decomposed in a last step to form

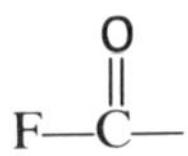

groups and cesium fluoride by heating.

## Intramolecular Cyclizations

Cyclic oligo ethers can be produced utilising the template effect of the cesium cation (Scheme 6).[16]

Especially synthesis of *N*-containing crown ethers could be accomplished via the cesium method (Scheme 7).[17]

Lactone formation of $\omega$-halogeno-carboxylic acids with high yields especially for the large sized rings is carried out with $Cs_2CO_3$ (Scheme 8):[18,19]

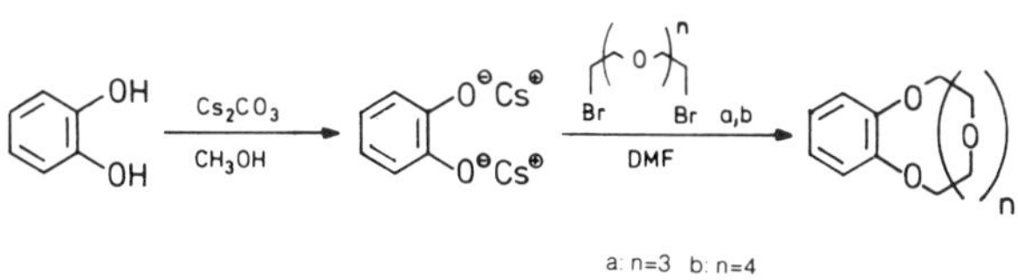


**Scheme 6.**

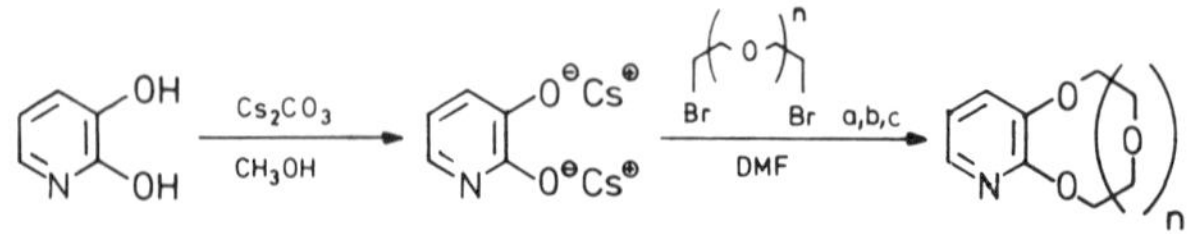

a: n=3 b: n=4 c: n=5

**Scheme 7.**

## Fluorinations

Among the ionic fluorides used in the preparation of organofluorine compounds,[20,21] only cesium fluoride shows sufficient solubility even in aprotic solvents. It can be used in halogen exchange reactions as a reactive source of nucleophilic fluorine either alone or as a catalyst to accelerate KF reactions[22] or in combination with calcium fluoride as a supported reagent.[22]

Clark[22] has given a typical example for a halex reaction:

$$C_6H_5CH_2Br \xrightarrow[\text{sulpholane}]{\text{CsF—CaF}_2} C_6H_5CH_2F$$

Reactivities of various cesium fluoride products compare as follows according to Scheme 9.[22]

Although cesium fluoride is available as a spray-dried powder with a bulk density of 1·0 g/ml, its high reactivity allows for its use in the crystalline form. This has a bulk density of 2·6 g/ml thus providing up to 2·5 times the molar amount of nucleophilic fluorine in the same volume as compared to spray-dried potassium fluoride (0·4 g/ml).

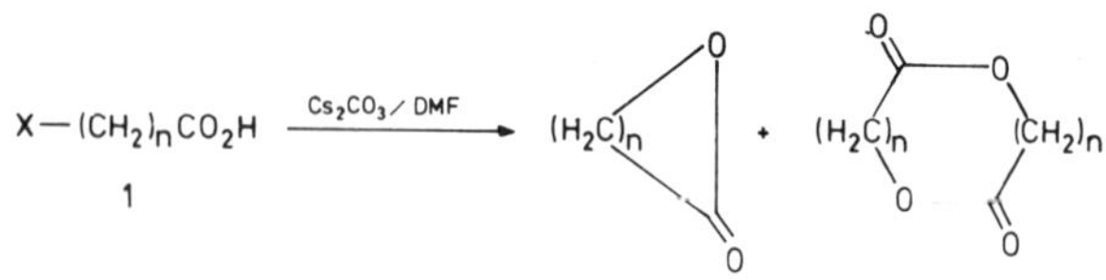

| 1 | n | X | 1 | n | X |
|---|---|---|---|---|---|
| a | 4 | I | f | 11 | I |
| b | 5 | Br | g | 12 | I |
| c | 8 | Br | h | 13 | I |
| d | 9 | I | i | 14 | I |
| e | 10 | I | j | 15 | I |

**Scheme 8.**

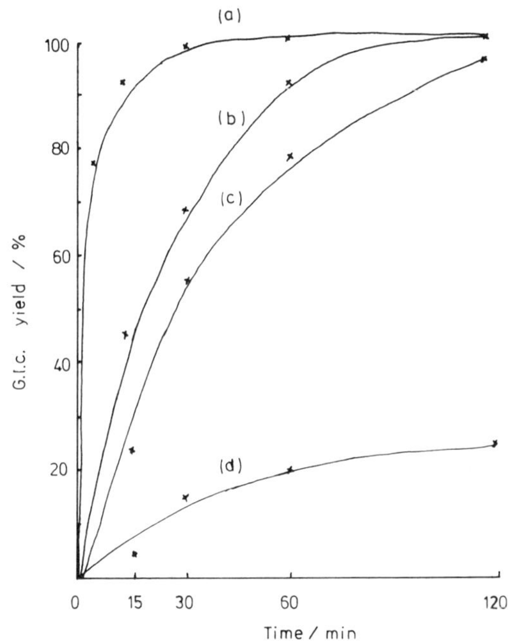


**Scheme 9.** Rate of production of $PhCH_2F$ from $PhCH_2Br$ in MeCN at 80°C using (a) CsF—$CaF_2$ (dried at 80°C for 1 h under reduced pressure); (b) CsF (dried at 80°C for 1 h under reduced pressure); (c) CsF with added $CaF_2$ (untreated laboratory reagents) and (d) CsF (untreated laboratory reagent).

# Summary

It was possible to compile this survey of applications because the commercial availability of cesium compounds has encouraged several industries to make use of their unique properties on a large scale. Since potential applications may require volume supply, Chemetall will always provide sufficient production capacity for cesium compounds to support industrial uses.

# References

1. Glover, D. M. (ed.), *DNA Cloning,* Vol. 1, 38. IRL Press, Oxford, Washington, 1985.

2. Hayami, R. & Terai, R., *Physics and Chemistry of Glasses*, **13** (1972) 102.
3. Rasper, J. & Ross, L., Schott Glaswerke, German Patent 35 01 898.
4. Kubota, S., Sakuragi, S., Hashimoto, S. & Ruan, J., *Nuclear Instruments and Methods in Physics Research*, **268** (1988) 275.
5. Emsely, J., *J. Chem. Soc.*, A (1971) 2511.
6. Clark, J. A., *Chem. Rev.*, **80** (1980) 429.
7. Clark, J. H. & Miller, J. M., *Tetrahedron Lett.* (1977) 599.
8. Horiki, K., Igano, K. & Inouye, K., *Chem. Lett.* (1978) 165.
9. Ogilvie, K. K. & Beaucage, S. L., *J. Chem. Soc., Chem. Commun.* (1976) 443.
10. Christen, H. R., *Grundlagen der Organischen Chemie*. Frankfurt am Main, 1975, p. 567.
11. *Encyclopedia of Polymer Science and Engineering*, 2nd edn, Vol. 12, 32ff.
12. Graham, D. P., *J. Org. Chem.*, **31** (1966) 955.
13. Madison, N. L., British Patent 1 141 265, 1969.
14. Madison, N. L., US Patent 3 324 145, 1964.
15. Hill, J. F. & Garabectian, M. E. of E I DuPont de Nemours, German Patent Appl. 2 147 102, 1979.
16. Klieser, B., Rossa, L. & Vögtle, F., *Kontakte*, **1** (1984) 3.
17. von Keulen, B. J., Kellogg, R. M. & Piepers, O., *J. Chem. Soc., Chem. Commun.* (1979) 285.
18. Krnizinga, W. H. & Kellogg, R. M., *J. Chem. Soc., Chem. Commun.* (1979) 286.
19. Krnizinga, W. H. & Kellogg, R. M., *J. Am. Chem. Soc.*, **103** (1981) 5183.
20. Gerstenberger, M. & Haas, A., *Angewandte Chemie, Int. Ed. Engl.*, **20** (1981) 647.
21. Prescott, W., *Chem. Ind.*, London (1978) 56.
22. Clark, J. H., Hyde, A. J. & Smith, D. K., *J. Chem. Soc., Chem. Commun.* (1986) 791.

## Bibliography

Comprehensive reviews on general aspects of the chemistry mentioned above are:

Bick, M., *Ullmann's Encyclopedia of Industrial Chemistry*, Vol. A6, pp. 153–6, 1986.

Clark, J. H., *Chem. Rev.*, **80** (1980) 429–52.

Klieser, B., Rossa, L. & Vögtle, F., *Kontakte*, **1** (1984) 3–17.

Yakobson, G. G. & Akhmetova, N. E., *Synthesis*, **3** (1983) 169–84.

# Catalysed oxidations using peroxygen reagents

**K. M. Dear**

*Interox, Widnes, UK*

Catalytic oxidation is an important process widely used in the chemical industry. Peroxygen reagents, particularly hydrogen peroxide, find increasing use and suitability for this type of reaction. This arises from the versatility of activation of hydrogen peroxide (Fig. 1). Direct activation, catalysis, peroxy acid and conversion to other compounds are the main routes. Catalysis by metal species or acid/base catalysis offer distinct commercial advantages over conventional peroxy acid routes. Reduction in capital costs, operational costs, improved safety and effluent control are all targets of research chemists in hydrogen peroxide catalytic chemistry.

Catalytic hydrogen peroxide reagents have now been applied to a wide range of transformations[1] (Table 1). Indeed, the versatility now matches the peroxy acids and advantage will be taken increasingly of these clean, environmentally favourable processes. Application areas to be covered are epoxidation,[2] alcohol oxidation, ketone oxidation, aromatic hydroxylation, sulphur oxidations[3] and nitrogen oxidations.

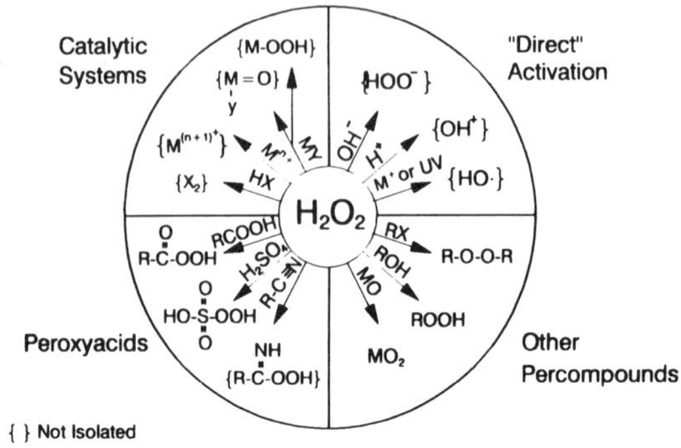


**Fig. 1.** Activation of $H_2O_2$.

# Mechanisms

Catalytic oxygen transfer reactions are broadly represented by the equation shown in Fig. 2. The active catalyst reacts with the substrate to give the oxidised product and the reduced catalyst. In many instances, the hydrogen peroxide role is to oxidise the reduced catalyst to the active form. The important parameters to be considered for maximum selectivity are the metal catalyst, oxygen donor and the ligand attached to the metal in question.

**Table 1.** Application of peroxygen systems

| | Acidic $H_2O_2$ | Alkaline $H_2O_2$ | Catalytic $H_2O_2$ | Peracids | TBH catalytic |
|---|---|---|---|---|---|
| Olefins | | | | | |
| Epoxidation | | + | + | + | + |
| Hydroxylation | | | + | + | |
| Oxidative cleavage | | | + | + | + |
| Aromatics | | | | | |
| Hydroxylation | + | | + | + | + |
| Oxidative cleavage | | | + | + | |
| Side chain oxidation | | | + | | |
| Other carbon compounds | | | | | |
| Alcohol oxidation | | | + | + | + |
| Aldehyde oxidation | | + | + | + | |
| Ketone oxidation | | + | + | + | |
| Sulphur compounds | + | + | + | + | + |
| Nitrogen compounds | | + | + | + | + |

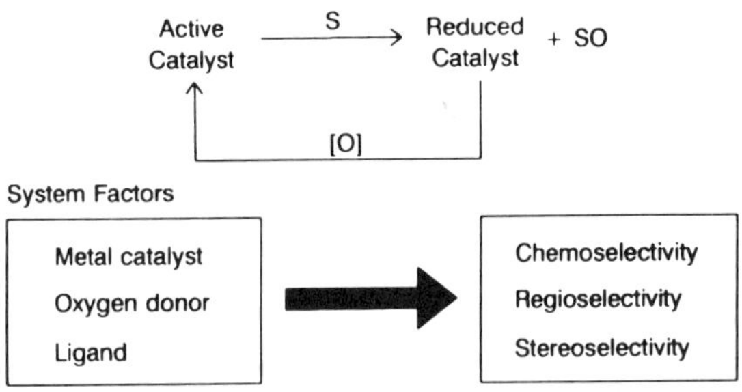


**Fig. 2.** Catalytic oxygen transfer.

The metal catalyst can achieve oxidation by either one-electron processes or two-electron processes. One-electron processes (Fig. 3) are effected by production of the hydroxyl radical which can then react with the substrate to produce a variety of products. Two-electron processes (Fig. 4) produce peroxometal or oxometal intermediates as the active catalyst. Typical examples of the former are arsenic and selenium whereas manganese, chromium and iron exhibit oxometal mechanisms (Fig. 5).

# Epoxidation

The most significant industrial use of peroxygen reagents is the epoxidation of olefins (Fig. 6). Although at present performic or peracetic acids are generally used, there is an increasing call for

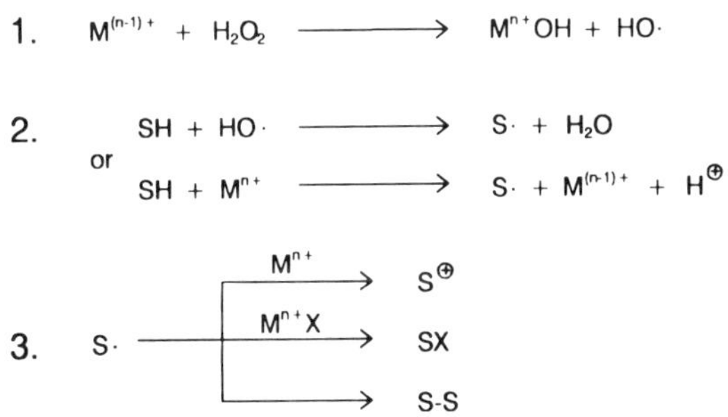


**Fig. 3.** Catalytic oxygen transfer—effect of metal catalyst (one-electron processes).

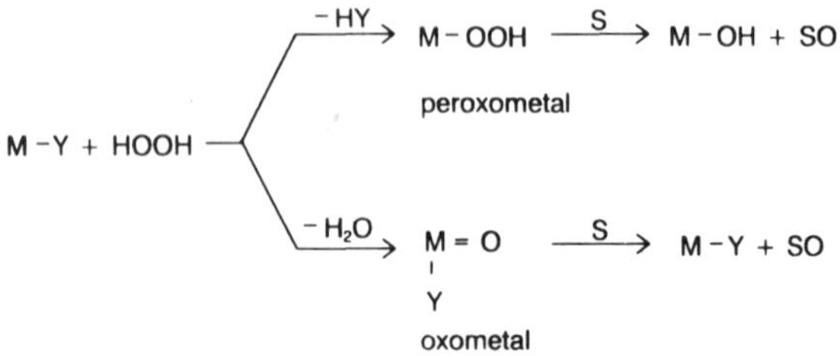


**Fig. 4.** Catalytic oxygen transfer—effect of metal catalyst (two-electron processes).

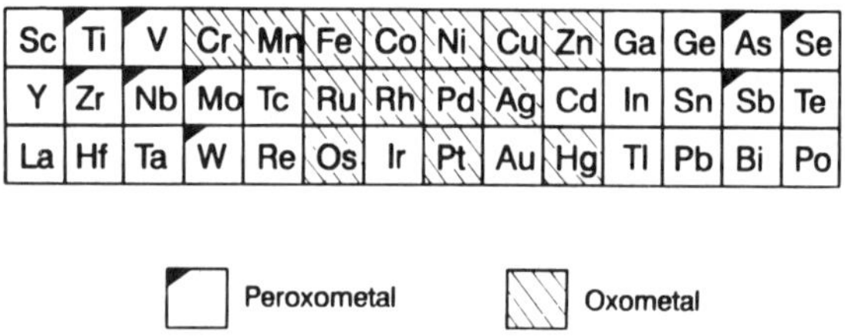


**Fig. 5.** Oxygen transfer catalysts.

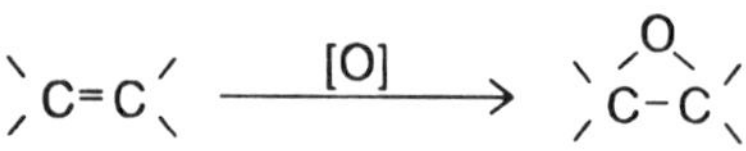


| Systems | Scope |
|---|---|
| $RCO_3H$ | Wide |
| $H_2O_2$ / Metal | Medium |
| ROOH / Metal | Medium |
| $H_2O_2$ / Nitrile | Limited |
| $H_2O_2$ / $HO^{\ominus}$ | Limited |

**Fig. 6.** Epoxides.

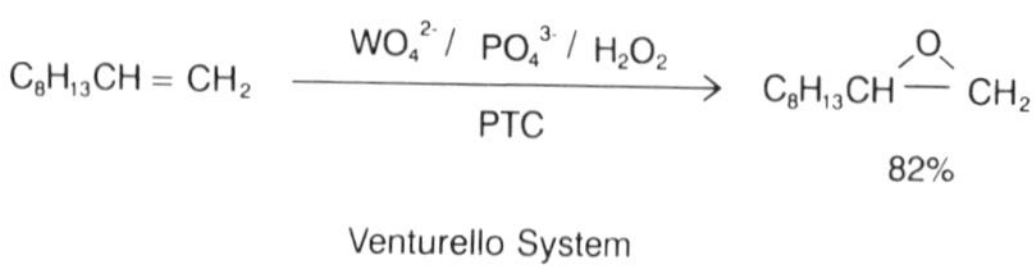


**Fig. 7.** Tungstate catalysed epoxidations.

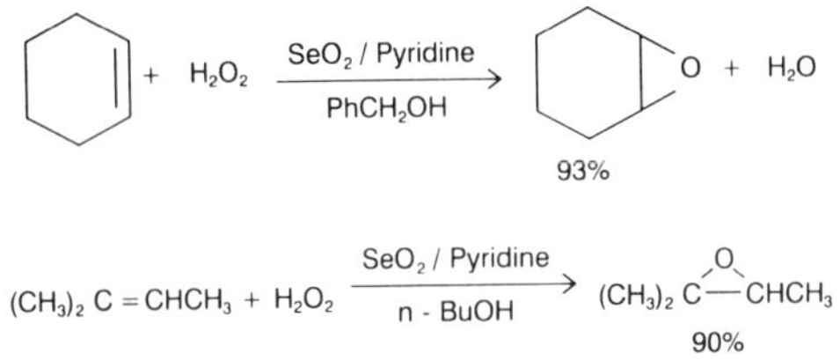


**Fig. 8.** Selenium catalysed epoxidations.

environmentally clean processes. Modern and future epoxidation reagents are likely to be based on simple catalytic hydrogen peroxide processes. Molybdates and tungstates have long been used as hydrogen peroxide catalysts for the epoxidation of activated water soluble olefins like allyl alcohol.[4] More recently, Venturello[5] has adapted such catalysts for epoxidation of non-activated olefins and preparation of acid sensitive epoxides (Fig. 7). Important advances have been made more recently by the titanium zeolite catalysts developed by Enichem.[6] The main advantage in epoxidation of the latter are their excellent efficiencies with terminal olefins.

Our own work with selenium catalysis has shown good performance with internal olefins (Fig. 8). The proposed mechanism (Fig. 9) involves formation of a peroxometal intermediate (hydroxy hydroperoxide). In this case the reduced catalyst must be dehydrated by azeotropic distillation before treatment with hydrogen peroxide.

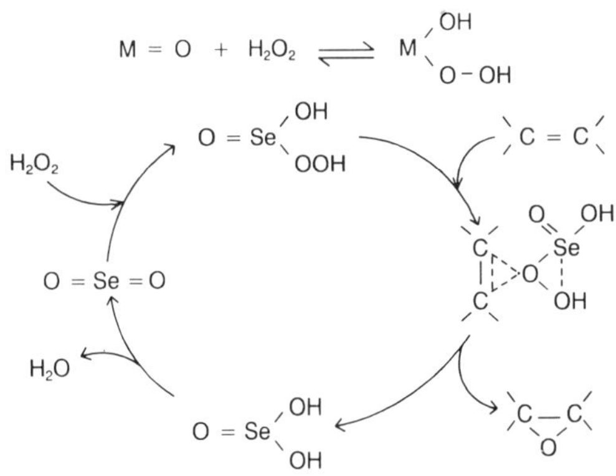

**Fig. 9.** Peroxometal mechanisms.

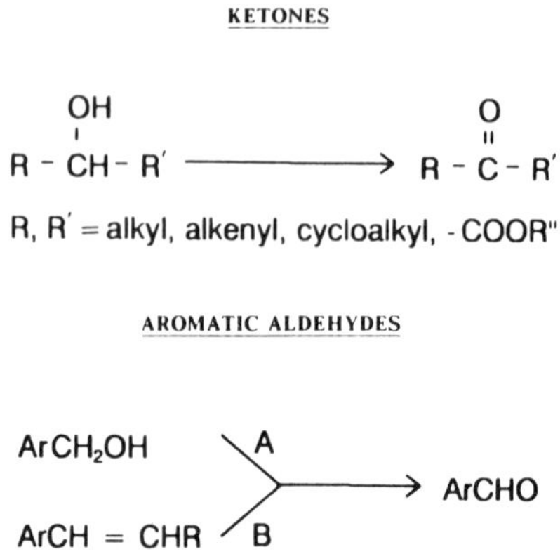


**Fig. 10.** Alcohol oxidation.

# Alcohol oxidation

The oxidation of primary and secondary alcohols to aldehydes, ketones and carboxylic acids is a useful chemical process. New methods have been developed in the recent past[7] (Fig. 10) involving catalysed hydrogen peroxide routes. Tungstates and molybdates,[8] titanium zeolites[9] and hydrogen bromide[10] have been utilised as effective catalysts for this oxidation of alcohols to carbonyl compounds. Typical results are shown in Table 2. High yields and selectivities are produced using hydrogen peroxide and ammonium molybdate. Normal reaction conditions are 85–90°C, two-phase reac-

**Table 2.** M-peroxo catalysed alcohol oxidations using $H_2O_2$/molybdate at 90°C. $H_2O$/organic with phase transfer catalyst.

| Alcohol | Carbonyl | % Conversion | % Selectivity | % Yield |
|---|---|---|---|---|
| 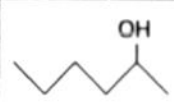 | 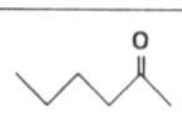 | 96 | ~100 | 96 |
| 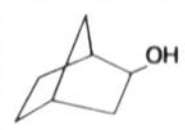 | 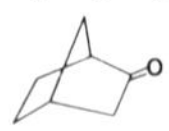 | 85 | ~100 | 85 |
| 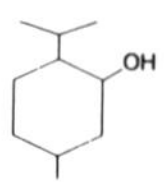 | 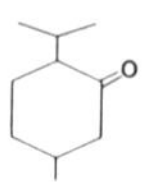 | 96 | 96 | 92 |
| 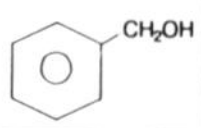 | 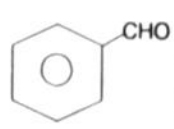 | 92 | ~98 | ~90 |

Reaction:

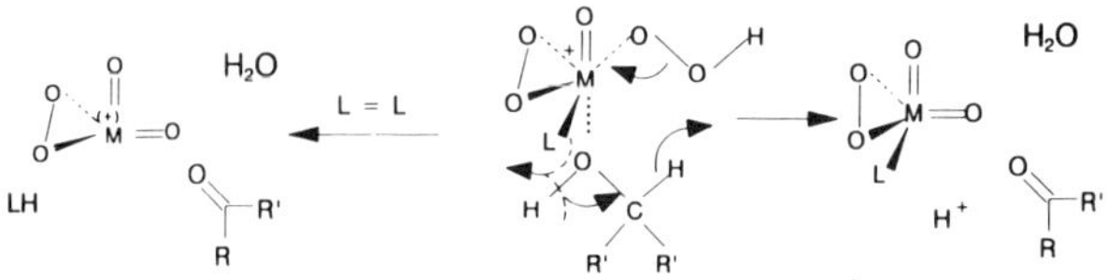


**Fig. 11.** Alcohol oxidation scheme.

tion with or without solvent with a total reaction time of 3–4 hours. The mechanism involves the formation of a metal peroxo complex with molybdenum (Fig. 11).

Oxidation of alcohols or aldehydes to the carboxylic acid often requires alternative catalysts to those quoted above (Fig. 12). With aromatic or heterocyclic aldehydes a careful choice of catalysts is required as the results may be affected by the 'Cannizaro reaction' giving acids and alcohols.

## Ketone oxidation

The oxidation of ketones to esters and lactones by peroxygen systems is the well-known 'Baeyer Villiger' reaction. Numerous catalysts based on peroxometal mechanisms have been published[11] (Fig. 13). *E*-Caprolactone is an important Interox product and we have investigated a wide range of catalysts[12] (Fig. 14). Effective catalysts are generally Lewis acids, best selectivities being obtained with metal fluorides of antimony, arsenic or boron. The processes have greater

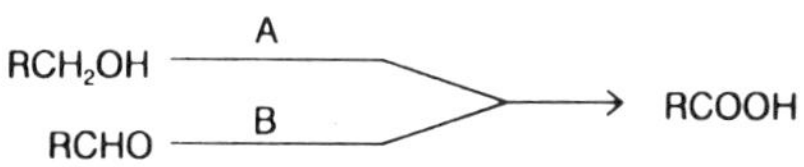


**Fig. 12.** Carboxylic acids.

**Fig. 13.** Lactones.

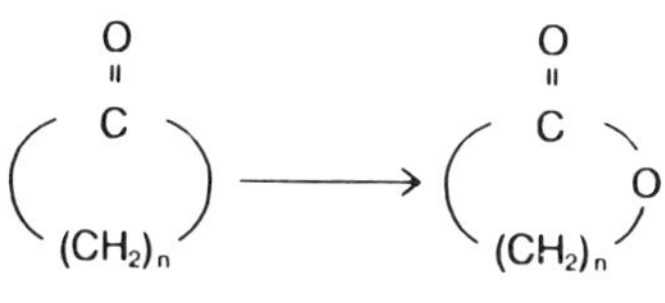

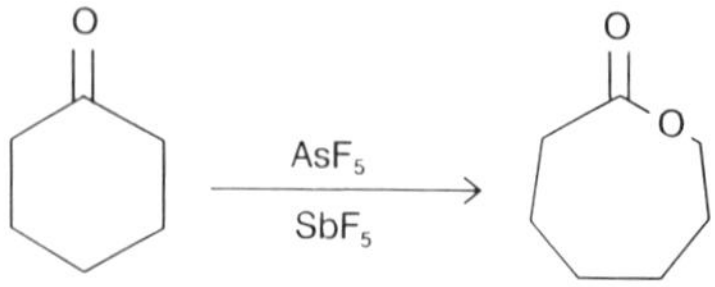


**Fig. 14.** Oxidation of cyclohexanone.

than 90% selectivity, but it is necessary to use high strength hydrogen peroxide and continuous azeotropic distillation to remove water during the reaction.

## Aromatic hydroxylation

The direct hydroxylation of aromatic compounds is a high capacity industrial process. Rhône Poulenc and Enichem produce catechol and hydroquinone from phenol by catalytic hydrogen peroxide processes (Fig. 15). Acid catalysis,[13] transition metal ions[14] and titanium zeolites[15] are all effective methods for phenol hydroxylation. Acid catalysis proceeds by an ionic mechanism involving an intermediate hydroxonium ion ($H_3O_2^+$) whereas transition metal ions promote the formation of hydroxyl radicals to effect substitution.

## Sulphur oxidation

Hydrogen peroxide is an important reagent in the oxidation of a wide variety of sulphur compounds. Thiols and thioethers are generally the substrates of main interest. The nucleophilicity of the sulphur atom governs the choice of reagent systems. We have made a detailed study

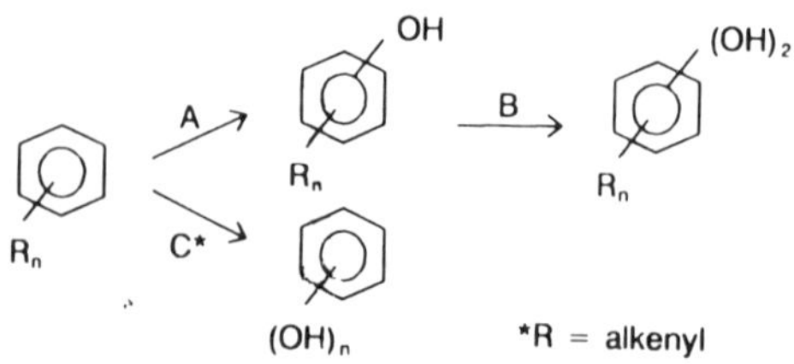


**Fig. 15.** Phenols—direct hydroxylation.

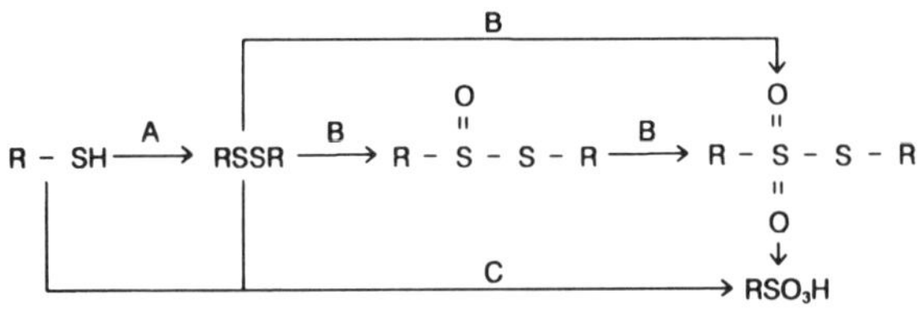

Systems

| A. | B. | C. |
|---|---|---|
| $H_2O_2/H^{\oplus}$ | $RCO_3H$ | $H_2O_2$/AcOH (opt. Fe, Cu) |
| $H_2O_2/HO^{\ominus}$ | | $H_2O_2$/Mo, V, W |
| $H_2O_2$/AcOH (opt. Fe, Cu) | | $RCO_3H$ |

**Fig. 16.** Thiol oxidation products.

of the use of various catalyst systems to effect the required transformations. Important considerations are matching reagent strength to nucleophilicity without over-oxidation.

Thiols are first oxidised to disulphides and then through various intermediates to sulphonic acids, the final oxidation product (Fig. 16). Mild catalysts with hydrogen peroxide are generally effective for disulphide preparation. A variety of more active catalyst systems are required for sulphonic acids. We have found that tungstate catalysis is effective for aliphatic thiols and iron catalysed free radical systems for aromatic thiols.

The oxidation of thiophenol (Fig. 17) shows good selectivities to disulphide formation with a number of catalysts, copper or tungsten, generally being the most effective. With excess hydrogen peroxide the best yields of phenyl sulphonic acid were achieved with Fe(II) catalysis.

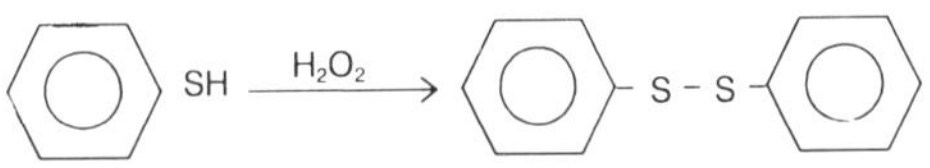

| Oxidant System | % Conversion of RSH | % Selectivity to RSSR |
|---|---|---|
| Cu II / $H_2O_2$ / HOAc | 84 | 89 |
| Fe II / $H_2O_2$ / HOAc | 72 | 88 |
| W / $H_2O_2$ / EtOH (Natural pH) | 56 | 86 |
| W / $H_2O_2$ / EtOH (pH 4-6) | 71 | 93 |
| V / $H_2O_2$ / EtOH (Natural pH) | 47 | 70 |

**Fig. 17.** Thiol oxidation.

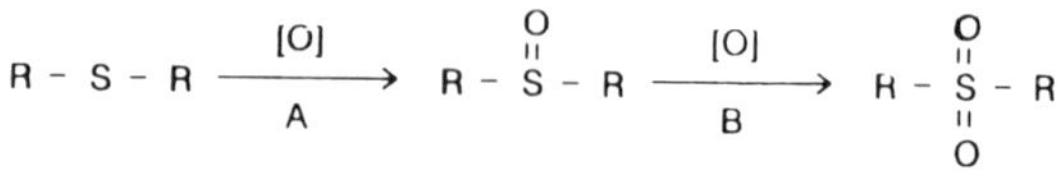

**Fig. 18.** Sulphoxides and sulphones.

Thioethers or heterocyclic sulphur compounds are oxidised to sulphoxides or sulphones (Fig. 18). Such molecules find extensive use in pharmaceutical and agrochemical products. The purity of the final product is generally of high importance, closely affecting the choice of peroxygen system. A wide range of catalysts has been examined to ascertain suitability of sulphoxide or sulphone production. Typical systems are illustrated for thiophene oxidation (Fig. 19). Useful catalysts for sulphoxide and sulphone production are peroxometal catalysts, i.e. tungstates, vanadates or molybdates.

# Nitrogen oxidations

Similar catalysis investigations have been applied to the oxidation of tertiary amines (Fig. 20). The nucleophilicity of the nitrogen atom has a prominent effect on choice of catalyst. Simple aliphatic tertiary

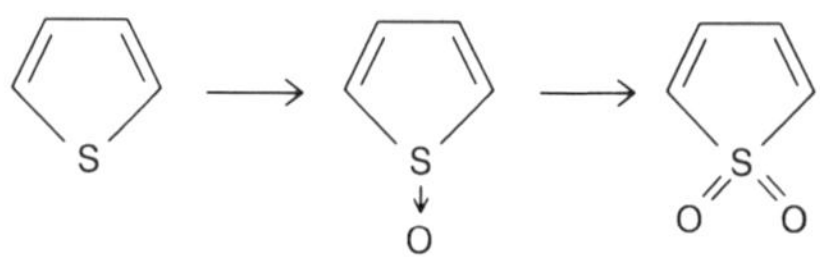

| Oxidant System | STAGE 1 % sulfoxide | STAGE 2 % sulfone |
|---|---|---|
| $H_2O_2$ / HOAc | 100 | <10 |
| $H_2O_2$ / W / EtOH<br>$H_2O_2$ / Mo / EtOH<br>$H_2O_2$ / V / EtOH | 100 | >95 (at 1% w/w catalyst and 0.1% w/w) |
| MMPP | 100 | >93 |

**Fig. 19.** Thiophene oxidation.

$$R_3N \xrightarrow{[O]} R_3N\rightarrow O$$

**Fig. 20.** Amine oxides—oxidation of tertiary amines.

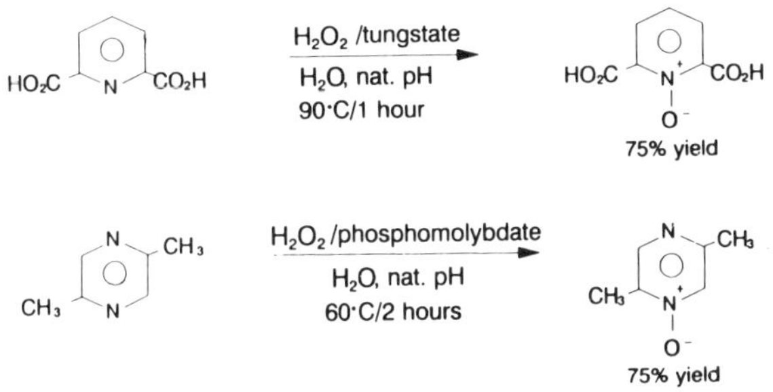

**Fig. 21.** *M*-peroxo catalysed *N*-oxidations.

amines are generally easily converted to amine oxides by uncatalysed hydrogen peroxide.[16] Aromatic and heteroaromatic systems generally give better results in the presence of catalysts. Selenium compounds[17] and tungstates and molybdates[18] have been reported. Typical results are illustrated for dipicolinic acid and dimethyl pyrazine (Fig. 21).

## Summary

We can see that the scope of catalysed hydrogen peroxide reactions is extension. Similar versatility to the widely used peroxycarboxylic acids has been developed. Catalytic oxidation is an increasingly important issue and considerable investigation continues in academic and industrial laboratories.

## References


1. Interox publication 'Peroxygen compounds in organic synthesis'.
2. Interox publication 'Epoxidation'.
3. Interox publication 'Sulphur compounds'.
4. Beg, M. A. & Ahmad, I., *J. Cat.*, **39** (1975) 260.
5. Bortolini, D., Conte, V., DiFuria, F. & Modena, G., *J. Org. Chem.*, **50** (1985) 2688.
6. Enichem, Eur. Patent Appl. 230 949 AZ, 1987.
7. Interox investigations.
8. Bortolini, D., DiFuria, F. D., Modena, G. & Seraglia, R. *J. Org. Chem.*, **51** (1986) 2661.
9. Anic, EP 102 655, 1984.

10. Interox investigations.
11. Union Carbide, US Patent 4 160 769, 1977; Allied Chemical, US Patent 4 213 906, 1980.
12. Interox, US Patent 4 353 832, 1982.
13. Rhône Poulenc, DE Patent 2 658 545, 1977.
14. Mobil Oil, US Patent 4 578 521, 1986.
15. Anig, DE Patent 3 309 669, 1983.
16. Höechst, DE Patent 3 545 152, 1987.
17. Interox, Eur. Patent 224 662, 1987.
18. Montedison, Eur. Patent Appl. 201 934, 1986.

# Applications of hydrides in chemical synthesis


**Edward A. Sullivan**

*Morton International, Danvers, Massachusetts, USA*


Chemical synthesis has long relied on reduction as one of its principal tools in building commercially useful organic compounds. Most organic reductions involve saturation of an unsaturated functional group, although some are substitution reactions. These include conversion of aldehydes and ketones to either alcohols or hydrocarbons, reduction of acyl halides and esters to alcohols, saturation of olefins to hydrocarbons, and dehalogenation to hydrocarbons. Among nitrogen-containing functional groups, reduction converts imines and quaternary iminium salts to amines, and amides, nitriles and nitro compounds to the corresponding amine (see Fig. 1).

Historically, organic reductions generally employ one of three major types of reactions (Fig. 2). The most common of these is catalytic hydrogenation, which can be applied to a variety of functional groups and is usually preferred when the compound contains only the functional group to be reduced. Typical hydrogenation catalysts are nickel, palladium, platinum and copper chromite, each of which can be made by several methods depending on the specific catalytic hydrogenation desired.

Another category of widely used traditional reducing agents is the reactive metals, including zinc, magnesium and iron. These usually

| Substrate | Product |
|---|---|
| Olefins<br>Dehalogenation<br>Aldehyde / ketone | to hydrocarbon |
| Aldehyde / ketone<br>Acyl halide<br>Ester | to alcohol |
| Amide<br>Nitrile<br>Nitro / Nitroso<br>Imine<br>Quaternary iminium | to amine |

**Fig. 1.** Typical organic reductions.

require activation by acid or by formation of an amalgam. For example, the Clemmensen reduction makes use of zinc amalgam with acid to convert aldehydes and ketones to the corresponding hydrocarbon. Another common active metal system is the Birch reduction, in which metallic lithium or sodium in liquid ammonia reduces aromatics to alicyclics. The Bouveault–Blanc reaction, which reduces esters to alcohols, also uses metallic sodium.

The third category of classical reducing systems is metal alkoxides, of which the most common is aluminum isopropoxide and isopropanol. This is known as the Meerwein–Ponndorf reduction and converts aldehydes and ketones to the alcohol. The alkoxide undergoes rapid metal–hydrogen exchange, allowing its use in catalytic quantities, with the alcohol as the source of hydrogen. This reaction usually needs forcing conditions such as acetone distillation.

A new category of reducing agents came on the scene in the 1950s when complex metal hydrides became commercially available. Lithium aluminum hydride, the first to be introduced, is a powerful reducing agent for a wide variety of organic functional groups. This reagent quickly found use in the synthesis of pharmaceutical products because of its broad applicability under very mild conditions, such as at room temperature. In many cases, however, its strong reducing power became a disadvantage when only one functional group was to be

- Catalytic hydrogenation
  Ni, Pt, Pd, $CuCr_2O_4$
- Active metals
  Li, Na, Zn, Mg, Fe
- Metal alkoxides
  $Al(Oi\text{-}Pr)_3/i\text{-}PrOH$

**Fig. 2.** Traditional reducing agents.

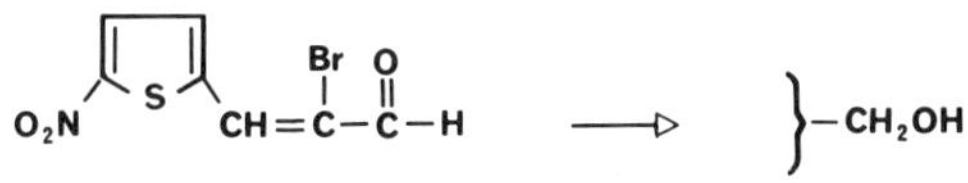


**Fig. 3.** Selectivity of $NaBH_4$ reduction: only the aldehyde group is reduced.

reduced, in a compound containing several different reducible functionalities.

Sodium borohydride became available soon thereafter and was quickly put to good use because it is a milder reducing agent than $LiAlH_4$, and because of its chemoselectivity, that is its ability to reduce aldehydes and ketones in the presence of other reducible groups in the same molecule. This selectivity is best exemplified by $\alpha$-bromo-(5-nitro-2-thienyl) acrolein (Fig. 3), in which the aldehyde group is reduced exclusively, despite the presence of nitro and halogen groups and both aliphatic and conjugated heterocyclic unsaturation.[1]

In addition to being selective in its reductions, $NaBH_4$ is also convenient and economical to use. Convenience results from its solubility in a broader range of solvents than $LiAlH_4$ and from its greater resistance to solvolysis (see Fig. 4). $NaBH_4$ is extremely stable in aqueous solution to sufficient alkalinity; this property is unique for metal hydrides which usually must be protected from water because of their rapid hydrolysis, generating hydrogen gas. With $NaBH_4$, a choice of product forms is available: powder, pellets and stabilized water solution.

The economic advantages of $NaBH_4$ are a consequence of its high efficiency. In reducing aldehydes and ketones, one mole of $NaBH_4$ can reduce four moles of the carbonyl compound in good to excellent yield and under mild conditions (see Fig. 5). The reagent's low molecular weight (37·84) and the favorable 4:1 stoichiometry result in the need

| $LiAlH_4$ | $NaBH_4$ |
|---|---|
| – Powerful reducing agent | – Mild, selective |
| – Vigorous reaction with water ⟶ $H_2$ | – Stable in aqueous alkali |
| – Soluble in ethers | – Soluble in water and alcohols, ammonia and amines, glycol ethers, $CH_3CN$, DMF and DMSO |

**Fig. 4.** Convenience of $NaBH_4$ compared to $LiAlH_4$.

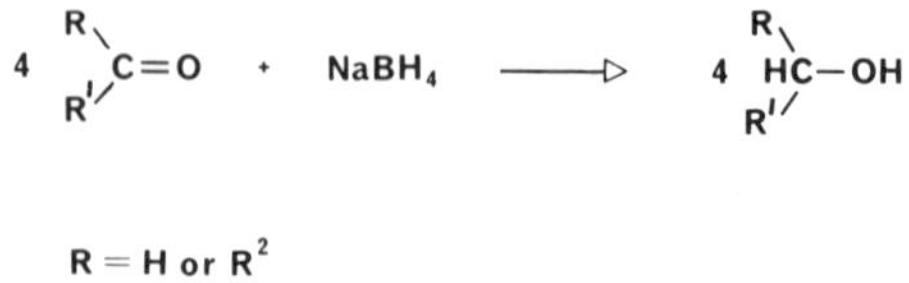


**Fig. 5.** Stoichiometry of $NaBH_4$ reduction of aldehydes or ketones.

for only small amounts to carry out a desired reduction with favorable economics.

A few examples of commercial synthetic uses of $NaBH_4$ will show its practical utility. Vitamin A or retinol has been made by the $NaBH_4$ reduction[2] of the corresponding (Fig. 6) aldehyde. Note that the conjugated double bonds are not affected during $NaBH_4$ reduction.

Another commerical aldehyde reduction in the pharmaceutical area is the manufacture of the antimicrobial agent dihydrostreptomycin[3] (Fig. 7). The broad spectrum antibiotic chloramphenicol[4] is also made using $NaBH_4$ (Fig. 8). In this case an acyl azide is reduced to the

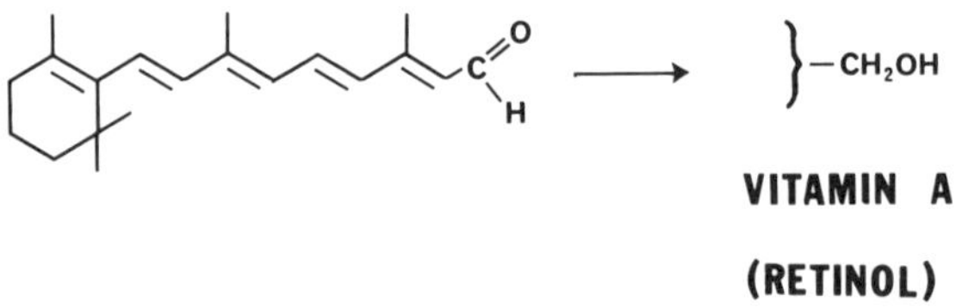


**Fig. 6.** Commercial uses of $NaBH_4$: Vitamin A synthesis.

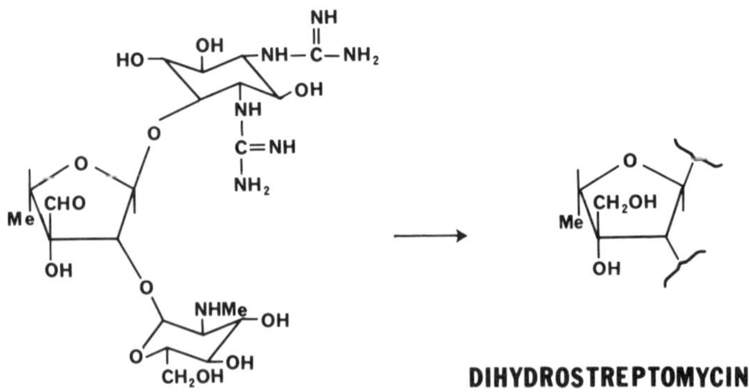


**Fig. 7.** Commercial uses of $NaBH_4$; Dihydrostreptomycin synthesis.

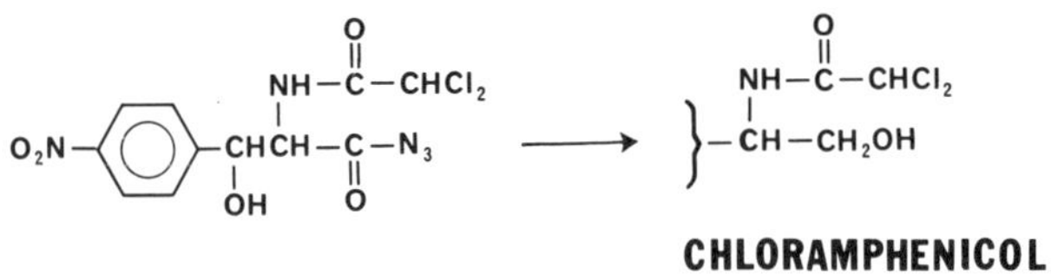


**Fig. 8.** Commercial uses of $NaBH_4$: Acyl azide reduction in chloroamphenicol synthesis.

corresponding alcohol. Note also that the nitro and amide groups are not reduced by $NaBH_4$.

A further example is provided by the analgesic or aspirin substitute, Talwin®; during its manufacture a 1,2-dihydropyridine is reduced by $NaBH_4$ to a 1,2,5,6-tetrahydropyridine,[5] constituting reduction of a carbon–carbon double bond (Fig. 9).

A last example of practical synthetic applications of $NaBH_4$ is nonmedicinal and is found in the stabilization of hops extracts used in brewing beer.[6] Hops, or their extracts, are used in brewing to impart the characteristic slightly bitter flavor of beer. This flavor is due to natural hops constituents called isohumulones. The isohumulones are very sensitive to both sunlight and artificial light, resulting in their decomposition and rapid formation of a mercaptan. Even a few parts per million of the mercaptan gives an undesirable 'skunky' odor and flavor to beer. By reducing the ketone group in isohumulones to the alcohol, they become insensitive to light, preventing this problem (Fig. 10).

Some of the above examples point out an important fact—that $NaBH_4$ does reduce functional groups other than aldehydes and ketones. Although more limited in scope than aldehyde and ketone

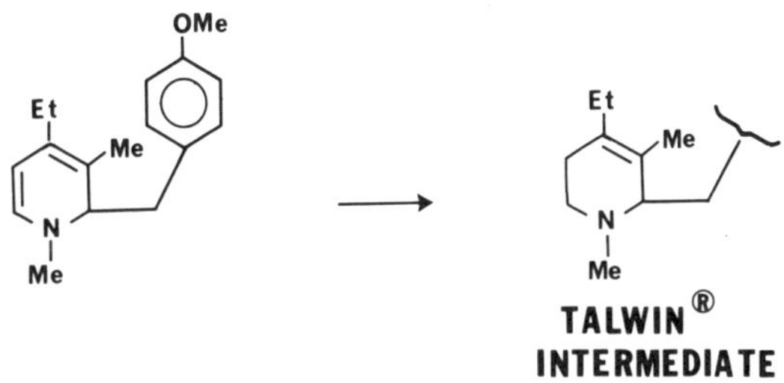


**Fig. 9.** Commercial uses of $NaBH_4$: Carbon–carbon double bond reduction in making the analgesic Talwin®.

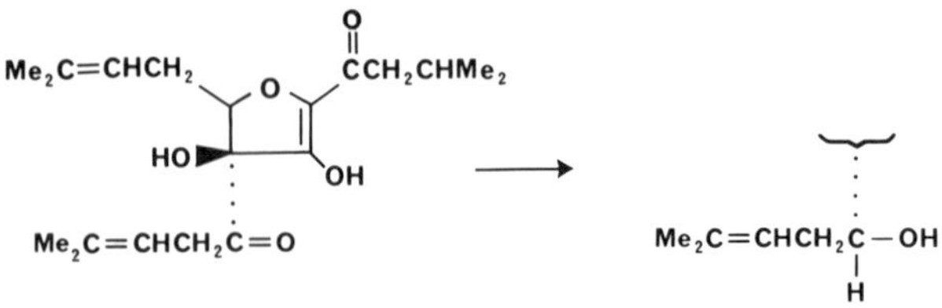

**ISOHUMULONE**

**Fig. 10.** Commercial uses of $NaBH_4$: Light-stabilization of isohumulones, used in brewing beer, by keto group reduction.

reductions, they are nevertheless useful. Imines[7] and quaternary iminium compounds[8] are generally reduced by $NaBH_4$, saturating the carbon–nitrogen double bond (see Fig. 11). Whereas alkyl and aryl halides are usually inert to $NaBH_4$, under solvolytic conditions, secondary and tertiary alkyl halides which can form stable carbonium ions can be dehalogenated to the hydrocarbon by $NaBH_4$.[9]

In general, the $NaBH_4$ reduction of esters is too slow to be of practical use, but in the presence of a lithium halide or a calcium halide, which forms $LiBH_4$[10] or $Ca(BH_4)_2$ *in situ*, ester reductions proceed readily in tetrahydrofuran, ethanol or diglyme. More difficult reducible groups, e.g. amides, nitriles or nitro compounds, are not reduced by $NaBH_4$ alone, but they can be reduced effectively by $NaBH_4$ in the presence of transition metal salts such as cobalt or nickel, forming amines.[11]

Such reactions involving the use of a co-reagent or catalyst with $NaBH_4$ are a tribute to organic chemists whose creativity led them to work out $NaBH_4$ systems for reducing functional groups not reducible with $NaBH_4$ by itself. As a further example, we return to aldehydes and ketones, which are normally reduced by $NaBH_4$ to alcohols.

| | |
|---|---|
| – Imine: | $>C{=}N{-}$ to $>CH{-}NH{-}$ |
| Quaternary iminium: | $>C{=}\overset{+}{N}{<}$ to $>CH{-}N{<}$ |
| – Dehalogenation | via stable carbonium ion |
| – Ester | *in situ* $LiBH_4$ or $Ca(BH_4)_2$ |
| – Amide<br>Nitrile<br>Nitro/nitroso | } metal catalyzed |

**Fig. 11.** $NaBH_4$ reduction of other functional groups.

Aldehydes or ketones

$$4\,RCH(=O) + NaBH_4 + 2\,H_2O \longrightarrow 4\,RCH_2OH + NaBO_2$$

Peroxides

$$4\,R_2CHOOH + NaBH_4 \longrightarrow 4\,R_2CHOH + NaBO_2 + 2\,H_2O$$

Metal cations

$$4\,M^{2+} + NaBH_4 + 2\,H_2O \longrightarrow 4\,M^0 + NaBO_2 + 8\,H^+$$

**Fig. 12.** Borohydride reduction of impurities suspected of causing quality problems.

However, in many instances, what is desired is the corresponding hydrocarbon rather than the alcohol. Formation of a tosylhydrazone intermediate and subsequent $NaBH_4$ reduction[12] forms the hydrocarbon. The net result is equivalent to the earlier mentioned Clemmensen reduction.

A different type of organic reduction which uses $NaBH_4$ consists of the removal of trace amounts of impurities from bulk organic chemicals. More than a billion pounds of organics are being purified each year with $NaBH_4$ to reduce low levels of carbonyls, peroxides or trace metals (Fig. 12). Alcohols and glycols, esters, alkoxylates, amines and amides, olefinic monomers and surfactants are among the bulk organics being treated (see Fig. 13). Addition of 50–500 ppm $NaBH_4$ to the bulk material does the job without the need for special equipment or process modifications, and the cost per pound of purified product is only one-tenth to one cent. Typical results for alcohols and ethylene glycol are shown in Table 1; special grades of these products are used in cosmetics, surfactants and prepolymerization ester manufacture, to name a few. Normal impurities are carbonyl groups such as the corresponding aldehyde in primary alcohols, and they are usually present at about 100 ppm. If not removed or reduced, they could lead to color, odor or stability problems in finished products. $NaBH_4$ treatment is an easy and economical solution to the problem, and does not require energy-intensive processing such as distillation.

— Alcohol and alkoxylates
— Amine and amides
— Ester and sulfates
— Quaternary ammonium compounds

**Fig. 13.** Classes of compounds purified with sodium borohydride.

**Table 1.** Carbonyl reduction.

| | Carbonyl level (ppm) | |
|---|---|---|
| | Initial | After SBH purification |
| Ethanol | 35 | 3 |
| 2-Ethylhexanol | 210 | 15 |
| $C_{13}$ alcohol | 120 | 40 |
| $C_{18}$ alcohol | 570 | 200 |
| Ethylene glycol | 280 | 20 |

Another metal hydride that finds significant use in organic synthesis, viz. sodium hydride, should also be mentioned briefly. While not a reducing agent for organics, sodium hydride acts as a strong base in base-catalyzed condensation and alkylation reactions.[13] Sodium hydride is used commercially in aldol, Claisen, Darzens, Dieckmann and Stobbe condensations. Alkylations with NaH frequently involve amines such as phenothiazine, but can also take place at carbon, such as in active methylene compounds.

Sodium hydride offers several advantages in comparison to alternative bases such as sodium amide, metal alkoxides or metallic sodium. It is a stronger base and therefore allows shorter reaction times under mild conditions. Unlike the alternatives, a large excess of the reagent is not required; this provides better economics. Side reactions are eliminated because forcing conditions are not needed, and the hydrogen evolved during NaH condensations and alkylations provides a convenient measure of the extent of reaction.

In addition to traditional methods of reducing organic compounds, metal hydrides, and particularly $NaBH_4$, offer a practical alternative for reducing numerous functional groups in good yield, selectively and under mild conditions. Furthermore, $NaBH_4$ has proved useful for removing trace impurities from bulk organic chemicals, thus preventing potential color, odor or stability problems in finished products. Finally, NaH has found significant use in base-catalyzed condensations and alkylations, offering significant advantages over competitive reagents.

## References

1. Carrara, G., Ettorre, R., Fava, F., Rolland, G., Testa, E. & Vecchi, A., *J. Am. Chem. Soc.*, **76** (1954) 4391.

2. Hawks, O. D., British Patent 778 753, 10 July 1957, to Eastman Kodak Co.
3. (a) Kaplan, M. A., US Patent 2 790 792, 30 April 1957, to Bristol Labs, Inc.
(b) Makepeace, E. H., US Patent 3 397 197, 13 August 1968, to Chas. Pfizer and Co., Inc.
4. Ehrhart, G., Siedel, W. & Nahm, H., *Chem. Ber.*, **90** (1957) 2088.
5. Archer, S., US Patent 3 250 678, 10 May 1966, to Sterling Drug Inc.
6. Koch, G. H., Herwig, W. C. & Kissel, T. L., US Patent 3 044 879, 17 July 1962, to Miller Brewing Co.
7. Billman, J. H. & Diesing, A. C., *J. Org. Chem.*, **22** (1957) 1068.
8. Whaley, W. M. & Robinson, C. N., *J. Am. Chem. Soc.*, **75** (1953) 2008.
9. (a) Brown, H. C. & Bell, H. M., *J. Org. Chem.*, **27** (1962) 1928.
(b) Bell, H. M. & Brown, H. C., *J. Am. Chem. Soc.*, **88** (1966) 1473.
10. Nystrom, R. F., Chaikin, S. W. & Brown, W. G., *J. Am. Chem. Soc.*, **71** (1949) 3245.
11. (a) Satoh, T., Suzuki, S., Suzuki, Y., Miyaji, Y. & Imai, Z., *Tetrahedron Lett.* (1969) 4555.
(b) Egli, R. A., *Helv. Chim. Acta*, **53** (1970) 47.
(c) Nose, A. & Kudo, T., *Chem. Pharm. Bull.*, **29** (1981) 1159.
12. Caglioti, L., *Tetrahedron*, **22** (1966) 487.
13. Banus, M. D. & Hinckley, A. A., Manufacture, handling and uses of sodium hydride. In *Handling and Uses of the Alkali Metals*. American Chemical Society, Washington, 1957, pp. 106–17.

# The Zinin reduction—a simple and highly selective process

**J. F. Hamlin & I. M. McRobbie***

*A. H. Marks and Company Limited, Bradford, UK*

## Introduction

The reduction of nitro compounds to the corresponding amines is performed regularly in both the laboratory and on the industrial scale. These conversions can be undertaken via a number of different reagents. This paper will concentrate on the use of anionic divalent sulphur compounds (in the form of ammonium, sodium or potassium sulphides, hydrosulphides or polysulphides). The first reported reduction of a nitroarene by sulphide was made by the Russian chemist, Zinin, in 1842,[1] who reduced nitrobenzene (I) to aniline (II) using ammonium hydrosulphide (Fig. 1).

Although this technique was subsequently superseded by other methods such as catalytic hydrogenation and iron and hydrochloric acid, it is still used industrially with sensitive compounds that are either incompatible with acids or are capable of being further reduced. Nevertheless, the Zinin reduction is not well known and is rarely mentioned in the current literature, or reviewed.[2]

* To whom correspondence should be addressed.

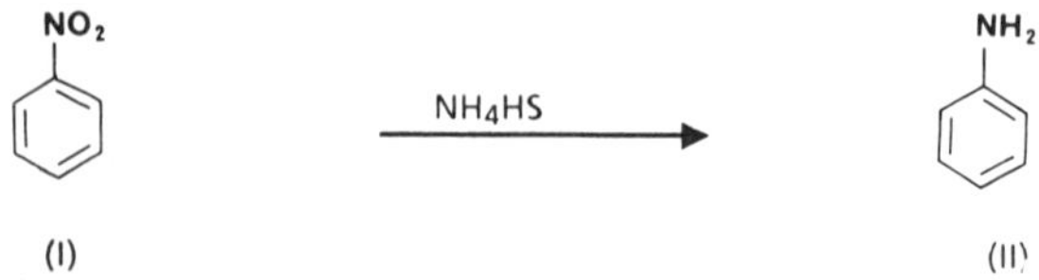


**Fig. 1.** The Zinin reduction—the first example.

# Scope

The utility of the reaction is due to the relative mildness of alkaline sulphide reductions compared with typical metal–acid methods or catalytic hydrogenation. This is due to a lower reduction potential and a narrower range of electromotive force. Consequently, it is often possible to achieve a high degree of control over the rate and selectivity by using sulphides at different pH. Indeed, the high selectivity of this technique is such that, in virtually every case, a nitro group is reduced even in the presence of other substituents which could potentially be reduced. Some examples are given in Table 1.

The process is so selective that only one nitro group is preferentially reduced in polynitro compounds. However, it is difficult to always predict which nitro group will be reduced as the mechanism(s) is not clearly understood. Nevertheless, steric hindrance and the relative reduction potentials of the different nitro groups are likely to be important parameters. Resonance and hypercongugation factors have also been used to explain the outcome of certain reactions.[9]

Many cases of the selective reduction of polynitro compounds are cited in the literature. For example, 2,4,6-trinitrotoluene (III) is reduced by sodium hydrosulphide to give 4-amino-2,6-dinitrotoluene (IV) in 72% yield (Fig. 2).[10]

**Table 1.** Selective reduction of nitro groups.

| Nitro compound | Product | Ref. |
|---|---|---|
| *p*-Nitrobenzaldehyde | *p*-Aminobenzaldehyde | 3 |
| *p*-Nitrobenzoic acid | *p*-Aminobenzoic acid | 4 |
| *p*-Nitrobenzamide | *p*-Aminobenzamide | 5 |
| *p*-Nitrobenzene sulphonic acid | *p*-Aminobenzene sulphonic acid | 6 |
| 1-Nitroanthraquinone | 1-Aminoanthraquinone | 7 |
| 4-Nitroazobenzene | 4-Aminozoabenzene | 8 |

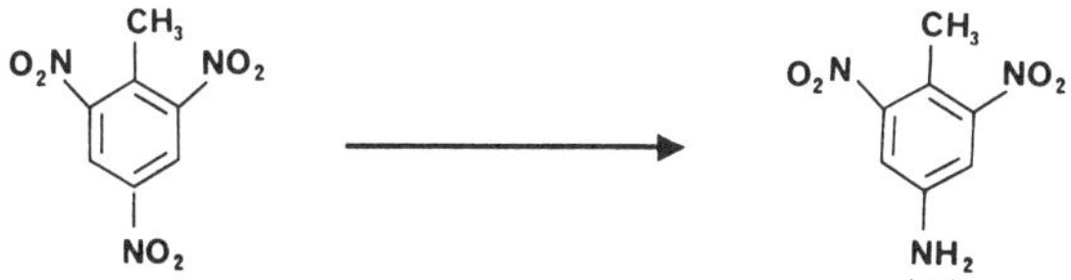

(III) (IV)

**Fig. 2.** Reduction of TNT.

Other instances include the reduction of 2,3-dinitrotoluene to 2-nitro-3-aminotoluene[11] and 2,4-dinitrotoluene to 2-nitro-4-aminotoluene.[12] Of note, 2,4-dinitro-*m*-xylene is reduced to 2-nitro-4-amino-*m*-xylene (VI) (see Fig. 3), whereas 2-nitro-*m*-xylene is not reduced.[13(a),(b)]

With the polynitrophenols, electronic considerations override steric effects and the more hindered *o*-nitro groups are reduced. It is believed that this is due to sulphide attack on the corresponding ortho quinoid resonance contributor (VII) which is less stable than the para contributor (VIII) (see Fig. 4).[14]

Perhaps the most commercially important case is the reduction of picric acid (IX) to sodium picramate (X) which is extensively used as a dyestuff for wool and leather (see Fig. 5).[15]

Other examples include anisole derivatives such as 2,4-dinitroanisole which is reduced to 2-amino-4-nitroanisole.[3]

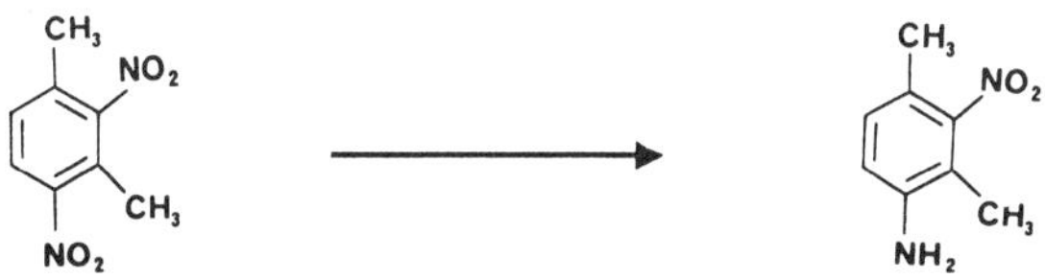

(V) (VI)

**Fig. 3.** Reduction of 2,4-dinitro-*m*-xylene.

(VII) (VIII)

**Fig. 4.** Resonance contributors of the 2,4-dinitrophenate ion.

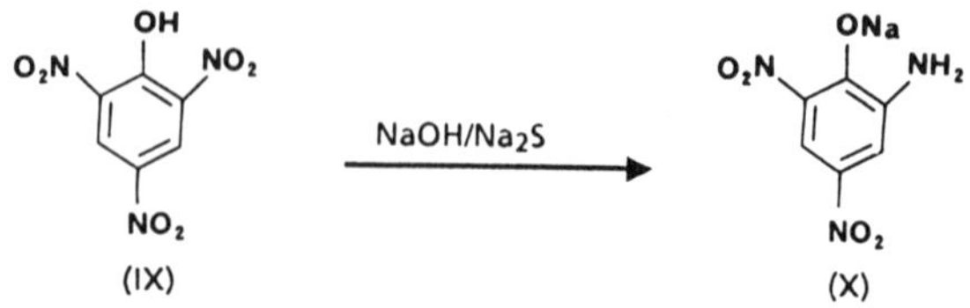


Fig. 5. Synthesis of sodium picramate.

Nitroso compounds are readily reduced to the corresponding amine and many examples are quoted in the literature.[2]

A number of side reactions are known but it must be emphasised that these very rarely occur. The most significant are as follows:[2]

Dehalogenation (ArX→ArH)
Formation of a sulphonic acid ($ArNO_2 \rightarrow ArSO_3H$)
Elimination of a sulphonic acid group ($ArSO_3H \rightarrow ArH$)
Hydroxylation (ArH→ArOH)
Hydroxylamine formation ($ArNO_2 \rightarrow ArNHOH$)
Replacement of halogen by a mercapto group (ArX→ArSH)
Thiosulphonic acid formation ($ArSO_3 \rightarrow ArSO_2SH$)
Decarbonylation (ArCOCOAr→ArCOAr)

Two side reactions deserve special mention as they give unexpected products in good yield. Thus, reduction of *o*-nitrophenylazo compounds give benzotriazoles (Fig. 6).[16(a),(b)]

The other notable reaction is the reduction of *p*-nitrotoluene (XI) in ethanol to give *p*-aminobenzaldehyde (XII) in 75% yield (Fig. 7).[17(a),(b)]

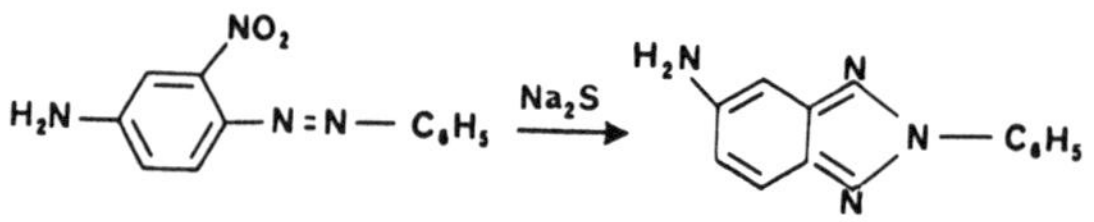


Fig. 6. Benzotriazole formation.

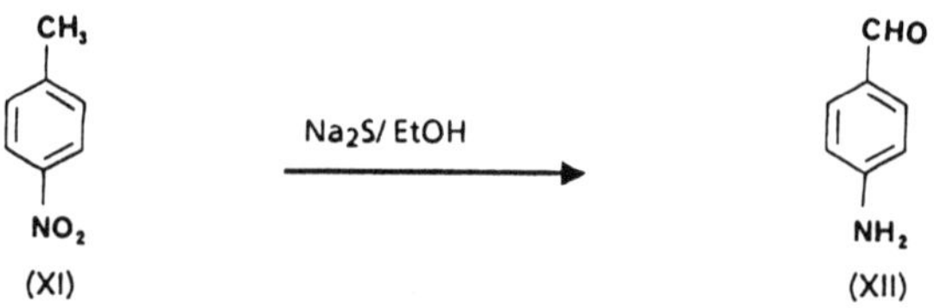


Fig. 7. Synthesis of *p*-aminobenzaldehyde.

## Mechanism

Depending upon the nature of the reducing species, the stoichiometry of the reaction may be depicted as follows:

Sulphide
$4ArNO_2 + 6S^{2-} + 7H_2O \rightarrow 4ArNH_2 + 3S_2O_3^{2-} + 6OH^-$
Disulphide
$ArNO_2 + S_2^{2-} + H_2O \rightarrow ArNH_2 + S_2O_3^{2-}$
Hydrosulphide
$4ArNO_2 + 6HS^- + H_2O \rightarrow 4ArNH_2 + 3S_2O_3^{2-}$

In actual practice, however, the stoichiometry described above is not usually observed owing to various competing side reactions, which lead to oxidative losses of the anionic sulphur species. Typically, 1·8–4·0 moles of sodium sulphide or hydrosulphide and 1·6–3·0 moles of ammonium sulphide are used per mole of nitro compound.[18]

Kinetic studies[19] indicate that the disulphide ion is more effective than the sulphide ion and that when sufficient alkali is present to keep the equilibrium

$$S_2^{2-} + H_2O \rightleftharpoons HS_2^- + OH^-$$

on the left-hand side, then the rate of reaction is first order for both the disulphide ion and nitro compound.

Increasing the concentration of water and the presence of electron-withdrawing groups is reported to rapidly increase the reaction rate.[19(a)] It is postulated that the actual reduction goes via the following route:

$$ArNO_2 \rightarrow ArNO \rightarrow ArNHOH \rightarrow ArNH_2$$

Later work[20] has shown that although the disulphide ion is the most effective reducing species for certain Zinin reductions, it is not universally so. A more plausible explanation of the widely varying reagents, solvent, pH and temperature conditions required to optimise specific reductions is via consideration of relative reduction potentials. In other words, the reduction potential of the sulphide species is tailored to match the appropriate nitro group potential.

## Applications

The Zinin reduction, although more expensive than the iron–hydrochloric acid procedure, has a wide range of application, particu-

**Table 2.** Commercially available products made via the Zinin reduction.

| Product | Use |
|---|---|
| Picramic acid (2-amino-4,6-dinitrophenol) | Dyestuff and in cosmetics (hair dye) |
| Sodium picramate | Dyestuff |
| 2-Amino-4-nitrophenol | Dyestuff |
| 2-Amino-6-nitro-*p*-cresol | Dyestuff |
| 6-Amino-2,4-dichlorophenol | Intermediate for oxyclozanide (an anthelmintic) |
| 2,6-Dichloro-4-nitroaniline | Dyestuff |

larly in the areas of partial reduction and the reduction of anthraquinones. Key examples include the preparation of nitroamines from polynitro compounds, reduction of nitrophenols and nitroanthraquinones and the manufacture of aminoazo compounds from the corresponding nitro derivatives.

Commercial applications are centred on the manufacture and use of nitroamines and reduced nitrophenols. Many of these compounds are used as dyestuffs, whilst others are attracting interest as 'building blocks' for pharmaceutical and veterinary products. Some specific examples are presented in Table 2.

## Manufacture

A number of factors have been found to influence the yield and quality of the finished product. These include pH, temperature, choice of reductant, quantity of reductant used, choice of solvent and solvent concentration.

Kinetic studies[2] suggest that:

(i) disulphide ion be used in preference;
(ii) as high a concentration of water be used as is consistent with complete or partial solution of the nitroarene in the reaction mixture;
(iii) excess alkali be used to ensure that almost all the reductant is present as the disulphide or sulphide ion;

(iv) excess reductant be used to ensure that the reaction does not stop at an intermediate stage.

The reagents used are generally ammonium, sodium or potassium sulphides, hydrosulphides or polysulphides. Manganous sulphide and ferrous sulphide have also been used but have not yet found widespread application. In some cases, where the presence of excess alkali is harmful, sodium sulphide is used with the addition of magnesium salts which remove the sodium hydroxide as it is formed by precipitation of magnesium hydroxide.[20]

The preferred solvent is water in which the inorganic reagents are very soluble. However, the addition or water miscible solvents such as the lower alcohols and dioxane are sometimes effective in improving yields.

No general rules concerning the method of carrying out the reaction can be given, since the best order of addition of reagents varies from compound to compound and the optimum temperature ranges from room temperature to reflux. However, to illustrate the simplicity of performing the Zinin reduction on the industrial scale, the manufacture of sodium picramate is described below.

## Sodium picramate

Sodium picramate, the sodium salt of 2-amino-4,6-dinitrophenol (X), has been produced at A. H. Marks for over 60 years. Although many improvements have been made, particularly in the area of materials handling, the process is still essentially the same as the classic procedure developed by Zinin. Thus water is charged to the glass-lined reactor (see Fig. 8). Picric acid is then added, followed by excess aqueous ammonia. The resulting deep yellow slurry is stirred whilst a 20% aqueous sodium sulphide solution is added. The reaction is appreciably exothermic and the temperature is kept below 80°C by cooling and by controlling the sodium sulphide addition rate. The pH is kept neutral throughout the reaction by the addition of sulphuric acid. During the reaction, the reaction mixture changes from yellow to deep red as the product gradually precipitates as fine red crystals of the monohydrate.

After the sodium sulphide addition is complete, the reactor contents are stirred for a further hour and then cooled to 40°C. The product is

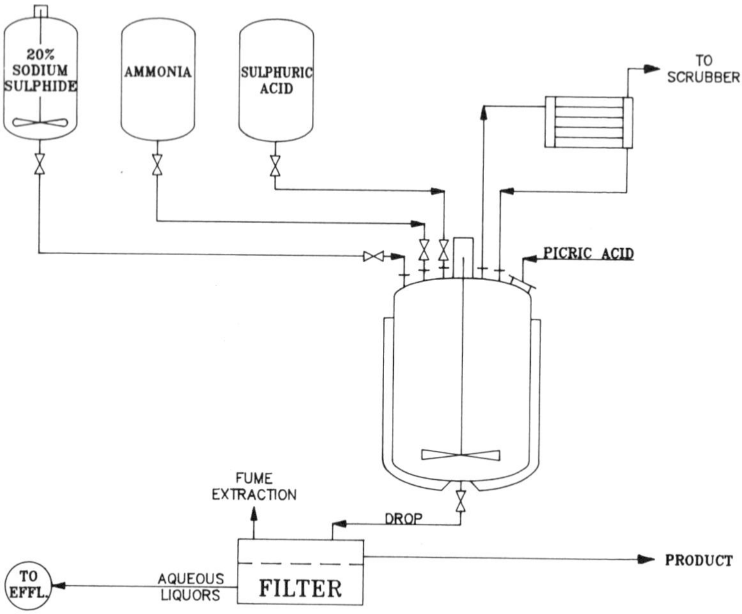


**Fig. 8.** Sodium picramate plant.

isolated by filtration and is obtained as a deep red solid containing ca. 65% active ingredients and ca. 35% water. The product is not dried as it is an explosive material. The presence of heavy metals leads to rapid decomposition and as a result steps must be taken to ensure that any such impurities are precluded at all times.

A flowsheet outlining the processing steps is depicted in Fig. 9.

## Alternative selective reductions

A large number of reduction methods have been used to convert nitro compounds to the corresponding amines. These include:

—*Catalytic.* This method involves the use of hydrogen (or hydrogen generated *in situ*) and a catalyst such as nickel, copper, platinum or palladium.

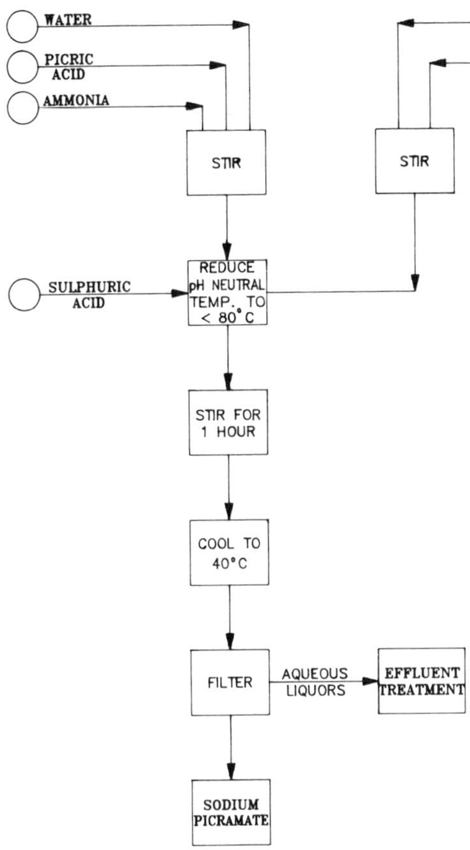


**Fig. 9.** Flowsheet for the manufacture of sodium picramate.

—*Metal and acid.* Iron and hydrochloric acid (Béchamp method[21]) is a classic example. Other metals such as tin and zinc have also been used. The advantages of the Béchamp process is that catalytic amounts of acid (less than 2% of theory) are used.

—*Metal and alkali.* This method generally gives azoxy, azo and hydrazo compounds. However, ferrous sulphate in alkaline solution does convert nitroarenes to the corresponding amines.

—*Metal hydrides.* These include lithium aluminium hydride and sodium borohydride. Yields vary from moderate to good but the use of such reagents is costly.

—*Sodium hydrosulphite.* This method is rarely used because of the relatively high cost.

—*Sulphite.* Sodium sulphite (or sodium bisulphite) is used to convert nitro compounds to the corresponding amines. However, there is a tendency for a sulphonic acid group to be introduced at the same time.

All of the above techniques have been used to selectively reduce polynitroarenes. For example, $Sn^{2+}$ has been used to reduce 2,4-dinitro-*N,N*-dimethylaniline to 2-amino-4-nitro-*N,N*-dimethylaniline.[22] Triethylammonium formate in the presence of a palladised charcoal catalyst is effective in selectively reducing dinitrophenols and anilines to the corresponding *o*-amino derivatives in yields varying from 24 to 57%.[23] However, 2,4-dinitrotoluene is reduced to 2-nitro-*p*-toluidine in 92% yield. Similarly, hydrogenation of 2,4-dinitrotoluene with a Raney copper catalyst is reported to give 2-nitro-*p*-toluidine in quantitative yield.[24] Various catalysts in the presence of hydrazine have been shown to give high selectivity in the reduction of polynitro compounds.[25]

Despite these other techniques showing signs of selectivity they all suffer from a number of major drawbacks (when compared with the Zinin reduction) and these include:

(i) lack of general applicability;
(ii) lack of selectivity when the polynitroarenes contain other substituents susceptible to reduction;
(iii) often expensive reagents are used which are either irrecoverable or recoverable with difficulty.

## Conclusion

The Zinin technique provides a simple, rapid and effective means of selectively reducting polynitroarenes. Furthermore, the high selectivity of the process is such that the presence of other substituents which are normally susceptible to reduction are left intact.

The reaction conditions are extremely mild so little decomposition of the starting material is observed and, as a result, high yields are common. Water is mainly used as the solvent, which consequently reduces the potential for hazardous and environmentally unacceptable occurrences. In terms of actual manufacture, these factors lead to the use of relatively simple and easily operable plant and equipment.

The advantages of the Zinin reduction are such that it is virtually always chosen as the route for selectively reducing polynitroarenes on the industrial scale.

## References


1. Zinin, N., *J. Prakt. Chem.* **27** [1], (1842) 149.
2. Porter, H. K., *Organic Reactions,* **20** (1973) 455.
3. Blanksma, J. J., *Rec. Trav. Chim. Pays-Bas,* **28** (1909) 107.
4. Fischer, G. *Ann.,* **127** (1863) 142.
5. Beilstein, F. & Reichenbach, E., *Ann.,* **132** (1864) 144.
6. Limpricht, H. *Ann.,* **177** (1875) 79.
7. Farben, I. G., German Patent 492 466, 1930.
8. Charrier, G. & Beretta, A., *Gazz. Chim. Ital.,* **54** (1924) 981.
9. Hodgson, H. H., *J. Soc. Dyers Colour,* **62** (1946) 114.
10. Purdue Research Foundation, US Patent 2 669 584, 1950.
11. Limpricht, H., *Ber.,* **18** (1885) 1402.
12. Beilstein, F. & Kuhlberg, A., *Ann.,* **155** (1870) 11.
13. (a) Blanksma, J. J., *Rec. Trav. Chim. Pays-Bas,* **28** (1909) 94.
    (b) Shell Development Company, US Patent 2 430 421, 1943.
14. Astle, M. J. & Cropper, W. P., *J. Am. Chem. Soc.,* **65** (1943) 2395.
15. Brand, K., *J. Prakt. Chem.* **74** [2], (1906) 491.
16. (a) Chakrabarty, B. & Barat, K. K., *J. Indian Chem. Soc.,* **5** (1928) 585.
    (b) Chakrabarty, B. & Barat, K. K., *J. Indian Chem. Soc.,* **5** (1928) 558.
17. (a) Hodgson, H. H. Davies, R. R. & Beard, H. G., *J. Chem. Soc.,* **4** (1944).
    (b) Campaigne, E., Buddle, W. M. & Schaefer, G. F., *Org. Synthesis,* **4** (1963) 31.
18. Groggins, P. H., *Unit Processes in Organic Synthesis.* McGraw-Hill, New York, 1958, pp. 186–90.
19. (a) Hojo, M., Tagaki, Y. & Ogata, Y., *J. Am. Chem. Soc.,* **82** (1960) 2459.
    (b) Hashimoto, S., Sunamoto, J. & Sato, K., *Bull. Chem. Soc., Japan,* **40** (1967) 2860.
    (c) Hashimoto, S. & Sunamoto, J., *Bull. Chem. Soc., Japan,* **40** (1967) 2867.
    (d) Hashimoto, S. & Sunamoto, *J. Bull. Chem. Soc., Japan* **41** (1968) 499.
    (e) Bhave, R. R. & Sharma, M. M., *J. Chem. Tech. Biotechnol.,* **31** (1981) 93.
20. Kirk, R. E. & Othmer, D. F., *Encyclopedia of Chemical Technology,* 3rd edn, Vol. 2, John Wiley, New York, p. 368.
21. Béchamp, A. J., *Ann. Chim. Phys.* **42** [3], (1854) 186.
22. Ayling, E. E., Gorvin, J.H. & Hinkel, L. E., *J. Chem. Soc.,* (1942) 775.

23. Terpko, M. O. & Heck, R. F., *J. Org. Chem.*, **45** (1980) 4992.
24. Jones, W. H., Benning, W. F., Davis, P., Mulvey, D. M., Pollak, P. I., Schaeffer, J. C., Tull, R. & Weinstock, L. M., *Ann. N.Y. Acad. Sci.*, **158** (2) (1969) 471.
25. Proc. 5th Conf., 1975., *Catalysis in organic synthesis*. Academic Press, New York, 1976, pp. 273–85.

# New reagents for old processes

**T. W. Bastock**

*Contract Chemicals Limited, Prescot, UK*

**J. H. Clark**

*Department of Chemistry, University of York, UK*

## Introduction

Pressure from legislative and environmental bodies together with a growing awareness within the chemical industry in developed countries has led to a search for new environmentally friendly processes to replace unacceptable outdated reactions. Clearly the dumping of hydrated aluminium chloride from Friedel–Craft reactions and oxidation by-products containing nitric acid, hexavalent chromium and permanganate is nearly at an end. The problem often is not the inorganic product but the organic residue it contains. Progress towards environmentally acceptable processes has been slow, but recent research and development work in the area of supported reagents and their application in organic synthesis is beginning to come to commercial fruition and promises much for the future.

## Supported reagents

Supported reagents are materials based on insoluble inorganic or organic solids onto which a reagent has been adsorbed. This article will focus on inorganic materials that usually have a high surface area (up to 1000 $m^2 g^{-1}$), such as silica, alumina, clay or charcoal, as the support.

The supported reagents can be prepared in many ways but the most popular method involves the slow evaporation of a well-agitated slurry of a solution of the reagent and the support. With good reagent dispersion, an increase in effective surface area of the reagent of two orders of magnitude can be achieved, leading to a concomitant increase in reagent reactivity. A favourable change in the entropy of activation of reaction due to restricted molecular movement over a solid surface is also likely to enhance reaction rates although the surfaces of supported reagents tend to be highly irregular with cracks and steps as well as microcrystalline reagent clusters.

The potential advantages of using supported reagents in organic synthesis are numerous and significant:

* improvement in reagent reactivity;
* improvement in reagent selectivity (resulting from microporous constraints on molecular geometry or intra/intermolecular adsorption selectivity);
* ease of handling and use (supported reagents are usually non-corrosive free-flowing powders and often have low toxicity);
* ease of product and reagent recovery (the supported reagents can be filtered from the reaction mixture and may be re-used);
* good thermal and mechanical stabilities;
* reactions are usually carried out in non-polar solvents (since reaction takes place on the surface of the solid; the solvent only plays the part of a heat transfer and working fluid).

Some of these advantages (improved reaction selectivity, lack of reagent effluent, use of non-polar solvents) are especially important in today's 'green' technological climate.

Supported reagents have been known since Fetizon's original article describing silver carbonate–celite in 1968.[1] The first comprehensive reviews on their applications in synthesis appeared in 1979[2] and the first entire textbook devoted to the subject was published in 1987.[3] The last few years have seen major breakthroughs in the subject, particularly with regard to the discovery of truly catalytic supported reagents, the effective use of modern analytical techniques to probe the surfaces of the materials and to help optimise their performance in chemical reaction, and the successful scale-up of supported reagents to an industrial scale and their first significant use in the manufacture of organic chemicals. The application of supported reagents to some of the most important types of chemical reaction will be used to help illustrate these important developments.

## Friedel–Craft reactions and solid acids

Freidel–Craft reactions are usually accompanied by two major disadvantages:

1. Large volumes of hydrated aluminium chloride catalysts exist as reaction residues needing disposal.
2. Di- or polysubstituted by-products together with polymeric tars can be formed requiring separation and disposal and causing problems with product purity.

Solid acids that give the desired level of activity but which can be easily removed from the reaction mixture with no residual inorganic contamination of the organic products offer obvious advantages over existing methods. Considerable progress has been made in achieving these goals through recent developments in supported reagent chemistry.

Conventional Friedel–Craft reactions are shown in Figs 1 and 2. In the acylation reaction in Fig. 1, aluminium chloride is used in stoichiometric proportions. A complex is formed with the acid chloride which then substitutes the aromatic nucleus, giving the ketone product, which in turn forms a complex with the Lewis acid. This product complex must be broken down, typically with aqueous acid, to release the benzopherone, but also creating large volumes of corrosive and toxic effluent. The suspected links between aluminium and the debilitating Altzheimer's disease exacerbates the problem of waste disposal.

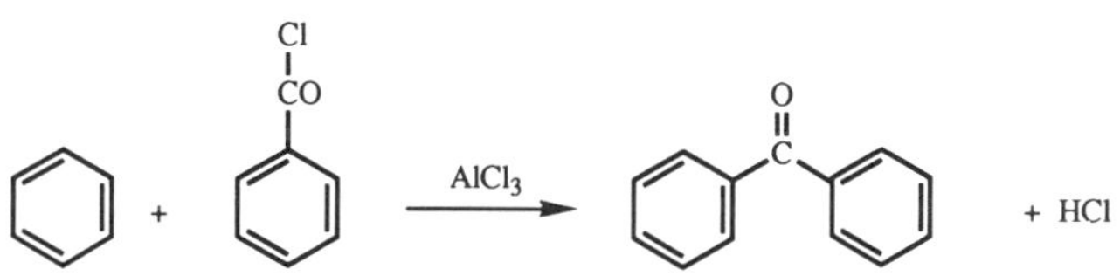


**Fig. 1.** A major drawback is the large amount of aqueous effluent from this reaction.

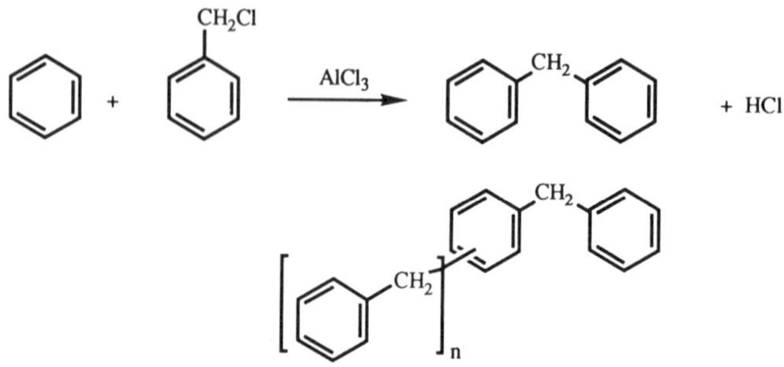


**Fig. 2.** The benzylation of benzene using aluminium chloride.

Figure 2 shows a typical alkylation reaction which although requires only catalytic amounts of aluminium chloride leads to di-substitution and polymeric by-products.

The Freidel–Crafts acylation of aromatic compounds with carboxylic acids has been performed over cation-exchanged montmorillonites, as shown in Fig. 3, as well as modified zeolites, although stoichiometric quantities of the metal cation were required and high product yields were only obtained with the more reactive longer chain carboxylic acids. Similar reactions and results have been obtained using modified zeolites.[4] Truly catalytic benzoylation and sulphonylations using supported reagents including those shown in Figs 4 and 5 have recently been announced.[5]

Ion-exchange montmorillonites[6] and montmorillonite-based supported reagents[7] have been shown to be capable of catalysing Friedel–Crafts alkylation reactions, although there are significant differences in their reactivity. In the case of the benzylation of benzene shown in Fig. 6, reaction times for ion-exchanged montmorillonites are 0·25–14 h at reflux temperatures whereas the fastest

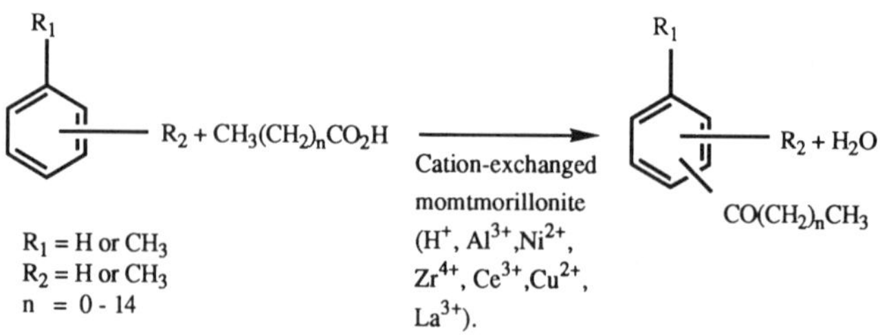


**Fig. 3.** The acylation of aromatic substrates using montmorillonites.

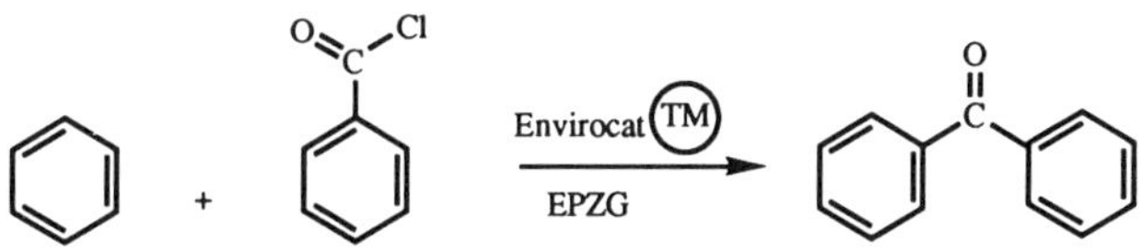

**Fig. 4.** The catalytic benzoylation of benzene using an Envirocat.

montmorillonite supported reagent, 'Clayzic', catalyses the reaction in minutes at room temperature.

The latter now forms the basis of an entirely new industrial process for the manufacture of diphenylmethane and related products. The technology has been successfully transferred from the bench to the plant, giving a more environmentally acceptable and more cost effective manufacturing process.

Friedel–Crafts alkylation using the supported reagent Envirocat™, EPZ10, has been successfully extended to a wide range of reactions including the preparation of substituted diphenylmethanes, the use of formaldehyde as a source of '$CH_2$' and alkylations using alkyl halides and alcohols (Fig. 7).[5]

Figure 7(a) shows the formation of benzylbiphenyls from the benzylation of biphenyl. The reaction is fairly typical of the use of supported reagents in that it is usually run in the absence of solvent, although a bulk inert solvent can be employed where preferred. The high specificity of this reaction to produce the 4-isomer cannot be matched using conventional Lewis acids and the reaction has been successfully scaled up to enable production of commercial quantities of

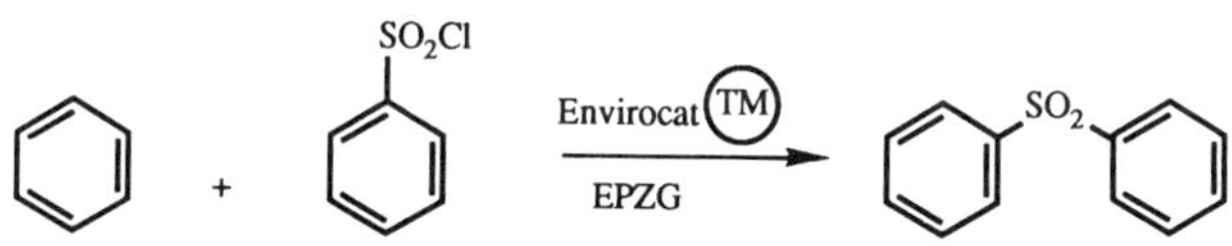

**Fig. 5.** The catalytic sulphonylation of benzene using an Envirocat.

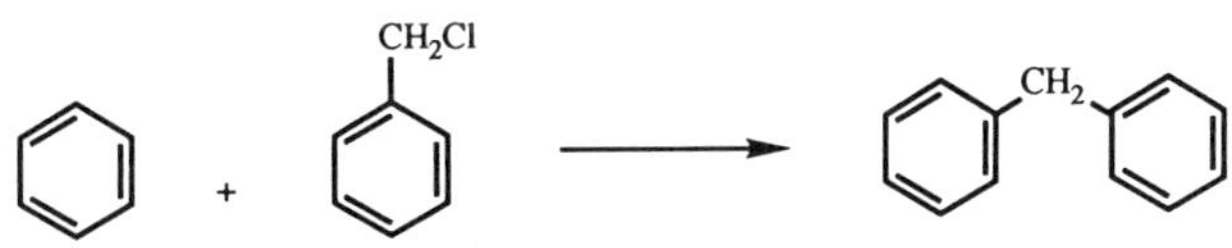

**Fig. 6.** The benzylation of benzene to give diphenylmethane.

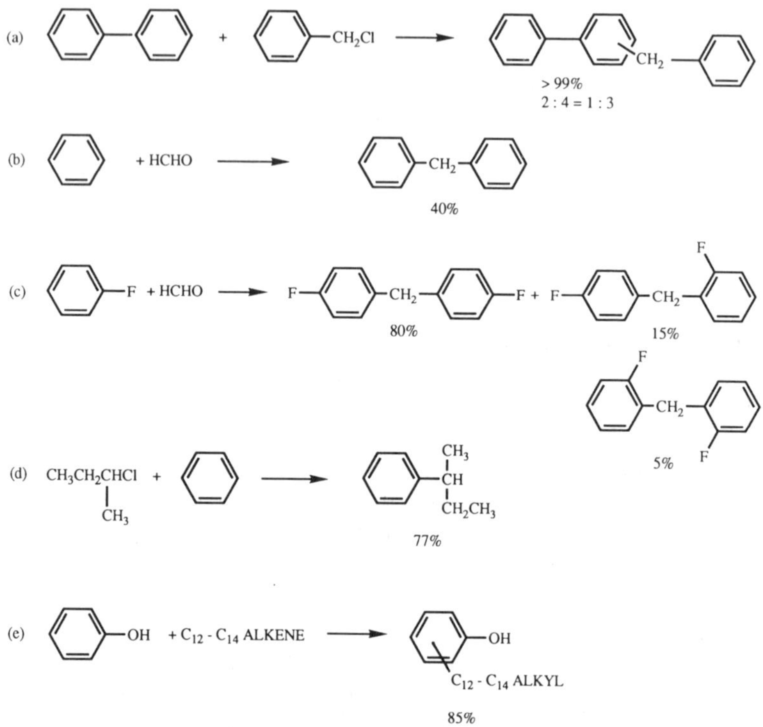


**Fig. 7.** Catalytic alkylations using an Envirocat.

the desired 4-isomer, which is an important product in the paper industry.

Figure 7(b) shows the reaction of benzene and formaldehyde over EPZ10 to give diphenylmethane. The yield is approximately 40% with respect to formaldehyde. One by-product appears to be formic acid, suggesting that the support has oxidative powers. Some polymeric products are formed by further substitution of DPM. On further investigation it was found that $CO_2$ was liberated during the reaction, possibly by the following route:

$$HCHO + O_2 \rightarrow CO_2 + H_2O$$

Carrying out this reaction under nitrogen did not alter the reaction. At present the mechanism of reaction is not understood. A benzyl alcohol species is the obvious intermediate for the reaction mechanism but this product has not been identified in the reaction mixture. Attempts to carry out the reaction with benzyl alcohol and benzene under similar conditions gave only slow reaction (6% conversion after 8 h reflux). This result would suggest that the intermediate is formed on the reagent surface and never released until the reaction is complete. The very specific substitution pattern shown in reaction (c) may result from this intermediate formation. Reaction (c) shows the preferential formation of 4,4′-difluorodiphenylmethane, possibly because of alignment of the reacting molecules in this conformation on the surface of the support. Reactions such as (d) also raise the interesting question of competitive isomerisation when using supported reagents. In many cases such as this it does seem that the use of these new catalysts may well lead to little side chain isomerisation as well as reduced multiple substitution. The latter may well be due to the molecular constraints imposed by the microporous catalysts.

## Other applications of solid acids

Many other supported reagent 'solid acids' have been reported and several of these show promise as industrial catalysts. Treatment of certain inorganic oxides with aluminium chloride in refluxing carbon tetrachloride results in the evolution of 1 mol of hydrogen chloride for every mole of aluminium chloride used, suggesting the formation of $(OAlCl_2)$ species on the surface. These tamed versions of the powerful Lewis acid can be used as cracking catalysts.[8] The most active of these materials is based on silica gel and is, unusually, an intense red colour when protected from the atmosphere. The $SiO_2$–$AlCl_3$ supported reagent compares well to $AlCl_3$ itself in the cracking of *n*-hexadecane: it is more selective, does not form high molecular weight tars, is not corrosive to the reactor and does not leach inorganics into the product solution.

More commonly, aluminas and clays are used as supports for preparing solid acids. Fluorinated aluminas have been used as solid catalysts for typically acid-catalysed reactions including cracking, polymerisation, isomerisation and disproportionation for many years. An interesting more recent development of these areas has been the proposed application of such catalysts to CFC (chlorofluorocarbon) technology, especially with regard to CFC destruction and to fluorine–

$$CHF_3 + CHClF_2 + CHCl_2F \longrightarrow CHF_3 \longrightarrow HF + CO_2$$

**Fig. 8.** The acid-catalysed destruction of CFCs.

for-chlorine exchange (e.g. Fig. 8) so as to reduce their ozone-depleting properties.[9]

A large volume of literature has been devoted to clay based supported reagents and their use in reactions normally requiring acidic conditions. Perhaps the best known of these solid acids are 'Clayfen' (ferric nitrate supported on montmorillonite) and 'Claycop' (cupric nitrate supported on montmorillonite) discovered by Laszlo and extensively applied by his group and others to a variety of organic reactions.[10] One of the most appealing reaction targets for these reagents is aromatic nitration. Existing technology for the nitration of aromatic hydrocarbons is old and fraught with difficulties, such as the use of aggressive mineral–acid mixtures and the production of unwanted isomers and side-products. Claycop has been shown to be capable of giving improved regioselectivity in the nitration of aromatic hydrocarbons under Menke conditions, as shown in Fig. 9.[11] The advantages of using a solid acid coupled with improvements in reaction regioselectivity are obvious and the further improvement of such methodology to make the reaction catalytic with respect to the supported reagent may well bring such chemistry to commerical fruition.

A similar story can be told of regioselective aromatic chlorinations.

| | Regioselectivity | |
|---|---|---|
| | Claycop/$AC_2O$ | Homogeneous ($HNO_3$ + $AC_2O$) |
| R = $CH_3$ | 2 - 6 | 0.33 |
| R = s - $C_3H_7$ | 11 - 9 | 1.17 |
| R = t - $C_4H_9$ | 21 - 5 | 3.07 |

**Fig. 9.** The regioselective nitration of aromatic substrates.

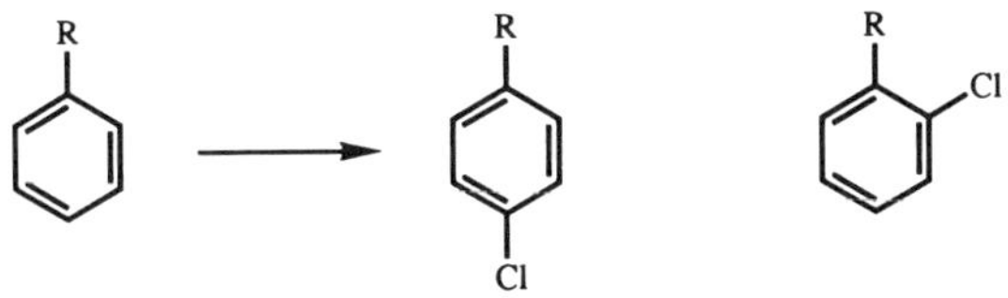

| | major | minor |
|---|---|---|
| R = $CH_3$ | 82 | 18 |
| R = $\underline{i}$- $C_3H_7$ | 80 | 20 |
| R = Cl | 97 | 3 |

**Fig. 10.** The chlorination of aromatic substrates using t-butyl hypochlorite $-H^+$, $Na^+$, faujasite X.

Traditional electrophilic chlorination reactions of aromatic substrates using reagents such as elemental chlorine in ethanoic acid are tedious and messy. It has been reported that several organic chlorine-containing compounds become active electrophilic chlorinating agents in the presence of silica and zeolites. Supported reagents such as *t*-butyl hypochlorite–silica chlorinates alkylbenzenes, naphthalene and anisole readily at 25°C,[12] but perhaps the most significant results have been obtained using *t*-butyl hypochlorite–$H^+$, $Na^+$ faujasite X (a partially proton-exchanged form of zeolite X).[13] The latter gives para-selective monochlorination of alkyl-, phenyl- and halobenzenes under mild conditions (Fig. 10). As with the use of supported reagents in aromatic nitration, the only major drawback with this new chemistry is that the reactions are at best stoichiometric with respect to the supported reagent.

## Solid bases

Less attention has been given to the development and application of supported reagent solid bases although such materials have been shown to be useful for a variety of important reactions including carbon–carbon bond forming reactions and oxidations. Progress has also been made with regard to various technical aspects of their application including the demonstration of truly catalytic behaviour, high turnover efficiency and the use of packed columns.

A solid superbase with a basicity (H—) of greater than 37 has been prepared by successive treatment of Ψ-alumina with sodium hydroxide

$$RCH_2NO_2 + R'CH{=}CHCOR^2 \longrightarrow R'CH(RCH_2)CH_2COR^2 \quad \text{(ref 18)}$$

$$PhCH_2Ph + O_2 \longrightarrow PhCOPh \quad \text{(ref 17)}$$

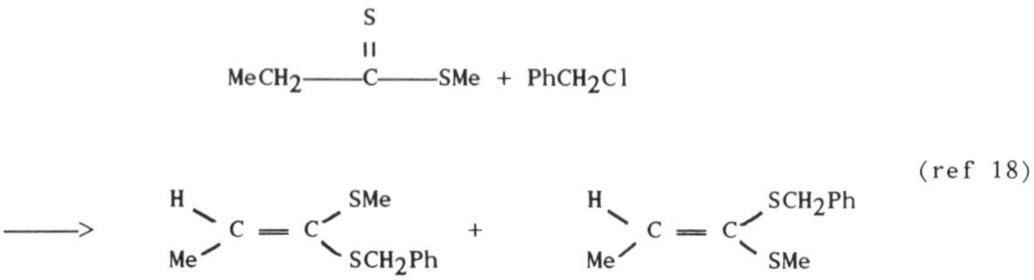

| Base | Selectivity |
|---|---|
| $^tBuOLi$ | 68 : 32 |
| Alumina | 53 : 47 |
| KF–Alumina | 85 : 15 |

**Fig. 11.** Some reactions catalysed by KF-alumina.

and sodium.[14] The base was shown to be highly effective in the catalytic isomerisation of 5-vinylbicyclo[2.2.1]hept-2-ene to 5-ethylidenebicyclo [2.2.1] hept-2-ene. The latter is an important third comonomer of ethylene–propylene synthetic rubber. It would be hoped that with such a high basicity, solid bases of this type might find wider application in reactions of hydrocarbons.

Perhaps the most widely studied solid base is potassium fluoride supported on alumina which provides the best compromise between activity and cost of all the supported fluorides.[15] The base has been used for many carbon–carbon bond forming reactions, most notably the Michael reaction, for which it appears to be especially useful in terms of activity, efficiency and the ability to undergo multiple turnovers.[16] It can be used with or without a bulk solvent. Some of the reactions which KF–alumina can catalyse including highly selective alkylations are shown in Fig. 11.

## Oxidations

Apart from the use of supported reagent solid bases, which will catalyse reactions via organic anions or radical anions and which should prove useful for the oxidation of relatively acidic hydrocar-

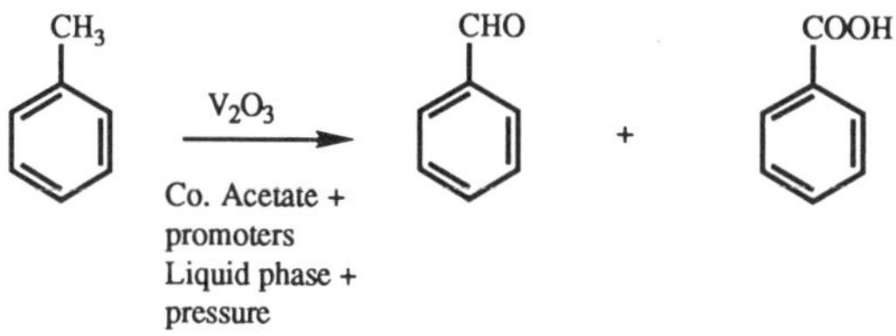


**Fig. 12.** The side chain oxidation of toluene.

bons, supported metal oxidants might well be considered as complementary reagents since the mechanism of their action is likely to involve organic radical cation intermediates.

Conventional oxidative processes to produce simple molecules can have considerable drawbacks. The normal methods using oxygen and a catalyst in the gas phase to oxidise toluene to benzaldehyde and benzoic acid (Fig. 12) have serious disadvantages, including low conversion of the substrate, expensive plants associated with pressure gas phase reactions, by-products and poor yields.

Other classical oxidation reactions (Fig. 13)[19] using permanganate or dichromate reagents have a similar list of drawbacks. Inevitably they are becoming environmentally unacceptable because of heavy metal effluent and other hazardous waste.

It has been shown (Fig. 14)[20] that the use of conventional oxidants such as potassium dichromate supported on silica or alumina can selectively oxidise alkyl halides and alcohols to the corresponding aldehyde. The advantages of this method are higher conversions and more selective reaction. The oxidant turns from an orange to a brown colour as the reaction takes place. It can be filtered off at the end of the reaction rather than forming an intractable sludge. The reagent is not, however, catalytic.

A new generation of truly catalytic supported reagent oxidants have been recently reported.[21] These supported chromium and manganese

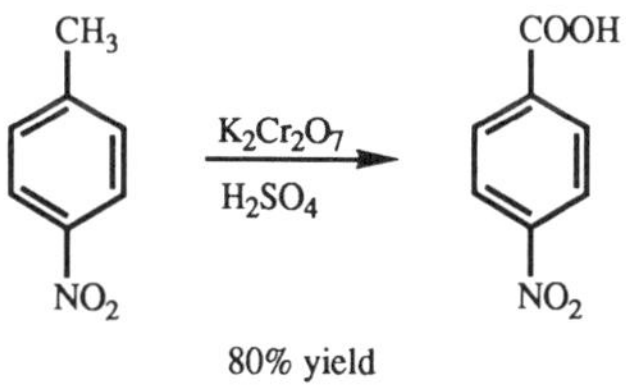


**Fig. 13.** The dichromate oxidation of p-nitrotoluene.

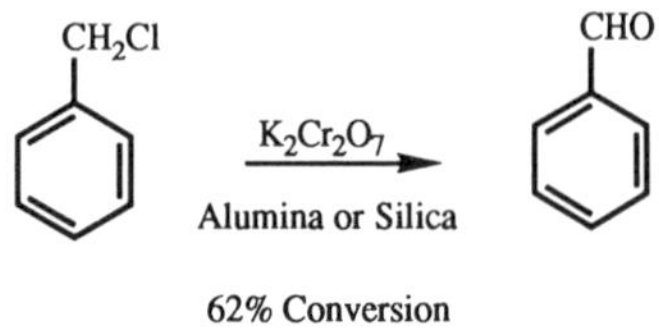


**Fig. 14.** The oxidation of benzyl chloride using supported $K_2Cr_2O_7$.

reagents show a number of interesting and important properties:

1. The actual catalytic sites in active materials are present at a very low surface concentration: excess metal oxidant sites can only act stoichiometrically.
2. Active reagents show outstandingly high turnover efficiencies and only require air as the consumable feedstock (other than the substrate).
3. The method of preparation of the reagents and the way in which they are employed in oxidation reactions are critical to their performance.

The model reaction used to test the activity of these new catalysts was the oxidation of diphenylmethane to benzophenone (Fig. 15). Conversion of 70% (bulk substrate) at rates up to 1·3% $h^{-1}$ with isolated yields of 60% have been achieved.

Commercial 'Envirocats™' catalysts based on materials of this type have now been developed and applied to an increasing number of organic oxidations including those shown in Fig. 16. Reactions may be run in a suitable inert bulk solvent although the use of neat substrate is again preferred.

# Summary

After twenty years as academic curiosities relying mostly on serendipitous discoveries, supported reagents now seem set for a promising

**Fig. 15.** The catalytic oxidation of diphenylmethane using supported chromium and manganese.

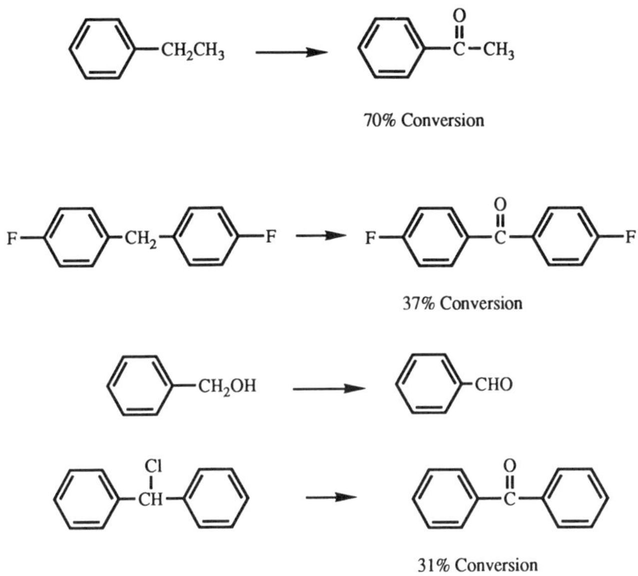

**Fig. 16.** Catalytic oxidations using an Envirocat.

future in industrial chemistry. Their easy use and recovery, high activity and often improved selectivity in reactions have always been attractive features, but the real breakthroughs have come in their successful transition to truly catalytic activity and the optimisation of their performance in commercially important areas. There is little doubt that the success of these new materials will in part be due to the growing awareness of the unacceptability of much of today's chemical process technology. We should not lose sight, however, of the underlying chemistry which is exciting, innovative and imaginative.

## References

1. Fetizon, M. & Golfier, M., *C. R. Hebd. Seances Acad. Sci.* (*Ser. B*), **B267** (1968) 900.
2. McKillop, A. & Young, D. W., *Synthesis* (1979) 401 and 481.
3. Laszlo, P. (ed.), *Preparative Chemistry Using Supported Reagents.* Academic Press, San Diego, 1987.
4. Chiche, B., Finiels, A., Gauthier, C., Geneste, P., Graille, J. & Pioch, D., *J. Mol. Cat.*, **42** (1987) 229; *J. Org. Chem.*, **51** (1986) 2128.

5. Envirocats™, 1990., Supported Reagents, Contract Catalysts, Knowsley, Merseyside.
6. Laszlo, P. & Mathy, A., *Hel. Chim. Acta.*, **70** (1987) 577.
7. Clark, J. H., Kybett, A. P., Macquarrie, D. J., Barlow, S. J. & Landon, P., *Chem. Commun.* (1989) 1353; Eur. Patent Appl.
8. Drago, R. S. & Getty, E. E., *J. Am. Chem. Soc.*, **110** (1988) 3311; US Patent 4 719 190, 1988.
9. Hegde, R. I. & Barteau, M. A., *J. Catal.*, **120** (1989) 387.
10. Laszlo, P., *Acc. Chem. Res.*, **19** (1986) 121.
11. Laszlo, P. & Vandormael, J., *Chem. Lett.* (1988) 1843.
12. Smith, K., Butters, M. & Paget, W. E., *Synthesis* (1985) 1155.
13. Smith, K. & Butters, M., *Synthesis* (1985) 1157.
14. Suzukamo, G., Fukao, M. & Minobe, M., *Chem. Lett.* (1987) 585.
15. Ando, T., Brown, S. J., Clark, J. H., Cork, D. G., Hanafusa, T., Ichihara, J., Miller, J. M. & Robertson, M. S., *J. Chem. Soc. Perkin II* (1986) 1133.
16. Laszlo, P. & Pennetreau, P., *Tetrahed. Lett.* (1985) 2645.
17. Kybett, A. P., D. Phil. thesis, University of York, 1988.
18. Villemin, D., *Chem. Commun.* (1985) 870.
19. Kamm, O. & Matthews, A. O., *Org. Syn. Coll.*, **1** (1941) 392.
20. Santaniello, P., *Nouv. J. Chimie* (1980) 279.
21. Clark, J. H., Kybett, A. P., Landon, P., Macquarrie, D. J. & Martin, K., *Chem. Commun* (1989) 1355; Eur. Patent Appl. EP 0 336 768; EP 0 352 878.

# Polymer-bound pyridine-containing reagents and catalysts as process aids

**G. L. Goe, E. F. V. Scriven & E. E. Sowers**
and in part **T. D. Bailey, G. F. Hillstrom, J. G. Keay, D. W. McQuigg & C. R. Marston**

*Reilly Industries, Inc., Indianapolis, Indiana, USA*

## Polyvinylpyridine (REILLEX$^{TM}$ polymers)

Pyridine itself is well established as the solvent of choice for many chemical reactions.[1] This derives from its solvating power for polar molecules and the resistance of its ring carbon atoms to attack by electrophiles and nucleophiles under normal conditions. Pyridine has found extensive use as an acid scavenger, and as both a catalyst and solvent for acylation reactions. Pyridine and its derivatives form complexes with a great many metals. The above properties guided us in the development of crosslinked polyvinylpyridine polymers (REILLEX$^{TM}$ polymers). These polymers are insoluble in all solvents, which offers the advantage of easy separation from the product after completion of reaction. They are remarkably stable to heat (Table 1), have the advantages of high capacity, resistance to oxidation, and they can be recycled many times. Two polymers are currently offered in commercial quantities by Reilly Industries, Inc. They are REILLEX$^{TM}$ 402 polymer and 425 polymer (poly-4-vinylpyridine 2% and 25% crosslinked with divinylbenzene respectively). REILLEX$^{TM}$ polymers are essentially pyridine in solid form, and they undergo typical

**Table 1.** Typical properties of REILLEX™ polymers

| Property | REILLEX™ 402 polymer | REILLEX™ 425 polymer |
|---|---|---|
| Physical form | Off-white granular powder | Off-white beads |
| Particle size | ca. 60 mesh | 18–50 mesh |
| Apparent bulk density ($g/cm^3$) | 0·45 | 0·29 |
| ($lb/ft^3$) | 28 | 18 |
| Skeletal density ($g/cm^3$) | 1·15 | 1·14 |
| Particle density ($g/cm^3$) | — | 0·6 |
| Surface area ($m^2/g$) | ca. 0·5 | ca. 90 |
| Moisture retained upon filtration (% by weight) | 36–39 | 50–60 |
| Approximate p$K_a$ | 3–4 | 3–4 |
| Hydrogen ion capacity in water (meq/g) | 8·8 | 5·5 |
| Swelling from free base to hydrochloride form, in water (%) | 100 | 52 |
| Swelling from dry state to solvent-saturated state | | |
| Methanol | 70–75 | 28–32 |
| Acetone | 30–35 | 32–36 |
| Water | 33–37 | 12–16 |
| Isopropanol | 13–17 | 28–32 |
| Toluene | 8–12 | 18–22 |
| Ethyl acetate | 3–6 | 32–36 |
| Hexane | 0 | 12–16 |
| Temperature stability, maximum 225°C | | |

pyridine reactions, i.e. protonation, alkylation, complexation with metals, and *N*-oxide formation.

These properties have been exploited in a number of applications. The polymers have been used as catalysts,[2] reagents, and reagent supports[3] in organic chemistry. Reports of their use in metal recovery[4] and for pollution control[5] are growing.

## Applications as Catalysts and Reagents

REILLEX™ 402 polymer has been used as an acid scavenger in the reaction of alcohols with $\alpha,\beta$-unsaturated acid chlorides.[6] Hydrochloric acid which causes competing polymerization can be removed easily

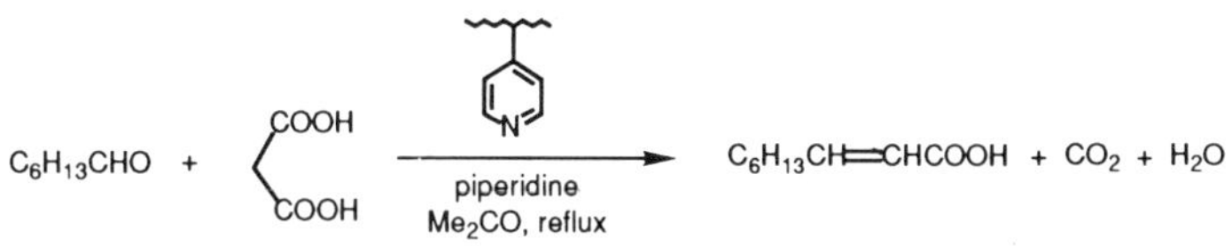


**Fig. 1.**

by addition of REILLEX™ 402 polymer. We have found that REILLEX™ 402 polymer (with a trace of piperidine) acts as a basic catalyst in the Knoevenagel reaction (Fig. 1).[6]

REILLEX™ 425 polymer in the hydrochloride salt form acts as a mild acid catalyst which is effective for ketal,[6] enamine,[6] and tetrahydropyranyl derivative[7] formation (Fig. 2). This polymeric catalyst is not only as effective as its monomeric counterpart, but also has the

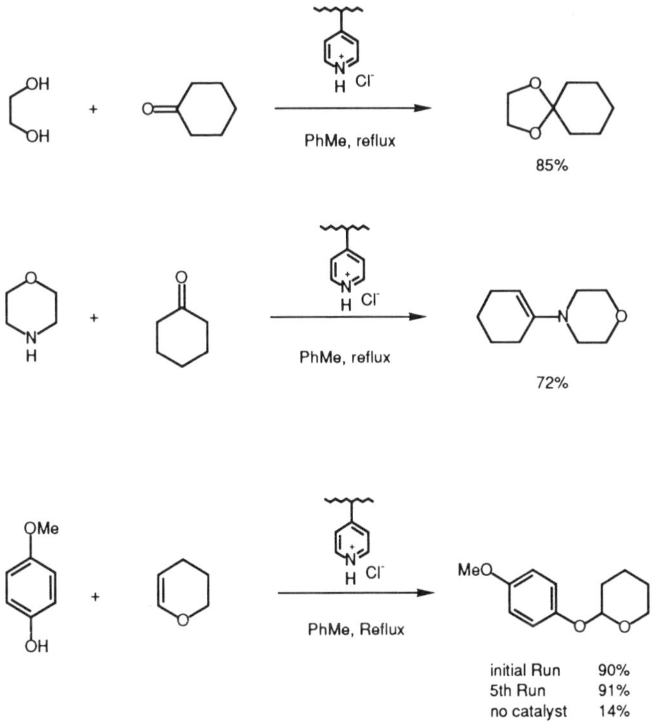


**Fig. 2.**

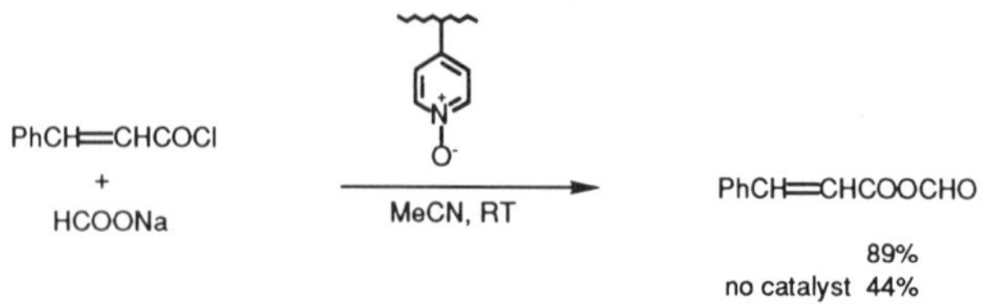


**Fig. 3.**

advantage of ease of removal from products and the option of recycle. REILLEX™ 402 polymer hydrochloride salt has been recycled five times in the tetrahydropyranylation of 4-methoxyphenol without loss of activity.

REILLEX™ 425 polymer *N*-oxide, prepared by treatment of PVP with 30% hydrogen peroxide in glacial acetic acid, is a good catalyst for the preparation of mixed formic anhydrides (Fig. 3).[8]

The superior thermal and mechanical properties of REILLEX™ polymers coupled with their ability to support metals has led to their use as catalyst supports. Metals may be supported on REILLEX™ polymers in three ways (Fig. 4).

Complexes (type A) can be made by adding an aqueous solution of

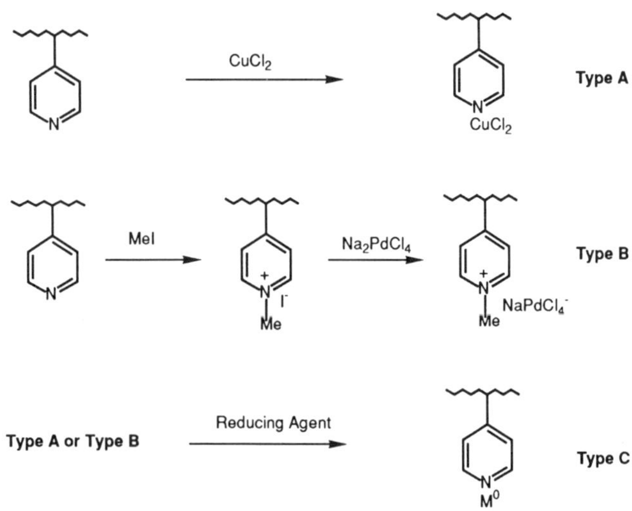


**Fig. 4.**

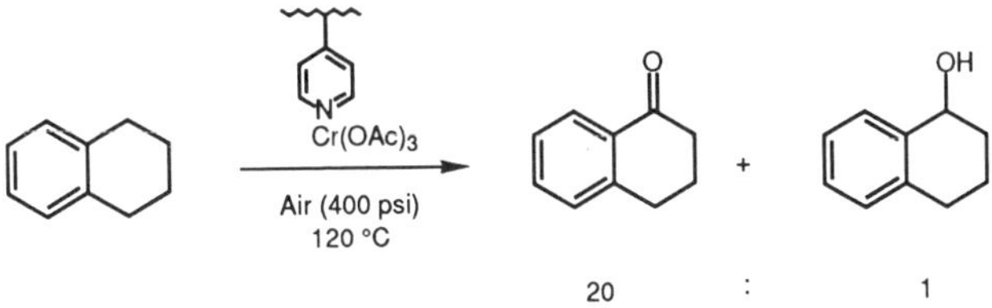


**Fig. 5.**

an inorganic salt to a slurry of the degassed resin. Most metals that can form stable complexes with tertiary amines can be supported by coordination on REILLEX™ 402 polymer (powder) and 425 polymer (beads). Polyvinylpyridinium salts that contain the metal in the anion (type B) are made either by prior or *in situ* quaternization of the pyridine and subsequent anion exchange. Treatment of type A or B catalyst with reducing agents (e.g. hydrazine, formic acid, diborane, or hydrogen) yields type C catalyst in which zero valent metal atoms are distributed throughout the polymer network.

1,2,3,4-Tetrahydronaphthalene-2-one is produced by liquid phase oxidation to tetrahydronaphthalene using chromium (III) acetate as a homogeneous catalyst (Fig. 5).[9] We have carried out this reaction with a heterogeneous catalyst, 2% chromium (III) acetate on a REILLEX™ polymer (type A catalyst).[10] This heterogeneous catalyst works well and is easy to recover and recycle. Use of REILLEX™ 425 polymer beads offers the opportunity to run this reaction in a fixed or fluid bed process.

Acetic acid is now usually manufactured by the Monsanto process, which involves the homogeneous carbonylation of methanol in the presence of a rhodium species and methyl iodide.[11] The catalyst used in this reaction may be 'heterogenized' by adding REILLEX™ 425 polymer (19·6 g) to a typical homogeneous reaction mixture of rhodium chloride trihydrate (0·4 g), methyl iodide (28·8 g) and methanol (100 g) in acetic acid (198 g) (Fig. 6). The heterogeneous catalyst is formed as the reaction vessel is raised to the reaction temperature (180–185°C) and pressurized with carbon monoxide (65–80 bar).[12]

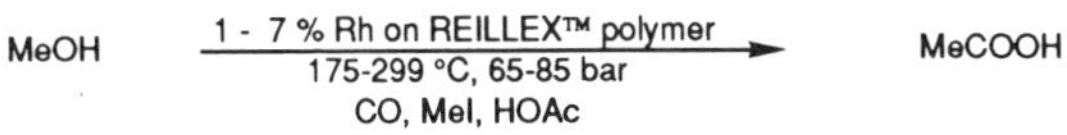


**Fig. 6.**

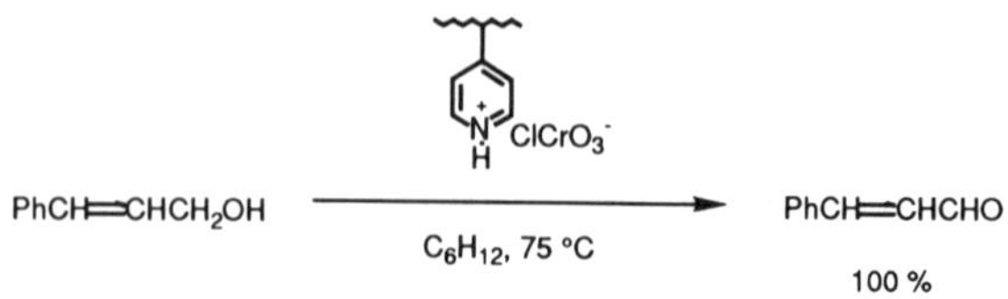


Fig. 7.

When acetic acid is chosen as the solvent, the catalyst can be recovered by filtration and used again without further treatment. Rates using the heterogeneous catalyst are four times faster than those observed under comparable homogeneous conditions, and conversion up to 99% is commonly achieved. No problem with rhodium leaching was encountered. Previous work using anion exchange resins as catalyst supports for acetic acid preparation at low temperatures and pressures gave much slower rates and rhodium leaching occurred.[13]

Zero valent metals supported on polyvinylpyridines (type C catalysts) have found use for hydrogenation,[14] hydroformylation of olefins,[15] and hydrolysis of nitriles.[16]

Pyridinium chlorochromate was introduced as an oxidizing agent by Corey[17] in 1975, and it is now in general use.[18] Polyvinylpyridinium chlorochromate can be made easily be treating an aqueous slurry of PVP with chromium trioxide and hydrochloric acid. It has been used as an oxidizing agent (in excess) for the conversion of benzylic alcohols to aldehydes (Fig. 7).[19]

Reducing agents may also be supported on polyvinylpyridines. PVP borane has been used (in equimolar amount) to reduce ketones to alcohols (Fig. 8).[20]

Polyvinylpyridine, like monomeric pyridine, can coordinate bromine and form a hydrobromide perbromide. Both of these have proved useful as brominating agents (Fig. 9).[21,22]

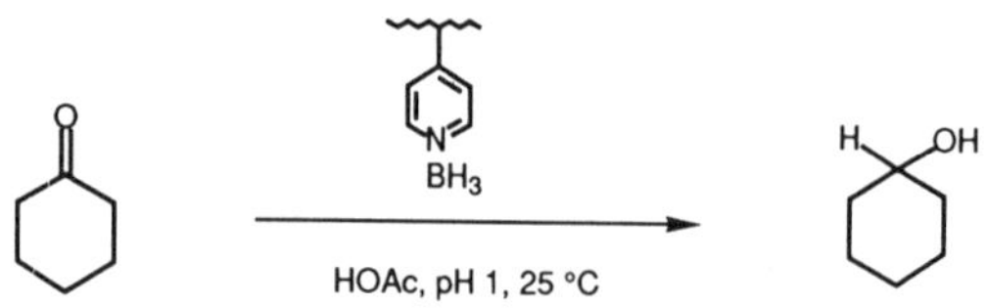


Fig. 8.

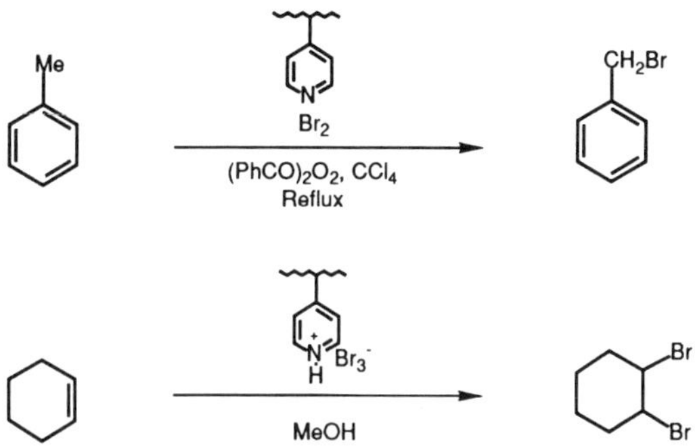


**Fig. 9.**

## Metal Recovery

REILLEX™ polymers act as weak base anion exchange resins for extraction of metals from acidic solutions. A type B complex is formed and the metal can be released from it by neutralization with aqueous sodium hydroxide (Fig. 10).[23] The extraction of vanadium from salt-roast leach solutions, the recovery of chromic acid wastes, and palladium from electroplating wastewater have been studied among others.[4]

## Pollution Control

The ability of pyridine to form stable tightly bound complexes with phenols is well known.[24] REILLEX™ polymers also readily form complexes with phenols even in high salt media. This provides an advantage over amine base resins which are usually only effective for removing phenols in salt-free aqueous media.

REILLEX™ polymers can be regenerated by elution with small volumes of an organic solvent, such as methanol. Therefore, REILLEX™ polymers are suitable for recovering phenols from

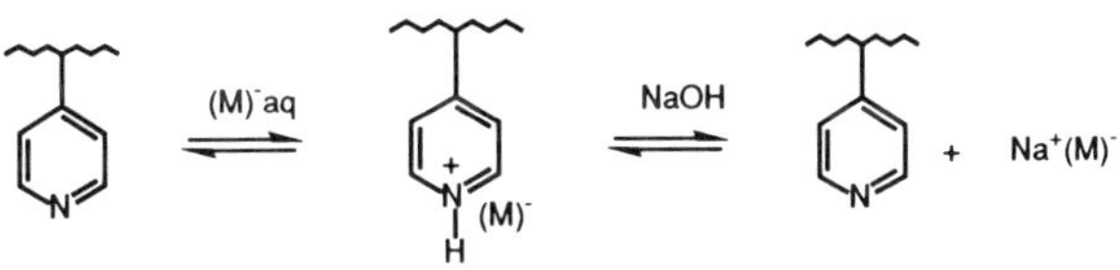


**Fig. 10.**

wastewater. Carboxylic and sulfonic acids and iodine are also readily adsorbed by REILLEX™ polymers.[5]

### Enzyme Immobilization

*N*-Cyanopolyvinylpyridiniums can immobilize proteins.[25] Reaction takes place by a ring opening–ring enclosure route (Fig. 11). This approach has been used for enzyme immobilization,[26] and for synthesis of catalytic polyelectrolytes suitable for ester hydrolysis.[27]

## Polymer-bound dialkylaminopyridine

The advantages of using 4-dimethylaminopyridine (DMAP) as a catalyst for many synthetically important acylations, particularly those of sterically hindered hydroxyl groups in steroids and other natural products, were first reported by Steglich of the University of Bonn and his collaborators at Schering AG in the late 1960s. At about the same time Litvinenko, in the USSR, observed a remarkable rate enhancement when studying the kinetics of the benzoylation of deactivated anilines when he used DMAP instead of pyridine as the catalyst. This and later work has been reviewed.[28] Now, 4-dimethylaminopyridine is the catalyst of choice for many acylations carried out on an industrial scale. The potential advantages in process efficiency and environmental control that would be offered by DMAP supported on an insoluble polymer have been recognized and Reilly Industries now offer POLYDMAP™ polymer as a commercial product. The properties and applications of this product are compared with DMAP itself and some polymers that have been reported in the chemical literature.

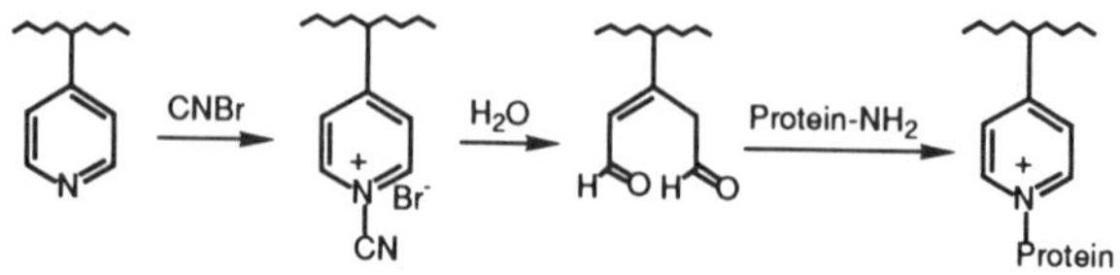


**Fig. 11.**

## Synthesis of Polymer-Bound 4-Dialkylaminopyridines

Klotz and coworkers were the first to report the attachment of a dialkylaminopyridine to a polymer.[29] This soluble polymer (**1**) was made by reaction of an acid-functionalized dialkylaminopyridine with a partially laurylated polyethyleneimine of 60 000 molecular weight using the carbodiimide method. This and other similar polymers effectively catalyzed the hydrolysis of *p*-nitrophenyl caproate. Later Verducci and coworkers[30] produced an insoluble polymer (**2**) by condensing 4-(4′-carboxyethyl)piperidinopyridine with the amino form of a Merrifield resin (1% divinylbenzene crosslinked polystyrene).

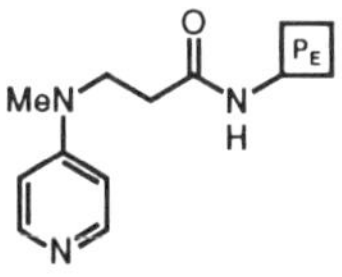


**1**

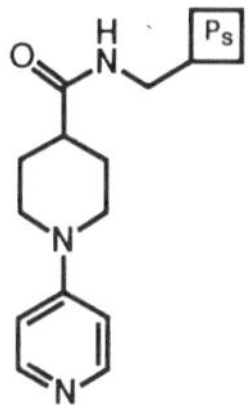


**2**

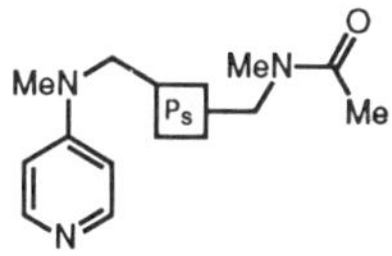


**3**

$P_E$ - polyethyleneimine

$P_S$- 1% divinylbenzene crosslinked polystyrene

Shinkai and coworkers avoided an amide linkage in the connection between the dialkylaminopyridine and the polymer backbone (used in **1** and **2**) in order to avoid the risk of hydrolysis during a caustic regeneration cycle.[31] They reacted a methylaminopolystyrene (4% crosslinked) with 4-chloropyridine; however, this reaction took 4 days at 150°C. The product was then acetylated to give a polymer (**3**) that contained 1·6 meq/g of DMAP and acted as a catalyst in the esterification of benzoic acid with methanol.

Fréchet and Tomoi with their respective groups have made comprehensive studies of the synthesis and structural requirements for catalytic activity of insoluble crosslinked dialkylaminopyridine catalysts. Two synthetic approaches have been compared (Fig. 12). Tomoi[32] prefers the route based on copolymerization of preformed monomer, whereas Fréchet[33] has found that crosslinked chloromethylated polystyrenes can be modified very effectively.

The groups of Fréchet,[34] Tomoi,[35] and Verducci[36] also have studied the effects of variation of the ratio of DMAP to styrene units, variation in the crosslinking, and variation of the length and nature of the spacer arm ($x$) (**4**).

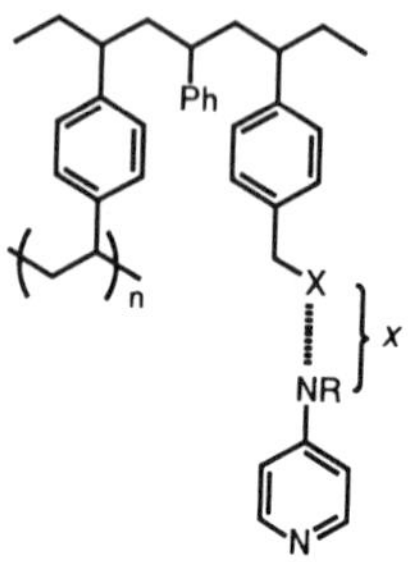


**4**

4-Pyrrolidinopyridine (PPY) is another excellent acylation catalyst and it too has been prepared linked to polymers. One method[37] of preparation developed by Mathias and his group involves the cyclo-polymerization of 4-diallylaminopyridine with free radical initiators

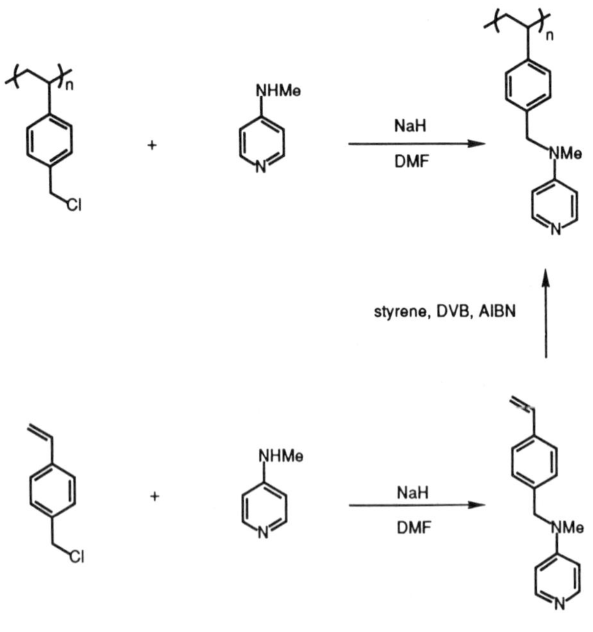


**Fig. 12.**

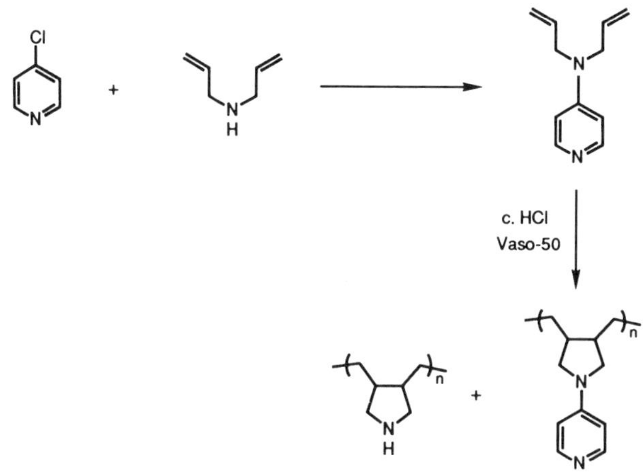


**Fig. 13.**

(Fig. 13). The other, by Narang,[38] involves condensation of 4-aminopyridine with an anhydride (Fig. 14).

# POLYDMAP$^{TM}$ polymer

We have synthesized a crosslinked polystyrene gel bead polymer,[39] insoluble in all solvents, that has the physical properties listed in Table 2. The polymer is supplied swollen and wet with toluene, as toluene is one of the most useful solvents in which to carry out DMAP catalyzed reactions.

## Conversion of Dimedone to Trione (5)

We chose to study of conversion of dimedone to trione (**5**) as an example of a reaction which is used in a key step in the production of

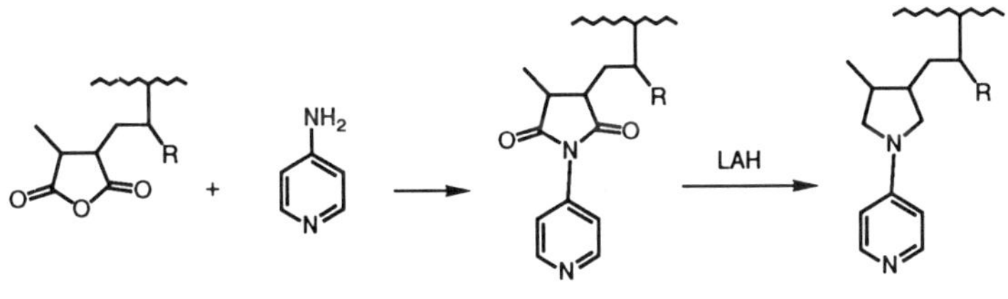


**Fig. 14.**

**Table 2.** Typical properties of POLYDMAP™ polymer

| | |
|---|---|
| Physical form | Crosslinked gel beads swollen in toluene |
| Toluene content | 45–47 wt % |
| Particle size | (Dry state) |
| | >80% between 30 and 80 US mesh |
| | >80% between 170 and 600 μm |
| Apparent bulk density | 0·56 kg/l (toluene wet state) |
| | 35·0 lb/ft$^3$ |
| DMAP equivalence | 1·4 meq/g (dry state) |
| | 0·4 eq/l (toluene wet state) |
| | 12·0 eq/ft$^3$ (toluene wet state) |
| Swelling in toluene | 92 vol % (from dry state) |
| Relative swelling | Toluene = 1·0 |
| Cyclohexane | 0·50 |
| Acetonitrile | 0·60 |
| Dioxane | 0·90 |
| Tetrahydrofuran | 1·05 |
| Pyridine | 1·10 |

several important herbicides (**6–8**).[40] As DMAP and polymer-supported aminopyridines have been reported to catalyze reactions of this type, we considered this an ideal reaction in which to compare the effectiveness of DMAP and POLYDMAP™ polymer. Our polymer catalyzes the dimedone–trione conversion (Fig. 15) with great facility, and it can be recycled many times without regeneration (Table 3).

A slight decrease in isolated yield of trione was observed after 20

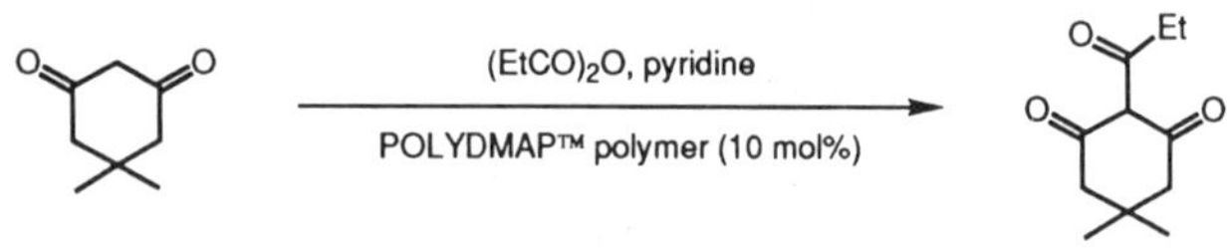


**5**

| Catalyst | % Isolated yield | % Relative activity |
|---|---|---|
| REILLEX™ polyDMAP™ polymer (10 mol %) | First cycle 86·5 | (97·2) |
| | 10th cycle 86 | (96·6) |
| | 30th cycle 78 | (87·6) |
| No catalyst | 50·5 | |
| DMAP (10 mol %) | 89 | |

**Fig. 15.**

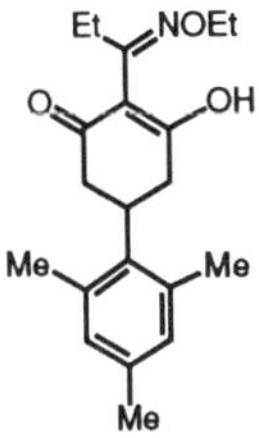

**6**

PP-604
Tralkoxydim
GRASP®

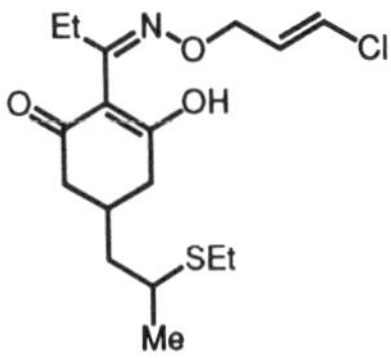

**7**

RE-45601

SELECT®

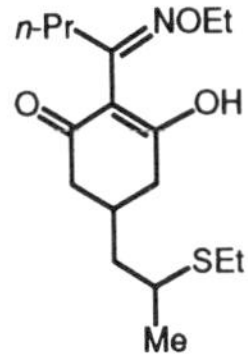

**8**

BAS-9052

POAST®

uses of our polymer at a 10 mol% catalyst loading, with no regeneration. The yield of trione may be maintained by increasing reaction time. Not only is the sustained catalytic activity of the polymer remarkable, but the ease of handling the gel beads (which do not break up) from run to run without mechanical loss is also noteworthy. On further recycle the catalyst showed a slow decline in activity. It was

**Table 3.** Conversion of dimedone to trione (**5**) with propionic anhydride and POLYDMAP™ polymer (10 mol%) in pyridine at 100°C—effect of number of catalyst cycles on yield of trione (**5**)

| Number of cycles | % Conversion to **5** after 1·5 h (relative activity)[a] | % Conversion to **5** after 4 h (relative activity) | % Isolated yield of **5** after 4 h |
|---|---|---|---|
| 1 | 96·0 (97·5) | 99·5 (100) | 90·5 |
| 5 | 89·5 (90·9) | 98·5 (99·0) | 90·0 |
| 10 | 83·0 (84·3) | 97·0 (97·5) | 87·5 |
| 15 | 77·5 (78·7) | 95·0 (95·5) | 86·0 |
| 20 | 70·0 (71·1) | 93·0 (93·5) | 86·0 |
| 25 | 64·5 (65·5) | 89·5 (90·0) | 82·0 |
| 30 | 60·5 (61·4) | 87·0 (87·4) | 82·0 (6 h)[b] |
| No catalyst | 20·7 | 47·5 | — |
| DMAP (10 mol %) | 98·5 | 99·5 | — |

[a] Relative activity compared with DMAP

$$= \frac{\%\ \text{conversion using polymer}}{\%\ \text{conversion using DMAP}} \times 100\%.$$

[b] Isolated yield after 6 h at 100°C.

found that regeneration of the catalyst, after 25 cycles, by treatment with ammonium hydroxide did afford a slight increase (3–5%) in isolated yield. Even after 25 uses the relative activity of the polymer after 4 h was 90% compared with both DMAP and its first use—this is a useful measure of reaction completion. We took percent conversion at 1·5 h as the point to compare the effect of the polymer and DMAP on rate of reaction. On the first recycle the relative activity of the polymer is 94·5%; this falls to 90% (cycle 5) and 61·4% (cycle 30). We feel that this fall-off in the rate of reaction is more than counterbalanced by other advantages offered by the polymer.

## Acetylation of 1-Methylcyclohexanol

The acetylation of 1-methylcyclohexanol, a difficult acylation, has been selected by many workers to test the effectiveness of various acylation catalysts including DMAP and other polymeric forms of DMAP.[28,30,33,35,41] We have used this reaction to compare the relative activity of POLYDMAP™ polymer to DMAP and to determine the mechanical stability of the polymer beads after 30 days at 80°C (Fig. 16).

Our polymer gives (relative activity 96·6% after 10 cycles) a superior performance to that of a polymer reported by Verducci and coworkers[30] which gave a 30% loss in relative activity over 10 cycles in this reaction under similar conditions. The relative activity after 30 cycles (87·6%) is comparable to that in the dimedone–trione conversion (87·4%). The mechanical properties of POLYDMAP™ polymer are still excellent after cycle 30 (720 h at 80°C in acetic anhydride). We have obtained 1-methylcyclohexyl acetate in 95% isolated yield from this reaction using designed preparative conditions.

The conversion of linalool to linalyl acetate has been well studied using DMAP as a catalyst.[32] In this reaction we have found POLYDMAP™ polymer to perform similarly to DMAP (Fig. 17).

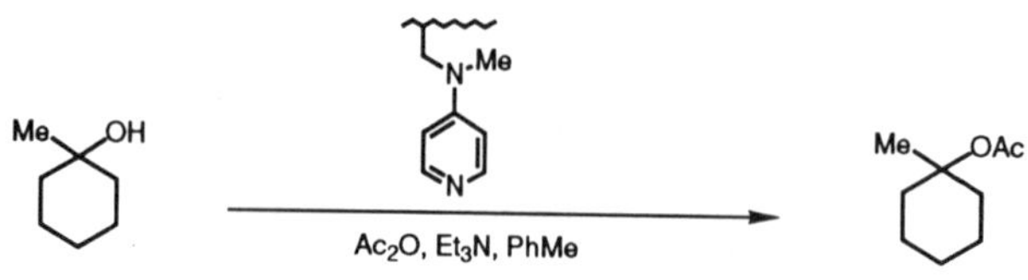


**Fig. 16.**

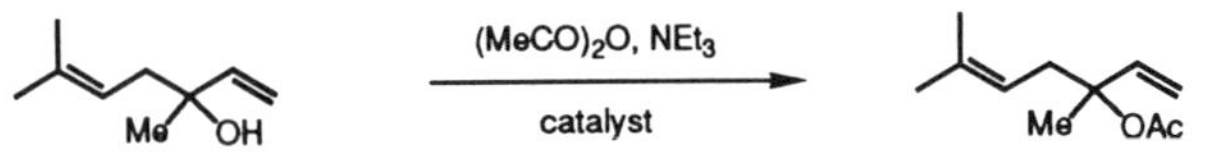


| | |
|---|---|
| POLYDMAP™ polymer (10 mol%) | 90.5% |
| DMAP (10 mol%) | 93.0% |

**Fig. 17.**

# Conclusions

Commercially available pyridine-containing polymer applications have been highlighted in this article. They offer in some cases an advantage over traditional methods, particularly where product isolation and purification is difficult. Process research teams developing synthetic routes that involve reactions of the types described should always consider the option of using a polymer-supported reagent or catalyst.

# References


1. Goe, G. L., Pyridine and pyridine derivatives. In *Kirk–Othmer: Encyclopedia of Chemical Technology*, Vol. 19, 3rd edn, 1982, pp. 454.
2. REILLEX™ Report 3, Catalysts. Reilly Industries, Inc., 1986.
3. REILLEX™ Report 4, Polymeric reagents. Reilly Industries, Inc., 1986.
4. REILLEX™ Report 5, Metal recovery. Reilly Industries, Inc., 1986.
5. REILLEX™ Report 6, Pollution control. Reilly Industries, Inc., 1989.
6. Hallensleben, M. L. & Wurm, H., *Angew. Chem. Intl. Edn. Engl.*, **15** (1976) 163; Reilly Industries, Inc., Pyridine functionality in polymer form. Indianapolis, 1980.
7. Johnston, R. D., Marston, C. R., Krieger, P. E. & Goe, G. L., *Synthesis* (1988) 393.
8. Fife, W. K. & Zhang, Z., *J. Org. Chem.*, **51** (1986) 3744; Fife, W. K. & Xin, Y., *J. Am. Chem. Soc.*, **109** (1987) 1278.
9. Coon, R. W., US Patent 4 473 711, 1984.
10. Goe, G. L., Bailey, T. D. & Beadle, J. R., US Patent 4 753 911, 1988.
11. Paulik, F. E., Hershman, A., Knox, W. R. & Roth, J. E., US Patent 3 769 329,, 1973.
12. Goe, G. L. & Marston, C. R., Eur. Patent Appl. 277 824, 1988.
13. Drago, R. S., Nyberg, E. D., El A'mma, A. & Zombeck, A., *Inorg. Chem.*, **20** (1981) 641; Drago, R. S. & El A'mma, A. US Patent 4 328 125, 1982.
14. Carlock, C. T., US Patent 4 178 314, 1979.
15. Carlock, J. T., US Patent 4 178 312, 1979.
16. Watanabe, Y., US Patent 3 980 662, 1976.

17. Corey, E. J. & Suggs, J. W., *Tetrahedron Lett.* (1975) 2647.
18. Piancatelli, G., Scettri, A. & D'Auria, M., *Synthesis* (1982) 245.
19. Fréchet, J. M. J., Warnock, J. & Farrall, M. J., *J. Org. Chem.*, **43** (1978) 2618.
20. Domb, A. & Avny, Y., *J. Macromol.-Sci. Chem.*, **A22** (1985) 167, 183.
21. Sket, B. & Zupan, M., *J. Org. Chem.*, **51** (1986) 929.
22. Fréchet, J. M. J., Farrall, M. J. & Nuyens, L. J., *J. Macromol. Sci.-Chem.*, **A11** (1977) 507.
23. Kennedy, D. C., *Chemical Engineering*, **106** 16 June 1980.
24. Bramley, A., *J. Chem. Soc.*, **109** (1916) 434.
25. Pittner, F., Miron, T., Pittner, G. & Wilcheck, M. *J. Am. Chem. Soc.*, **102** (1980) 2451.
26. Ishimatsu, Y., Shigesada, S. & Kimura, S., US Patent 3 915 797, 1975; Okubo, T., US Patent 3 983 097, 1976; Ishimatsu, Y., Shigesada, S., Suzuki, H., Kitagawa, H. & Kimura, S., US Patent 4 205 128, 1980.
27. Iso, N., Okubo, T., Kitano, H. & Kunugi, S., *J. Am. Chem. Soc.*, **97** (1975) 2882.
28. Höfle, G., Steglich, W. & Vorbrüggen, H., *Angew. Chem. Int. Ed. Engl.*, **17** (1978) 569; Scriven, E. F. V., *Chem. Soc. Rev.* (1983) 129; Neises, B. & Steglich, W., *Org. Synth.*, **63** (1984) 183.
29. Hierl, M. A., Gamson, E. P. & Klotz, I. M., *J. Am. Chem. Soc.*, **101** (1979) 6020; Delany, E. J., Wood, L. E. & Klotz, I. M., *J. Am. Chem. Soc.*, **104** (1982) 799.
30. Guendouz, F., Jacquier, R. & Verducci, J., *Tetrahedron Lett.*, **25** (1984) 4521.
31. Shinkai, S., Tsuji, H., Hara, Y. & Manabe, O., *Bull. Chem. Soc. Jpn.*, **54** (1981) 631.
32. Tomoi, M., Akada, Y. & Kakiuchi, H., *Makromol. Chem., Rapid Commun.*, **3** (1982) 537.
33. Deretani, D., Darling, G. D., Horak, D. & Fréchet, J. M. J. *Macromolecules*, **20** (1987) 767.
34. Fréchet, J. M. J., Darling, G. D., Itsuno, S., Lu, P., de Meftahi, M. V. & Rolls, W. A., *Pure and Appl. Chem.*, **60** (1988) 353; Fréchet, J. M. J., Rolls, W. A. & Darling, G. D., *Future Trends Polym. Sc. Technol.*, ed. E. Martuscelli, C. Marchetta & L. Nicolais. Technomic Lancaster, Pa., 1987, pp. 118–33; Deratani, A., Maraldo, T., Darling, G. D. & Fréchet, J. M. J., *Reactive Polymers*, **9** (1988) 19.
35. Tomoi, M., Ishigoki, S., Arita, Y. & Kakiuchi, H., *J. Polym. Sc.*, **26** (1988) 2251.
36. Guendouz, F., Jacquier, R. & Verducci, J., *Tetrahedron*, **44** (1988) 7095.
37. Mathias, L. J. & Cei, G., *J. Am. Chem. Soc.*, **20** (1987) 2645.
38. Narang, S. C. & Ramharack, R., *J. Polym. Sc., Lett. Edn.*, **23** (1985) 147.
39. Reilly Patents pending.
40. Iwataki, I., Shibuya, M., Ishikawa, H. & Kawana, T., US Patent 4 249 937, 1981; Bird, B. G. & Karge, R. L., Australian Patent 52 992, 1988.
41. Goe, G. L., Huckstep, L. H. & Scriven, E. F. V., *Chem. Ind.* (*London*) (1982) 722.

# Organic reactions in liquid ammonia solution

**O. Meth-Cohn***

*Sterling Organics UK Limited, Cramlington, UK*

## Unique properties of liquid ammonia

As shown in Table 1, liquid ammonia is capable of operation at atmospheric pressure from −33 to −78°C but, because of its favourable critical temperature (above which it is inevitably a gas), it can be used under pressure safely even in excess of 100°C. Its ready miscibility with organic solvents and ability to dissolve a variety of metals underlies its wide application.

Some of the reactions conducted in liquid ammonia and exemplified in the remainder of the text are:

- reductions
- alkylations
- aryne chemistry
- aminations
- acetylene chemistry
- oxidations
- dehydrohalogenations
- rearrangements

* Present address. Sunderland Polytechnical College, Sunderland, UK.

**Table 1.** Ammonia $NH_3$

| | |
|---|---|
| Boiling point | −33°C |
| Melting point | −78°C |
| Critical temperature | 132°C |
| Heat of vaporisation | 5·6 kcal/mole |
| Excellent solvent for metals | |
| Miscible with most organic solvents | |
| Typical price | 23 pence/kg<br>18–19 cents/lb |

Laboratory operations in liquid ammonia are easily conducted using a solid carbon dioxide condenser while plant scale would require a suitable refrigeration plant.

# Dissolving metal chemistry

## Reductions in Liquid Ammonia

Probably the single most valuable process conducted in liquid ammonia is reduction, and of this class the Birch reduction is of unique importance. When a group 1 or 2 metal (e.g. lithium, sodium, potassium or calcium) is dissolved in ammonia a blue solution ensues due to the formation of a solvated electron, the key to the reduction process (Fig. 1). Lithium is more potent than sodium in these reductions and potassium still less effective.

$$Na\cdot + \text{liquid } NH_3 \rightarrow Na^+(NH_3)_n + \cdot(NH_3)_m \qquad \text{Deep turquoise solution}$$

**Fig. 1.** Reductions—metal solvation in ammonia.

The process usually requires the presence of an alcohol or an ammonium salt which behaves as an acid (i.e. a source of a proton) (Table 2).

In the Birch reduction the electron is added to a benzene ring to give an intermediate which sequentially picks up a proton, another electron and finally a proton to give a dihydrobenzene (Fig. 2). The dihydro compound is in most cases a 1,4-derivative, an otherwise inaccessible reduction pattern. Thus when benzoic acid is reduced (or any derivative of benzene bearing an electron-withdrawing group) a

**Table 2.** Reduction methods

| | |
|---|---|
| Metal | Li > Na > K |
| Solvent | $NH_3$ with or without co-solvent (ether) |
| Other reactant | An alcohol: EtOH, PrOH, *t*-BuOH<br>or an ammonium salt<br>(these are *acids* in liquid $NH_3$) |

high yield of 1,4-dihydrobenzoic acid is produced (Fig. 3). Anisole, typical of electron-releasing derivatives, required lithium for reduction and results in the 2,5-dihydro derivative, since the charge from electron addition would be repelled by the substituent. Similarly naphthalenes such as $\alpha$-naphthol are reduced in the non-substituted ring, phenylglycine gives the 2,5-dihydro-derivative (Fig. 3) and anthrancene can be reduced even to the hexahydro stage (Fig. 4). One of the exceptions to the rule is dimethylaniline which gives a 2,3-dihydro product (Fig. 4).

The reduction is not limited to aromatic systems. Although isolated double bonds are stable, conjugated double bonds can be reduced (e.g. butadiene gives but-2-ene) and, when the reduction step is conducted in the absence of an alcohol, can be transformed into useful dimers (Fig. 5). Esters are readily reduced to alcohols, whereas acids are unaffected, allowing transformations such as that shown in Fig. 6 to be achieved.

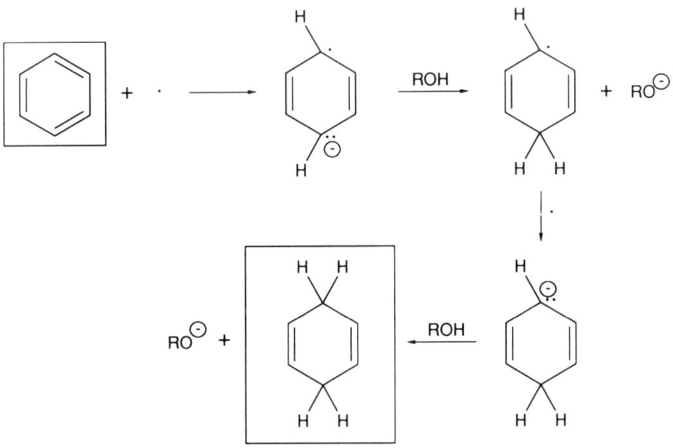


**Fig. 2.** The Birch reduction.[1]

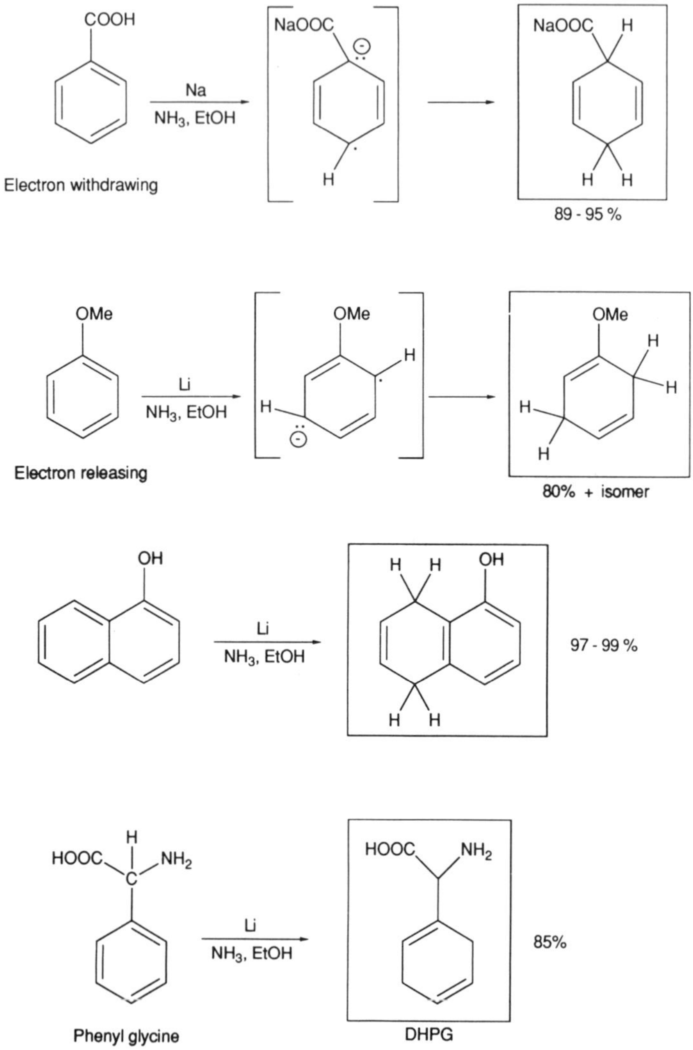


**Fig. 3.** Further examples of the Birch reduction.

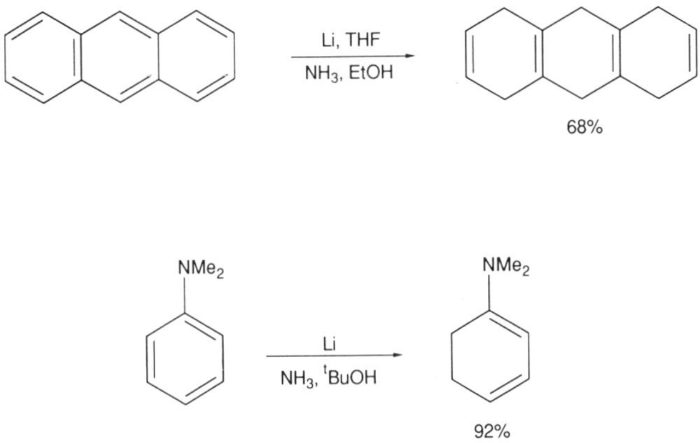


**Fig. 4.** Unusual Birch reactions.

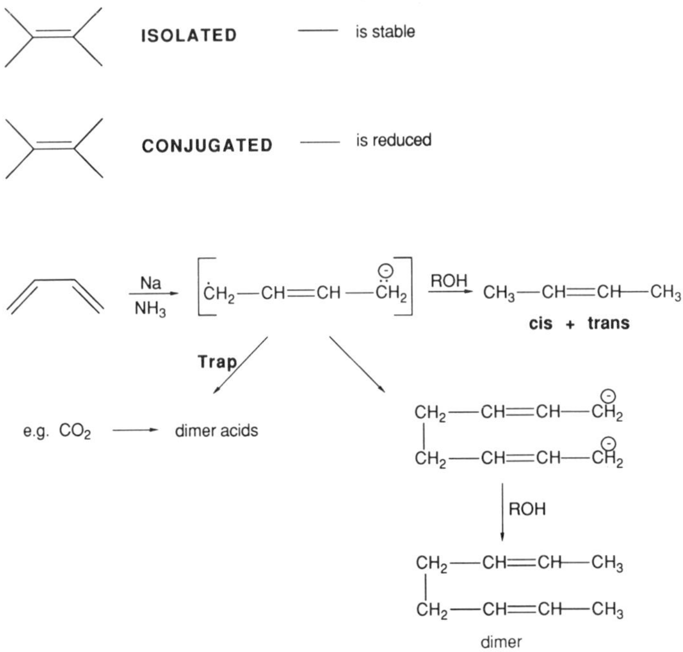


**Fig. 5.** Alkene reductions.

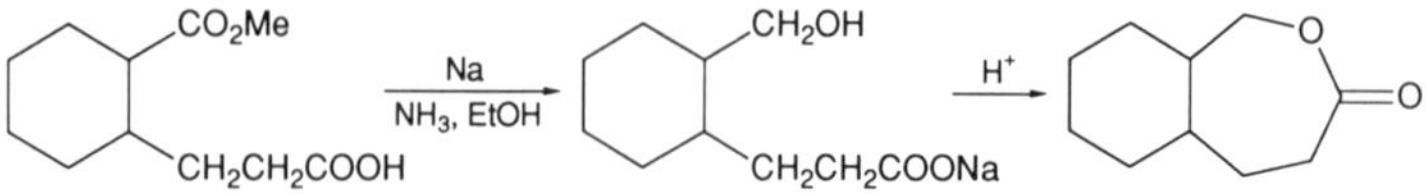


**Fig. 6.** Ester reduction.

# Reactions using metal amides

## Alkylations

When sodium is dissolved in ammonia containing a catalytic amount of a ferric salt (usually ferric nitrate), hydrogen is liberated and a solution of sodamide, a very strong base, is formed. This may be used as such or isolated for use at higher temperature in another solvent (e.g. for the Chichibabin amination of pyridines—see later). Examples of the use of sodamide for alkylation are shown in Fig. 7 and the particular

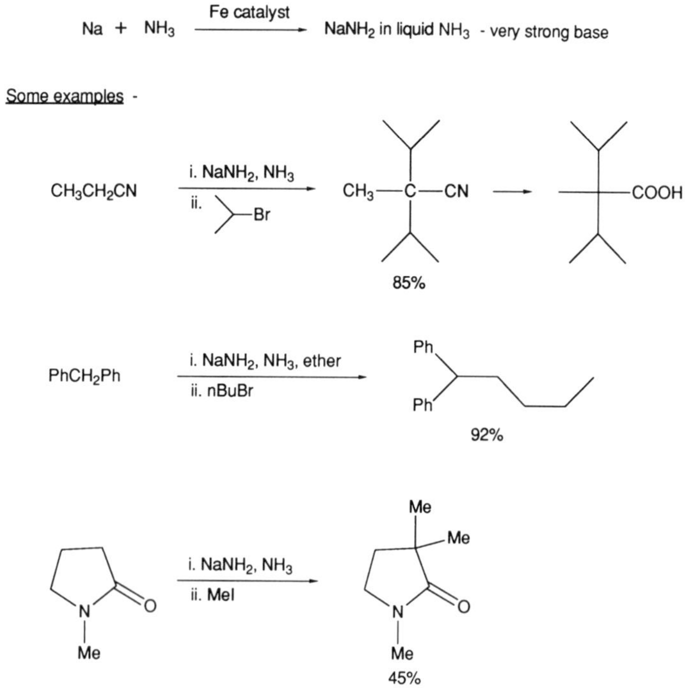


**Fig. 7.** The use of sodamide in alkylation.

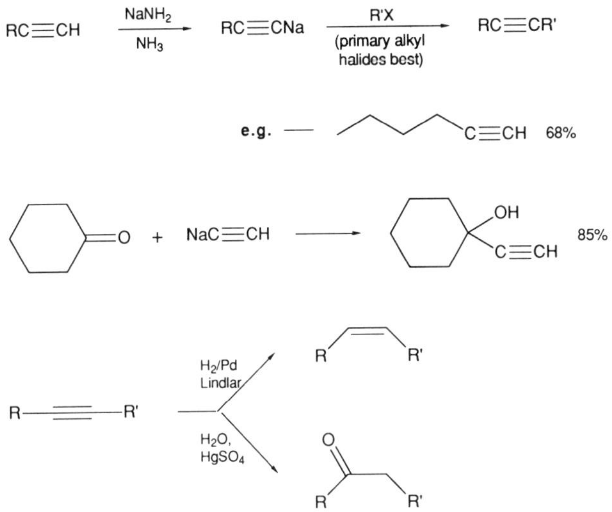


**Fig. 8.** Alkyne alkylation.

case of alkyne alkylation in Fig. 8. Since acetylene can be mono- or dialkylated and since the products are, for example, ready sources of *cis*-alkenes or ketones (Fig. 8), this chemistry is of particular versatility.

## Benzyne-mediated reactions

When a halobenzene is treated with, for example, potassium amide in liquid ammonia, hydrogen halide is eliminated and a highly reactive benzyne is produced. In the absence of other nucleophiles this intermediate reacts with ammonia to give an aniline (Fig. 9) which may be differently substituted compared to the halobenzene precursor. The mode of elimination (e.g. whether the *ortho*- or *para*-hydrogen is eliminated from a *meta*-halo-substituted benzene (Fig. 9)) and subsequent mode of attack of the benzyne by a nucleophile (indicated by ↯ in Fig. 9) is totally predictable on the basis of whether the substituent, R, is electron-withdrawing (EWG) or electron-releasing (ERG). A useful way of harnessing the reaction preparatively is to allow the nucleophile (e.g. a carbanion) also to be generated by the metal amide, either intermolecularly or intramolecularly as illustrated in Fig. 10.

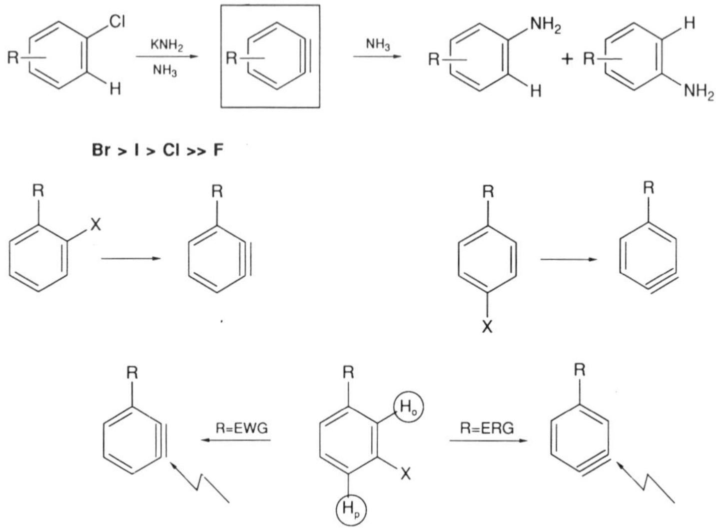


**Fig. 9.** Benzyne and related intermediates.[2]

## Aminations

Less accessible anilines are also usefully made in this way (Fig. 11) and the isolated sodamide is used in high boiling hydrocarbon solvents to aminate pyridines in the well-known Chichibabin reaction. Invariably $\alpha$-amination is preferred and dependent upon temperature. 2-Amino, 2,6-diamino and even 2,4,6-triaminopyridines can be produced.

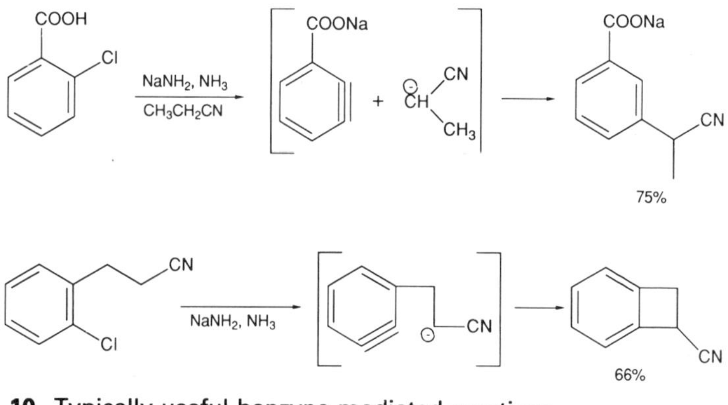


**Fig. 10.** Typically useful benzyne-mediated reactions.

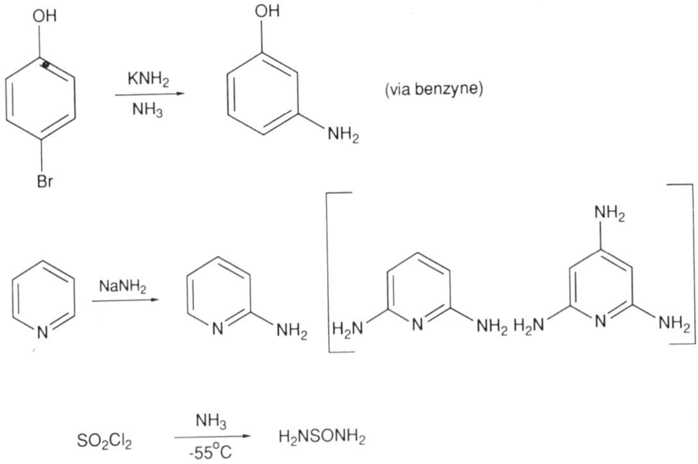


**Fig. 11.** Aminations.[3]

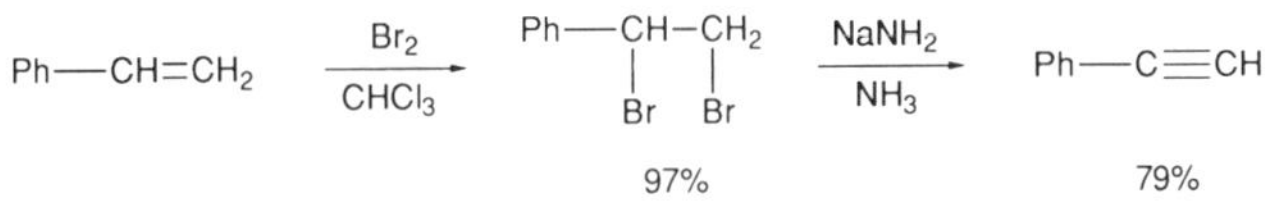


**Fig. 12.** Dehydrohalogenation.

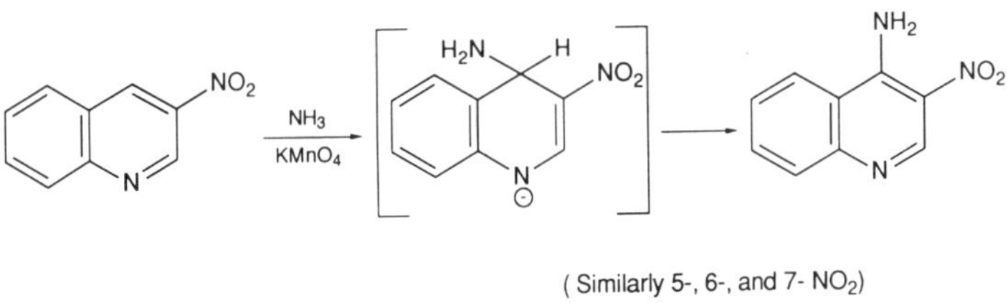


**Fig. 13.** Oxidation with $KMnO_4$ in $NH_3$.[4]

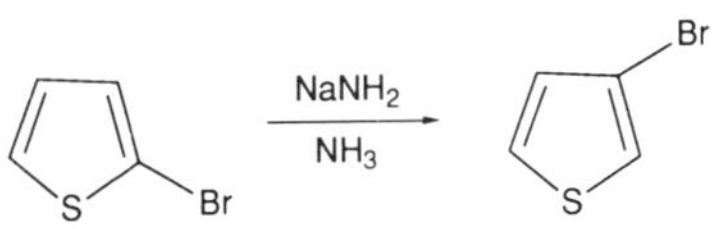


**Fig. 14.** Aromatic halogen arrangement.

### Dehydrohalogenation

The powerful base strength of sodamide allows efficient elimination of the bromine adduct of styrene, giving ready access to phenylacetylene (Fig. 12).

### Miscellaneous

Recent lesser exploited applications of liquid ammonia involve, surprisingly, the efficient oxidation with potassium permanganate to allow the synthesis of nitroaminoquinolines (Fig. 13) and the rearrangement of 2-bromothiophen to the difficultly made 3-bromo-isomer in high yield by brief treatment with sodamide in liquid ammonia (Fig. 14).

## References

1. March, J., *Advanced organic chemistry,* 3rd edn, John Wiley, 1985, p. 700.
2. ibid, p. 586.
3. *Comprehensive heterocyclic chemistry,* Volume 2, ed. A. J. Boulton & A. McKillop. Pergamon Press, 1984.
4. Wozniak, M., Baranski, A., Nowak, K. & Van der Plas, H. C., *J. Org. Chem.,* **52** (1987) 5643.

# A new route to acetal compounds

**M. Yamashita & H. Asada**

*UBE Industries Limited, London, UK*

## Introduction

Alkylnitrite (RONO) is well known as a useful agent for synthesizing oximes and diazo compounds by nitrosation of active methylene groups, but no studies using alkylnitrites together with palladium catalyst in a synthetic reaction have been reported.

The unique reactivity of the alkylnitrite/palladium catalyst combination was discovered by UBE in the course of studies to develop a new process for production of oxalic diester from carbon monoxide, and resulted in the full commercialization, in 1978, of a unique process for oxalic diester production.

An extensive R&D programme has also resulted in other new oxidation reactions using the Pd/RONO system.

In 1985, an oxiacetalization of acrylontrile was successfully commercialized to produce an intermediate of Vitamin $B_1$ and in 1986 a multi-purpose plant was constructed to be able to proceed with several kinds of reaction using the Pd/RONO system. A certain volume of carboxylic diesters and acetals are now ready for prompt production.

In this paper, we would like to introduce such new organic syntheses using the Pd/RONO system and also several useful compounds, derived from the products through these synthetic methods.

# Synthesis of carboxylic diesters

## Synthesis of Oxalic Diesters in a Liquid Phase

Oxalic diester is synthesized on a laboratory scale by reacting carbon monoxide, alcohol and oxygen under water-free conditions in the presence of the $PdCl_2/CuCl_2$ catalyst:

$$2CO + 2ROH + \tfrac{1}{2}O_2 \xrightarrow{PdCl_2/CuCl_2} (COOR)_2 + H_2O \tag{1}$$

However, in this reaction the water formed inhibits further reaction, and it had been necessary to add dehydrating agents to the reaction mixture.

In our study to overcome this problem, it was found that nitric acid changed to RONO in a reaction system containing nitric acid:

$$CO + HNO_3 + ROH \xrightarrow{Pd} RONO + CO_2 + H_2O \tag{2}$$

and further study under higher RONO concentration revealed the important role of RONO in this oxidation reaction:

$$2CO + 2RONO \xrightarrow[ROH]{Pd/C} (COOR)_2 + 2NO \tag{3}$$

Production of RONO from the NO by-product in the presence of $O_2$ was also checked:

$$2NO + 2ROH + \tfrac{1}{2}O_2 \rightarrow 2RONO + H_2O \tag{4}$$

Commercial operation of our oxalate plant started in 1978 and has operated well since. In this plant, the reaction condition is as follows:

ROH: *n*-BuOH
Pressure: 100 kg/cm$^2$ G
Temperature: 90°C

The overall reaction scheme in the system is

$$2CO + 2BuOH + \tfrac{1}{2}O_2 \rightarrow (COOBu)_2 + H_2O \tag{5}$$

## Synthesis of Oxalates in the Vapour Phase

It is a very important point to maintain the activity of the catalyst by operating reactions (1) and (2) separately in two reactors in order to get a higher yield and selectivity of reactions. In other words, reaction (3) should always be operated in a moisture-free condition.

In our development, these problems have been solved satisfactorily by the use of a vapour phase reaction system using methyl nitrite. Figure 1 shows a vapour phase oxalate production process.

Metallic Pd alone or metallic Pd systems as DMO synthesis catalysts have also been reported. These catalysts have superior activity, selectivity and life. Production processes for ethylene glycol or oxamide by hydrogenation or amidation of DMO:

$$(COOCH_3)_2 + 4H_2 \rightarrow HOCH_2CH_2OH + 2CH_3OH \tag{8}$$

$$(COOCH_3)_2 + 2NH_3 \rightarrow (CONH_2)_2 + 2CH_3OH \tag{9}$$

have been developed, but are not discussed in this paper.

Another reason why the economical process of this reaction using RONO could be established is because reaction (7) is carried out with high selectivity and high efficiency. NO produced by the main reaction is quantitatively recycled under the correct conditions to regenerate RONO for use in the oxidation reaction.

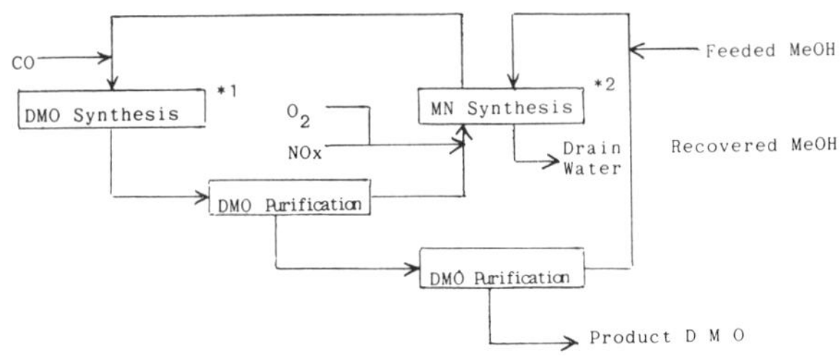


$$1^{*}\quad 2CO + 2CH_3ONO \longrightarrow (COOCH_3)_2 + 2NO \tag{6}$$

$$2^{*}\quad 2NO + 2CH_3OH + {}^{1}/{}_{2}O_2 \longrightarrow 2CH_3ONO + H_2O \tag{7}$$

DMO = Dimethyl oxalate MN = Methyl nitrite

**Fig. 1.** Dimethyloxalate (DMO) synthesis.

### Synthesis of Malonates

Malonates are synthesized from ketene, carbon monoxide and RONO in either the vapour or liquid phase in the same way as dialkyl oxalate:

$$CH_2=CO+CO+2RONO \rightarrow \begin{matrix} COOR \\ | \\ CH_2 \\ | \\ COOR \end{matrix} + 2NO \tag{10}$$

### Synthesis of Succinate

Dialkyl succinates are synthesized by using ethylene as the raw material:

$$CH_2=CH_2+2CO+2RONO \rightarrow \begin{matrix} CH_2COOR \\ | \\ CH_2COOR \end{matrix} + 2NO \tag{11}$$

## Synthesis of acetals and ketals

The Wacker process for acetoaldehyde production by oxidation of olefin using the $PdCl_2/CuCl_2$ catalyst is well known as a commercial process:

$$C_2H_2+PdCl_2+H_2O \rightarrow CH_3CHO+Pd+2HCl \tag{12}$$

$$2CuCl_2+Pd \rightarrow 2CuCl+PdCl_2 \tag{13}$$

$$\tfrac{1}{2}O_2+2CuCl+2HCl \rightarrow 2CuCl_2+H_2O \tag{14}$$

---

$$C_2H_4+\tfrac{1}{2}O_2 \rightarrow CH_3CHO \tag{15}$$

The same reaction with the Pd/RONO catalyst produces a corresponding acetal or ketal from some olefins (Table 1). The reaction is quite similar to the Wacker process from a commercial value point of view. The superiorities of the Pd/RONO catalytic reaction are:

1. The olefin does not directly come into contact with oxygen.
2. Acetals and ketals can be produced with high selectivity without the presence of water.

**Table 1.** Oxyacetalization and ketalization examples

| Starting olefin | Products |
|---|---|
| $CH_2{=}CH_2$ | $CH_3CH(OCH_3)_2$ |
| $CH_2{=}CHCH_3$ | $CH_3C(OCH_3)_2CH_3$ |
| $CH_2{=}CHCOOCH_3$ | $(CH_3O)_2CHCH_2COOCH_3$ |
| $CH_2{=}CHCN$ | $(CH_3O)_2CHCH_2CN$ |
| $R{-}C_6H_4{-}CH_2CH{=}CH_2$ | $R{-}C_6H_4{-}CH_2C(OCH_3)_2CH_3$ |

This nitrite process is also superior to the Wacker process for direct acetal synthesis and oxidation of olefins containing electron withdrawing groups such as nitrile, etc.

$$CH_2{=}CHX + 2CH_3ONO \rightarrow \underset{OCH_3}{\overset{OCH_3}{\overset{|}{\underset{|}{C}}}}H_2CH_2X + 2NO \qquad (16)$$

$$2NO + 2CH_3OH + \tfrac{1}{2}O_2 \rightarrow 2CH_3ONO + H_2O \qquad (17)$$

$$CH_2{=}CHX + 2CH_3OH + \tfrac{1}{2}O_2 \rightarrow \underset{OCH_3}{\overset{OCH_3}{\overset{|}{\underset{|}{C}}}}H_2CH_2X + H_2O \qquad (18)$$

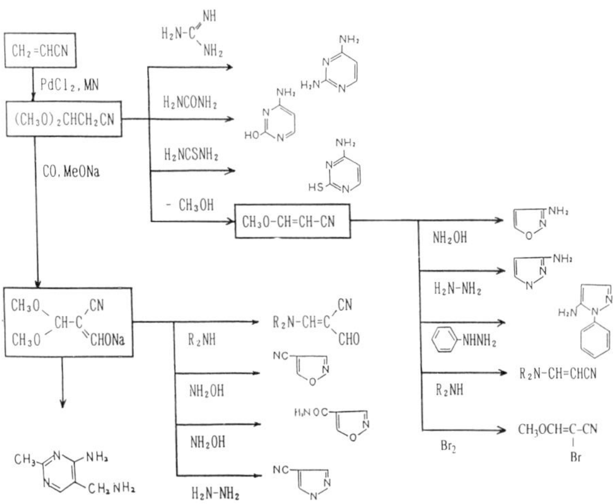

**Fig. 2.** Derivatives of acrylonitrile.

# Derivatives of acetals

### Derivatives of Acrylonitrile

These are shown in Fig. 2.

### Derivatives of Acrylates

These are shown in Fig. 3.

# Conclusion

In this paper we have reported 'oxidation reaction using the Pd/alkylnitrite system', focusing on the commercialized process.

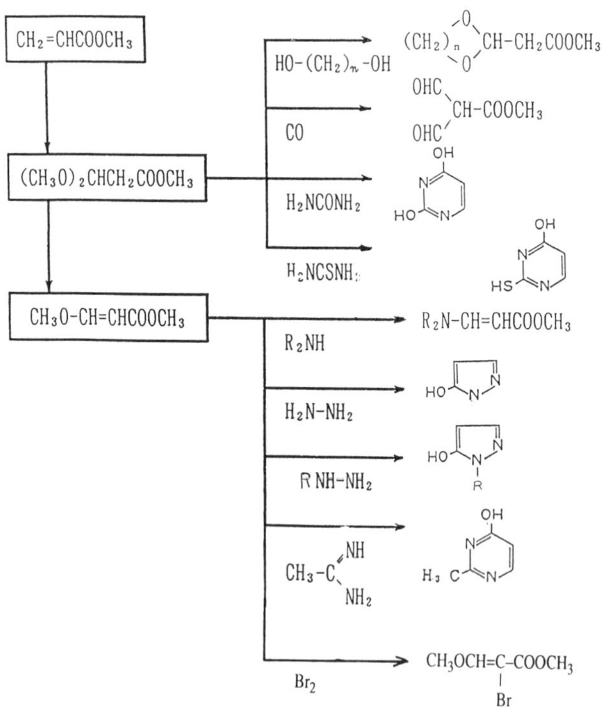


**Fig. 3.** Derivatives of acrylates.

However, we have not yet established the mechanisms of these reactions. We hope that many related compounds will be produced from continuous activity in research.

We also hope that many compounds obtained from these reactions will be useful as raw materials of fine chemicals, such as pharmaceuticals and agrochemicals.

# Novel amine and carboxyl protecting group

**Roy Valentine***

*La Porte Organic Chemicals Division, Middlesbrough, UK*

## Introduction

A new protecting group based on bis-(4-nitrophenyl)ethane (Fig. 1) will be described with particular reference to peptide synthesis.

There is currently a great interest in peptide synthesis, particularly in the fields of immunology and virology, where short chain peptides are showing promise as a new generation of vaccines.

Peptides are short chains of amino acids and could be considered to be synthesised by condensing the amino group of one amino acid with the carboxylic acid group of another (Fig. 2). However, this simplistic approach does not work in practice as the reaction is thermodynamically unfavourable. As there is no differentiation between the reactivity of the functional groups of one amino acid and the other, mixtures would also be formed.

Chemists have developed methods for controlled peptide synthesis; the essential features of which are outlined below (see Fig. 3):

1. protection of amino group of one amino acid
2. protection of carboxy group of the other amino acid

* Present address: High Force Limited, Durham, UK.

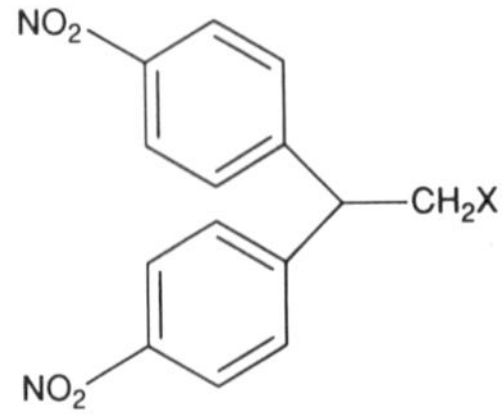


**Fig. 1.**

3. protection of side chain functional groups
4. activation of carboxy group
5. coupling of both amino acids
6. removal of one protecting group
7. further coupling with another protected amino acid.

Although peptide chains can be grown from either end it has been found advantageous to grow the chain from the amino end of the chain. Protecting group chemistry has evolved gradually, taking into account the conflicting reactivities of the functional groups present.

## Protection Group Requirements

There are three groups to protect (Fig. 4). The reagents used for cleavage of the temporary protecting groups must be compatible with the semi-permanent protecting groups, as must the reagents for activation and coupling.

The semi-permanent protecting groups are put into place early in the synthesis and are removed at the end.

The temporary protecting groups are removed after each coupling stage. The reagents used for this deprotection must not have any adverse effects on the growing peptide. It is this criteria that causes

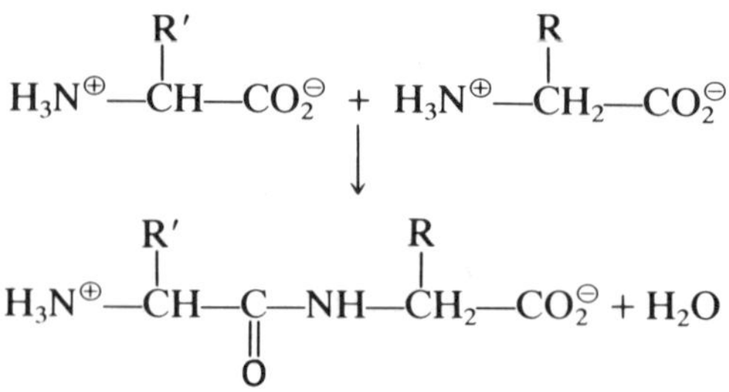

**Fig. 2.**

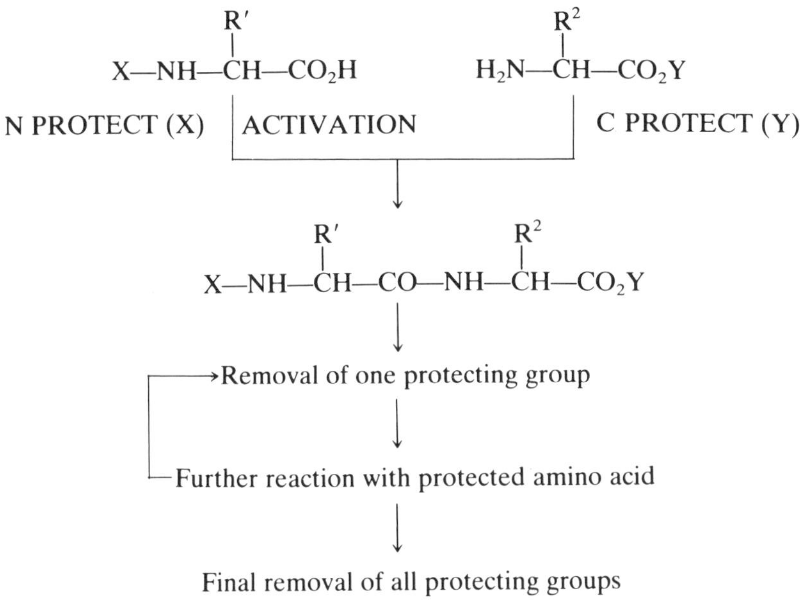


**Fig. 3.**

problems as strong acids, e.g. trifluoroacetic acid, or extended reaction times lead to unwanted by-products.

Although many hundreds of compounds have been proposed and used as protecting groups only a few have gained prominence

### *Acid Labile Types, e.g. t-Butoxycarbonyl (BOC)* (Fig. 5)

#### *Advantages*

(i) relatively inexpensive
(ii) soluble in many solvent systems.

#### *Disadvantages*

(i) deprotection with trifluoroacetic acid places constraints on side chain protection
(ii) gives rise to carbocations which attack growing peptide and form unwanted by-products.

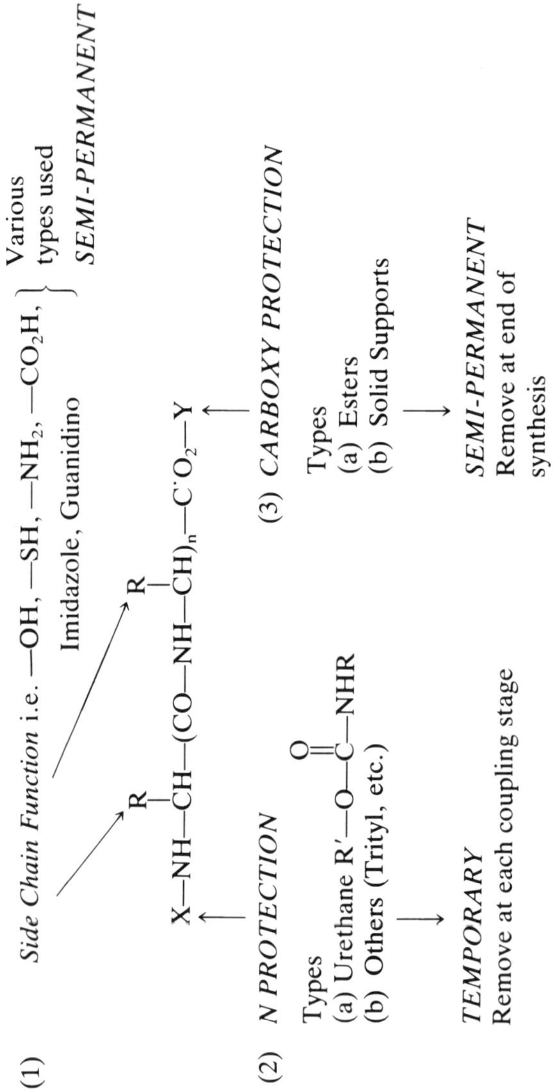

(1) Side Chain Function i.e. —OH, —SH, —NH$_2$, —CO$_2$H, Imidazole, Guanidino
Various types used
SEMI-PERMANENT
R
R
X—NH—CH—(CO—NH—CH)$_n$—CO$_2$—Y
(2) N PROTECTION
Types
(a) Urethane R′—O—C(=O)—NHR
(b) Others (Trityl, etc.)
TEMPORARY
Remove at each coupling stage
(3) CARBOXY PROTECTION
Types
(a) Esters
(b) Solid Supports
SEMI-PERMANENT
Remove at end of synthesis


**Fig. 4.**

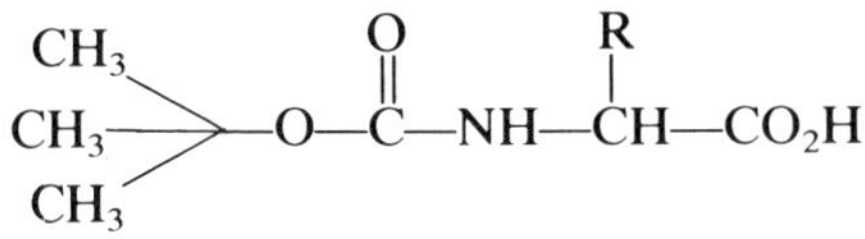

**Fig. 5.**

*Base Labile Types, e.g. Fluorenylmethoxycarbonyl (FMOC)* (Fig. 6)

*Advantages*

(i) compatible with acid labile side chain protecting group
(ii) no intermediate carbocations formed and so there is no need for scavengers to mop up these reactive intermediates.

*Disadvantages*

(i) expensive
(ii) very susceptible to mild base cleavage; basic solvent impurities can give rise to side reactions.

# Bis-(4-Nitrophenyl)ethane derivatives

These have been developed as new base labile protecting groups which offer advantages over FMOC (presently the most popular base labile protecting group) and also the acid labile groups (see Fig. 7).

### BNPE Amine Protection

The succinimidyl derivative (prepared via the chloroformate) has been found to derivatise amine groups efficiently (Fig. 8).

The BNPEOC protected amino acids are good crystalline materials which makes purification easy. This is an important point when peptide chains of any length are required as amino acid purity is paramount.

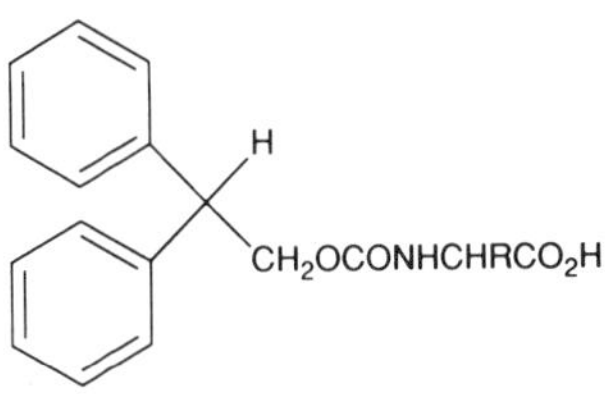


**Fig. 6.**

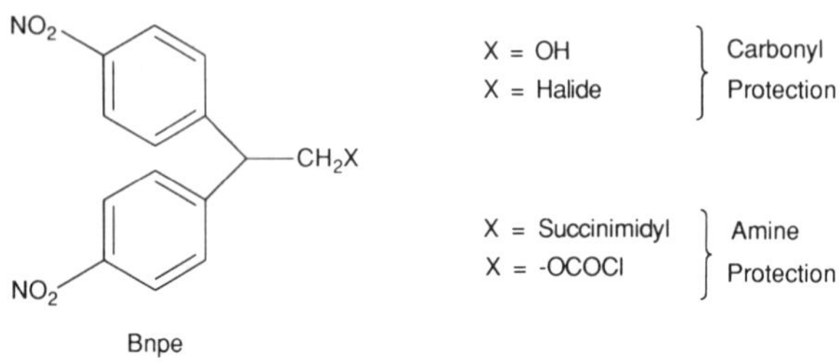


**Fig. 7.**

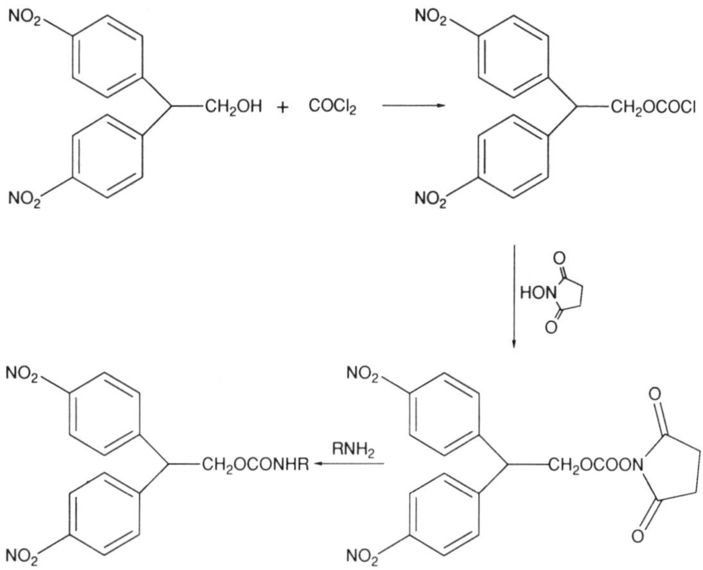


**Fig. 8.**

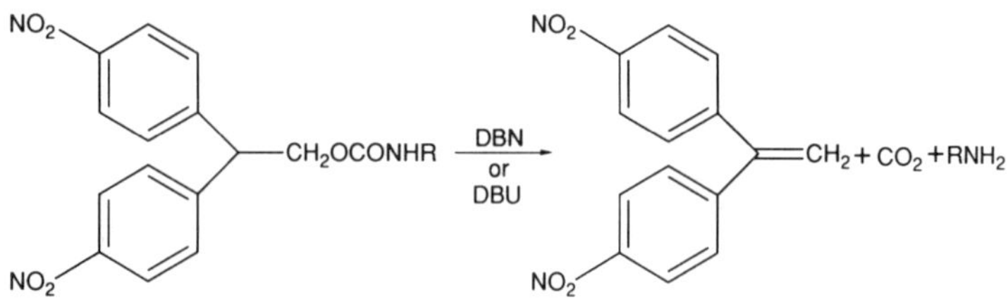


**Fig. 9.**

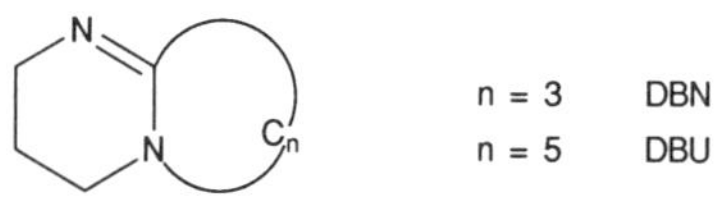

Fig. 10.

## Removal of BNPEOC Group

As with FMOC, removal is via $\beta$-elimination (Fig. 9). This has been found to proceed via the E1cb mechanism, i.e. prior ionisation. It is a fast deprotection and very specific.

The reagents used are (Fig. 10)

1,5-Diazabicyclo-(4,3,0)non-5-ene (DBN)
1,8-Diazabicyclo-(5,4,0)undec-7-ene (DBU).

These reagents are strong bases but, importantly, they are weak nucleophiles so the chance of unwanted side reactions is considerably reduced. They can be used in a buffered system with acetic acid, thus avoiding extremes of pH.

The olefin produced in the elimination appears to be unreactive in the system and the peptide can be separated easily.

The BNPEOC protected amine is stable to the following:

trifluoroacetic acid, HF/methanol, acetic acid,
NaOH, pyridine, triethylamine, *N*-methylmorpholine,
*N*-ethylpipiridine, piperdine

These are common reagents used in peptide work and stability to these materials is very important.

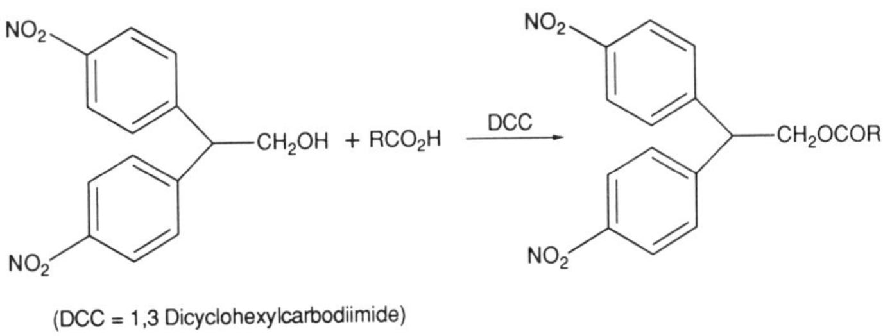


Fig. 11.

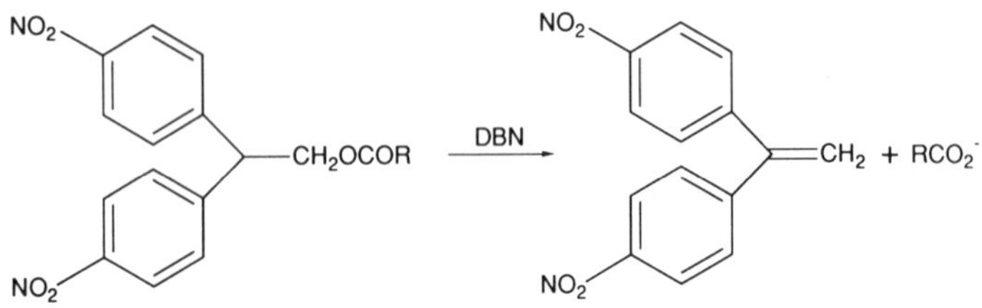


**Fig. 12.**

## BNPE Carboxy Protection

Although not as important as amine protection some methodologies could take advantage of this property.

The hydroxy or halide derivatives of the BNPE group are suitable for carboxy protection (Fig. 11).

Again this can be removed quickly with DBN or DBU (Fig. 12).

It could be useful if an acid labile protecting group were to be removed from a side chain as BNPE is stable to acid conditions.

## Preparation of an *N*-protected Amino Acid

Most of the naturally occurring amino acids have been derivatised without problem using the BNPEOC group (Fig. 13). Dimethyl formamide is the preferred solvent; 1·1 equivalents of the amino acid

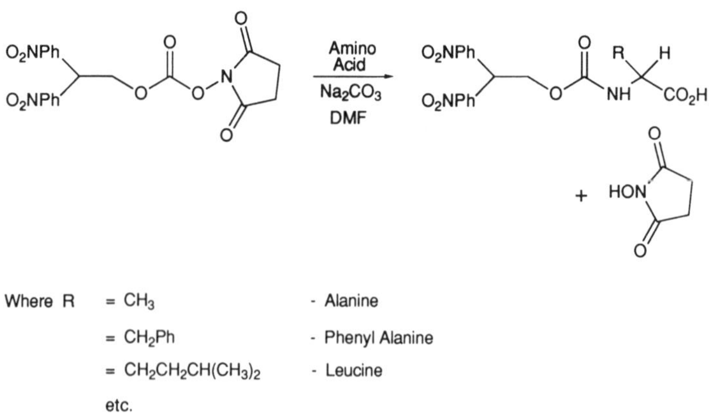


**Fig. 13.**

is dissolved in DMF with 2 equivalents of $Na_2CO_3$. The BNPEOC/*N*-succinimide is added at 0°C and reacted for 15 min. Work-up is done from the diluted aqueous solution by extraction with ethyl acetate. The protected amino acids are stable crystalline products which can be separated and purified easily prior to coupling.

## Preparation of a Tripeptide

The scheme in Fig. 14 illustrates the coupling of BNPEOC protected alanine with a dipeptide protected at the carboxy end as its methyl ester. The BNPEOC group is removed in approximately 30 minutes using an equimolar quantity of DBN in acetic acid. The methyl ester is hydrolysed to give the free tripeptide.

Other amino acids with side chain functional groups have been successfully protected:

| *Amino acids* | *Side chain functinality* |
|---|---|
| Glutamine | $CO_2H$ (*t*-butyl ester) |
| Glutamic acid | $CO_2H$ (*t*-butyl ester) |
| Methionine | Thioether |
| Tryptophan | 2° Amine |
| Tyrosine | Phenolic |

This shows the versatility of the BNPEOC groups and compatibility with the common side chain functional groups often encountered.

## Solid Phase Peptide Synthesis

By attaching the C-terminal end of an amino acid to a suitably functionalised polymer resin, a peptide chain can be grown from the N-terminal end.

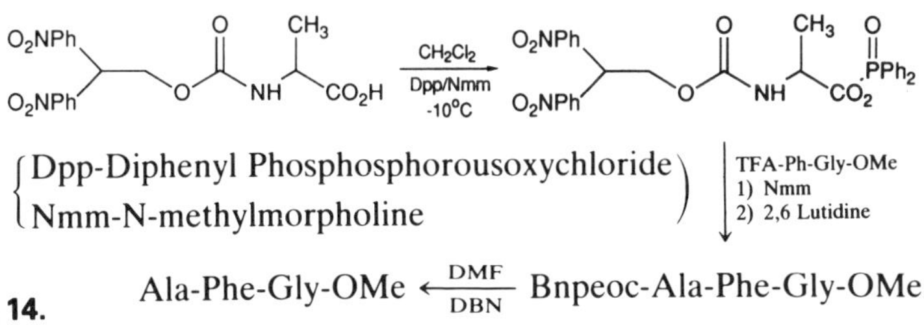


**Fig. 14.**

This method has the advantage that the polymer-bound peptide can be easily separated from the reagents used for protection, coupling and deprotection stages of the synthesis, and it also lends itself to automation. In addition large excesses of reagents can be used to achieve high yields at each stage with the excess being easily washed off at the end.

The BNPEOC group has been used successfully in this fashion using chemistry similar to that outlined previously.

### Other Uses of BNPEOC Protecting Group

The uses of BNPEOC groups are not restricted to peptide synthesis but are equally applicable to other areas. Two areas under study are nucleic acid synthesis and $\beta$-lactam synthesis.

In the nucleic acid area the BNPEOC could be used in place of the structurally similar protecting group: *p*-nitrophenethyl. This is presently used for protecting the reactive functionalities of the heterocyclic bases used, e.g. thymine, cytosine, adenine and guanine. It is only removed slowly whereas the BNPEOC group should be superior in this respect.

The BNPEOC group should also prove to be a useful and versatile protecting group for $\beta$-lactam synthesis.

## Conclusion

This short paper has outlined the problems encountered with the use of protecting groups in peptide synthesis and has described a new, novel group which has a number of unique advantages over existing compounds. Work is continuing and it is hoped that the BNPEOC group will develop commercially in the future.

## Acknowledgements

The author would like to thank Professor R. Ramage and co-workers of Edinburgh University for their ideas and dedicated effort which has laid the foundation in this new and important area of synthetic chemistry.

# Bibliography

Atherton, E., Logan, C. J. & Sheppard, R., Peptide synthesis using FMOC. *J.C.S. Perkin I* (1981) 538–546.

Bodansky, M., *Principles of Peptide Synthesis*. Springer-Verlag, 1984.

McOmie, J. F. W., Review of protecting groups and strategy. *Chemistry and Industry*, **18** (1979) 603–27.

Merrifield, R. B., Solid phase synthesis. *Angewante Chemie*, **24**(10) (1985) 799–892.

# Developments in sonochemistry

**Peter Martin* & Terry Goodwin‡**

*AEA Technology, Harwell, UK*

## Introduction: benefits and mode of action of ultrasound in chemistry

Sonochemistry is the study of the effects of ultrasound on chemical reactions, ultrasound being the name given to sound having a frequency beyond that perceptible to human hearing (about 16 kHz). It is a branch of chemistry that has experienced a rapid growth in both interest and effort over the past ten years, with the result that in 1990 for the first time a European conference was devoted to it. The effects of ultrasound are well documented in the scientific literature.[1–3] These effects include increased rates of reaction,[4] reduced induction times, improved yield[5] and/or selectivity,[6] and on occasion changes to the course of reaction.[7] In many cases ultrasound allows reactions to be performed at greatly reduced temperature or pressure,[8,9] compared to those usually required, with obvious potential benefits to industrial practice.

What is sonochemistry? In one sense it can be seen as an addition to an already well-established family of 'special' techniques for targeting

* To whom correspondence should be addressed.
‡ Present address: British Steel, Welsh Laboratory, Port Talbot, UK.

energy into reacting systems: these have included photo-, radio- and microwave chemistry, each of which leads to specific effects which straightforward thermal excitation does not. The ways in which sound energy acts are, however, very different from those associated with the electromagnetic spectrum, and must be understood primarily in terms of transient mechanical effects in the liquid reaction medium.

The available spectrum of sound energy ranges from the low-frequency, long-wavelength sound used in sonar (modest resolution, long range) to the megahertz ultrasound used in diagnostics (good resolution, short range), and made so familiar in obstetric 'scans'. Most practitioners of sonochemistry, however, use the 20–50 kHz 'near ultrasound' range which is optimal for creating cavitation in the reaction medium, and it is cavitation which is crucial to the chemical and physicochemical effects observed. As an aside, certain classes of sonochemistry do nonetheless occur in the megahertz region—much vintage work in this area has been done by Henglein at the Hahn–Meitner Institut in Germany—and the implications for the safety of healthcare diagnostics formed a subject of hot debate at the 1989 Ultrasonics International Conference in Madrid.

The cavitation that occurs in an insonated medium is no different in essence from that created by a poorly designed marine propeller or a maloperated pump; a comprehensive study has been provided by Young.[10] Provided a critical amplitude is exceeded, waves of compression and rarefaction sweep across a liquid of limited compressibility, tearing open voids of gas or vapour, which subsequently implode with devastating effects on the microscopic scale. The precise way in which this localised release of energy leads to chemical effects is the subject of considerable discussion, but more than one of the proposed mechanisms probably applies in many cases. Theories can be divided into the *physical*, the *electrical* and the *mechanical*.

According to the *physical* or 'hot-spot' theory, reactions are modified by the local extremes of temperature and pressure created by cavitation bubble collapse. Calculation suggests transient temperatures in excess of 1000°C and pressures of thousands of atmospheres, though direct measurement at this microscopic scale is of course impossible. Indirect evidence is to hand, however, in the sonochemical performance of ultra-high pressure reactions such as this Diels–Alder cycloaddition of dienes to *o*-quinones[11] (Fig. 1).

This reaction, which usually requires 11 kbar (*sic*) pressure, can be performed at atmospheric pressure in the presence of ultrasound.

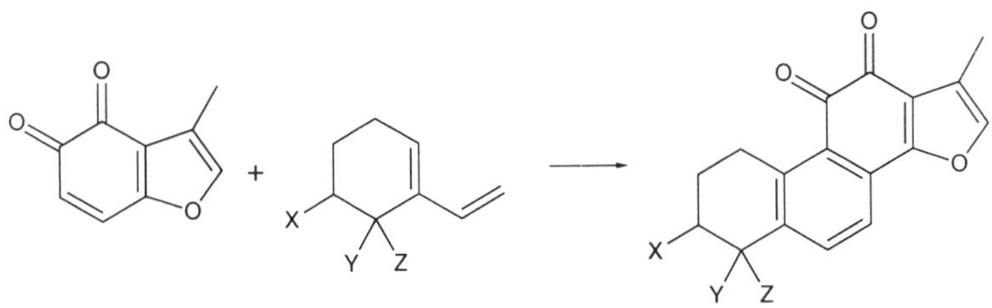

**Fig. 1.**

Many other examples are known of organic syntheses normally needing reflux conditions which can be run sonochemically at room temperature. The advantage to reactor design, operation and cost of replacing macroscopic high temperature and pressure conditions with microscopic ones are obvious: cheaper materials of construction, fewer pressure vessels, lower energy costs, etc.

By contrast with the 'hot-spot' theory, a number of workers, notably Margulis of the Moscow All Union Institute of Organic Synthesis, favour an electrical approach to ultrasonic acceleration of reactions. The existence of the well-known phenomenon of sonoluminescence would seem to support this view, and it is easy intuitively to imagine static charge separation induced by the creation and collapse of cavitation bubbles. Evidence for the involvement of direct electrical action comes from the effects of ultrasound in promoting single electron transfer processes. For instance, Dickens and Luche[12] found that in the reaction of *p*-nitrobenzyl bromide and lithium 2-nitropropanate, ultrasound favoured a single electron transfer mechanism yielding the *C*-alkylated product in preference to the normally preponderant route via *O*-alkylation (Fig. 2).

The *mechanical* effects of ultrasound in sonochemistry, by contrast, are visible, dramatic and well documented. Just as a pump operating in a cavitating regime soon suffers serious erosional damage, so in heterogeneous sonochemistry powerful microjects induced by bubble implosion are able to break through surface layers. This mechanical surface attack is particularly useful in heterogeneous organometallic syntheses, where oxidised layers on the surface of, for example, magnesium, lithium, zinc or copper can prevent or delay the onset of reaction. Ultrasound therefore enables rapid initiation of reaction in many such cases and permits the use of dirtier (and cheaper) reagents or wet solvents. Numerous examples can be cited in the preparation of

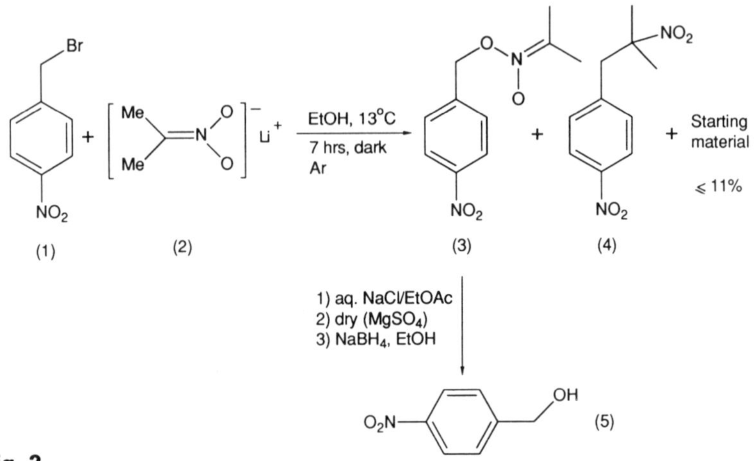

Fig. 2.

Grignard reagents,[13] in Barbier[14] and Reformatsky[15] reactions and in Ullmann coupling.[16]

The mechanical effects of ultrasonic cavitation can help too in the break-up of friable solid or liquid–liquid reagents, and in assisting desorption and mass transfer. A particularly valuable application is in phase transfer catalysis,[17] where the intimate contact given by ultrasound can sometimes eliminate the requirement for a catalyst altogether.

The examples given above have been of heterogeneous reactions. In homogeneous (i.e. single liquid phase) systems ultrasound acts by the creation of active species such as free radicals. A number of examples are available in the field of homogeneous organometallic chemistry,[18] and an important area of potential application is in the initiation of polymerisation without the use of chemical initiators.

## Methods of introducing ultrasound into a reacting system

The key element in introducing ultrasound to a reaction vessel is the transducer, which transforms (by piezoelectric or magnetostrictive means) an electrical signal into ultrasonic vibration of the reaction

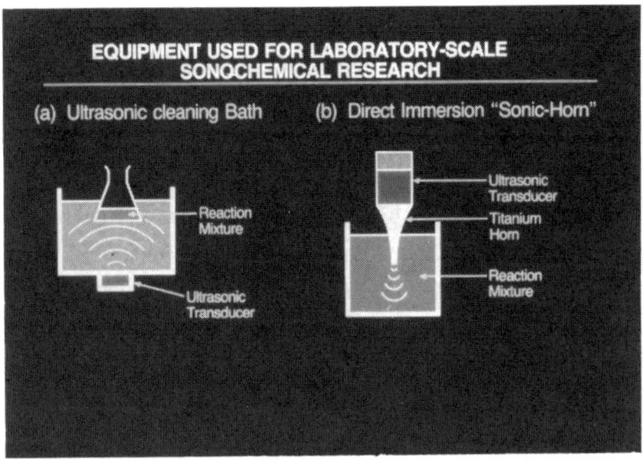


**Fig. 3.** Equipment used for laboratory-scale sonochemistry.

medium. The two types of device used in laboratory (i.e. small-scale) practice are the *cleaning bath* and the *sonic probe* or *horn* (Fig. 3).

The cleaning bath is normally of the common laboratory type, consisting of one or more transducers mounted directly on the base of a stainless steel bath. Ultrasonic energy is transmitted from the vibrating floor of the bath through the water to the reaction flask, which can be of any conventional type. Though used by many experimentalists, this arrangement has a number of drawbacks. The main deficiency lies in the low power and intensity transmitted to the reagents: the modest power of the bath is dissipated over a considerable volume, of which the reaction vessel occupies only a small portion, and further attenuation and refraction occurs at the glass wall of the flask. The presence of standing waves, with zones of maximum and minimum energy, further complicates the situation and makes it difficult realistically to compare results obtained by different workers, since the positioning of the reaction flask can be crucial.

If by contrast a direct-immersion sonic horn is used, the ultrasonic intensity in the reaction mixture at the probe tip can exceed that obtained in a cleaning bath by one or two orders of magnitude. The volume so insonated, however, is quite small, and the depth and shape of the vessel combined with the circulation patterns induced by streaming can profoundly affect the results obtained. Still more important, when thinking ahead to industrial applications, is the cavitational damage suffered by the (usually titanium) probe tip: under conditions of high intensity the probe life is limited to a few tens of hours, raising serious questions when considering the long maintenance-free operation required of industrial equipment.

It can be seen, then, that the simple-minded scaling-up of either of these two types of equipment for practical use would be attended by major difficulties. Indeed, attempts of this sort have been made, and some examples are given below; for successful application of ultrasound at high intensities and large scales a good deal of subtlety is called for. Only in the last five years and predominantly through the work of the Harwell Sonochemistry Development Club has a truly integrated approach been taken to the sonochemical engineering design problem. We discuss below some of the general principles and challenges involved before describing the types of solution that have been employed.

## Methods and equipment in large-scale sonochemistry

### Challenges of Scale-up

It can be seen that the scale-up of sonochemistry from the laboratory invites a series of important challenges, which involve both space (i.e. treating a larger volume) and time (the achievement of equipment that will give maintenance-free operation over long periods). In applying the integrated approach alluded to above there are a series of logical steps in applying electrical power derived from the mains to the ultimate effects of enhanced reaction performance in a chosen reaction volume; with each step may be involved an energy loss and a resulting inefficiency.

The first step involves the ultrasound *generator* where mains power is converted into a signal to the transducer; inevitably some energy will be lost as heat. Next there is the *transducer* itself, where there will be further losses both as heat and sound energy. Further inefficiencies will be involved in *coupling* the transducer to the load, by whatever means is chosen. Finally, because it is unlikely (and actually undesirable) that the whole reaction volume should be insonated perfectly uniformly, the *contacting and circulation pattern* of the reagents within the reactor must be considered. There are problems here for the ultrasonic engineer, the chemical engineer and the chemist.

## Early Techniques

It is important first to consider the types of reaction mostly targeted for sonochemical treatment. From the examples already quoted it is seen that we are concerned primarily with fine, as opposed to bulk, chemicals, for the production of which batch reactors of a few hundreds or at most a few thousands of gallons volume are used. Early techniques sought to adapt these reactors in three basic ways: (i) as large cleaning baths; (ii) by means of immersible transducers; and (iii) by the use of probes.

### *'Cleaning Bath' Reactors*

Multiple transducers can be mounted on the (preferably flat) walls of existing reactor vessels to produce reactors that are in essence large cleaning baths (Fig. 4). The number and configuration of transducers would need to be optimised for best effect in a reactor of given shape and volume, though there is little evidence of systematic investigation of this question. This type of equipment has been proposed for solid/liquid reactions such as that of Simmons–Smith:[19]

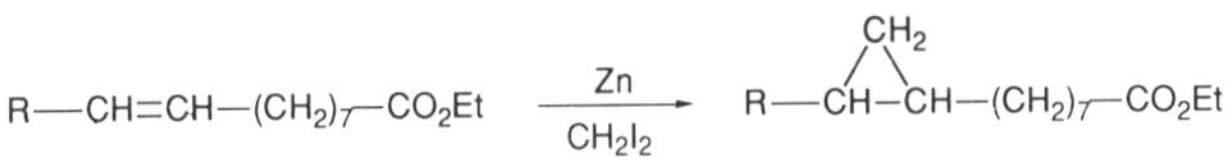

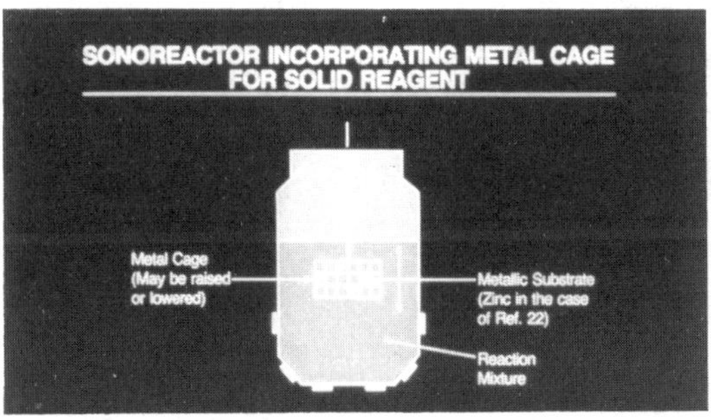


**Fig. 4.** 'Cleaning bath' sonoreactor.

The solid reagent, in this case zinc, can be contained in a steel cage near the centre of the reactor and raised or lowered to control the rate and optimum ultrasonic geometry. Other applications involving bulk lithium or magnesium can be envisaged.

This arrangement has the advantage that an existing vessel could in principle be used, and elevated pressures tolerated without the need for special seals around the transducers. There are, however, a number of disadvantages:

(i) The ultrasonic field is of relatively low intensity and its spatial distribution is poorly defined.
(ii) Attenuation within the reaction medium, and upper limits on the size and power of transducers, mean that very large vessels will contain an uninsonated core. This restriction of active insonation to a limited region not too far from the wall limits the ultimate size of the equipment.
(iii) The means of attachment of transducers to the vessel is critical, and surface matching, particularly to curved surfaces, must be carefully considered. There are implications too for transducer replacement.
(iv) The fact that transducers are externally mounted means that acoustic protection for the operators must be provided.

### *Immersible Transducers*

Transducers have been developed for the cleaning industry which are immersed in the cleaning fluid.[20] Similar schemes could in principle be

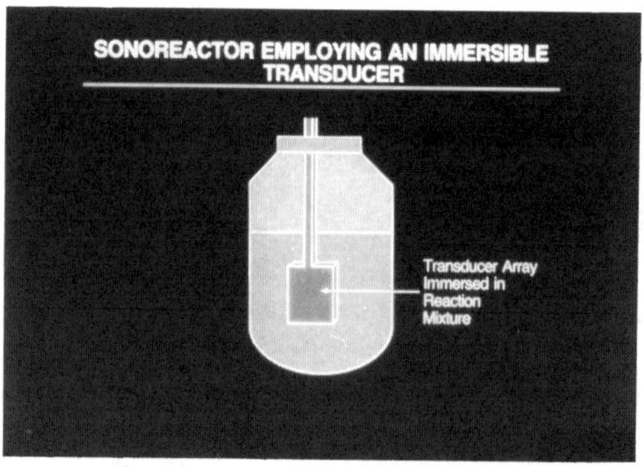


**Fig. 5.** Immersible transducers.

applied to chemical reactors and the transducers inserted and withdrawn through a port (Fig. 5). There are obvious disadvantages, however, from the point of view of maintenance in the mutual protection of transducer element and aggressive reaction liquor. Nonetheless, this type of scheme has been used in high-performance electroplating baths.

### *Probe Transducers in Industrial and Pilot Plant*

The use of direct insonation by titanium probes gives very high local intensities, and, due to streaming, desirable circulation with a reactor vessel (Fig. 6). The two major drawbacks are the inevitable erosion of the probe tip, which makes it unlikely that such an arrangement would even be used industrially at reasonable intensities, and limitations on the size of the reactor (unlike the 'cleaning bath' type, it is now the periphery of the vessel that is poorly insonated), unless the probe is used in a flow cell. A number of workers have used such a flow cell arrangement in laboratory and small-scale pilot work.[9]

## Recent Developments

In the last few years industrial interest has prompted a number of research groups to take a more serious, systematic, approach to the design of reaction schemes at pilot and larger scale. In the United Kingdom, the Harwell Sonochemistry Development Club was able to draw together a group of some 20 companies with the specific aim of

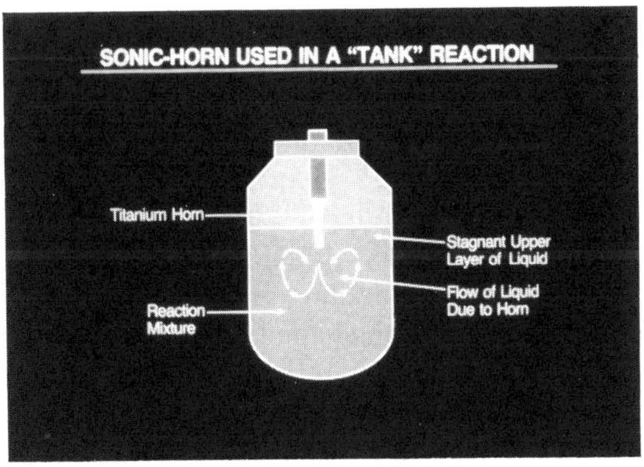


**Fig. 6.** Ultrasonic probe or horn used in a tank reactor.

demonstrating, if possible, the feasibility of application initially at the 10–100 litre scale: and in Toulouse and Milan respectively Berlan and Ragaini have made considerable progress, partially in collaboration with individual companies. In this work the Harwell group has perhaps had the edge, by virtue of using a small multi-disciplinary team of ultrasonic engineers, chemists and chemical engineers working in close cooperation and in conjunction with a specialist manufacturer.

One of the key developments of recent years has been the recognition that in scaling-up it is desirable to exercise tight control on the flow pattern of the reagents, so that the best and most uniform contact is obtained between reagents and the ultrasonic field. This could be characterised as bringing the reaction to the ultrasound rather than the ultrasound to the reaction, as was the case in the schemes indicated above. By way of illustration, in Fig. 6 the streaming induced by the horn circulates the reagents to some extent so that the same package of reaction mixture is not in the high-intensity zone continuously; this circulation pattern might be improved (for instance by the addition of a draft tube or mixing impeller) but 'dead zones' and poorly insonated portions would inevitably exist.

Deliberate recirculation is therefore desirable, and each of the three groups mentioned above have used through-flow reactors of various types, with the option of recirculation. A number of through-flow insonators are available from manufacturers, and have been tested by industrialists with modest results. Examples of off-the-shelf equipment in this broad category are: whistle reactors (in which ultrasound is generated by fluid flow impinging on and vibrating a fixed blade) and multi-transducer ducts such as the reverberatory ultrasonic mixer (RUM) and cylindrical or prismatic ducts available from, for example, the Maysonics or Branson companies. The chief drawbacks found have been the relatively low intensities available and/or operational difficulties. For this reason each of the development groups mentioned has opted for home-made equipment.

Berlan[21] has described a loop reactor of half to one litre total volume featuring a home-made insonating device (Fig. 7). The liquid reagents are circulated from a holding tank through a cylindrical vessel arranged axially in a water bath shaped as a regular hexagonal prism, and having a transducer mounted in the centre of each of its six lateral faces. This geometry echoed to a degree the pentagonal duct transducer available for some time from the Branson Corporation but

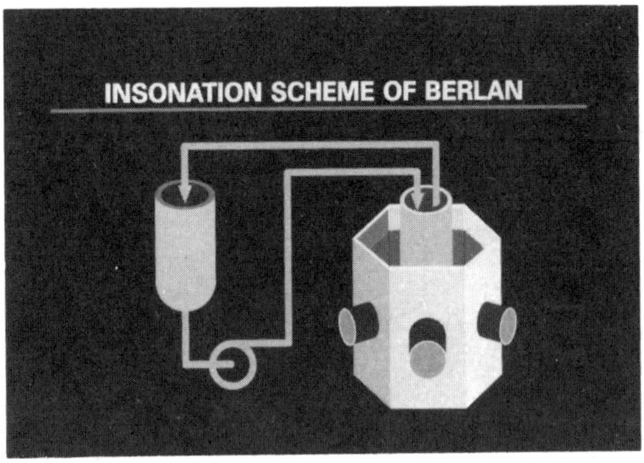


**Fig. 7.** Laboratory loop reactor of Berlan.

subsequently apparently withdrawn from the market. Berlan gave no details of the power and intensity of the ultrasonic field.

Ragaini[22] and others have looked at a number of small laboratory recycle schemes, and have also applied for a patent on the catalytic through-flow reactor shown in Fig. 8. In this equipment the liquid reagents flow through an annular duct of a few centimetres diameter, which also can contain catalyst or solid reagent granules. The transducers are arranged around the duct inside a protective cover, while a central tube carries temperature control fluid. A diagram of a small pilot plant incorporating three such reactors in series was shown.

Recycle reactors have been a feature of the Harwell sonochemical engineering programme since its inception. Early examples consisted simply of a 15–20 litre flow circuit insonated by a single probe

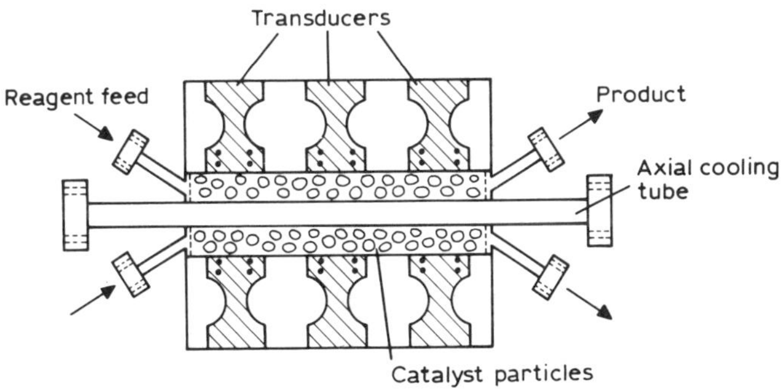


**Fig. 8.** Tubular reactor of Ragaini and others.

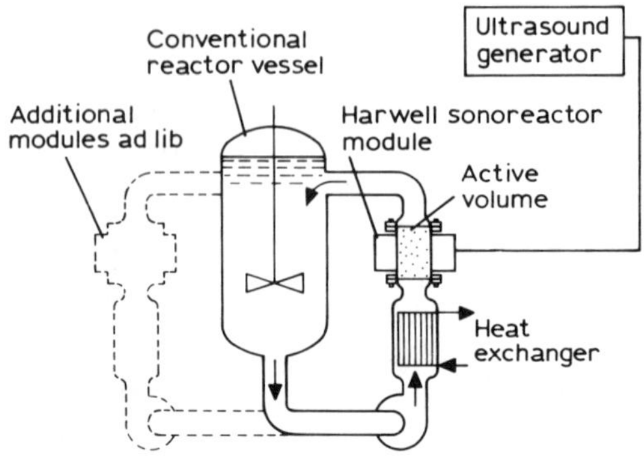


**Fig. 9.** Harwell-type pumped-loop reactor insonated by a sonoreactor module.

transducer, but this was soon followed by a reactor circuit featuring the unique Harwell sonoreactor module (Fig. 9). Note that in describing equipment of this type either the insonator device itself or the ensemble of insonated zone, holding volume and pumped circuit can be referred to as the sonoreactor. The Harwell insonating device consists functionally of a tube of 13 cm diameter and 30 cm long, and virtually the whole of this volume is intensely insonated.

Considerable effort was dedicated to the insonator design by power ultrasonic engineers before a satisfactory solution was reached, and this is now the subject of a patent application. Though it was apparent from the start that a multiple-transducer duct was required, it soon became clear that due to the non-linear behaviour of systems vibrating at high amplitudes theoretical approaches were of limited use: design is inevitably subject to a degree of trial and error. By careful tricks of transducer geometry no part of the reactor is subjected to cavitational damage. It is this combination of high intensity and minimal erosion that is unique to the Harwell equipment.

It is a feature of the insonating module that it can be used either as a laboratory batch reactor or as part of a flow circuit, and it is offered in either form. The 'guts' of these two variants are interchangeable, enabling data gained in the laboratory to be directly translated in terms of a loop reactor of hundreds of litres.

Extensive trials have been undertaken using this reactor loop on the Barbier reaction (Fig. 10) and also on an example of a two-liquid-phase hydrolysis reaction. Most encouragingly, early results indicate a significant improvement in the efficiency of energy usage as scale-up proceeds.

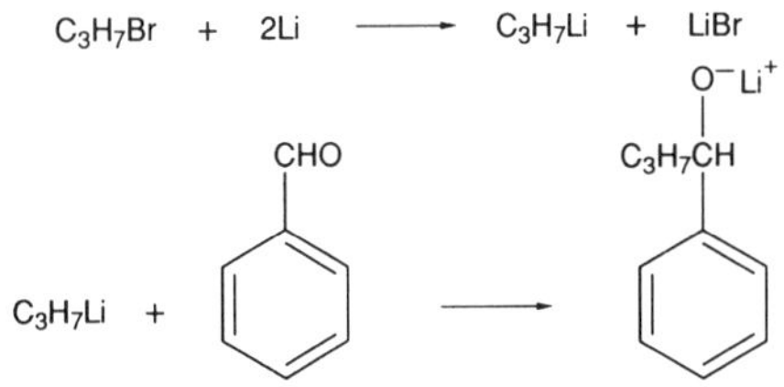


**Fig. 10.**

# Conclusion

In recent times, and particularly during the second half of the 1980s, the science and art of sonochemistry have advanced very significantly. Not only has the number of examples of exciting new reactions continued to grow unabated, but a better grasp has been obtained with regard to fundamental mechanisms. On top of this the application of sonochemistry to pilot and industrial scale reactions has been seen to be both desirable (to judge from the level of interest from manufacturers) and feasible. The technology and practice of sonochemistry is therefore set to become established as an industrial technique.

# References


1. Bremner, D., *Chemistry in Britain* (1986) 633.
2. Mason, T. J., *Ultrasonics,* **24** (1986) 245.
3. Boudjouk, P., *J. Chem. Ed.,* **63** (1986) 247.
4. Moon, S., Duchin, L., & Cooney, J. V., *Tetrahedron Letters,* **20** (1979) 3917.
5. Ando, T., Kawate, T., Ichihara, J. & Hanafusa, T., *Chem. Lett.* (1984) 725.
6. Horton, A. M., Hollinshead, D. M. & Ley, S. V., *Tetrahedron,* **40** (1984) 1737.
7. Ando, T., Sumi, S., Kawate, T., Ichihara, J. & Hanafusa, T., *J. Chem. Soc. Chem. Comm.* (1984) 439.
8. Han, B.-H. & Boudjouk, P., *Organometallics,* **2** (1983) 769.
9. Moulton, K. J., Koritata, S. & Frankel, E. N., *J. Am. Oil Chem. Soc.,* **60** (1983) 1257.
10. Young, F. R., *Cavitation,* McGraw-Hill, 1989.
11. Lee, J. & Snyder, J. K., *J. Am. Chem. Soc.,* **111** (1989) 1522.
12. Dickens, M. & Luche, J.-L., *Proc. 1st Meeting Eur. Soc. Sonochemistry* (1990) 98.

13. Brown, H. C. & Racharla, U. S., *Tetrahedron Letters,* **26** (1985) 4311.
14. Luche, J.-L. & Damiano, J. C., *J. Am. Chem. Soc.,* **102** (1980) 7926.
15. Han, B.-H. & Boudjouk, P., *J. Org. Chem.,* **47** (1982) 5030.
16. Suslick, K. S., Casandonte, D. J. & Doktycz, S. J., *Chem. Mater.,* **1** (1989) 6.
17. Ezquerra, J. & Alvarez-Builla, J., *J. Heterocycle Chem.,* **25**(3) (1988) 917.
18. Nakamura, E., Machii, D. & Inubushi, T., *J. Am. Chem. Soc.,* **111** (1989) 6849.
19. Repic, Q., Lee, P. G., & Gigers, U., *Org. Preps & Procedures Int.,* **16** (1984) 25.
20. Kerry Ultrasonics, Hitchin, Herts, UK.
21. Berlan, J., Scale-up in practice—the solutions. In *Proc. Symp. Applications of High Intensity Ultrasound,* Leatherhead, 1990.
22. Ragaini, V., *Proc. 1st Meeting Eur. Soc. of Sonochemistry* (1990) 106.

# Selectivity in electroorganic synthesis


## J. E. Toomey

*Reilly Industries, Inc., Indianapolis, Indiana, USA*


## Introduction

The ability to transform chemicals in a selective fashion is not a modern goal. However, achieving selectivity is of even greater concern today. One driving force for high selectivity in electrochemical processes comes from the inherent competition with conventional reaction schemes. In order to produce commercial products economically, electrolysis must provide some economic advantage over the alternatives; one such advantage is increased selectivity. Another driving force is that electrochemistry usually provides a broader spectrum of redox chemistries for a given substrate than conventional chemistry does; that is, the anticipated product can be subject to either incomplete reaction or over-reaction if electrolysis conditions are not controlled to prevent it. Consider, for instance, many of the early reports of electrochemistry of pyridines. Oxidation of alkylpyridines to pyridinecarboxylic acids was often accompanied by ring oxidation.[1,2] Under certain conditions, degradation of the ring could be extensive. In a similar fashion, reduction of alkylpyridines led to significant amounts of tetrahydro compounds (incomplete reduction) as well as the anticipated piperidines.[3,4] When functional groups attached to the pyridine ring were reduced, ring reduction or cleavage of carbon–heteroatom bonds was often a side reaction, occasionally the predominant process.[5–8] Fortunately modern techniques take selectivity into account, and address the ways of avoiding over-reduction and over-oxidation of organic molecules. This research explores the means available to the electrochemist for improving selectivity through study of three research goals from pyridine electrochemistry.

## Selectivity concepts

Selectivity ($S$) is a criterion-of-merit for raw material utilization much like chemical yield and current efficiency. It measures the molar yield ($Y$) corrected for converson ($C$) of the starting material:

$$S = Y/C \qquad \text{for } S > Y$$

Obtaining a high selectivity of concurrent with high conversion is the usual goal of electrochemical process development. While there are conventional processes that operate at low conversion and high selectivity, facile separation of product and recycle of starting material are generally necessary. Since many electroorganic reactions are not readily amenable to easy recycle of the starting material, the best case would seem to be high conversion of the starting material.

The cost of production of a given chemical determines the utility of that industrial technology being considered. There are extremely few, truly unique technologies for the production of an organic chemical. Alternative processes almost always exist. Therefore, electroorganic processes compete with conventional technologies for doing the same thing in most instances. If there is a value in electrolysis, it must lie in cost reduction over these alternatives, be it capital costs or operating costs. High process selectivity can reduce both types of costs.[9] These production costs are an inverse function of selectivity; thus, when one plots 'costs' versus increasing 'selectivity', a hyperbolic curve is obtained (Fig. 1). There are two notable features of these hyperbolic curves. One is that costs are reduced by an increase in selectivity at all levels. The other is that above a certain level of selectivity (the 'knee' of the curve), the rate of decrease in costs is moderated, usually becoming quite small. Therefore, knowing where the process is on the curve can be helpful in deciding whether further research is justified,

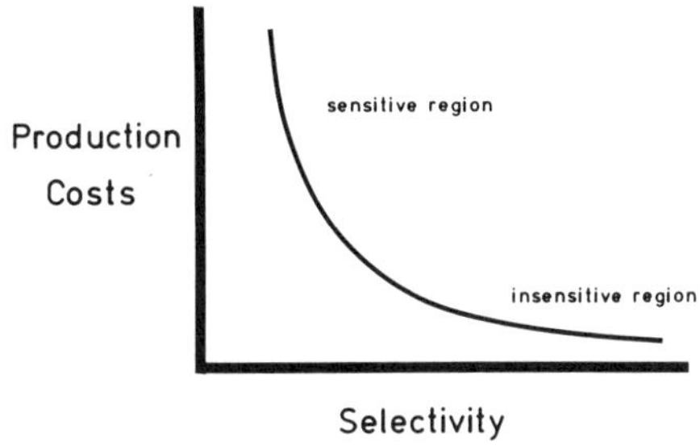


**Fig. 1.** Total production costs vs selectivity (Reilly Industries).

given the anticipated cost reduction by an incremental improvement in selectivity.

As was mentioned in the first section, a possible advantage of electrochemical technology is an increase in selectivity. There are sound reasons for suspecting that this could be achieved in practice. The electron potential can be varied, if not controlled, to allow at least one selective electron transfer to, or from, the substrate. In cases where the product distribution is effected by that step, a degree of selectivity can be achieved. However, potential control need not lead to selectivity.[10] Electrolyses are usually done near room temperature and thus, avoid the thermal by products seen in some conventional technologies. As an example, the catalytic hydrogenation of 2-cyanopyridine to 2-picolylamine was a technology that Reilly Industries had practiced for many years on a production scale. Thermal and other byproducts were always a problem. Electrochemical reduction led to high isolated yield and selectivity at virtually complete conversion. Since raw material utilization is one of the largest cost factors in the production of speciality products, increased efficiency in this utilization can result in significant reduction in the cost of production. In addition, high selectivity has secondary or inherent advantages besides the obvious one of economical production stated above. When a product is produced more selectively, the product is usually easier to isolate. Separation of by products from the desired product can be much more efficient. The space–time yield of the chemical reactor goes up and less time is spent in making the product.

## Selectivity parameters

The means to achieve selective electrochemical transformations are numerous and the interactions between them are complex. Thus, the study and development of these reactions can be difficult at the research level. Our researches have identified twelve primary parameters with up to five or six additional secondary ones. The research reported here only covers primary selectivity parameters. Those are ones that are the most influential and generally exert their influence directly. These twelve parameters can be subdivided into three classes: (i) operation parameters, (ii) materials selection parameters, and (iii) cell design parameters.

## Operation parameters

1. Current density
2. Flowrate
3. Charge passed
4. Temperature

There are four operation variables that can be changed during electrolysis to alter selectivity.

Changes in current density are reflected in variations of junction potential at the electrode surfaces. This in turn can cause the product distribution to change and alter selectivity; this distribution can be a sensitive function of current density.

Alterations in electrolyte flowrates or flow patterns can effect mass transfer to the electrode. The mechanistic pathway taken can be affected and cause a different product distribution also. However, the sensitivity of product distribution to changes in electrolyte flowrate is normally small, at least above some minimum flowrate that is a function of the cell design and electrode form used.

The degree of charge passage can have a material effect on selectivity for quite apparent reasons. In certain cases, the control realized using this variable can offer some practical advantages.[11]

Finally, varying the electrolyte temperature will usually cause small changes in the selectivity.

## Materials selection parameters

1. Starting material
2. Electrode

3. Solvent
4. Supporting electrolyte
5. Catalyst

The choice of five different materials for a contemplated synthesis has been found to change selectivity patterns. One of the lesser-used variables is the choice of starting material. Of course, many times there is simply a small selection available in practical terms; however, one should always give due consideration here as quite large changes in processing costs can be made.

The choice of electrode materials-of-construction has been recognized as having a strong bearing on process viability.[12] In fact, the product distribution may be most sensitive to this variable.

The choice of solvent has also been recognized as being important for a variety of reasons. Not only does it change how molecules are solvated, but it can also affect adsorption, mass transfer rates, proton availability, and other factors.

In a similar fashion, the supporting electrolyte choice has a bearing on these secondary parameters, sometimes causing gross alterations in selectivity for reasons that are not always understood.

Lastly, the presence or absence of either a soluble metal ion (homogeneous catalysis) or one affixed to the electrode surface (heterogeneous catalysis) can have profound effects on process results. These catalysts can occupy a position of prominence equal to the electrode material choice.

## Cell design parameters

1. Electrolyte distribution
2. Electrode form
3. Ion-exchange membrane

The last group of parameters may not be widely accepted as such, but a recognition of the true character of these variables is slowly coming to light. Cell design can alter selectivity in three ways. The electrolyte distribution patterns within the cell have obvious influences on mass transfer and junction potential distributions.

The coupling of electrode processes with electrolyte distribution is somewhat difficult to study and the real causality can only be hinted at.

In addition to the flow pattern of electrolyte, the form that the electrode takes has a strong influence on the type of products made. In this respect, three-dimensional electrodes have a beneficial effect on most reactions with regard to making them more selective. Some electrochemists find this result puzzling since the junction potential distribution is liable to be larger in such electrode forms than planar ones. Therefore, product selectivity should be lower in the three-dimensional electrode. The issue has no simple answer because experiment seems to be at variance with that concept, at least for certain types of three-dimensional electrodes. Certainly, there are three-dimensional forms that would seem to be incompatible with achieving a high selectivity, particularly those with small pores and ultra-high specific surface area (>50 $cm^2/cm^3$). Moreover, the coupling of fluid dynamics with the electrode form is expected to be large and a treatment of these two parameters as independent variables may very well be incorrect.

Finally, the choice of ion-exchange membrane can determine how the concentrations of species within the electrolytes vary. Almost all divided cell syntheses proceed with changes in mass across the physical divider. Just the ion transport process alone can transfer a surprising quantity of solvent within the solvation shell of the ions. Consider the transformation of one mole of substance requiring 4 F/mol. This would necessitate the transfer of four moles of singly charged ions through the membrane. The usual rule of thumb is that six to twelve solvent molecules are transferred with each ion. Hence the electrolysis would entail the transfer of 24–48 moles of solvent! Naturally, the gross changes in concentration can affect the product distribution. Whether the membrane is anion-selective or cation-selective can determine the direction and magnitude of these changes.

## Experimental apparatus

Since developmental-scale, industrial research installations (>0·5 kg/day) are rarely described in the literature, the apparatus used for this work will be depicted. The equipment shown in Fig. 2 was used for most of the work, although some work was done in installations with fewer than four cells. Electrolyte flow was in parallel and current flow was in series. The reservoirs were 4 liter capacity and most of the

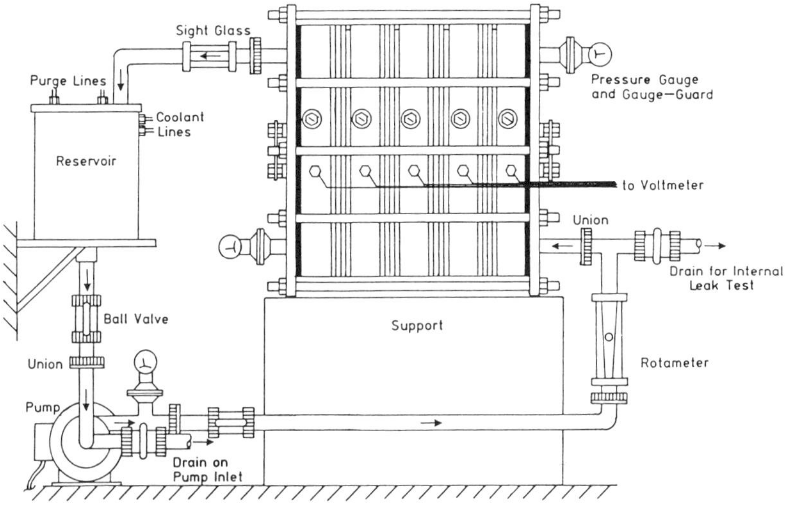


**Fig. 2.** Laboratory equipment (Reilly Industries).

pipework was 0·5 in schedule 80 polypropylene. The pumps were centrifugal, magnetic-driven ones capable of delivering 5–6 liters/min at 10–15 lb/in$^2$ gage, although normal operating pressures and flowrates were well below these maximum values. The cell design used is shown in Fig. 3.[13]

# Selectivity examples

## Control of the Product from Reduction of Pyridinecarboxamides

This work was aimed at producing the three isomeric carbinols (**2**) or isomeric amines (**3**) and not the aldehydes (**4**), as that form of selectivity from amine reductions had already been studied.[14] All three isomers of the monoamides (**1**) were reduced.[15] The major product was determined by control of six of the twelve parameters described earlier. Choice of starting material was important because the isomeric carbinols (**2**) could also be prepared from the corresponding isomeric aldehydes (**4**)[14,16] and the isomeric amines (**3**) could be prepared from the corresponding nitriles (**5**).[17]

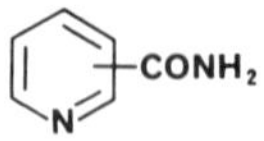

(1)

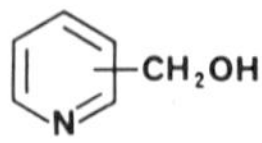

(2)

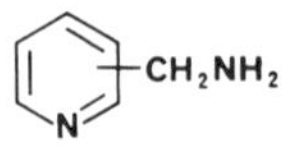

(3)

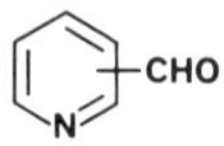

(4)

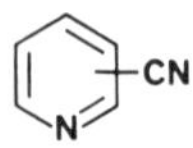

(5)

Certainly, having a single, readily available precursor to both products would be an advantage if selectivity to each product could be made to be high. Control of pH was known to be important from Lund's work[14] so the choice of solvent and/or supporting electrolyte could be critical to selectivity when trying to avoid aldehyde as the major product. We found pH was indeed a factor in controlling selectivity to either (**2**) or (**3**), albeit a minor one when at least two moles of acid were used per mole of amide. Limiting the amount of charge passage to that needed for nearly complete conversion (4 F/mol theoretical $Q$) prevented over-reduction of the products. However, small amounts (<5%) of picolines could always be detected even when limiting charge passage. Current density had an effect on the product distribution, selectivity being at a maximum between 12 and 30 mA/cm$^2$. The use of a packed-bed cathode made from lead shot (3·2 mm average diameter) was a critical factor in obtaining the best results. Certainly, the use of an electrode with a relatively high specific surface area (8–11 cm$^2$/cm$^3$) made low current densities practical and led to improvements in product selectivity. Finally, the addition of $Ti^{4+}$ salts had the most dramatic effect, but not on the selectivity of

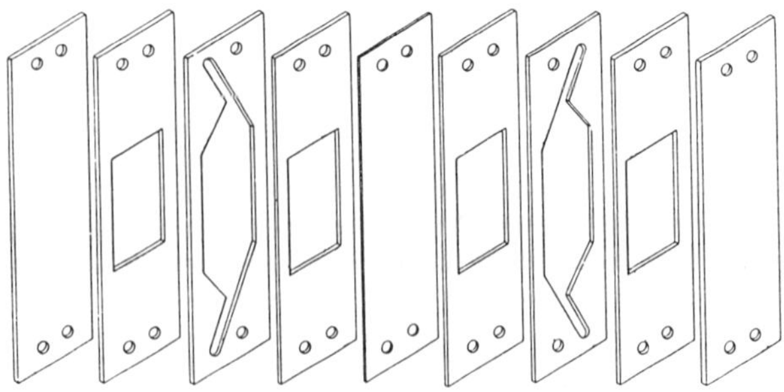

**Fig. 3.** Cell design (US Patent 4 589 968, Reilly Industries).

**Table 1.** Reduction of the isomeric pyridinecarboxamides

| Amide | Titanium sulfate (wt%) | Major product | Selectivity (%) | Yield (%) | Current efficiency (%) |
|---|---|---|---|---|---|
| 2-(**1**) | None | 2-(**2**) | 84 | 81 | 95 |
| 2-(**1**) | None | 2-(**2**)[a] | 78 | 48 | 97 |
| 2-(**1**) | 0·1 | 2-(**3**) | 74 | 70 | 85 |
| 3-(**1**) | None | 3-(**2**) | 68 | 65 | 70 |
| 3-(**1**) | 0·1 | 3-(**3**) | 77 | 75 | 60 |
| 4-(**1**) | None | 4-(**2**) | 91 | 89 | 100 |
| 4-(**1**) | 0·1 | 4-(**3**) | 73 | 72 | 100 |

[a] No aldehyde was detected.

carbinol (**2**) production. The absence of these salts meant that (**2**) was always the major product, the actual value of the selectivity being determined by the other five parameters. The presence of these salts altered the major product to amine (**3**), the other five parameters again exerting their own effect but never reducing (**3**) to the status of a minor product. Furthermore, although (**3**) could be a sizable side-product in the absence of $Ti^{4+}$ salts, the amine (**3**) was never the major product regardless of the choice of the other five parameters. It was the major product only in the presence of $Ti^{4+}$

In Table 1, yields are mole ratio of isolated product to material charged to the reactor; conversions were based on analysis after workup; selectivities were calculated from yield and conversion data combined with measured charge passed.

During reductions without added $Ti^{4+}$ salts, the major side-product was the dimer (**6**). Presumably, this material arose from radical coupling of intermediates from the 'aldehyde' stage, although no actual aldehyde could be detected by analysis. Dimer (**6**) was formed as a roughly equal mixture of diastereoisomers and represented as much as 20% of the product distribution. Dimer formation was inhibited with added $Ti^{4+}$ salt and neither (**6**) nor the known corresponding imine dimer (**7**) could be detected (<1%).

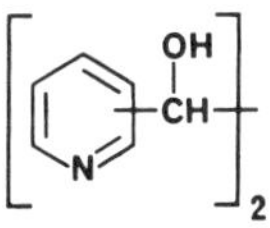


(6)

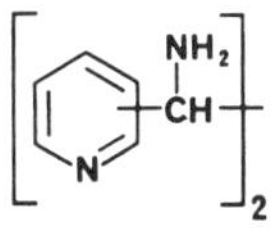


(7)

## Control of the Product from Reductive Dimerization of Pyridinium Quaternary Salts

The dimerization of pyridinium salts by electrolysis has been known since 1909;[18] however, the reduction has a tendency not to proceed to high conversions[19] unless an extraction solvent (ex-cell) is used in a planar electrode cell design where special hydrodynamic conditions are maintained.[20] With this reaction, cell design is an important parameter group. The use of high surface area electrodes was found to allow complete conversion of pyridinium salt without the use of extracting solvents or planar electrode cell designs.[21] An additional advantage of these three-dimensional cell designs was an increase in the space–time yield of the reactor. As one would expect, electrode material-of-construction was an important variable. Here, Pb and its alloys were best. Limiting charge passage to that needed for complete conversion prevented over-reduction of the product. Choice of starting material structure could be termed trivial in this case since only pyridinium quaternary salts are known to dimerize by electrolysis. That is, hydrodimerization of pyridine bases occurs to only a minor extent, the major products of reduction of the free base being the corresponding piperidine and piperideine.[3,4] Within the manifold of different pyridinium salts, structure seemed to play a minor role in altering current efficiency; selectivity appeared to be largely insensitive to structure alterations as long as some other functional group beside H was on the nitrogen atom.

When 1-methylpyridinium chloride (**8**) was reduced in alkaline solution at a Pb cathode, the tetrahydrobipyridine (**9**) was produced in 100% selectivity.

Reducing 1-acetylpyridinium chloride (**10**) under essentially the same conditions gave dimer (**11**) in 98% selectivity.

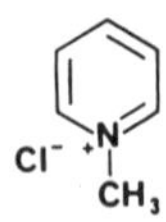

(8)

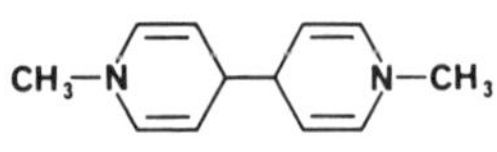

(9)

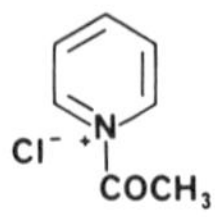

(10)

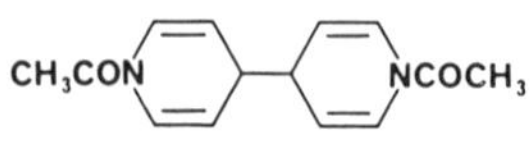

(11)

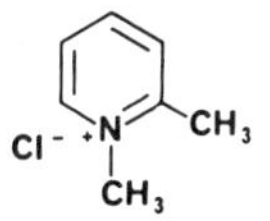

(12)

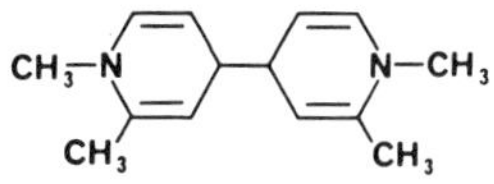

(13)

Electrolysis of 1,2-dimethylpyridinium chloride (**12**) gave dimer (**13**) in 98% selectivity also.

In all of these cases, current density and concentration of starting material had little effect on selectivity. The product could be easily isolated as an upper-layer oil because it was not soluble in the slightly basic aqueous electrolyte. Moreover, since the starting material was very soluble in the aqueous system, the process could be run continuously by addition of pyridinium salt to the electrolyte. Product could be withdrawn by separation of layers.

In Table 2, yields are mole ratio of isolated product to material charged to the reactor; conversions were based on analysis after workup; selectivities were calculated from yield and conversion data combined with measured charge passed.

## Control of the Product from Direct Oxidation of Alkylpyridines

Historically, Reilly Industries has been involved with anodic reactions on pyridines since the late 1940s. The first electrochemical plant built at Reilly was for the production of isonicotinic acid from 4-picoline. This intermediate was needed for the manufacture of Isoniazid, an antitubercular drug. The plant was operated in the 1950s for that purpose and had a nameplate capacity of 1000 Mt/year. Because the equipment that was available at that time was so poor, this technology was eventually replaced by catalytic oxidation, but the replacement of technology was recognized as equipment limited only, not inherently

**Table 2.** Reduction of pyridinium salts

| Pyridinium salt | Dimer | Selectivity (%) | Yield (%) | Current efficiency (%) |
|---|---|---|---|---|
| (**8**) | (**9**) | 100 | 95 | 86 |
| (**10**) | (**11**) | 98 | 93 | 91 |
| (**12**) | (**13**) | 98 | 91 | 77 |

limited. It was only a chronological disability. Today the occurrence of tuberculosis is rare, fortunately, and the demand for Isoniazid is small.

The ability to achieve selectivity is most important for oxidative technologies, irrespective of whether electrolysis is involved or not. Moreover, electrochemistry holds more of a hope for accomplishing this than some other technologies do, for the reasons outlined earlier. In the normal sense of the word 'selective', the application of selectivity to direct oxidation of alkylpyridines is virtually unknown. In fact, it is a technology that has been exclusively practiced in beaker cell designs for the production of exhaustively oxidized products, pyridine carboxylic acids, in preference to lower oxidation state products, such as aldehydes, ketones, or carbinols.[22–26] Contrary to certain opinions, the ring itself is not insusceptible to degradation, and even 'exhaustive' oxidation to a carboxylic acid is selective in the sense that the pyridine ring is still intact.[1,2] Those who are familiar with permanganate oxidation of alkylpyridines, especially polyalkylpyridines, can readily attest to the >15–20% loss in yield due to ring oxidation.

There are really three meanings of selective, direct oxidation in the alkylpyridine series. The first would be the one just referred to, an oxidation of all the alkyl groups to carboxylic acids, leaving the ring intact. This would be exemplified by oxidation of 2,6-lutidine (**14**) to dipicolinic acid (**15**). The second would be complete oxidation of one or more alkyl groups to carboxylic acids, leaving at least one alkyl group in a pristine state. Oxidation of (**14**) to 6-methylpicolinic acid (**16**) would be an example of that type. Lastly, partial oxidation of one or more alkyl groups to form aldehydes, ketones, or carbinols constitutes the third type. Oxidation of (**14**) to the dicarbinol (**17**) would be an example of that type.

The approach taken here was to explore anodic reactions in a given system to observe how selective these reactions could be made.[11] The system was direct oxidation in aqueous sulfuric acid at a packed-bed

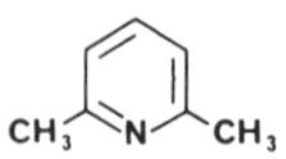

(14)

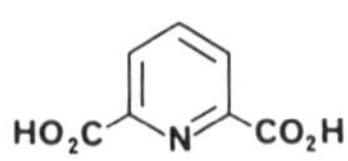

(15)

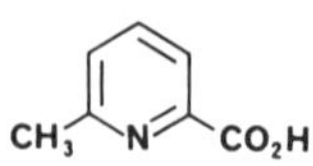

(16)

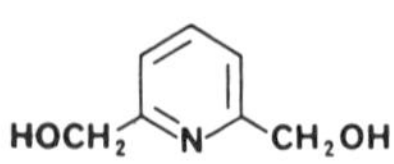

(17)

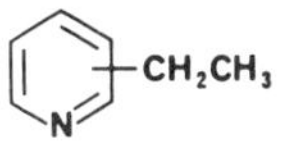

(18)

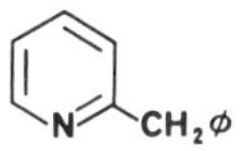

(19)

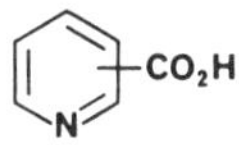

(20)

$PbO_2$ anode (on Pb substrate) using a flow cell divided with a cation-exchange membrane. A series of isomeric ethylpyridines (**18**) and 2-benzylpyridine (**19**) were used as the substrates. There is only one previous report on these substrates, oxidation of 4-(**18**) to isonicotinic acid, 4-(**20**) in 75% yield, but using a noble metal anode, beaker cell, and nitric acid electrolyte.[26] As such, while a very good selectivity was observed (>75%), the system was impractical for large-scale synthesis.

As Table 3 shows, the selectivities for the isomeric acetylpyridines (**21**) were modest, but excellent selectivity was observed in the case of 2-benzoylpyridine (**22**). The major byproduct for the isomeric (**21**) was the corresponding carboxylic acid. The control of selectivity will obviously be structure dependent. Current density was found to be a factor, the selectivity being best in the range of 25–70 $mA/cm^2$. The electrode materials-of-construction and electrode form had a large bearing on changing selectivities, being best in packed-bed, $PbO_2$ anodes. Charge passage strongly influenced the product distribution and the results shown in Table 3 are approximately optimum. Finally, pH had a minor role if at least one mole of sulfuric acid per mole of pyridine base was used. When less was used, conversions became poor.

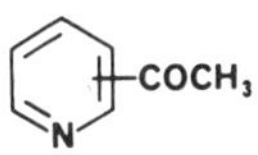

(21)

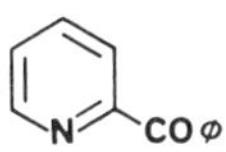

(22)

**Table 3.** Oxidation of alkylpyridines

| Alkylpyridine | Ketone selectivity (%) | Ketone yield (%) | Carboxylic acid yield (%) | Ketone current efficiency (%) |
|---|---|---|---|---|
| 2-(**18**) | 49 | 40 | 30 | 27 |
| 3-(**18**) | 53 | 45 | 35 | 30 |
| 4-(**18**) | 65 | 39 | 15 | 26 |
| 2-(**19**) | 91 | 78 | — | 62 |

In Table 3, yields are mole ratio of isolated product to material charged to the reactor; conversions were based on analysis after workup; selectivities were calculated from yield and conversion data combined with measured charge passed.

## Conclusions

This research has shown that there are a number of complicated factors involved in controlling or altering selectivity in electroorganic syntheses. However, it also indicates that these parameters can be studied on a practical level and are amenable to treatment by reasonable, relatively simple conceptualization. Not all of the factors are independent of one another; quite the contrary, many are interdependent. The sheer number of important variables combined with their confounded nature can make process identification an extremely difficult undertaking. These variables can be classified according to their origin, and the magnitude of the effect on selectivity can be discussed in general terms, in spite of the complexity.

These selectivity concepts can be applied to synthetic situations, as has been shown by the three examples of pyridine electrochemistry. Not all of these synthetic systems could be developed into high selectivity processes, especially some of the direct oxidations, but nevertheless a remarkable number could be. This high selectivity made electrolysis a more viable technology for the production of industrial chemicals. The increase in technology viability was sufficient to make some cathodic reductions more cost effective than the available conventional counterparts, even catalytic hydrogenation. Prior to this work, most industrial electrochemists had concluded that electrolysis simply could not compete with catalytic conventional technologies. This does not have to be so.

## References

1. Yokoyama, M. & Yamamoto, K., *Bull. Chem. Soc. Jpn.*, **7** (1934) 28.
2. Rivkind, E. P., Mityania, K. G., Grinblat, E. I. & Derbysheva, E. K., *Vopr Tekhnol. Ulavlivaniya Pererab, Prod. Koksovaniya*, **2** (1974) 102.
3. Ferles, M., Havel, M. & Tesarova, A., *Coll. Czech. Chem. Commun.*, **31** (1966) 4121.

4. Ferles, M., *Coll. Czech. Chem. Comm.*, **24** (1959) 1029.
5. Sorm, F., *Coll. Czech. Chem. Comm.*, **13** (1948) 57.
6. Wibault, J. P. & Boer, H., *Rec. Trav. Chim.*, **68** (1949) 72.
7. Ferles, M. & Pyrstas, M., *Coll. Czech. Chem. Comm.*, **24** (1959) 3326.
8. Ferles, M. & Tesarova, A., *Coll. Czech. Chem. Comm.*, **32** (1967) 1631.
9. Jansson, R. E. W. in *Electrochemical Cell Design,* ed. R. E. White. Plenum Press, New York, 1984, pp. 175–95.
10. Pletcher, D., *J. Electroanal. Chem. Interfacial Electrochem.*, **179** (1984) 263.
11. Toomey, J. E., US Patent 4 482 439, 1984.
12. Tilak, B. V., Sarangapani, S. & Weinberg, N. L., in *Techniques of Chemistry,* ed. A. Weissberger, Vol. 5, *Technique of Electroorganic Synthesis,* eds B. V. Tilak & N. L. Weinberg, Part III. John Wiley & Sons, New York, 1982, p. 196.
13. Toomey, J. E., US Patent 4 589 968, 1986.
14. Lund, H., *Acta Chem. Scand.*, **17** (1963) 2325.
15. Toomey, J. E., US Patent 4 695 352, 1987.
16. Zuman, P., *J. Org. Chem.*, **46** (1981) 1906.
17. Toomey, J. E., US Patent 4 482 437, 1984.
18. Emmert, B., *Ber.*, **42** (1909) 1997.
19. Colchester, J. E. & Entwisle, J. H., US Patent 3 478 042, 1969.
20. Misumi, T., Furuhashi, S. & Shiga, M., US Patent 4 176 020, 1979.
21. Toomey, J. E., US Patent 4 670 111, 1987.
22. Yokoyama, M., *Bull. Chem. Soc. Jpn,* **7** (1932) 69; Ochiai, E. & Okuda, S., *J. Pharm. Soc. Jpn.*, **70** (1950) 156.
23. Kulka, M., *J. Am. Chem. Soc.*, **68** (1946) 2472.
24. Kruglikov, S. S. and Khomyakov, V. G., *Tr. Mosk. Khim-Teknol. Inst.*, **32** (1961) 194.
25. Alamelin, J., Talitha, K. S., Rahakrishnaumurthi, K., Chidambaram, S., Pathy, M. S. V. & Updupa, H. V. K., *Trans. Soc. Advan. Electrochem. Sci. Technol.*, **6** (3) (1971) 97.
26. Ito, A. & Kawada, K., *Ann. Rept Takamine Lab.*, **5** (1953) 14.

# The industrial synthesis of optically active compounds

**R. A. Sheldon***

*Andeno BV, Venlo, The Netherlands*

## Introduction—chirality and biological activity

In the last two decades there has been growing public awareness and concern with regard to the exposure of man and his environment to an ever-increasing number of chemicals. The benefits arising from the use of therapeutic agents, pesticides, food and feed additives, etc., are manifold. However, in addition to the desired biological effects these products can exhibit undesirable side-effects. Consequently, a proper assessment of their potential benefits and risks requires a thorough knowledge of their interactions with biological organisms. Only then can their efficacy be optimized to provide maximum benefits with minimum undesirable side-effects.

This is particularly relevant where the use of chiral compounds is involved. Many biologically active substances contain one or more chiral centres and consist, therefore, of two or more optical isomers which should be considered as different substances. Thus, in general, only on the enantiomers (the eutomer) of a racemic mixture is

* Present address: Technical University of Delft, The Netherlands.

responsible for the desired biological activity towards the target organism.[1] At best the other isomer (the distomer) constitutes unnecessary ballast, the removal of which would significantly reduce the chemical burden on the environment.

However, more often than not the distomer inhibits the desired effect of the eutomer and/or exhibits undesirable side-effects. A much-quoted example is the drug thalidomide which was marketed in the 1960s as a sedative and administered as a racemic mixture. Unfortunately, it was not known at that time that although the *R*-enantiomer is an effective sedative the *S*-enantiomer is one of the most potent teratogens known.

Many other less dramatic but nevertheless significant examples are known. For example, $\beta$-adrenergic blocking agents ($\beta$-blockers) characteristically have the aryloxypropanolamine structure (**1**) containing one chiral centre. In general $\beta$-blocking activity resides virtually completely in the *S*-enantiomer. Similarly, it is the *S*-enantiomer of the $\alpha$-arylpropionic acid (**2**) class of anti-inflammatory drugs that is responsible for the desired therapeutic effect.

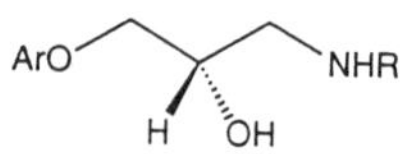


(1)

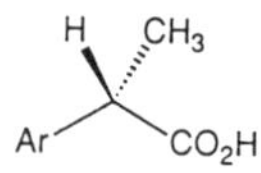


(2)

When one considers that biological activity is generally a result of selective binding of the substance in question to a receptor site or to the active site of an enzyme in the target organism it is not so surprising that the two enantiomers of a racemate display completely different properties. In fact it was suggested[2] as long ago as 1956 that the potency of a drug is directly proportional to the so-called eudismic ratio (the activity ratio of the eutomer compared to the distomer). Simply stated, Pfeiffer's rule says that the potency of the eutomer is determined by how well it fits into the active site and if both enantiomers are able to bind it is unlikely to be an exclusive fit.

Moreover, in comparing the biological properties of two enantiomers it is not just a question of comparing their respective activities *per se*. For a proper comparison it is essential to study the effect of enantioselectivity on the rates of absorption, transport, accumulation and especially biotransformation (metabolism) of the two enantiomers *in vivo*.

As noted above, this phenomenon of enantioselectivity is not restricted to pharmaceuticals but is characteristic of all biologically active agents. Another market segment that is being subjected to increasing scrutiny in this context is agrochemicals. For example, it is well known that the herbicidal activity of the $\alpha$-phenoxypropionic acid (**3**) group of plant protection agents resides predominantly in the *R*-enantiomer. Consequently, there is an increasing trend to market these compounds as single enantiomers, thereby reducing significantly their ecological impact.

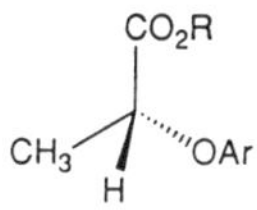


(3)

In short, the ever-increasing demand for more selective drugs, pesticides, etc., which are more targeted in their action, show less side-effects and are more environmentally acceptable, is providing an important stimulus to companies to market these products as pure enantiomers. Consequently, there is an increasing demand for efficient, economical methods for the industrial-scale synthesis of optically active compounds.

## Synthetic methodology

Synthetic routes to optically active compounds can be conveniently divided into three groups, on the basis of the type of raw material used (see Scheme 1).

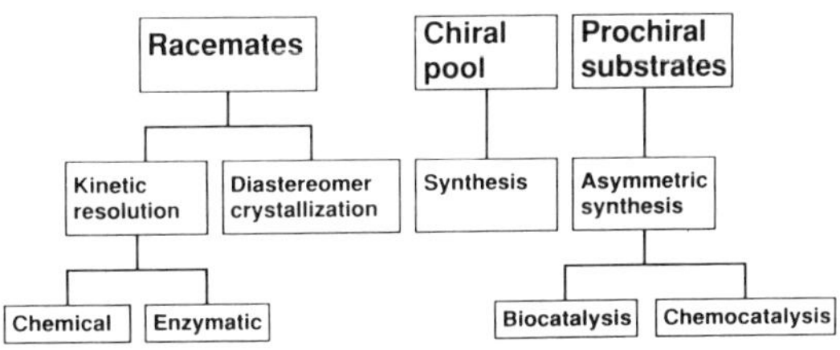


**Scheme 1.** Optically active compounds—methods of preparation.

## The Chiral Pool

The chiral pool refers to inexpensive, readily available natural products (and their derivatives) such as carbohydrates, amino acids and lactic acid. These substances can be transformed into synthetic products via chemical manipulation which may involve retention or inversion of configuration or chirality transfer.

For example, the optically active $\alpha$-phenoxypropionic acid herbicides referred to earlier can be prepared from the appropriate enantiomer of lactic acid as shown in Scheme 2.

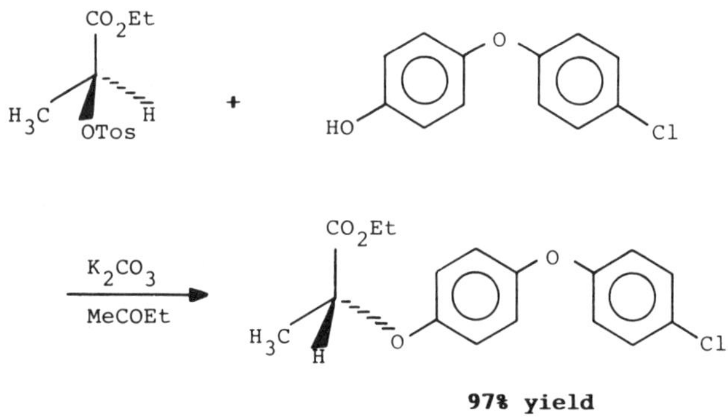


**Scheme 2.**

## Racemate Resolution via Crystallization

The resolution of racemates constitutes the main method for the industrial synthesis of pure enantiomers. Methods for resolution can be divided into three types (see Scheme 3): direct preferential crystallization, crystallization of diastereomeric salts and kinetic resolution.

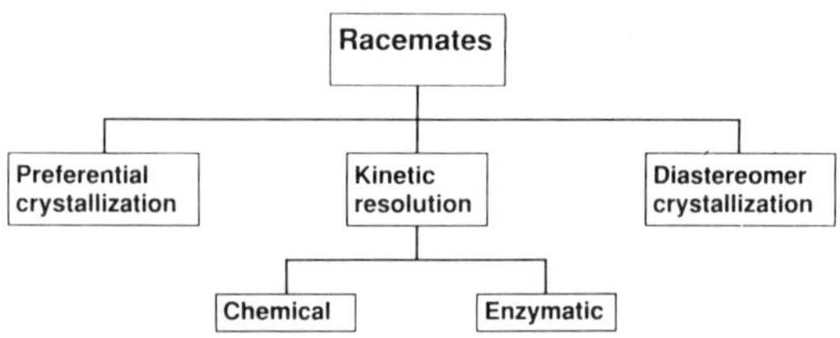


**Scheme 3.** Resolution of racemates.

Direct preferential crystallization of the required enantiomer (also referred to as resolution by entrainment) is possible only with racemates that are so-called conglomerates in the solid state, i.e. they are the mechanical mixtures of crystals of the two enantiomers. Preferential crystallization is dependent on differences in rates of crystallization of two enantiomers. By seeding with crystals of one enantiomer it is possible to achieve preferential crystallization. Resolution by entrainment is widely used on an industrial scale, for example in the manufacture of α-methyl-L-Dopa[4] and Chloramphenicol.[5] It is a particularly attractive method when it is accompanied by spontaneous, *in situ* racemization of the unwanted enantiomer which allows for a theoretical yield of 100%.[6]

However, more than 80% of racemates are true racemic compounds, i.e. a homogeneous solid phase of the two enantiomers co-existing in the same unit cell, and cannot be separated by preferential crystallization. Racemic compounds may be separated by diastereomer crystallization which generally involves reaction of the racemate with an optically pure acid or base (the resolving agent) to give a mixture of two diastereomeric salts whose physical properties are different.

Thus, when a racemic acid A is combined with an optically pure base B a mixture of two diastereomeric salts is formed which is separated by crystallization:

$$\underset{\text{Racemate}}{(\text{DL})\text{-A}} + \underset{\text{Resolving agent}}{(\text{L})\text{-B}} \rightarrow \underset{\text{Diastereomers}}{(\text{D})\text{A}(\text{L})\text{B} + (\text{L})\text{A}(\text{L})\text{B}} \quad (1)$$

Diastereomer crystallization is widely used for the industrial manufacture of pure enantiomers. A typical example is the Andeno process (see Scheme 4) for the manufacture of (D)-(−)-phenylglycine, an antibiotic intermediate, using optically pure camphor sulphonic acid as the resolving agent. This process is used for the production of more than a thousand tons per annum of D-(−)-phenylglycine.

The theoretical, once-through yield of a resolution via diastereomer crystallization is 50%. In practice, chemical yields of >40% of material of high (>95%) optical purity, in a single crystallization, are considered to be good resolutions. In general, industrial processes involve racemization of the unwanted enantiomer, for obvious economic reasons.

It is possible to obtain once-through yields >50% when the diastereomer remaining in solution undergoes spontaneous epimeriza-

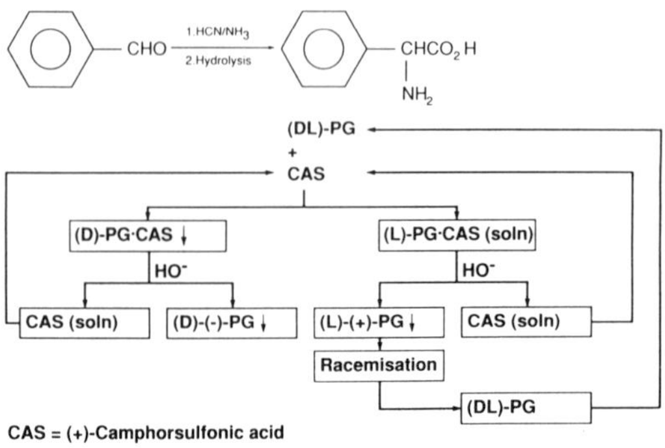


**Scheme 4.** Manufacture of D-phenylglycine DSM/Andeno.

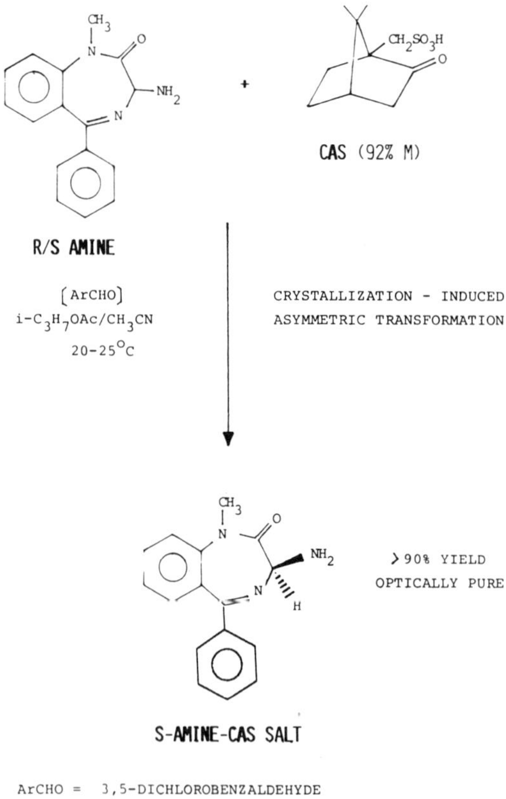


**Scheme 5.** Crystallization-induced asymmetric transformation.

tion, often referred to as diastereomer interconversion. Such a process is referred to as a crystallization-induced asymmetric transformation and allows for a theoretical yield of 100%. For example, crystallization of the mixture of diastereomeric salts formed by combination of the racemic amine shown in Scheme 5 with camphor sulphonic acid, in the presence of a trace of an aldehyde as epimerization catalyst, afforded the *S*-amine salt in 90% yield in a single crystallization.[7]

## Kinetic Resolution of Racemates

Kinetic resolution depends on the fact that the rates of reaction of two enantiomers with an optically active reagent are different. Preferably, the optically active reagent should function in catalytic quantities and may be an enzyme or a simple chemical catalyst. The so-called enantiomeric ratio *E* is a measure of the efficiency of a particular kinetic resolution (in this case the required enantiomer is the unreacted substrate):

$$E = \frac{\ln([1-\mathrm{c}][1-\mathrm{ees}])}{\ln([1-\mathrm{c}][1+\mathrm{ees}])} \tag{2}$$

where c = degree of conversion of substrate
ees = enantiomeric excess of remaining substrate

In favourable cases ($E > 50$), such as many enzymatic transformations, the rates of reaction of the two enantiomers are substantially different, resulting in a virtually enantiospecific reaction.

Enzymatic kinetic resolution using hydrolytic enzymes is a method (see later) that is becoming increasingly popular for the industrial scale synthesis of pure enantiomers. Their main advantages are their relative simplicity and the high substrate concentrations which are often possible. Moreover, the use of enzymes in organic solvents (or even with neat organic substrates), as pioneered by Klibanov and coworkers,[8] has significantly broadened their scope in organic synthesis. It is worth noting, however, that highly selective kinetic resolutions have also been achieved with relatively simple chiral catalysts (for an example see Ref. 9).

## Prochiral Compounds—Catalytic Asymmetric Synthesis

The enantioselective conversion of a prochiral substrate to an optically active product, by reaction with a chiral reagent, is referred to as an

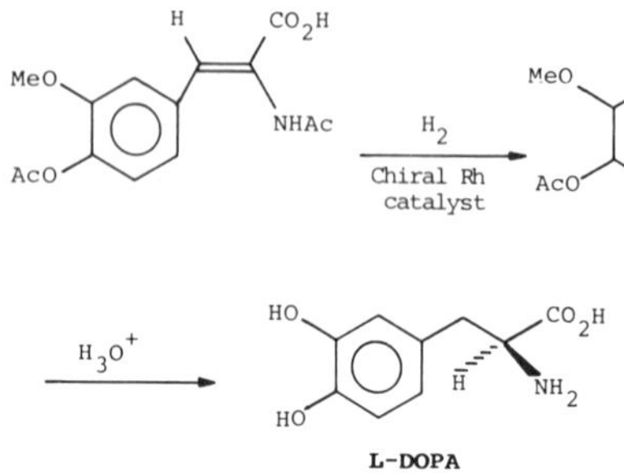


**Scheme 6.**

asymmetric synthesis. From an economic viewpoint the optically active reagent should function in catalytic quantities. Catalytic asymmetric syntheses can be divided into two groups based on whether they involve a simple chemical catalyst or a biocatalyst. An example of the former is the well-known Monsanto process for the manufacture of L-Dopa[10] by catalytic asymmetric hydrogenation (Scheme 6). An example of the latter is the Genex process[11] for the synthesis of L-phenylalanine by L-phenylalanine ammonia lyase (PAL)-catalysed addition of ammonia to *trans*-cinnamic acid (Scheme 7).

## Asymmetric Synthesis vs Kinetic Resolution

In considering the relative merits (economics) of different methods for the synthesis of pure enantiomers the question often arises: which is more attractive, asymmetric synthesis or kinetic resolution? This is a difficult question to answer. In the first place different raw materials are involved and the economics are obviously dependent on their relative price and availability.

Superficially it would seem that asymmetric synthesis is more attractive since it has a theoretical yield of 100% compared to 50% for kinetic resolution as shown in the hypothetical example (Scheme 8).

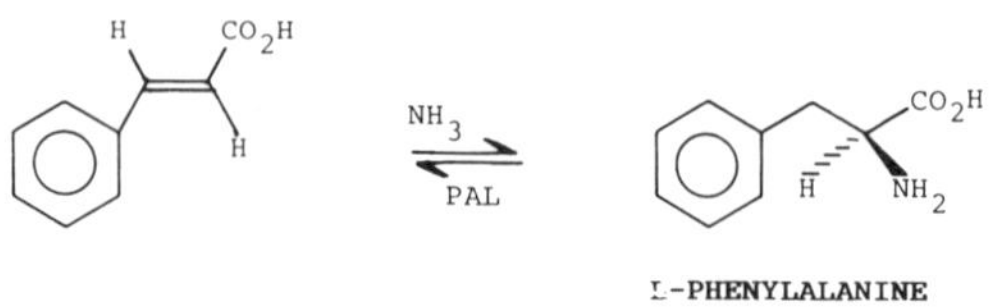


**Scheme 7.**

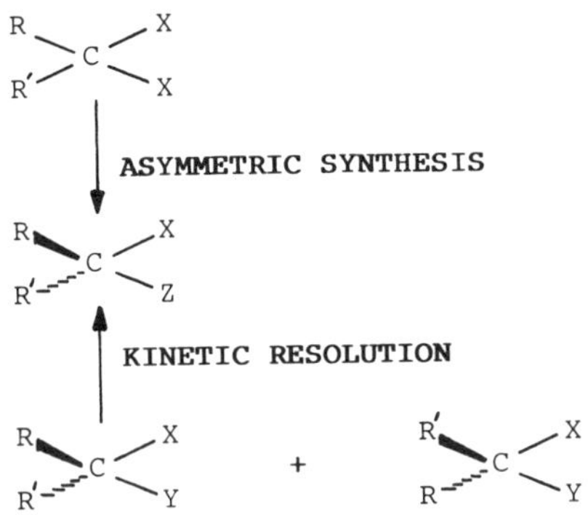


**Scheme 8.** Asymmetric synthesis vs kinetic resolution.

However, further inspection often reveals that kinetic resolutions have several advantages compared to asymmetric synthesis. One important advantage is the fact that the optical purity can in principle be tuned to any required value by adjusting the degree of conversion as illustrated in Fig. 1. This is a consequence of the fact that a kinetic resolution involves by definition reaction of a mixture (of enantiomers), the molar ratio of which varies with conversion, whilst asymmetric synthesis pertains to the reaction of a single (prochiral) compound and enantioselectivity is independent of conversion.

Moreover, kinetic resolutions tend in general to be simpler chemical processes than asymmetric syntheses. On the other hand, they suffer from the disadvantage that at least one extra step (racemization of the of the unwanted isomer) is generally involved. This disadvantage is

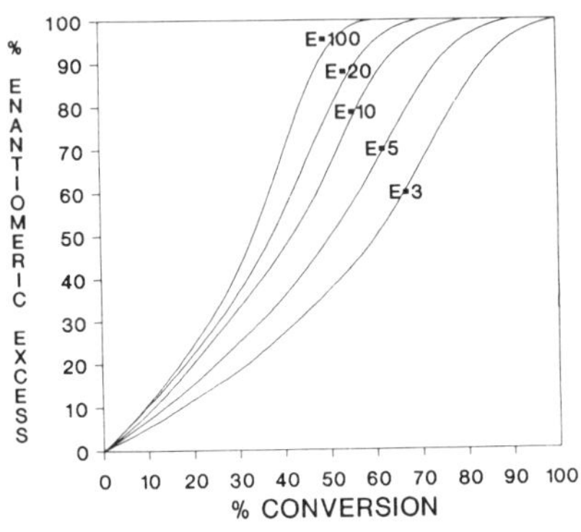


**Fig. 1.** Dependence of enantiomeric excess on relative rate (enantiomeric ratio, *E*) and conversion.

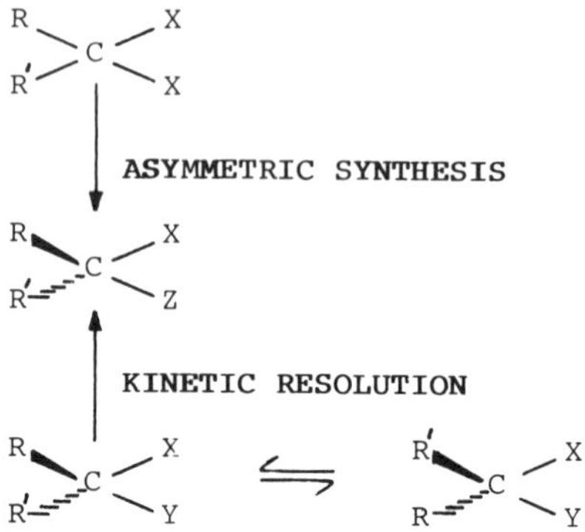


**Scheme 9.** Asymmetric synthesis vs kinetic resolution plus spontaneous racemization.

circumvented when conditions can be chosen where the unwanted enantiomer spontaneously racemizes, leading to a situation where the theoretical yield is 100% as shown in the hypothetical example in Scheme 9. In the context of industrial syntheses this is the situation that one is always looking (hoping) for.

## Enzymatic resolution of amino acids

Various methods have been reported in the literature[12] for the enzymatic resolution of amino acids. For example, several L-amino acids are produced on a commercial scale by acylase-catalysed enantioselective hydrolysis of *N*-acetyl-D,L-amino acids.[13]

At DSM a very efficient method has been developed[14] for the production of optically pure L- and D-amino acids via enantioselective hydrolysis of racemic amino acid amides mediated by an L-specific aminopeptidase (see Scheme 10). This process is currently used for the production of several D- and L-amino acids on a commercial scale.

The substrate is readily prepared from simple raw materials. Thus, Strecker reaction of the appropriate aldehyde with hydrogen cyanide and ammonia affords the corresponding amino nitrile. The latter is converted in high yield to the corresponding racemic amino amide by alkaline hydrolysis in the presence of a catalytic amount of acetone.[15]

The resolution step[16] is carried out with permeabilized whole cells of *Pseudomonas putida* ATCC 12633 resulting in a virtually enantio-

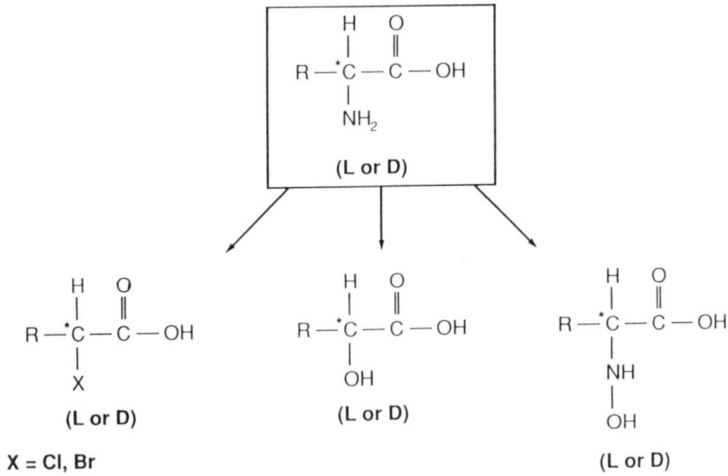

**Scheme 10.** DSM aminopeptidase process.

specific hydrolysis of the L-amino acid amide. A significant advantage of this process is its broad substrate specificity which allows for the resolution of a wide range of amino acids (see Table 1). Another advantage is the fact that the substrate is a precursor of the amino acid

**Table 1.** Substrate specificity of the amino peptidase from *Pseudomonas putida* ATCC 12633 (selected examples)

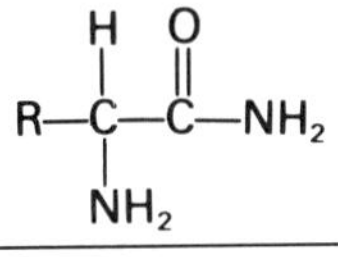

| R | R |
|---|---|
| Methyl | Benzyl |
| Ethyl | 2-Phenylethyl |
| *n*-Propyl | Phenyl |
| *i*-Propyl | *o*-Chlorophenyl |
| *i*-Butyl | *p*-Hydroxyphenyl |
| Allyl | *p*-Hydroxybenzyl |
| 2-Methylthioethyl | 2-Naphthyl |
| 3-Methylthioethyl | 2-Thienyl |

rather than a derivative. This means, in principle, that less steps are needed compared to processes which use a derivative as the substrate and involve the initial preparation of the racemic amino acid followed by derivatization.

The biocatalyst is used in soluble form in a batchwise process, thus allowing for the resolution of even poorly soluble amino acids. Recycling of the biocatalyst is in principle possible. A very simple and elegant method has been developed for the separation of the L-amino acid and D-amino acid amide products. Thus, addition of one equivalent of benzaldehyde (with respect to D-amino amide) results surprisingly in the formation of a water-insoluble Schiffs base of the D-amino amide which is readily separated.[17] Subsequent acid hydrolysis affords the optically pure D-amino acid.

Process economics dictate the recycling of the unwanted isomer. Racemization of the D-*N*-benzylidene amino acid amide (path A in Scheme 10) proceeds readily under mild conditions.[18] After removal of the benzaldehyde the racemic amino acid amide can be recycled to the enzymatic hydrolysis step.

A suitable method for the racemization and recycling of the L-amino acid (path B, Scheme 10) comprises conversion to the methyl ester followed by reaction with ammonia affording the amino amide. Addition of benzaldehyde and racemization at pH 13 gives the DL-amino amide. Although the synthesis of D-amino acid involves overall more steps than for the corresponding L-amino acid (referred to as substractive and attractive processes, respectively), the operations are relatively simple and virtually quantitative yields of D-amino acids are obtained.

Both D- and L-amino acids are widely used as pharmaceutical intermediates. For example, the earlier mentioned D-enantiomer of phenylglycine is a key intermediate in the manufacture of several semi-synthetic penicillins and cefalosporins.

Interest in optically pure amino acids as agrochemical intermediates is steadily growing. For example, D-valine is used in the synthesis of

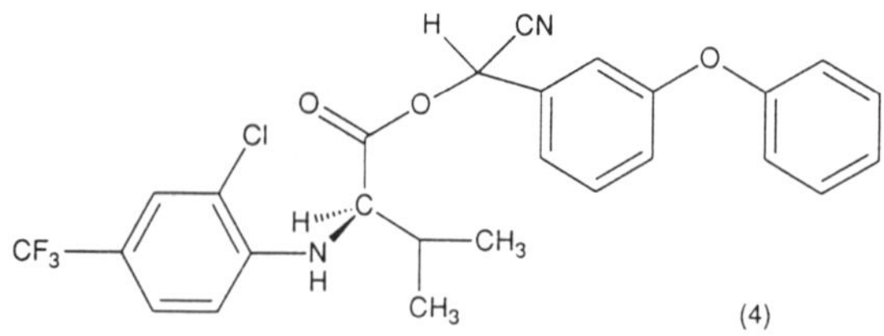


(4)

Fluvalinate (**4**), a broad spectrum pyrethroid insecticide developed by Zoecon.[19]

An essential structural feature for the *Pseudomonas putida*-derived aminopeptidase is the presence of an $\alpha$-hydrogen in the substrate. On the other hand, optically pure $\alpha$-disubstituted amino acids are also of commercial interest. To provide access to this class of amino acids a new biocatalyst derived from *Mycobacterium neoaurum* has been developed by DSM.[20] This L-specific aminopeptidase mediates the enantioselective hydrolysis of a range of $\alpha$-disubstituted amino acid amides (see Scheme 11). The latter are readily prepared by a Strecker reaction of hydrogen cyanide and ammonia with the appropriate ketone.

It is impossible to racemize $\alpha$-disubstituted amino acids directly due to the absence of the acidic $\alpha$-C—H bond. Recycling of the unwanted

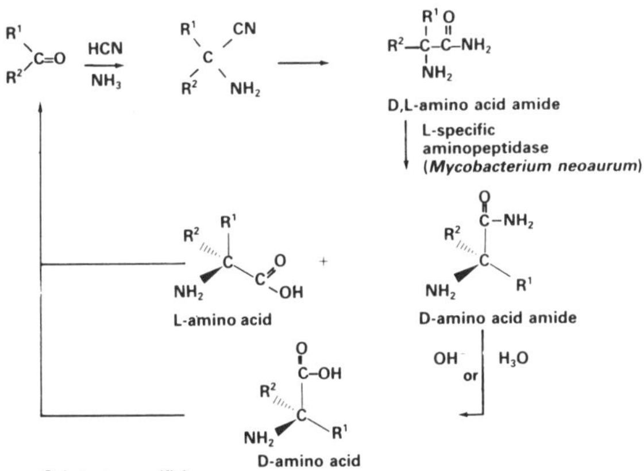


Substrate specificity

$R^2 = CH_3-$ $\quad R^1 = CH_3CH_2-$; $(CH_3)_2CH-$; $(CH_3)_2CHCH_2-$;

$C_6H_5-$; $4\text{-}H_3CO\text{-}C_6H_4-$; $3,4\text{-}(H_3CO)_2\text{-}C_6H_3$;

$C_6H_5CH_2-$

$R^2 = CH_3-CH_2-$ $\quad R^1 = C_6H_5CH_2-$

$R^2 = (CH_3)_2CH-$ $\quad R^1 = C_6H_5CH_2-$

**Scheme 11.** Enzymatic method for the optical resolution of racemic $\alpha$-disubstituted amino acids.

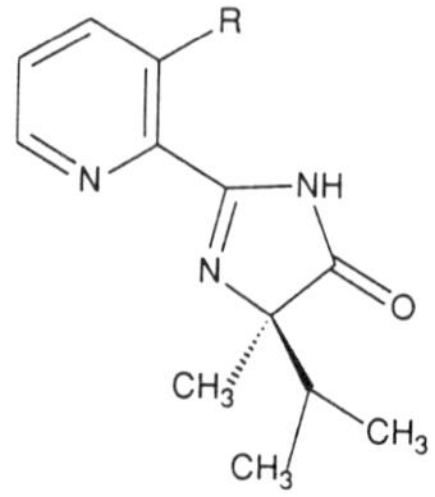

(5)

isomer is achieved, therefore, by conversion back to the starting ketone (see Scheme 11).

A possible commercial application of an optically pure α-disubstituted amino acid amenable via this technology is the use of D-α-methylvaline amide as an intermediate for the synthesis of the biologically active form of the broad spectrum herbicide Arsenal (**5**) developed by American Cyanamid.[21] It can be inferred from the patent literature that the D-enantiomer is significantly more active than the racemate in certain applications.

## Downstream derivatives of optically pure amino acids—ACE inhibitor intermediates

As mentioned above, readily available amino acids may be considered as part of the 'chiral pool' and as such are attractive starting materials for further elaboration to commercially interesting fine chemicals. For

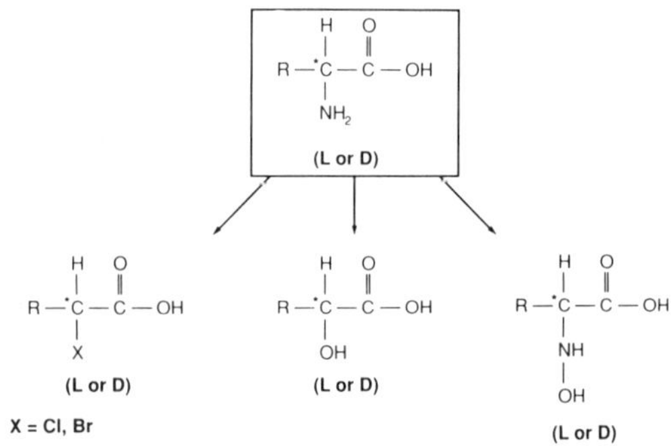

**Scheme 12.** Downstream derivatives of optically pure amino-acids.

example, efficient methods have been developed by DSM/Andeno for the conversion of D- or L-amino acids, available using the above described enzymatic resolution technology, to the corresponding optically pure $\alpha$-hydroxy acids, $\alpha$-haloacids and $\alpha$-hydroxyamino acids (see Scheme 12).

An example of an amino acid that has been successfully resolved on a commercial scale using the DSM technology is homophenylalanine. The substrate is readily prepared starting from benzaldehyde as shown in Scheme 13.

Both L- and D-homophenylalanine constitute commercially interesting starting materials for the production of a variety of Angiotensin Converting Enzyme (ACE) inhibitors, a relatively new class of antihypertensive agents. The ACE inhibitors[22] have found wide application in the treatment of hypertension since the market introduction of the first example, Captopril, by Squibb in 1981 (see Scheme 14 for structure).

Subsequent to the introduction of Captopril a variety of ACE inhibitors have been developed, many of which contain the homophenylalanine moiety as a common structural feature. The prime example of this group is Enalapril developed by Merck, Sharp and Dohme and introduced in 1984. Enalapril is currently manufactured by Merck via reductive amination of 4-phenyl-2-ketobutyric acid ethyl ester (**6**) with L-alanyl-L-proline (**7**) as shown in Scheme 15. The best results were obtained using a Raney nickel catalyst which gave an 80–90% yield of a product containing 87% of the required (*SSS*)-diastereomer together with 13% of the (*RSS*)-isomer.[23] The crude Enalapril was purified via crystallization of its maleic acid salt.

At Andeno/DSM we have developed[24] alternative commercially attractive routes to Enalapril starting from L- or D-homophenylalanine

$$\mathrm{PhCHO + CH_3CHO \xrightarrow{base} PhCH{=}CHCHO}$$

$$\mathrm{\xrightarrow[hydrogenation]{catalytic} PhCH_2CH_2CHO \xrightarrow[NH_3]{HCN} PhCH_2CH_2CH(NH_2)CN}$$

$$\mathrm{\xrightarrow[Acetone]{HO^-} PhCH_2CH_2CH(NH_2)CONH_2}$$

**Scheme 13.**

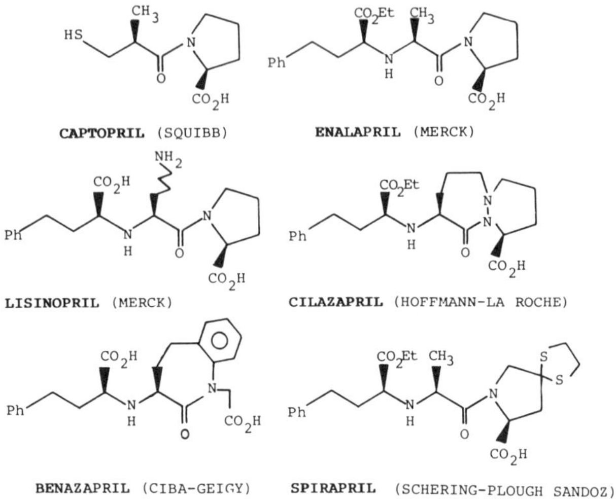


**Scheme 14.** Structures of various ACE inhibitors.

(see Scheme 16). Thus, coupling of L-homophenylalanine ethyl ester with an appropriate derivative of D(=R)-lactic acid, followed by deprotection, affords the key intermediate (**8**) in high yield and optical purity; (**8**) is a common intermediate for the synthesis of Enalapril and structurally related ACE inhibitors (see Scheme 14). For example,

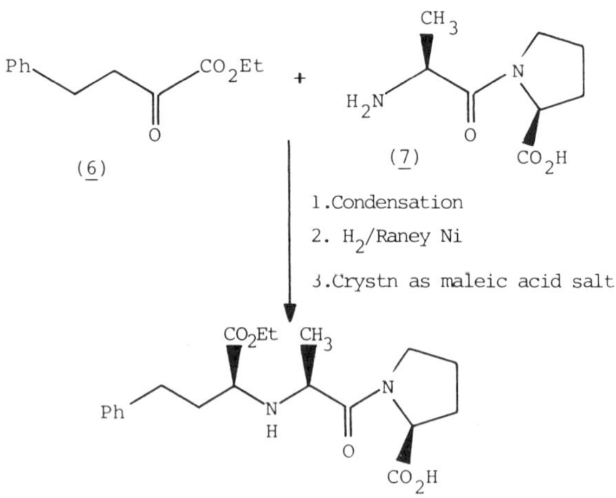


**Scheme 15.** Merck synthesis of Enalapril.

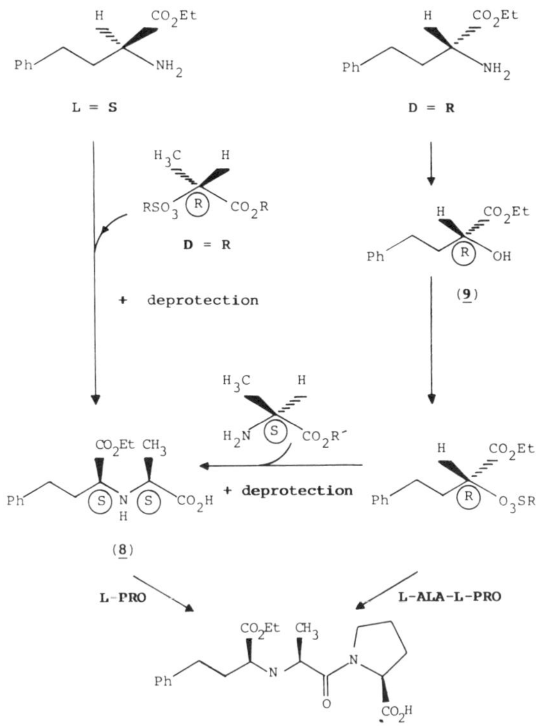


**Scheme 16.** Synthesis of Enalapril from L- or D-homophenylalanine.

reaction of the acid chloride hydrochloride salt of (**8**) with L-proline affords Enalapril in excellent yield.

Alternatively, D-homophenylalanine ethyl ester can be converted selectively, without significant loss of optical activity, to the corresponding $\alpha$-hydroxy compound (**9**). Treatment of the appropriate derivative of (**9**) with L(=*S*)-alanine ester followed by deprotection affords the key intermediate (**8**).

An alternative, shorter route to Enalapril, starting from D-homophenylalanine, involves reaction of the appropriate derivative of (**9**) with L-alanyl-L-proline. This route has the advantage that it is also applicable to the synthesis of other ACE inhibitors, such as Lisinopril, Cilazapril and Benazapril (see Scheme 14 for structures) which possess the homophenylalanine moiety but are not based on the key intermediate (**8**).

All of the above mentioned routes to ACE inhibitors starting from L- or D-homophenylalanine have one common advantage: the key resolution step is early in the overall synthesis. Indeed, this is an important factor in determining the economics of processes for the synthesis of optically active compounds in general. The resolution step should be as early as possible since this halves the amount of reagents, solvents, reactor volume, etc., required per kilo product in subsequent steps.

## Enzymatic resolutions with lipases—synthesis of optically active glycidyl derivatives

In recent years increasing attention has been placed on the use of lipases and esterases for enzymatic resolutions, especially in organic media.[25] Although the presence of a certain minimum amount of water is essential for the stability and function of enzymes, replacement of the rest of the water with an organic solvent should be possible without adversely affecting the enzyme.[8] Indeed, Klibanov and co-workers[8] have demonstrated for a number of enzymes that they can function in almost anhydrous (ca. 0·1% water) hydrophobic organic solvents. Moreover, they often exhibit greatly enhanced thermal stability and marked changes in substrate specificity compared to aqueous media. Furthermore, the use of organic solvents makes enzymatic processes more compatible with organic syntheses which generally involve water-insoluble compounds and it allows for the catalysis of reactions (e.g. transesterification) which could not be conducted in aqueous solution.

Commercial development of lipase-catalysed resolutions is in some cases well advanced. Chemie Linz (Austria) is reportedly operating a process for the lipase-mediated resolution of $\alpha$-halopropionic acids on a pilot-plant scale.[26] Optically pure $\alpha$-halopropionic acids are of interest as intermediates for the commercial synthesis of pure enantiomers of the $\alpha$-phenoxypropionic herbicides[3] discussed earlier.

Alternatively, the required optically pure $\alpha$-phenoxypropionic acid derivative can be obtained directly via lipase-mediated hydrolysis, esterification or transesterification of the corresponding racemic $\alpha$-phenoxypropionic acid (ester). In the example[27] depicted in Scheme 17 the unwanted *S*-isomer is racemized by heating with a catalytic amount of sodium methoxide.

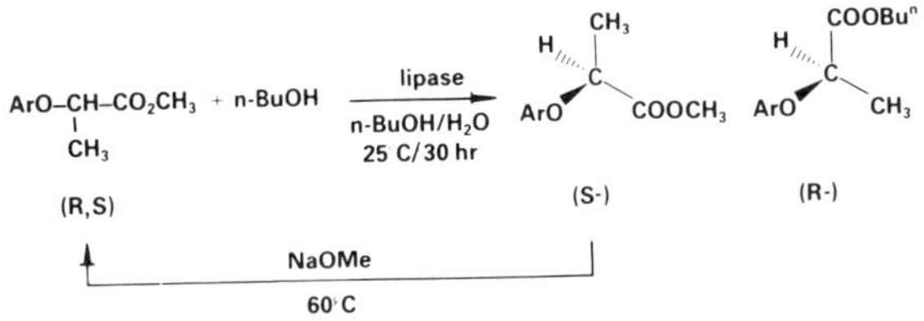


**Scheme 17.** *R*-$\alpha$-Phenoxypropionic esters via lipase-mediated transesterification.

Similarly, the therapeutically active *S*-enantiomers of $\alpha$-arylpropionic acid anti-inflammatory drugs (**2**) can be conveniently prepared by lipase-mediated hydrolysis of an appropriate ester. Thus, at Andeno we have developed a commercially attractive process for the synthesis of *S*-Ibuprofen (**2**: Ar = *p*-isobutylphenyl).[28]

At Andeno-DSM we have also developed a process for the commercial scale synthesis of optically pure glycidyl derivatives via lipase-mediated enantioselective hydrolysis of *R,S*-glycidyl butyrate (Scheme 18). This reaction was first reported by Ladner and Whitesides[29] using porcine pancreatic lipase. We have subsequently improved the original findings considerably to give *R*-glycidyl butyrate (**10**) and/or *R*-glycidol (**11**) with high enantiomeric excess. This process is currently being operated on a multi-ton scale at Andeno.

The *R*-glycidyl butyrate and *R*-glycidol products are readily converted to *R*- and *S*-glycidyl tosylate, respectively, of high enantiomeric purity. The high selectivity for nucleophilic attack at $C_1$ makes these glycidyl tosylates highly attractive intermediates for a wide range of commercially interesting products, such as optically active $\beta$-blockers (see earlier), phospholipids and many more (see Scheme 19).

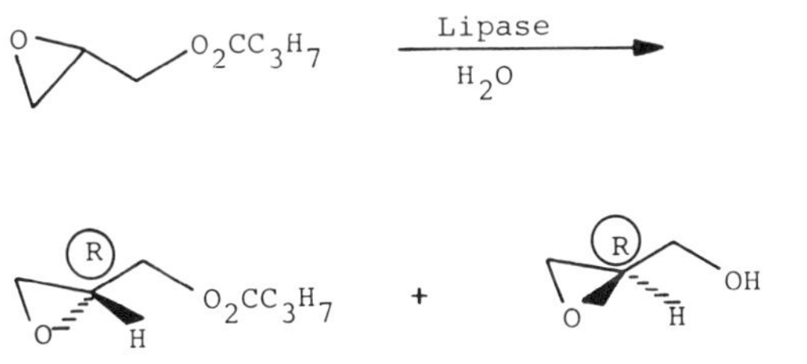


(5)

(10) (11)

**Scheme 18.**

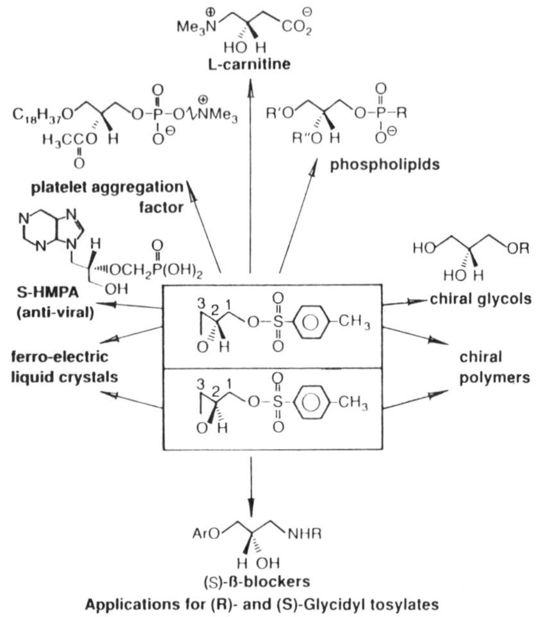


**Scheme 19.** Optically active glycidyl tosylates—versatile intermediates in organic synthesis.

Examples of $\beta$-blockers which are marketed as the single, biologically active *S*-enantiomer are *S*-Timolol (Merck), *S*-Penbutolol (Hoechst) and *S*-Levobunolol (Warner-Lambert). All of these products can be readily prepared starting from *R*-glycidol and/or *S*-glycidol tosylate.

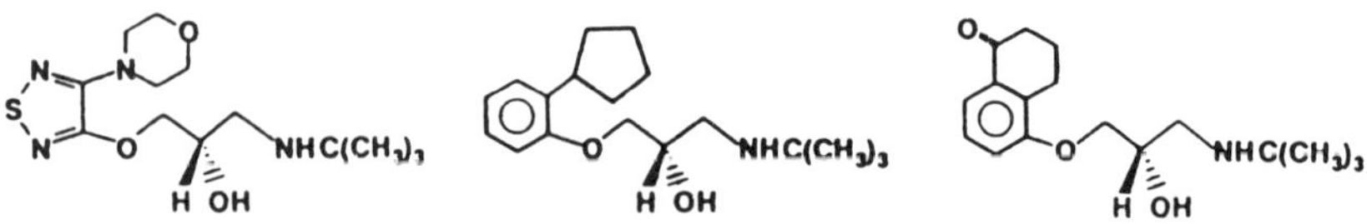

(*S*)-timolol (*S*)-penbutolol (*S*)-levobunolol

Alternative methods for the synthesis of optically pure glycidyl derivatives have been reported.[25] For example, the elegant method developed by Sharpless and co-workers[30] for catalytic asymmetric epoxidation of allylic alcohols can be used. In the original paper poor yields were reported for glycidol but improvements have been reported.[31] A method for the enantioselective microbial epoxidation of aryl allyl ethers to the corresponding aryl glycidyl ethers has been reported by IBIS workers.[32]

Both of these methods are asymmetric syntheses in contrast to the glycidyl butyrate process described above which is a kinetic resolution. Nevertheless, we consider the latter process to be commercially more attractive. It is simple, easy to scale up and can be performed at very high substrate concentrations, resulting in high volume yields.

## Recent developments in catalytic asymmetric synthesis

Progress in the area of chiral synthesis is by no means restricted to the enzyme-mediated kinetic resolutions mentioned above. In recent years much progress has been made in the area of catalytic asymmetric synthesis.

The catalytic asymmetric hydrogenation technology applied in the synthesis of L-Dopa (see earlier) was developed in the 1960s. Subsequently, much attention was focused on the area of catalytic asymmetric synthesis, but almost twenty years elapsed before Sharpless reported[30] the catalytic asymmetric epoxidation of allyl alcohols, the only example were the success of the asymmetric hydrogenation was really emulated.

Nevertheless, both the asymmetric hydrogenation and epoxidation reactions suffer from the same drawback: their scope (substrate specificity) is limited (to unsaturated $\alpha$-acylamino carboxylic acids and allylic alcohols, respectively). There is still a definite need, therefore, for chiral catalyst systems with a broader substrate specificity that are able to mediate the asymmetric hydrogenation, oxidation, carbonylation and isomerization of relatively simple molecules.

Recently, very promising results have been reported by Noyori and co-workers[33] who have employed chiral metal–Binap complexes in catalytic isomerizations and hydrogenations. These systems exhibit very high activities (substrate catalyst ratios often >1000) and enantioselectivities (>95%). Moreover, they appear to have a broad scope and are effective with relatively simple, largely unfunctionalized substrates.

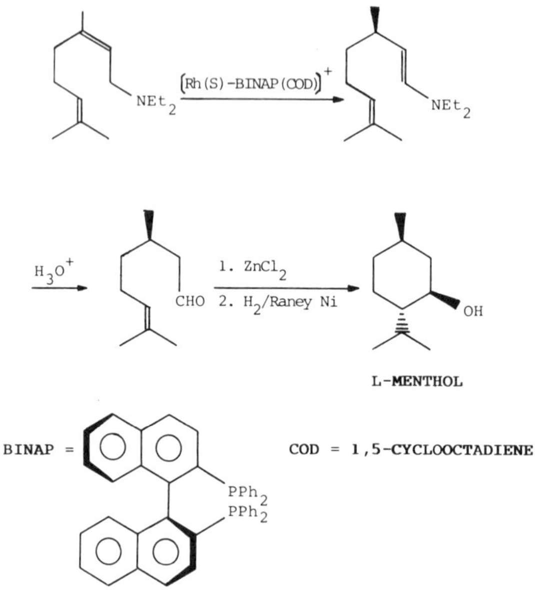


**Scheme 20.** Catalytic asymmetric synthesis of L-menthol.

This technology has reportedly been commercialized by Takasago in a new process for the production of L-menthol.[34] The key, chirality-inducing step in this elegant process is the isomerization of a prochiral allylic amine to a chiral enamine (see Scheme 20). This reaction, which employs a Rh-(*S*)-Binap catalyst at substrate/catalyst ratios of ca. 10 000:1, reportedly gives the required enamine in essentially quantitative yield and 98% enantiomeric excess.

# Concluding remarks

Revolutionary advances are currently being made in the industrial synthesis of optically active compounds. These developments are largely stimulated by the growing awareness of the importance of chirality in conjunction with biological activity.

With regard to the question of which method is the most economically attractive, e.g. enzymatic kinetic resolution, catalytic asymmetric synthesis or classical resolution via crystallization, there is no simple, all-encompassing answer. Numerous factors are involved[35] in determining the economics of different routes to a particular product and the method of choice will obviously vary from one product to another. One generalization, however, usually holds true and that is: the earlier in the synthesis the chirality is introduced the better. Finally, another general conclusion can be drawn: with the steadily growing arsenal of effective methods at our disposal there is no longer any excuse for marketing such products as racemic mixtures.

## Acknowledgements

The author wishes to acknowledge his gratitude to his co-workers at Andeno (L. A. Hulshof, V. H. M. Elferink, J. van Iersel, J. H. Roskam and H. J. M. Zeegers) and colleagues at DSM Research (E. M. Meijer, H. E. Schoemaker, J. Kamphuis, W. H. J. Boesten and M. Kloosterman). Without their numerous innovative contributions this article could not have been written.

## References

1. For leading references see: Ariëns, E. J., *Med. Res. Revs,* **6** (1986) 451; Ariëns, E. J., *Eur. Clin. Pharmacol.,* **26** (1984) 663–8,; Simonyi, M., *Med. Res. Revs.,* **4** (1984) 359–413; Ariëns, E. J., van Rensen, J. J. S. & Welling, W. (eds), *Stereoselectivity of Pesticides; Biological and Chemical Problems.* Elsevier, Amsterdam, 1988.
2. Pfeiffer, C., *Science,* July 1956, p. 29.
3. Nestler, H. J., *et al.,* British Patent 2 042 503, 1980, to Hoechst.
4. Anon., *Chem. Eng.,* **72** (1965) 247; see also Reinhold, D. F., Firestone, R. A. Gaines, W. A., Chemerda, J. M. & Sletzinger, M., *J. Org. Chem.,* **33** (1968) 1209–13.
5. Velluz, L., Amiard, G. & Joly, R., *Bull. Soc. Chim. France* (1953) 342–4; Amiard, G., *Experientia,* **15** (1959) 1–7.
6. For examples see: Sifniades, S., Boyle, W. J. & van Peppen, J. F., *J. Am. Chem. Soc.,* **98** (1976) 3738; Yamada, S., Hongo, C. & Chibata, I., *Chem. Ind. (London)* (1980) 539.
7. Reider, P. J., Davis, P., Hughes, D. L. & Grabowski, E. J., *J. Org. Chem.,* **52** (1987) 957–8.
8. Klibanov, A. M., *CHEMTECH,* February 1982, p. 116.
9. Martin, V. S., Woodward, S. S., Katsuki, T., Yamada, Y., Ikeda, M. & Sharpless, K. B., *J. Am. Chem. Soc.,* **103** (1981) 6237.

10. Vineyard, B. D., Knowles, W. S. & Sabacky, M. J., *J. Mol. Catal.*, **19** (1983) 159; Knowles, W. S., Sabacky, M. J., Vineyard, B. D. & Weinkauf, D. J., *J. Am. Chem. Soc.*, **97** (1975) 2567.
11. Hamilton, B. K., Hsiao, H. Swann, W. E., Anderson, D. M. & Delente, J. J., *Trends in Biotechnol.*, **3** (1985) 64–8.
12. For leading references see: Sheldon, R. A., Schoemaker, H. E., Kamphuis, J., Boesten, W. H. J. & Meijer, E. M., in *Stereoselectivity of Pesticides*; *Biological and Chemical Problems* ed. E. J. Ariëns, J. J. S. van Rensen & W. Welling. Elsevier, Amsterdam, 1988, pp. 409–51.
13. Chibata, I. & Tosa, T., *Ann. Rev. Biophys. Bioeng.*, **10** (1981) 197–216.
14. Meijer, E. M., Boesten, W. H. J., Schoemaker, H. E. & van Balken, J. A. M., in *Biocatalysts in Organic Syntheses,* ed. J. Tramper, H. C. van der Plas & P. Linko. Elsevier, Amsterdam, 1985, pp. 135–56.
15. Boesten, W. H. J., British Patent 1 548 032, 1976, to DSM/Stamicarbon.
16. Boesten, W. H. J., US Patent 3 971 700, 1976, to DSM/Stamicarbon.
17. Boesten, W. H. J., US Patent 4 172 846, 1979, to DSM/Stamicarbon.
18. Boesten, W. H. J., Schoemaker, H. E. & Dassen, B. H. N., Dutch Patent Appl. 8 501 093, 1985, to DSM/Stamicarbon.
19. *Farm Chemicals Handbook,* 72nd edn, Meister Publ Co., Willoughby, Ohio, 1986.
20. Boesten, W. H. J. & Peters, P. J. H., Eur. Patent 150 854, 1984, to DSM/Stamicarbon.
21. Eur. Patents 123 830, 1984, and 169 300, 1985, to American Cyanamid.
22. For leading references see: Natoff, I. L. & Redshaw, S., *Drugs of the Future,* **12** (1987) 475–83; Mackaness, G. B. *J. Cardiovascular Pharmacol.*, **7** suppl 1 (1985) 30–4; Ondetti, M. A., Rubin, B. & Cushman, D. W., *Science,* **196** (1977) 441.
23. Blacklock, T. J., Shuman, R. F., Butcher, J. W., Shearin, W. E., Budavari, J., & Grenda, V. J., *J. Org. Chem.*, **53** (1988) 836–44; see also Wyvratt, M. J., Tristram, E. W., Ikeler, T. J., Lohr, N. S., Joshua, H., Springer, J. P., Arison, B. H. & Patchett, A. A., *J. Org. Chem.*, **49** (1984) 2816–19.
24. Zeegers, H. J. M. & Hulshof, L. A., unpublished results.
25. For leading references see: Kloosterman, M., Elferink, V. H. M., van Iersel, J., Roskam, J. H., Meijer, E. M., Hulshof, L. A. & Sheldon, R. A., *Trends in Biotechnology,* **6** (1988) 251–6.
26. Kirchner, G., Scollar, M. P. & Klibanov, A. M., *J. Am. Chem. Soc.*, **107** (1985) 7072–6; Klibanov, A. M., US Patent 4 601 987, 1986; Dahod, S. K. & Siuta-Mangano, P., *Biotechnol. Biocng.*, **30** (1987) 995–9.
27. Cambou, B. & Klibanov, A. M., *Biotechnol. Bioeng.*, **26** (1984) 1449–54.
28. Elferink, V. H. M. & Hulshof, L. A., unpublished results.
29. Ladner, W. E. & Whitesides, G. M., *J. Am. Chem. Soc.*, **106** (1984) 7250–1.
30. Katsuki, T. & Sharpless, K. B., *J. Am. Chem. Soc.*, **102** (1980) 5976–80.
31. Klunder, J. M., Ko, S. Y. & Sharpless, K. B., *J. Org. Chem.*, **51** (1986) 3710–12.
32. Phillips, G. T., Robertson, B. W., Bertola, A. M., Koger, H. S., Marx, A. F. & Watts, P. D., Eur. Patent 193 228, 1986, to Shell and Gist-Brocades.

33. Tani, K., Yamagata, T., Akutagawa, S., Kumobayashi, H., Taketomi, T., Takaya, H., Miyashita, A., Noyori, R. & Otsuka, S., *J. Am. Chem. Soc.*, **106** (1984) 5208.
34. Tani, K., *Pure Appl. Chem.*, **57** (1985) 1845–54.
35. For a discussion see: Sheldon, R. A., Porskamp, P. A. & ten Hoeve, W., in *Biocatalysts in Organic Synthesis*, ed. J. Tramper, H. C. van der Pias & P. Linko. Elsevier, Amsterdam, 1985, pp. 59–80.

# The use of enzymes for the synthesis of fine chemicals and pharamceuticals

**S. West**

*Biocatalysts Limited, Pontypridd, Wales, UK*

## Introduction

Biotransformations (use of whole cells and enzymes as chemical catalysts) have been used many times during the past century for chemicals production. Past examples include the production of acetone and glycerol from starch (developed during the First World War) and the conversions of glucose into citrate and subsequently glutamate and lysine.

Two of the current biggest markets involving biotransformations are vitamin C and chemically modified penicillins. The current production of vitamin C involves a biotransformation step using the micro-organism *Acetobacter suboxydons* for the oxidation of sorbitol into sorbose.[1] Much work has been done on developing the whole process by biotransformation, but these cannot yet compete with the current mixed chemical/biochemical process. Another mixed chemical/biochemical process is modified penicillin production. The basic molecule is produced by fermentation. The next stage is to remove the benzyl (penicillin G) or phenoxy methyl (penicillin V) side chains. This leaves 6-aminopenicillanic acid (6-APA) which is ready for chemical modification. Removal of the side chains is currently achieved using the enzyme penicillin acylase from micro-organisms such as *Bacillus megaterium*, *Escherichia coli* and *Achromobacter* sp. Subsequent modifications are carried out chemically using acid chlorides, although modifications using the acylase enzyme in reverse have been studied.[2]

In the food ingredients area the major use of enzyme catalysts is for the production of fructose from starch. This is a good demonstration of the versatility of enzymes and the many ways they can be used. The first stage is carried out at 105°C and shows that not all enzymes are highly temperature sensitive; the second stage uses a batch process with a disposable catalyst and the third stage uses a continuous column reactor with an immobilised enzyme. This process is shown in Fig. 1.

The third stage is an isomerization process and this is very difficult to do chemically. However, the enzyme is not able to give a 100% conversion of glucose into fructose without continuously removing the fructose. If the fructose is not removed, then the reaction stops at about 45% conversion.

Other enzymes currently used in biotransformation processes are shown in Table 1.

The enzymes in Table 1 plus $\beta$-lactamases/penicillin acylase have an annual market value of about £15 million. This represents about 5% of the total industrial enzyme market which is still dominated by detergent enzymes. Growth in both the biocatalyst and industrial enzyme market has been slow over the past 10 years. However, 10

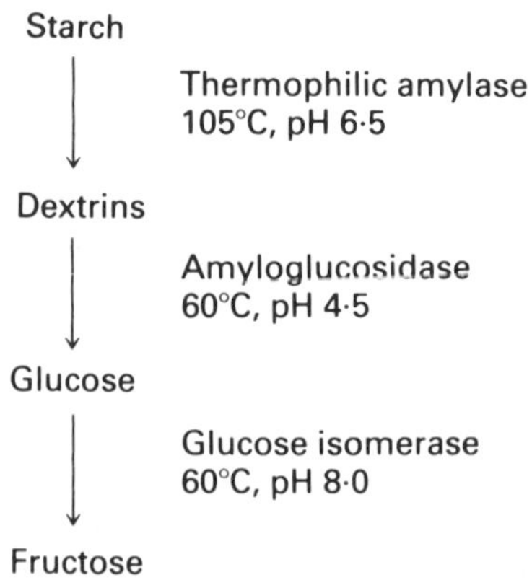


**Fig. 1.** Enzymic conversion of starch to fructose.

**Table 1.** Enzymes currently used in chemical processes

| Enzyme | Chemicals |
|---|---|
| Acylases | Resolved amino acids |
| Esterases/lipases | Ester synthesis/resolved hydroxy compounds, carboxylic acids |
| Hydantoinase | Resolved *p*-hydroxy phenylglycine |
| Peptidases | Peptide synthesis |
| Stereoid transforming enzymes | Steroid synthesis |
| Thermophilic proteases | Aspartame synthesis |

years ago very little research was being done on using biocatalysts for fine chemicals synthesis. This has now changed since the work in the early 1980s by Unilever on lipase catalysed transesterification[3] and the work by Klibanov on using enzymes in organic solvents.[4] Now over 10 publications a week are seen in the chemical literature using biocatalysts. The major single area of interest is still lipase catalysed reactions, with over 25% of the publications using lipases as biocatalysts. Almost a similar number appear using whole cells and the remainder are on non-lipase enzymes.

Whilst whole cells do have some advantages over enzymes, enzymes are generally seen as better and more versatile chemical catalysts. This article will concentrate on enzymes and in particular two classes, the lipases and secondary-alcohol dehydrogenases (*s*-ADH).

## Enzyme myths

Two myths have surrounded enzymes since their discovery. One has now been dispelled; it is now accepted that not all enzymes are fragile and temperature sensitive, and that many will function happily in non-aqueous solvents.

In addition many two-phase membrane systems have now been developed, where the enzyme is immobilised on the membrane. This technique has been used for many pharmaceutical resolutions and has been mainly pioneered by Sepracor Inc.

The second myth which has not yet been dispelled is that all enzymes are highly substrate specific. Whilst this is true of some enzymes, for example, phenylalanine dehydrogenase shows virtually no activity against any other amino acid, there are many enzymes that are only specific to functional groups and the remainder of the molecule is almost irrelevant. This is demonstrated later in Table 4 looking at *s*-ADH specificity.

## Lipases

There are over 20 different types of esterase/lipase on the market. Three types are of animal origin: pig liver esterase (PLE), porcine pancreatic lipase (PPL) and pre-gastric esterase (PGE). Both PLE and PPL have shown good stereoselectivity for chiral resolution work and PPL is used in the commercial process for the resolution of *R*,*S*-glycidyl butyrate. Microbial lipases have a very broad range of applications. A summary of these is shown in Table 2.

Outside the pharmaceutical industries it is mainly the conventional hydrolytic activities and the ability to synthesise esters that have made lipases useful. They are used for cheese and savoury flavour production, for fatty effluent treatment and grease traps and for the synthesis of natural flavour esters and emulsifiers.[5]

For the pharmaceutical industry it is the regio- and stereo-selectivity of lipases that make them valuable. This is demonstrated in Fig. 2.

**Table 2.** Sources and applications of lipases[a]

| Microbial source | Current applications/characteristics |
|---|---|
| *Alicaligenes* sp. | Asymmetric hydrolysis of symmetrical esters |
| *Candida cylindracea* (*rugosa*) | Resolution of D,L-chloropropionic acid esters; flavour ester synthesis; effective for primary and secondary alcohols |
| *Geotrichum candidum* | High specificity towards $C_{18}$ unsaturated fatty acids, resolution of 3-hydroxyalkanoic esters |
| *Mucor miehei* | Cocoa butter equivalent manufacture by trans-esterification; ester synthesis |
| *Penicillium cyclopium* | Monoglyceride synthesis; also exhibits stereo-selectivity |
| *Pseudomonas fluorescens* | Asymmetric hydrolysis of symmetric esters |
| *Rhizopus arrhizus* | Esterification, polyester hydrolysis |
| *Rhizopus japonicus* | Transesterification regio-selective acylation of *s*-hydroxy groups |

[a] All available from Biocatalysts Limited.

2a. (±) Endo-norbornenyl acetate ⟶ (+) Endo-norbornenol + (−) Endo norbornenyl acetate

2b. $HOCH_2CH(R)CH_2OH$ —Chemical acylation→ $AcOCH_2CH(R)CH_2OAc$

$HOCH_2CH(R)CH_2OH$ ↓ Enzyme acylation (ppl) → $HOCH_2CH(R)CH_2OAc$

$AcOCH_2CH(R)CH_2OAc$ ↓ Enzyme hydrolysis (ppl) → $AcOCH_2CH(R)CH_2OH$

**Fig. 2.** Stereo- (a) regio-selectivity (b) of lipases.

It can be said without doubt that during the 1990s more and more chiral drugs will come on to the market and that biocatalysts give the organic chemists one more technique for producing these optically active compounds. However, how successful enzymes will be and what percentage of these new chiral drugs will use a biotransformation for one or more of the production stages is still open to debate. More general information on lipases for pharmaceutical manufacture can be found in the papers by Schneider[6] and Kloosterman *et al.*[7]

## Secondary alcohol dehydrogenases

One problem with lipases as with all resolution processes is that unless the unwanted isomer can be recycled the yield at best will only be 50%. Hence the search is on for chiral synthesising agents rather than chiral resolving agents. Once again both chemical and biochemical routes exist. If we consider chiral hydroxy compounds then they can either be resolved using lipases or synthesised using secondary alcohol dehydrogenases (*s*-ADH). The advantages and disadvantages of these two enzyme types are summarised in Table 3.

Examples of the two process routes for the production of chiral 2-pentanol are shown in Fig. 3. (This is purely an illustration; it is not suggested that (*S*)-2-pentanol should be made in this way.)

Most of the work in the literature on *s*-ADH enzymes has used the enzyme isolated from horse liver (HLADH). However, microbial sources are available and these have much better potential, simply because of the economics of production.

**Table 3.** Advantages and disadvantages of *s*-alcohol dehydrogenases over lipases for chiral hydroxy compound production

| Factors | *s*-ADHs | Lipases |
|---|---|---|
| Process stages | One | Many |
| Theoretical yield | 100% | 50% |
| Enzyme cost | HIgh | Medium |
| Co-factor requirements | NAD, NADP | None |
| Substrate tolerance | Low–medium | High |
| Current availability | Milligram | Kilogram |

The specifications of *s*-ADH enzymes are very broad and details are given in Table 4 for two microbial sources. Table 4 shows that all secondary alcohols tested were susceptible to one or other of the enzymes. It should also be noted that whereas most *s*-ADH enzymes including HLADH are NADP dependent Biocat *s*-ADH 301 requires NAD as co-factor. This is a very important economic factor, for although the enzymes are expensive, the main current process cost comes from the co-factor requirement. Therefore the ability to use NAD which costs about one-tenth of NADP is very important economically.

*s*-ADH enzymes are still under commercial development, but of equal importance is the development of cheap co-factor recycle

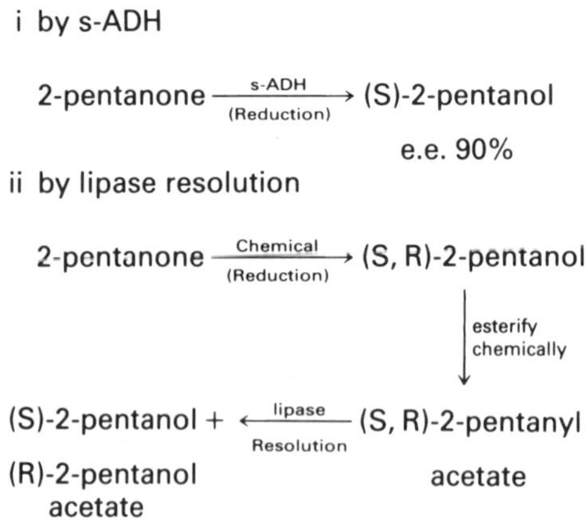


**Fig. 3.** Production of chiral 2-pentanol by *s*-ADH and lipase resolution.

**Table 4.** Relative activity of *s*-ADH on various substrates

| Substrates | Activity[a] | |
|---|---|---|
| | Biocat *s*-ADH 300[b] (Anaerobic thermophile) | Biocat *s*-ADH 301[c] (*Pseudomonas* sp.) |
| 2′-Alkanols | | |
| 2-Propanol | 100 | 100 |
| 2-Pentanol | 156 | 17 |
| 2-Hexanol | 134 | 11 |
| 3′-Alkanols | | |
| 3-Heptanol | 192 | 0 |
| Cyclic alcohols | | |
| Cyclohexanol | 77 | 21 |
| Cyclopentanol | 182 | 220 |
| Diols | | |
| 2,3-Butane diol | 8 | 72 |
| *trans*-1-2-Cyclohexane diol | 0 | 16 |
| *cis,trans*-1-4-Cyclohexane diol | 0 | 72 |
| Others | | |
| *s*-Phenethyl alcohol | 27 | 5 |
| 3-Methyl-2-butanol | 5 | 79 |

[a] Activity against 2-propanol taken as 100% in both cases.
[b] Work done at pH 8·5, 50°C using 10 nM alcohol and 0·8 mM NADP.
[c] Work done at pH 8·0, 50°C using 10 mM alcohol and 0·9 mM NAD.

**Table 5.** Ratio of initial velocities of reaction for *s*-alcohols to *r*-alcohols ($V_s/V_r$), Biocat *s*-ADH 300 and Biocat *s*-ADH 301

| Substrate | $V_s/V_r$ (at 50°C) | |
|---|---|---|
| | Biocat *s*-ADH 300 | Biocat *s*-ADH 301 |
| 2-Butanol | 1·1 | 0·7 |
| 2-Pentanol | 9·7 | 2·1 |

$V_s/V_r$ = ratio of reaction velocities against (*S*) and (*R*) isomers.

**Table 6.** Effects of temperature on the ratio of the specificity constant (kcat/km) for (*S*)- and (*R*)-2-pentanol

| Temperature | Specificity constant (*S*/*R*) | |
|---|---|---|
| | Biocat *s*-ADH 300 | Biocat *s*-ADH 301 |
| 10 | 70 | — |
| 15 | 75 | 5·4 |
| 20 | 81 | 5·1 |
| 25 | 87 | 4·8 |
| 30 | 94 | 4·4 |
| 35 | 103 | 4·0 |
| 40 | 114 | 3·4 |
| 45 | 128 | 2·6 |
| 50 | 143 | 1·4 |
| 55 | 162 | 0 |

systems and the synthesis of artificial co-factors. Great advances have been made in these areas in the past decade and the 1990s should see the first chemical process using a co-factor dependent enzyme. At the moment whole cell systems are favoured where co-factor requiring enzymes are needed.

Although the microbial *s*-ADH enzymes show broad chemical specificity they also show high stereo-selectivity. This is demonstrated in Table 5.

In particular, under the conditions used, Biocat *s*-ADH 300 shows a very high stereo-selectivity for (*S*)-2-pentanol. However stereo-selectivity is not absolute and is dependent upon the conditions used. In particular stereo-selectivity is dependent on temperature, and carrying out the reaction at different temperatures will alter the ratio of isomers obtained. This is shown in Table 6.

The results in Table 6 show that as the temperature is increased Biocat *s*-ADH 300 shows a higher and higher selectivity for (*S*)-2-pentanol whilst Biocat *s*-ADH 301 shows a decreasing activity. At 55°C the latter enzyme shows no stereo-selective preference at all.

## Conclusions

Whilst biocatalysts have had occasional successes in the chemical industry during the past century the level of research into novel biotransformations is now at an all-time high. The most popular biocatalysts are currently lipases and many commercial processes will appear during the 1990s. Enzymes such as secondary-alcohol dehydrogenases will probably become the favourite research biocatalyst of the 1990s and hence will start to be used in commercial processes in the twenty-first century.

## References

1. Holding, T., World Biotech report, in *Proceedings of Biotech 88*, London. Online Publications, May 1988, pp. 221–9.
2. Chetham, P. S. J., *Handbook of Enzyme Biotechnology,* ed. A. Wiseman. John Wiley & Sons, 1986, pp. 336–68.
3. Macrae, A. R., *Microbial Enzyme and Biotechnology,* ed. W. M. Fogarty. Applied Science Publishers, 1983, pp. 225–47.
4. Zaks, A. & Klibanov, A. M., *Science,* **224** (1984) 4654.
5. West, S. I., *Speciality Chemicals,* (April 1988) 410.
6. Schneider, M., *Performance Chemicals* (April 1989) 28.
7. Kloosterman, *et al., TIBTECH,* **6** (1988) 251.

# Biocatalytic preparation of chiral materials

**Neil W. Boaz & Kurt Laumen**

*Eastman Kodak Company, Rochester, New York, USA*

## Introduction

The number of organic molecules prepared via enzymatic catalysis has blossomed in the last several years. The enzymes that are used for these transformations can be in isolated form much as any other catalyst or as part of a whole cell system. These systems can promote a wide variety of transformations, but are especially useful for the preparation of optically active materials due to enzyme enantiospecificity. The widespread acceptance of this type of catalysis will not be realized until the general predictability of these transformations, which often still rely on trial and error biocatalyst–substrate matching, approaches that inherent in mainstream organic chemistry.

There are several methods to enhance the predictability of biocatalytic transformations. The most exact method uses the active site configuration as determined by X-ray analysis as a guide for substrate design and selection. This technique, however, is difficult and time-consuming. A more general approach involves the examination of various substrates that are structurally related to the natural substrate of the desired enzyme to define the steric requirements for successful enzymatic transformation. In this manner, even an early rudimentary active-site 'model' can produce excellent results while providing further refinement of the model. This type of approach can be applied to both isolated enzyme and whole cell transformations, and one of each will be discussed here.

## Enzymatic esterification of ferrocenylethanol[1]

For chemists who are often not accustomed to working with living systems, commercially available isolated enzymes provide simple and attractive biocatalysts alternatives, especially for induction of chirality due to enzyme enantiospecificity. Of these biocatalysts, hydrolases have been the most thoroughly explored due to the ready availability of the enzymes as well as the operational simplicity since no cofactors or additives are required.

In contrast, the marriage between organometallic and biocatalytic chemistry remains largely unexplored, though it affords great potential, especially for the preparation of optically active catalysts and ligands. Though prior work has verified the ability of organometallic species to undergo both microbial[2] and cell-free enzyme[3–5] transformations, the behavior of organometallic species compared to organic materials within an enzyme active site has remained unexplored. The investigation of the applicability of known enzyme active site models to an organometallic substrate such as 1-ferrocenylethanol would both extend the knowledge of the active site configuration and provide useful materials. The recent report by Wong and coworkers on the enzymatic esterification of this compound[4] prompts us to report similar investigations based on active site considerations that afforded improved results.

Reliable active site models are available for several hydrolase enzymes.[6] The lipase from *Pseudomonas fluorescens*[6(d),7] has a highly useful active site working hypothesis.[6(d),(e)] A schematic of this model that is effective for rationalizing both reactivity and enantioselectivity of acyclic compounds is shown below for $\alpha$-phenethyl acetate (**1a**).[6(d)] This active site model contains four points of interaction with the substrate: the catalytically active serine hydroxyl site (located such that the ester function must be two atoms from the substrate chiral center), a large hydrophobic pocket, a small hydrophobic binding site, and a niche that will accept only an $\alpha$-hydrogen. These interactions define the chiral discrimination and direct the *R*-enantioselectivity.

Our investigation sought to determine whether 1-ferrocenylethyl acetate would function as an organometallic analog (albeit a very different-shaped analog due to the large ferrocenyl substituent) of α-phenethyl acetate for this enzyme. α-Phenethyl acetate is the enantioselectivity benchmark substrate for this lipase,[6(d)] with an $E$ value (a directly proportional measure of the enantioselectivity of the reaction based on ratios of $K_m/V_{max}$ of each enantiomer)[8] of >1000. A high degree of enantioselectivity, the measure of success for this investigation, would indicate that the much larger ferrocenyl moiety performs similarly to the phenyl group in **1a** as an anchor in the hydrophobic pocket of the enzyme as depicted in **1b**. Thus 1-ferrocenylethanol provides a useful positive probe for active site interactions with an organometallic compound. In addition, a highly enantioselective kinetic resolution of a 1-ferrocenylethanol derivative followed by an efficient separation would provide an alternative to classical resolution[9] or chromatographic separation[10] for the preparation of chiral ferrocenyl bisphosphines such as BPPFA (**2a**) and BPPFOH (**2b**), highly useful ligands for homogeneous asymmetric catalysis.[11]

As reported by Wong *et al.*,[4] the documented lability of 1-ferrocenylethyl acetate to hydrolysis[12] rendered an enzymatic hydrolysis impossible. However, enzymatic esterification in 'anhydrous' organic solvents,[13] the reverse reaction of enzymatic hydrolysis, was still feasible. This normally unfavorable process is driven by both the lack of water and the use of a vinyl ester as the carrier of the acyl group (for esterification purposes), since upon formation of the acyl-enzyme intermediate from the serine hydroxyl and the vinyl ester,

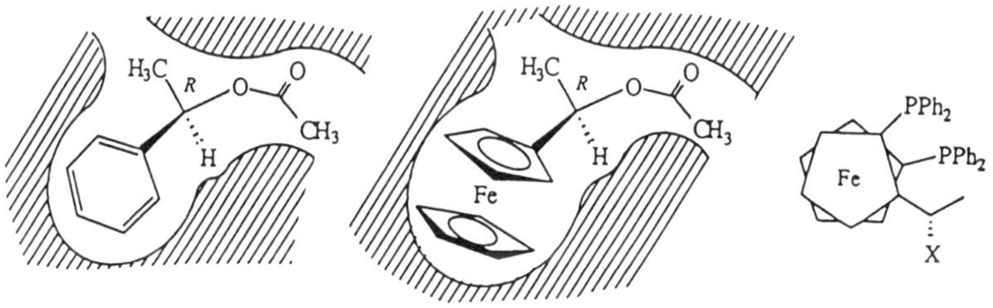


1a 1b 2

a: $X = NMe_2$
b: $X = OH$

**Fig. 1**

the vinyl alcohol liberated (which would normally compete with the alcohol substrate for the acyl-enzyme) isomerizes completely to the corresponding carbonyl compound.[14] This technique of enzymatic esterification has enjoyed rapid development due to the much enhanced solubility of organic substrates in organic compared to aqueous media. Indeed, the air- and water-sensitivity of many organometallic species suggest that this is the method of choice for these materials. Therefore, racemic 1-ferrocenylethanol (**3**) was submitted to enzymatic esterification (lipase SAM-II from *Pseudomonas fluorescens* (Amano International Enzyme Co.), 3 equiv vinyl acetate, organic solvent). The esterification reaction proceeded smoothly at room temperature, asymptotically approaching 50% conversion. This suggested high enantiomeric discrimination—i.e. only one enantiomer reacted (50% of the racemate). Indeed, high enantioselectivity (*E* value of 162) was observed for both the product *R*-acetate *R*-(−)-**4** ($[\alpha]_D^{25}$ −25·5° (c. 1·755, ethanol), lit. for *R*-**4**,[12] $[\alpha]_D^{25}$ −30·5° (c. 1·3, ethanol)) and recovered *S*-alcohol *S*-(+)-**3** ($[\alpha]_D^{25}$ +28·8° (c. 1·165, benzene), lit. for *S*-**3**,[12] $[\alpha]_D^{25}$ +32·0° (c. 1·2, benzene)) as shown in Table 1.[15] Most importantly, the high *E* value of the transformation and the absolute configuration of the reacting *R*-enantiomer indicates that the ferrocenyl moiety, though much larger, is a successful analog for the phenyl substituent. Therefore, a third dimension is available within the hydrophobic pocket to accommodate the ferrocenyl moiety in the active site.

Organic solvents reduce enzyme activity,[13] necessitating extended reaction time even in the optimal organic solvent (*t*-butyl methyl ether for this transformation) to achieve the desired 50% conversion to acetate **4** at room temperature. However, organic solvents also increase enzyme rigidity,[13] allowing the use of elevated temperatures to increase the esterification rate with little erosion of enantioselectivity (Table 1).[16]

**Table 1**

| Temperature | Time (h) | Percent conversion | % e.e. *R*-**4** | % e.e. *S*-**3** |
|---|---|---|---|---|
| RT | 294 | 50·0 | 96 | 92 |
| 40°C | 144 | 49·5 | 92 | 90 |
| 60°C | 96 | 50·5 | 92 | 90 |

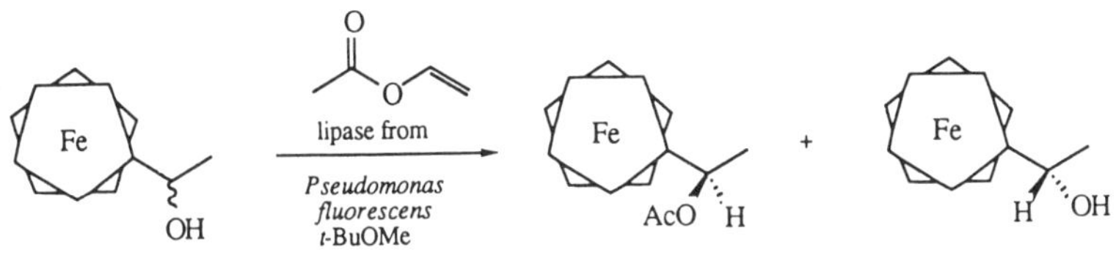


3 R-(-)-4 S-(+)-3

**Fig. 2**

A significant technical problem with enzymatic hydrolyses or esterifications of chiral alcohol derivatives is the separation of the different enantiomers, since the gross physical similarity of alcohol and ester precludes many attractive methods. Since chromatography, the most widely applicable separation technique, becomes cumbersome as the scale of the reaction is increased, other solutions are desirable. In this case the unusual configurational stability and facile formation of the 1-ferrocenylethyl carbonium ion[12,17] allowed selective replacement of the acetate function of **4** in the mixture of **3** and **4** with a dimethylamino substituent (aq. $HNMe_2$, methanol, room temperature) with retention of configuration.[12] This afforded the *R*-*N*,*N*-dimethylamino derivative *R*-(+)-**5** (45%) ($[\alpha]_D^{25}$ + 12·8° (c. 1·21, ethanol), literature for *S*-(−)-**5**,[9(a)] $[\alpha]_D^{25}$ − 14·1° (c. 1·6, ethanol)) and unreacted alcohol *S*-**3** (45%; total yield 90%) without racemization of either species.[12,15] A simple acid extraction (10% aqueous citric acid)

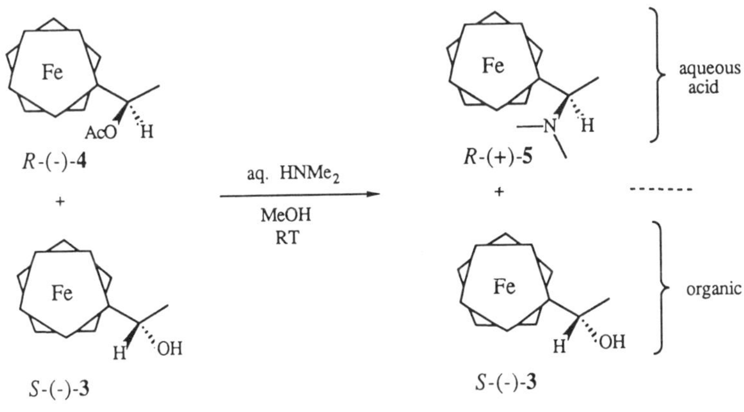


**Fig. 3**

separated the amine from the alcohol, completing the enantiomer separation. Advantageously, the alcohol *S*-**3** could be recrystallized to optical purity, affording maximum utility for this preparation.

Either the alcohol or amine can be elaborated to the optically active bisphosphines mentioned above,[11,18] affording a unique method for the preparation of these chiral ligands while demonstrating the utility of the application of active site considerations to the combination of biocatalytic and organometallic chemistry.

## A facile route to optically pure D-malic acid

Rational approaches to whole cell transformations are often more difficult due to the numerous processes occurring within the cell. Specificities and steric requirements based on studies using various substrates similar to that described above are much rarer, but have in certain instances been performed (e.g. for yeast, specifically yeast alcohol dehydrogenase). When little is known about the enzymes within the cell, a promising approach to the rational preparation of optically pure materials involves the depletion of the natural enantiomer (from a racemic mixture) of a chiral substrate via metabolism (or some side process). This concept of destructive resolution has been successfully applied to the preparation of D-*R*-malic acid (*R*-**6**).

D-Malic acid has been previously prepared by several methods including classical separation of diastereomeric salts of chiral amines[19] and multistep transformation of materials from the chiral pool such as L-asparagine.[20] in addition, some success has been reported on the enantioselective hydrolysis of the distal acid functionality of derivatives of racemic dimethyl malate.[21] However, none of these methods provides the potential for the preparation of reasonable quantities of D-malic acid inexpensively. The metabolism of L-malic acid by many microorganisms suggested that a destructive resolution might be possible. Unfortunately, there was no literature precedent or guidance for selecting a microorganism with the three requirements for success: high (absolute) specificity, ability to use L-malic acid as the sole carbon source, and rapid substrate turnover.

The approach used for the isolation of an appropriate microorganism involved searching for differential growth characteristics using D- and L-malic acid, respectively, as sole carbon sources. Microorganisms exhibiting selective growth on L-malic acid were then further investigated. More specifically, a sample of a microorganism, obtained either from a cell culture bank or from natural sources, was grown in media with D/L-malic acid as the carbon source. After 24 h of growth at 30°C, the media thus obtained was propagated on an agar plate (D/L malic acid as carbon source). After incubation, a small uniform colony was removed and grown on two sets of agar plates, one containing D-malic acid and the other with the L-enantiomer as the sole carbon source. A successful colony selection was indicated by robust growth on the L-plates and no growth (or minimal growth traceable to a small amount of yeast extract) on the plates containing D-malic acid. Two such organisms were found to work admirably, *Pseudomonas putida* ATCC 21244 and a soil isolate, *Acinetobacter calcoaceticus* ATCC 53927.

For preparative experiments, colonies thus isolated were incubated in minimal media with 1% D/L-malic acid (**6**) as the sole carbon source (checked against D-malic acid media control) at 30°C for 16–20 h. The optical purity of the D-malic acid depends upon the complete consumption of L-malic acid and thus continuously viable microorganisms. A potential threat was the rise in pH during the fermentation due to loss of the acid functions of the consumed L-malate, so the fermentation was run under pH control (pH 7) by the addition of aqueous acid. The consumption of acid was also a convenient measure of L-malate consumption, and corroborated with optical density increase (cell density) and dissolved oxygen content as indicators of residual (and consumption of) L-malate. Once growth had ceased (and the theoretical amount of acid had been consumed), the cells were removed by centrifugation and the supernatant was acidified to pH 2 with much gas evolution (presumably $CO_2$). The water was then removed *in vacuo* and D-malic acid ($[\alpha]_D^{20}$ + 8·4° (c. 1·20, $CH_3OH$))[22] was isolated in ca. 35% yield by extraction with acetone. The enantiomeric purity was determined to be >99% by first conversion to dimethyl malate (*R* – **7**) ($HC(OCH_3)_3$, HCl, $CH_3OH$), MTPA esterification of the free hydroxyl to afford **8** (MTPA-Cl, DMAP, $CH_2Cl_2$), and determination of the ratio of diastereomers by capillary vapor phase chromatography.

The *R*-(+) absolute configuration for the product (corroborated by the positive optical rotation) along with the high optical purity observed demonstrate the viability of the rational application of destructive resolution for the preparation of D-malic acid.

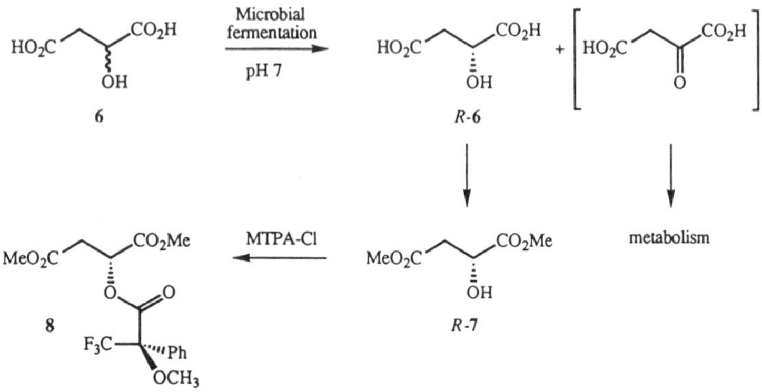

Fig. 4

# Conclusion

Rational approaches to the use of biocatalysis with either isolated enzymes or whole cells offers great potential for the future of these techniques in the preparation of organic molecules, especially those of high optical purity. Further demonstrations of rational design of substrates and systems should enhance the acceptability of biocatalysis to the organic community while providing useful information and materials.

# References and notes

1. Part of this work has been previously published in: Boaz, N. W., *Tetrahedron Letters*, **30** (1989) 2061.
2. (a) Johnson, R. A. & Murray, H. C., *J. Chem. Soc. Chem. Commun.* (1971) 989.
   (b) Ratajczak, A. & Misterkiewicz, B., *Chem. Ind. (London)* (1976) 903.
   (c) Sokolov, V. I., Troitskaya, L. L. & Rozhkova, T. I., *Gazz. Chim. Ital.*, **117** (1987) 525.
   (d) Gillois, J., Buisson, D., Azerad, R. & Jaouen, G., *J. Chem. Soc. Chem. Commun.* (1988) 1224.
   (e) Top, S., Jaouen, G., Gillois, J., Baldoli, C. & Maiorana, S., *J. Chem. Soc. Chem. Commun.*, (1988) 1284.
3. Hydrolysis: Alcock, N. W., Crout, D. H. G., Henderson, C. M. & Thomas, S. E., *J. Chem. Soc. Chem. Commun.* (1988) 746.

4. Esterification: Wang, Y.-F., Lalonde, J. J., Momongan, M., Bergbreiter, D. E. & Wong, C.-H., *J. Am. Chem. Soc.*, **110** (1988) 7200.
5. Oxidation: Yamazaki, Y. & Hosono, K., *Tetrahedron Lett.*, **29** (1988) 5769.
6. (a) Ohno, M., in *Enzymes in Organic Synthesis, Ciba Foundation Symposium III,* ed. R. Porter & S. Clark. Pitman, London, 1984; pp. 171–83.
(b) Ohno, M., Kobayashi, S. & Adachi, K., in *Enzymes as Catalysts in Organic Synthesis,* ed. M. P. Schneider. Reidel, Dordrecht, Holland, 1986, pp. 123–42.
(c) Lam, L. K. P., Hui, R. A. H. F. & Jones, J. B., *J. Org. Chem.*, **51** (1986) 2047.
(d) Laumen, K. E., PhD thesis, University of Wuppertal, Germany, 1987.
(e) Xie, Z.-F., Nakamura, I., Suemune, H. & Sakai, K., *J. Chem. Soc. Chem. Commun.* (1988) 966.
(f) Bjorkling, F., Boutelje, J., Gatenbeck, S., Hult, K., Norin, T. & Szmulik, P., *Tetrahedron,* **41** (1985) 1347.
(g) Boutelje, J., Hjalmarrsson, M., Szmulik, P., Norin, T. & Hult, K., in *Biocatalysis in Organic Media,* ed. C. Laane, J. Tramper & M. D. Lilly. Elsevier, Amsterdam, The Netherlands, 1987, pp. 361–8.
7. (a) Laumen, K. & Schneider, M. P., *J. Chem. Soc. Chem. Commun.* (1988) 598,
(b) Laumen, K., Breitgoff, D. & Schneider, M. P., *J. Chem. Soc. Chem. Commun.* (1988) 1459.
8. Chen, C.-S., Fujimoto, Y., Girdaukas, G. & Sih, C. J., *J. Am. Chem. Soc.*, **104** (1982) 7249.
9. (a) Marquarding, D., Klusacek, H., Gokel, G., Hoffmann, P. & Ugi, I., *J. Am. Chem. Soc.*, **92** (1970) 5389.
(b) Ratajczak, A. & Misterkiewicz, B., *J. Organomet. Chem.*, **91** (1973) 73.
10. (a) Armstrong, D. W., DeMond, W. & Czech, B. P., *Anal. Chem.*, **57** (1985) 481.
(b) Yakahashi, S. & Harada, A., Jpn Kokai Tokkyo Koho JP 61 205 293 (86 205 593), 11 September 1986.
11. (a) Hayashi, T., Yamamoto, K. & Kumada, M., *Tetrahedron Lett.*, (1974) 4405.
(b) Hayashi, T., Mise, T., Mitachi, S., Yamamoto, K. & Kumada, M., *Tetrahedron Lett.*, (1976) 1133.
(c) Hayashi, T., Mise, T. & Kumada, M., *Tetrahedron Lett.* (1976) 4351.
(d) Hayashi, T., Tajika, M., Tamao, K. & Kumada, M., *J. Am. Chem. Soc.*, **98** (1976) 3718.
(e) Hayashi, T., Kaysumura, A., Konishi, M. & Kumada, M., *Tetrahedron Lett.* (1979) 425.
(f) Brunner, H., Schonhammer, B., Schonhammer, B. & Steinberg, C., *Chem. Ber.*, **116** (1983) 3529.
(g) Brunner, H. & Kunz, M., *Chem. Ber.*, **119** (1986) 2868.
(h) Hayashi, T., Yamamoto, A., Hagihara, T. & Ito, Y., *Tetrahedron Lett.*, **27** (1986) 191.

(i) Ito, Y., Sawamura, M. & Hayashi, T., *J. Am. Chem. Soc.*, **108** (1986) 6405.
(j) Ito, Y., Sawamura, M. & Hayashi, T., *Tetrahedron Lett.*, **28** (1987) 6215.
(k) Ito, Y., Sawamura, M., Shirakawa, E., Hayashizaki, K. & Hayashi, T., *Tetrahedron Lett.*, **29** (1988) 235.
(l) Ito, Y., Sawamura, M. & Hayashi, T., *Tetrahedron Lett.*, **29** (1988) 239.
(m) Hayashi, T., Yamamoto, A. & Ito, Y., *Tetrahedron Lett.*, **29** (1988) 669.
(n) Ito, Y., Sawamura, M., Shirakawa, E., Hayashizaki, K. & Hayashi, T., *Tetrahedron,* **44** (1988) 5253.
(o) Hayashi, T., Kanehira, K., Hagihara, T. & Kumada, M., *J. Org. Chem.,* **53** (1988) 113.

12. Gokel, G. W., Marquarding, D. & Ugi, I. K., *J. Org. Chem.*, **37** (1972) 3052.
13. Klibanov, A. M., *Chemtech* (1986) 354.
14. Degueil-Castaing, M., De Jeso, B., Drouillard, S. & Maillazrd, B., *Tetrahedron Lett.*, **28** (1987) 953.
15. For *S*-**3**, the optical purity was determined by $^{1}$H NMR analysis in the presence of the chiral shift reagent Eu(tfc)$_3$ after separation from the ester. The acetate *R*-**4** was hydrolyzed (1:1 acetone/water) to alcohol *R*-**3** with retention of configuration[12] and analyzed as above.
16. A small amount of non-enzymatic esterification at the elevated temperatures as indicated by control reactions may account for part of the observed reduction in optical purities.
17. (a) Turbitt, T. D. & Watts, W. E., *J. Chem. Soc. Chem. Commun.* (1973) 182.
    (b) Turbitt, T. D. & Watts, W. E., *J. Chem. Soc. Perkin II* (1974) 177.
18. Hayashi, T., Mise, T., Fukushima, M., Kagotani, M., Nagashima, N., Hamada, Y., Matsumoto, A., Kawakami, S., Konishi, M., Yamamoto, K. & Kumada, M., *Bull. Chem. Soc. Jpn,* **53** (1980) 1138.
19. (a) Tanekawa, T., Hachiya, T., Yokokawa, Y. & Maruyama, H., Japan Kokai 77 77 011, 1977.
    (b) Nohira, H., Mishikawa, J. & Ishizaki, T., Jpn Kokai Tokkyo Koho 80 33 220, 1980.
    (c) Berning, W. & Siegel, H., Ger. Offen. DE 2 933 895, 1981.
    (d) Saigo, K., Kai, M., Yonezawa, N. & Hasegawa, M., *Synthesis* (1985) 214.
    (f) Sumitomo Chemical Co. Ltd, Jpn Kokai Tokkyo JP 57 56 439, 1982.
20. Ott, E., *Chem. Ber.*, **68** (1935) 1651.
21. Buibe-Jampel, E., Rousseau, G. & Salaun, J., *J. Chem. Soc. Chem. Commun.* (1987) 1080.
22. For comparison purposes, the optical rotation for commercially available L-malic acid is $[\alpha]_D^{20}$ − 9·0° (c. 1·20, $CH_3OH$).

# Advanced technologies for pharma peptide production

**C. Birr**

*Orpegen GmbH, Heidelberg, FRG*

There is an increasing appreciation of the broader applicability of less acid-stable temporary and permanent protecting groups in solid phase peptide synthesis strategies. 3,5-dimethoxyphenyl-2-propyl-2-oxycarbonyl-(Ddz)-amino acids in combination with *tert*-butyl type permanent protections for side functions allows much less aggressive chemical reaction conditions for deprotection and release of peptides from the solid phase[1] than does the Boc-/benzyl strategy.

In the Merrifield synthesis Ddz can be cleaved readily within 10–30 minutes by 1–5% (vol) trifluororacetic acid in dichloromethane.[2] On scaling-up of solid phase syntheses this halo-organic reagent can, for environmental reasons, be substituted by acetic acid or dioxane containing 1% hydrogen chloride gas. Ddz deprotection conditions do not harm side function protecting *tert*-butyl groups. Therefore, *tert*-butyl aspartate and glutamate as well as the *tert*-butyl ether of serine, threonine and tyrosine can be applied. Nitrogen side functions in lysine, arginine, histidine and tryptophane can be protected with Boc residues. For cysteine, *tert*-butylmercaptan is used to form a *S-S-tert*-butyl protecting function. Other Ddz derivatives of trifunctional amino acids like Ddz-Lys(Z), -Lys(Fmoc), -Arg(Mtr), -Arg(Tos), -His(Dnp), -Trp(Form), -Cys(Acm) and -Cys(Mbzl) are also available on bulk scale.[3]

Both the molecular characteristic of Ddz/*tert*-butyl moieties and the lower acid stress of the strategy on the immobilized protected peptides have positive implications on the final product quality and the scalability of solid phase syntheses.

- The lipophilicity and reactivity of commonly activated Ddz/*tert*-butyl amino acids are favourably compatible with protected peptide sequences both on polystyrene and polyacrylamide supports.
- There is a drastically diminished tendency for side reactions on $\beta$-*tert*-butyl aspartate, and also $\gamma$-*tert*-butyl glutamate, due to the steric hindrance of the compact *tert*-butyl ester moiety in the $\omega$-position.
- On cleavage of the temporary Ddz protecting group, this residue creates:

  —Its own scavenger, 3,5-dimethylphenyl-2-propylium cation.[4] This 3,5-dimethoxyphenyl-2-propylium cation has an increased reactivity towards itself and foreign carbonium ions. It neutralizes itself via aromatic substitution and ring closure reaction (Fig. 1).

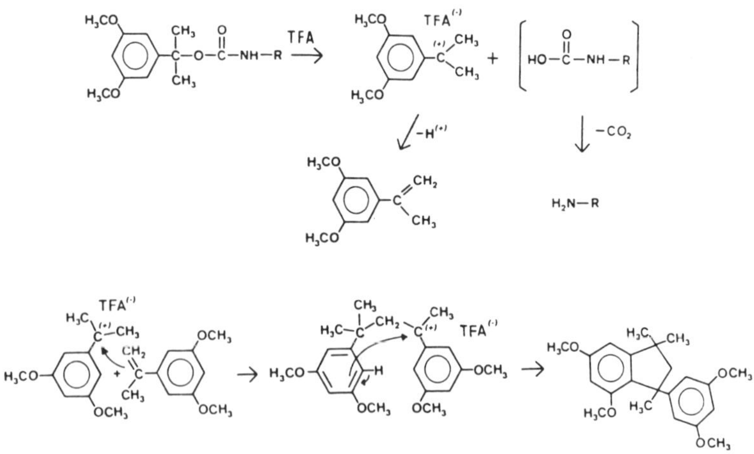

1 : Cleavage by trifluoroacetic acid of the Ddz-amino protecting group

2 : Formation of 1,3,3-trimethyl-1-(3,5-dimethoxy)-phenyl-5,7-dimethoxyindan

**Fig. 1.** The acid-labile 3,5-dimethoxyphenyl-2,2-propyloxycarbonyl-(Ddz) protecting group and its inherent scavenger principle.

—A photometric tool to determine the deprotection efficiency.[5] This is because Ddz has a very characteristic UV absorption pattern with maxima at 230, 276 and 282 nm. Thus the Ddz fission product allows a very precise determination both of the initial loading on the polymer phase of Ddz-amino acid derivatives and of the growing Ddz-peptide chain on the polymer. Also in this context, scale-up procedures with regard to amino acid concentrations in solution and wash-out efficiency have been controlled precisely in SPPS since 1972[6] (Figs 2 and 3).

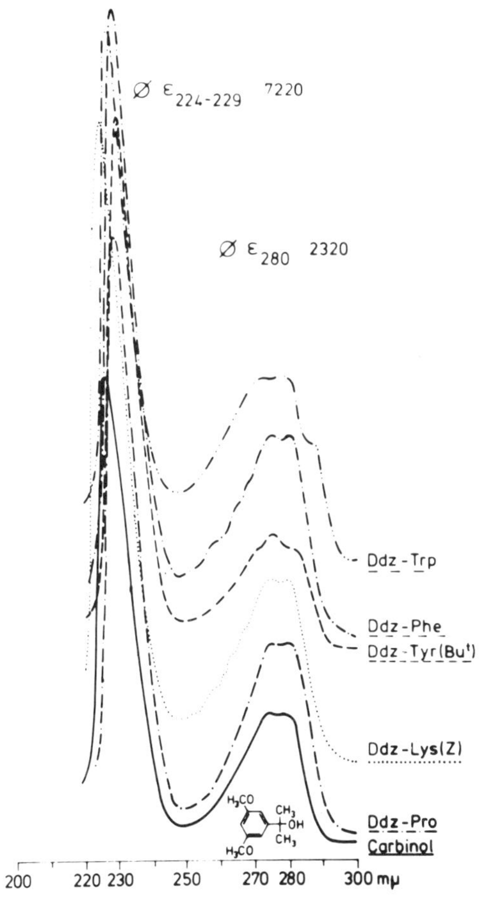


**Fig. 2.** Spectroscopic properties of Ddz-amino acids do not interfere with aromatic side functions.

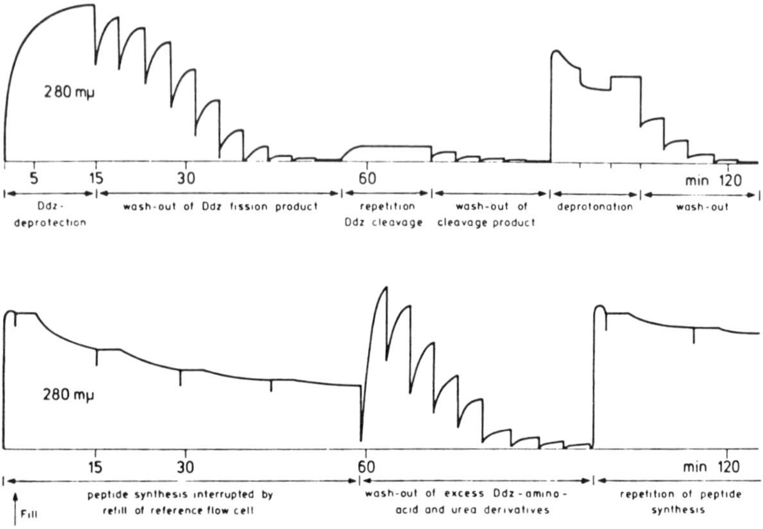


**Fig. 3.** Continuous flow photometric monitoring in SPPS with Ddz-amino acids.[5]

- For the C-terminal anchor and a linker moiety on solid support the Ddz/*tert*-butyl strategy allows a much higher chemical flexibility, as compared to the Boc/benzyl scheme. There are various procedures available for release of fully protected Ddz-peptides and fragments via hydrazinolysis, aminolysis or hydrolysis.[7] These possibilities very much enhance the variability of a synthesis plan for the production of pharma peptide analogues via modified fragments in the polymer phase (Fig. 4).

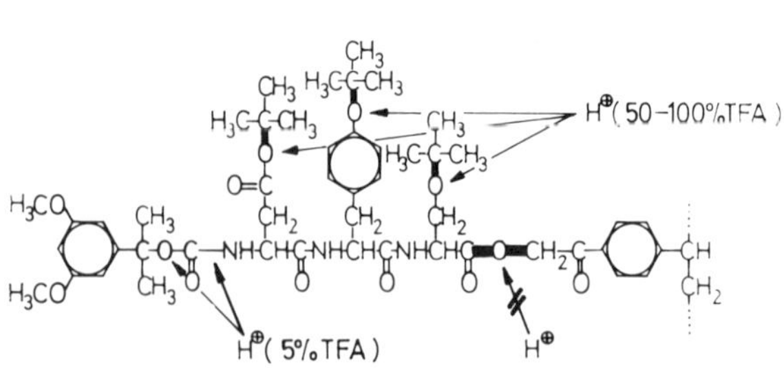


**Fig. 4.** The relation of bond strengths of Ddz/*tert*-butyl protecting groups with an electrophilic C-terminal linker function in SPPS.

A discussion on the scale-up potential with Ddz/*tert*-butyl in comparison to Boc/benzyl strategy has to consider the costs of starting materials vs final peptide raw product quality before purification.

It is our general experience that peptides made up from Ddz/*t*-butyl amino acids after detachment and deprotection have a much higher raw product quality than the same structures synthesized by the Boc/benzyl strategy on a regular Merrifield support. In other words, higher initial costs on starting materials are compensated for by less cumbersome purification methods and higher final yields of peptide.[8]

Competing with Ddz in terms of a solid phase strategy are the Fmoc-amino acids. Introduced by Carpino in 1972,[9] this urethane type of base-labile temporary protecting group was only finally accepted much later by solid phase chemists as a very useful alternative to the Boc residue in the Merrifield synthesis. Fmoc is stable to acid treatment and can be cleaved with 20–50% piperidine/dimethylformamide. It is not stable to hydrogenolysis. Therefore, *tert*-butyl type permanent protecting groups are advantageously applicable.

In solution phase synthesis, an occasional experience is that deprotection of Fmoc residues with base is accompanied by side reactions (which have not been fully elucidated). Although the general cleavage principle in Fmoc is a base driven $\beta$-proton elimination from the fluorenyl bridge carbon, followed by the release of the N-terminal carbaminate, there is a tendency for the cleavage in the opposite direction. This results in the transformation of the Fmoc-amino group into an isocyanate with the release of fluorenylmethanol. Of course, an isocyanate moiety can further react, resulting in several by-products[10] (Fig. 5).

A curious and critical investigator can identify the reality of Fmoc cleavage in solution simply by spotting the reaction mixture on to a TLC plate for separation and identification.

If an isocyanate intermediate were to be generated by Fmoc cleavage in solid phase reactions, it would be invisible in a reaction solution because of its immobilized nature. On the other hand, isocyanate by-products in the solid phase would reduce the amount of growing peptide chain. This, however, is not a regular experience in the operation of the Fmoc/*tert*-butyl strategy in Merrifield syntheses.

There is an increasing body of information on the successful application of this strategy for the preparation of sensitive pharmaceutical peptides. In particular, this success is due to the com-

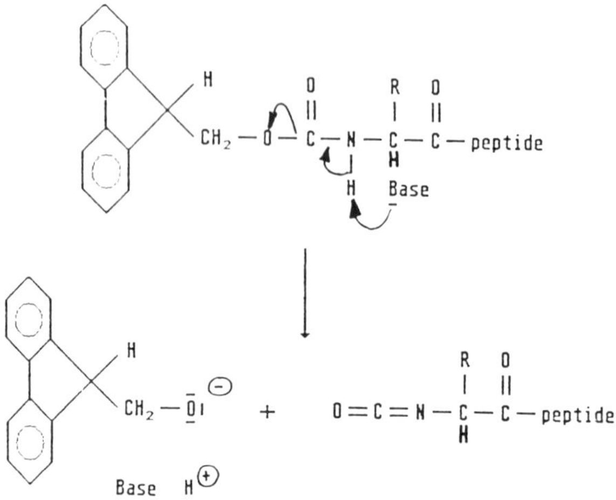


**Fig. 5.** Base catalysed isocyanate formation from Fmoc-aminoacyl peptides.

patibility of the basic Fmoc cleavage with bond stabilities at C-terminal anchoring functions and *tert*-butyl side chain protections.[11]

The base lability of the temporary Fmoc protecting group allows utilization of acid labile C-terminal linker moieties which can then be cleaved simultaneously with *tert*-butyl side chain protections. There is no need for hazardous chemicals like liquid hydrogen fluoride and trifluoromethanesulphonic acid to be used. Trifluoroacetic acid as used in the competing Ddz/*tert*-butyl strategy can be substituted by HCl gas in acetic acid or dioxane for final entire deprotection reactions. As with Ddz, the UV absorption characteristics of the Fmoc residue can be utilized to determine and control Fmoc-amino acid derivative concentrations in a solid phase synthesis.[12] Opposite to Ddz, however, the UV pattern of Fmoc interferes with absorption maxima of aromatic amino acids like phenylalanine, tyrosine, histidine and tryptophane. With Ddz, none of these aromatic amino acid side chains interact with the characteristic absorption maxima of the protecting group (*vide supra*).

On deprotection of Fmoc, the efficiency of the reaction can be controlled photometrically as already described for the Ddz residue. Loading capacities with Fmoc-amino acids on polymer or Fmoc-peptides on support can also be determined via Fmoc photometry.

Repetitive application of 20–50% piperidine in dimethylformamide for deprotection of Fmoc is, on the other hand, a definite permanent stress for the optical integrity of the $\alpha$-carbons of peptides. In addition, there is a tendency for base catalysed rearrangement with the Fmoc strategy. This latter aspect does not allow the use of reactive esters like benzyl, phenacyl, methyl and allyl for the protection of carboxylic side functions. Even $\beta$-*tert*-butyl aspartate shows an aspartimide rearrangement with alanyl or glycyl if they are C-terminally the next residues in the sequence. However, these base catalysed side reactions do not play a major role in the Fmoc/*tert*-butyl strategy and often can be neglected. In summary, the Fmoc advantages for the final deprotection of peptides and their acid sensitive detachment from polymer phase more than compensate for minor side reaction tendencies due to base catalysed rearrangements. We should bear in mind, however, that the optical purity of Fmoc-made peptides has to be checked carefully in the final product. It should be mentioned that the properties of Fmoc result in an interesting orthogonal protecting scheme for lysine derivatives, namely Boc-Lys(Fmoc) and Ddz-Lys(Fmoc).[3] Standard derivatives are Fmoc-Lys(Boc) and Fmoc-Lys(Ddz).[3]

Compared with the Boc/benzyl strategy, the overall quality of peptides made from Fmoc/*tert*-butyl amino acids more than compensates for the significantly higher costs of starting materials. As with Ddz/*tert*-butyl amino acids, the Fmoc-amino acid derivatives are all available in bulk quantities. The prices on the average are 25% lower than for Ddz derivatives. The ratios of costs over the range as averaged for all protected L-amino acid derivatives at kilogram quantities are:

$$\text{Boc}:\text{Fmoc}:\text{Ddz} = 1:3:4.$$

A major advance in peptide production has been made by the development of more advanced linker technologies in the solid phase process.

Compared to the costs and availability of the original chloromethylated polystyrene, the initial Merrifield resin, most of the more advanced linker-modified polymers are expensive, not available in bulk and sometimes exotic. Most commonly used is the benzhydrylamine-type polystyrene for the solid phase synthesis of peptide amides.[13] This linkage, however, requires the use of liquid HF or trifluoromethanesulphonic acid as detachment reagents, with all their damaging aggressiveness for those side functions like —OH,

—SH and —COOH, which are more sensitive to proton-induced reactions than a C-terminal amide.

The other most acid stable linker has 4-oxomethylphenylacetamidomethyl as a spacer grouping on polystyrene. This C-terminal benzyl ester-type linker (PAM resin[14]) on the one hand gives the polymer solvation properties better compatible with a peptide to be built up on the support, and on the other hand enforces the C-terminal ester against acidic attack followed by chain losses during repetitive cycles for Boc deprotections. For the final release of a peptide product from the polymer, application of liquid hydrogen fluoride or trifluoromethanesulphonic acid is required.

Because the hazardous chemicals these technologies use are inhibitory to large scale application of solid phase technology, other linkers were developed to overcome the limitations of the Boc/benzyl strategy with respect to the C-terminal polymer anchorage. In 1973, Wang presented a more acid labile benzyl ester-type linker[15] containing a phenoxy grouping. More recent developments on suitable supports combine the chemical characteristics of the linker molecule with Fmoc/*tert*-butyl strategy. Anchoring functions labile to acetic acid or low concentrations of trifluoroacetic acid help to release fully protected Fmoc/*tert*-butyl-peptides from the support as amides or free carboxylic acids. This then allows complete deprotection of the fully protected peptide after fragment condensation. All of the linker structures have as a general principle the introduction of methoxy or phenoxy functions into a benzyl alcohol or benzylamine groups to increase the acid lability at the C-terminal anchoring bond,whether it is of the benzyl ester, benzylamide or benzhydrylamide type. It is worth mentioning that all of these linker principles are not restricted to polystyrene and are applicable in polyacrylamide supports as well[16–18] (Fig. 6).

The most recent development by Kunz for a generally applicable linker principle is compatible with all current SPPS strategies.[19] It therefore has the best potential to revolutionize the utilization of the Merrifield synthesis for pharmaceutical production in the polymer phase. The new anchoring moiety, HYCRAM™,[20] contains a 4-hydroxycrotonoylamidomethyl grouping bonded to aminomethylpolystyrene or an aminomethylpolyacrylamide by spacer molecules like $\beta$-alanine, sarcosine, etc.

Any Boc-, Ddz- or Fmoc-amino acid caesium salt can be esterified with this 4-hydroxycrotonoylamidomethyl linker group via the 4-

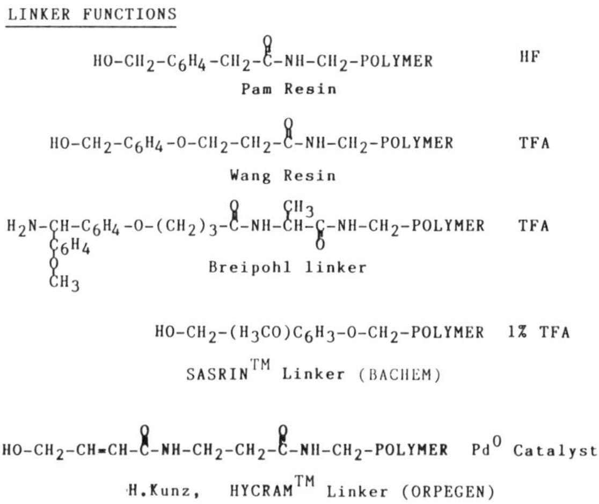


**Fig. 6.** Linker functions in SPPS. Only the HYCRAM™ on polymer needs no hazardous and environmentally difficult chemicals for the release of fully protected peptides.

bromocrotonoylamidomethyl pre-stage, forming an allylic ester on the polymer phase. Merrifield syntheses are then performed as usual, depending on the strategy selected. The allylic ester anchor on the polymer is stable to all acidic and basic conditions of the Boc/benzyl, the Ddz/*tert*-butyl and the Fmoc/*tert*-butyl protocols.[21]

For release of fully protected synthetic peptides from the HYCRAM™ support, the allylic linker polymer is activated in neutral conditions by the noble metal catalyst palladium (0) tetrakis (triphenyl-phosphane). The catalyst delocalizes the *p*-electrons of the allylic double bond. In this way a carboxylate anion at the C-terminus of the peptide is pushed out of the allylic ester group on the polymer, resulting in detachment of the still fully protected peptide. Additives like morpholine, dimedone or *N,N'*-dimethyl barbiturate as acceptor molecules take over the allylic group.[22] Solvents for this soft but complete deprotection technology are mixtures of 50% dimethylsulphoxide with dimethylformamide, *N*-methylpyrrolidone, toluene, dichloromethane, tetrahydrofuran and water. Oxygen has to be excluded during the noble metal catalysed detachment reaction. Even glycopeptides can be produced by the HYCRAM™ technology (Fig. 7).

Fully protected peptide carboxylates after separation from the

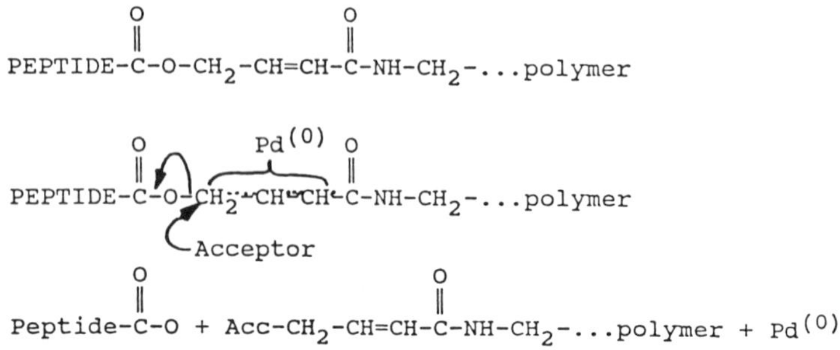

**Fig. 7.** The mechanism of the soft release of fully protected peptide acids from HYCRAM™ support. The peptide ester linkage becomes activated by Pa(O) catalyst followed by allyl acceptor mediated release of peptide at natural pH.

polymer phase can be converted into amides or other intermediates of interest by conventional solution methods. The fully protected peptides can also be used in subsequent fragment condensation strategies (Table 1).

With this new technology, environmentally difficult and hazardous chemicals are no longer needed to detach peptides from the polymer

**Table 1.** Computer assisted manufacturing (CAM) of pharmaceutical peptides in SPPS via HYCRAM^tM release technology

| Pharma peptide | Yield (%)[c] on HYCRAM™ resin | Peptide released from HYCRAM™ | |
|---|---|---|---|
| | | Raw (%)[c] | Final (%)[c,d] |
| LHRH[a] | 90 | 77 | 15 |
| Eledoisin[b] | 92 | 72 | 12 |
| Somatostatin[a] | 85 | 82 | 12 |
| Thymopentin[b] | 93 | 85 | 28 |
| ANP[b] | 85 | 78 | 8 |
| Thymosin-α 1[b] (13–28) | 80 | 68 | 24 |
| Calcitonin[b] | 78 | 70 | 3 |

[a] Synthesis strategy: Ddz/*tert*-butyl.
[b] Fmoc/*tert*-butyl.
[c] Yields based on initial load on polymer.
[d] Deprotected, purity >98% by HPLC.

phase. HYCRAM™ therefore fulfills all of the best requirements for pharamceutical production purposes, and is compatible with all available types of polymer phase and all current peptide synthesis strategies.

In a discussion on the scale-up potential of solid phase technology, a short view on instrumentation cannot be omitted.

Unfortunately, the most advantageous computer assisted manufacturing system for peptide synthesis has not yet been applied to production equipment. On the other hand, such equipment, operating on a chemical feedback principle would, on a production scale, be merely a tool to run the instrument automatically, not to optimize the process. Optimization has to be performed in the product development phase at small scale, resulting in precise specifications for each production cycle of the peptide to be produced.

All except one of the currently available commercial peptide synthesizers are tools for research and development. With respect to production potential, they are limited to the milligram to gram scale. All execute a preprogrammed cycle of operations in a Merrifield synthesis fully mechanically and operate more or less unattended. None of the instrumental systems on the market is fully automatic, because there is no direct chemical feedback from the ongoing peptide synthesis to modify the performance of the instrument. Various reactor types like VORTEX flasks, shakers and flow-through columns are used in the instruments.

Only one machine can deal with 100 grams and more of solid support and the corresponding volumes of solvents and reagent solutes. Its reactor is a large rocking round bottomed flask containing a sintered glass filter at the base.[24] Some of the instruments, introduced to the market more recently, are restricted exclusively to operation of the Fmoc/*tert*-butyl strategy and to use of polyacrylamide supports.

Although 100% transformations on the immobilized growing peptides are the categoric imperative of the Merrifield synthesis (*vide supra*) none of the currently marketed synthesizers has a final technical solution to this requirement. Nevertheless, for the pharmaceutical production of peptides, the solid phase technology that is performed by a mechanical system is very welcome.

Currently, several pharmaceutical companies produce their peptides in the solid phase with systems built up by in-house engineers. There is a major difference in the technical sophistication of commercial peptide synthesizers compared to the simplicity of pilot solid phase peptide production plants. It is most surprising that several pharmaceutical peptides are produced in a very conventional way by Boc/benzyl strategy, neglecting all the recent advances in methodology mentioned elsewhere in this article. Mainly, this is because documented drug master files required for GMP compliance do not readily allow alteration of a production procedure.

Another reason is the bulk availability of cheap Boc-amino acids and chloromethyl-polystyrene. Contrary to most of the scientific contributions, very highly loaded polymer supports are used in production plants, on average containing 2–4 mmol/g. This is due to the economic requirement that a minimum of volume should result in a maximum of product.[25]

From a more general point of view on the production of pharma peptides, mechanized Merrifield syntheses are very well supported by the GMP requirements of the FDA or its European counterpart, the EEC GMP guidelines. For a specific peptide product a mechanical solid phase synthesis can be more precisely recorded with regard to all of the technical operations than can a manual process. The validation of each chemical operation for production purposes fulfills the requirement for optimization at the same time. The less manual work that is involved, the better and easier a production process can be performed in compliance with GMP rules. Protocol recordings of all operational steps in combination with analytical control certificates satisfy these GMP requirements even though the mechanical procedures are not yet fully automatic.

There remains the need for computer assisted manufacturing (CAM) of solid phase peptides at industrial scale[26] (Table 1). Currently there is only one development that fits into this perspective. Computer assisted control of a solid phase peptide synthesis can be based on the chemical feedback from a dimethoxytrityl reaction with remaining free amino functions in polymer phase in an automated control system introduced by Horn in 1984.[27]

The controlling trityl reagent is applied in thousandfold excess related to the content of free amino groups in the polymer phase. The reagent binds covalently to primary amino functions. All excess and adsorbed trityl is washed off up to a photometrically zero level. Dimethoxytrityl is then released quantitatively by the action of 0·5% trichloroacetic acid in dichloroethane and measured quantitatively in a flow photometer. The computerized system compares the measured absorbance with a pre-set yield expectation, i.e. 99%. If this value is passed, the system allows the mechanical synthesizer to continue with the pre-programmed cycle of operations. If, however, the chemical control signals insufficient transformation, this is fed back into the synthesizer, which then repeats the uncompleted step. More chemical monitoring as above continues until the feedback system results in a photometric pass value. The maximum number of repetitions the system will perform is determined by the operator. With a yield expectation of 99·5%, a deviation of 0·5% from 100% transformation can be determined to a precision of ±0·05%.

## Summary and perspectives

The transition from laboratory to production scale SPPS is clearly marked by some specific achievements.

The Fmoc and Ddz/*tert*-butyl strategy in the solid phase have proved to be less prone to side reactions than with Boc/benzyl amino acid derivatives.

On-line records directly from production scale GMP operations can be obtained by UV monitoring of Ddz- or Fmoc-amino acids in the course of the progressing SPPS. The only production scale apparatus on the market[24] allows connection of a flow photometer for this purpose. All operational features of this instrument are very well suited for kilogram-scale peptide production. For example, metering of solvents and washing the resin can be performed at the same time when pre-activation is executed in a separate flask, resulting in more economic production times. For efficient recycling of solvents, different waste lines can be utilized to separate the various effluents.

The HYCRAM technology of peptide detachment from the polymer support by noble metal catalysed cleavage of an allylic ester bond at neutral pH can be executed directly in the SPPS reactor as a final pre-programmed step of the production process. This technology no longer requires any hazardous and environmentally difficult chemicals at pharmaceutical production scale. Fully protected peptide acids are harvested directly out of the process without additional manipulation of the solid phase. The peptides can be purified, characterized, converted into amides or other analogues and can be utilized for further fragment condensations before final deprotection of side functions.

In this respect the recent development of the TBTU reagent[28] for direct activation of protected amino acid derivatives and fragments represents another break-through in peptide production. This method of activation via HOBt esterification[29] *in situ* is reported to be entirely free from racemization. In addition, it allows direct recovery and recycling of excess amino acid derivatives or fully protected fragments from the waste filtrates of SPPS at production scale.

There is no doubt that soon these advances will, all together, result in pharmaceutical peptide products of reliable quality at reasonable cost, and consequently enhance national health care programmes.

## References


1. Birr, C., *Aspects of the Merrifield Peptide Synthesis.* Springer, Heidelberg, 1978.
2. Voss, C. & Birr, C., *Hoppe-Seyler's Z. Physiol. Chem.*, **162** (1981) 717.
3. Enquire at ORPEGEN GmbH, Czernyring 22, D-6900 Heidelberg, Germany.
4. Schmitt, B. & Birr, C., *Chem. Letters Jap.* (1980) 1005.
5. Birr, C., in *Peptides 1974,* ed. Y. Wolman. Wiley, New York, 1975, pp. 117–22.
6. Birr, C., *Liebigs Ann. Chem.*, **763** (1972) 162.
7. Birr, C. & Wengert-Müller, M., *Angew. Chem. Int. Ed. Engl.*, **18** (1979) 147.
8. Birr, C. & Pipkorn, R., *Angew. Chem. Int. Ed. Engl.*, **18** (1979) 536.
9. Carpino, L. A. & Han, G. Y., *J. Org. Chem.*, **37** (1972) 3404.
10. Birr, C., unpublished investigations.
11. Chang, C.D., Felix, A. M., Jimenez, M. H. & Meienhofer, J., *Int. J. Peptide Protein Res.*, **15** (1980) 485.
12. Sheppard, R.C., *Chem. Britain,* **19** (1983) 402.
13. Pietta, P. G. & Marshall, G. R., *J. Chem. Soc. D* (1970) 650.
14. Mitchell, A. R., Erickson, B. W., Ryabtsev, M. N., Hodges, R. S. & Merrifield, R. B., *J. Am. Chem. Soc.*, **98** (1976) 7357.
15. Wang, S.-S., *J. Am. Chem. Soc.*, **95** (1973) 1328.
16. Atherton, E., Caviezel, M., Fox, H., Harkiss, D., Over, H. & Sheppard, R. C., *J. Chem. Soc. Perkin Trans. I* (1983) 65.
17. Breipohl, G., Knolle, J. & Stüber, W., *Tetrahedron Letts,* **28** (1987) 5651.
18. Mergler, M., Tanner, R., Costeli, J. & Grogg, P., *Tetrahedron Letts,* **29** (1988) 4009.
19. Kunz, H. & Dombo, B. *Angew. Chem. Int. Ed. Engl.*, **27** (1988) 711; German patent Appl. P 37 20269.3, 18 June 1987.
20. Registered trademark ORPEGEN GmbH, Czernyring 22, D-6900 Heidelberg, Germany.

21. Becker, G., Nguyen-Trong, H., Birr, C., Dombo, B. & Kunz, H., in *Peptides 1988,* ed. G. Jung & E. Bayer. DeGruyter, Berlin 1989, pp. 157–9.
22. Birr, C., Becker, H., Nebe, C. T., Nguyen-Trong, H., Müller, T. & Schramm, M., in *Peptides. Proc Tenth Am. Peptide Symposium, San Diego,* ed. J. Rivier & G. R. Marshall. ESCOM, Leiden, 1989, pp. 1031–3.
23. Kunz, H., Dombo, B. & Kosch, W., in *Peptides 1988,* eds. G. Jung & E. Bayer. DeGruyter, Berlin, 1989, pp. 154–6.
24. Advanced ChemTech Model 400, Louisville, USA, for Europe: Zinsser Analytical, Frankfurt, Germany.
25. Epton, R., Wellings, D. A. & Williams, A., *Reactive Polymers,* **6** (1987) 143.
26. Horn, M., Novak, C. & Birr, C., in *Peptides 1988,* ed. G. Jung & E. Bayer. DeGruyter, Berlin, 1989, pp. 235–7.
27. Horn, M., US Patent 4 701 304, 1985; US Patent Appl. 06/87 513.
28. Knorr, R., Trzeciak, A., Bannwarth, W. & Gillessen, D., *Tetrahedron Letts,* **30** (1989) 1927.
29. König, W. & Geiger, R., *Chem. Ber.,* **103** (1970) 788.

# Poly(sulphur nitride)—a polymeric metal—and related conducting materials

**A. J. Banister**

*Department of Chemistry, University of Durham, UK*

Poly(sulphur nitride) holds a special place in chemistry because it is a true metal which contains no metal atoms.[1] Moreover it is a fibrous metal in which there is a weak but important bonding interaction between the polymer chains. This prevents the conductor/insulator ('Peierls') transition that normally occurs, on cooling, for other true one-dimensional conductors; the steady improvement in conductivity, towards eventual superconductivity at 0·27 K, is the property that shows it to be a genuine metal, and it is the odd-electron thiazene unit [S=N] that can be regarded as the pseudo-metal atom.

The discovery in 1975 of the metallic properties of $(SN)_x$ so stimulated further work that the first and third most cited chemistry references[2] during that year were the notes on its superconductivity and low temperature specific heat published by Greene and Street at IBM. The subsequent research revealed a quite remarkable combination of properties.

In texture it is a soft fibrous polymer like asbestos and rather 'springy' when deposited (by sublimation or electrolysis) on a soft material such as a platinum electrode; it then easily peels off as flakes. It adheres much better to rigid materials like glass, especially when thin (below ~1 $\mu$m). Unlike the tetramer $S_4N_4$, explosion is not a significant practical hazard. Detonation requires a hard blow with a hammer on a sample placed on a block of metal, and this detonation becomes sluggish with surface treatment. We have even made pellets of treated polymer in a press under 30 kbar (~30 000 atmospheres) pressure.[3]

Poly(sulphur nitride) is not only scientifically intriguing, it is a beautiful material. It is dull gold in colour and very thin films (less than ~0·2 μm) are a fine royal blue by transmitted light.[1] It has typical metallic reflectance, with a sharp cut-off (the 'plasma edge') at ~500 nm (at the red end of the visible spectrum), hence the gold colour. The linear chain structure of $(SN)_x$ (Fig. 1) leads to striking anisotropy in all its physical properties, for instance partially polarised light emerges after reflection because light polarised parallel to the fibres is reflected preferentially.[1] This results in the tiny domains in $(SN)_x$ layers (single crystal grains in which all constituent fibres are parallel) becoming visible under a polarising microscope.

We now turn to thermal and electrical properties. On warming *in vacuo* to 130°C (~110° for the powder), the polymer readily vaporises to $S_3N_3$ radicals.[4] Little energy is required for vaporisation and so this

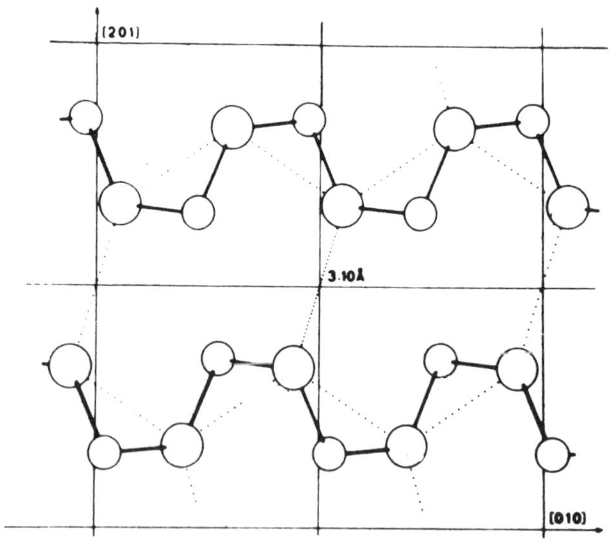


**Fig. 1.** Projection of $(SN)_x$ structure on [102] which shows layer-like packing of chains. (From Ref. 1.)

can be accomplished with a 'clean-cut' edge, using a low energy laser or electron beam.[3] Remarkably the $S_3N_3^{\cdot}$ radicals recombine on a water-cooled cold finger to regenerate $(SN)_x$ with an approximately 90% recovery. Since the chains grow parallel to the surface, even layers of polymer are readily obtained (so long as the underlying surface is clean) and the coating often penetrates under masking materials.[3]

Films of $(SN)_x$ obtained by the classical procedure[1] have a conductivity[3] of about 70 S $cm^{-1}$. The conductivity of crystals (usually 1000–3000 S $cm^{-1}$) increases tenfold between 1 atm and 20 atm pressure; this effect is unique for nontransition metals.[1]

Among the most remarkable and potentially useful properties of $(SN)_x$ are the high work function (ionisation energy) and the high electronegativity of 3·0,[5] contrasting with values 1·5 (Al) to 2·5 (Au) for other non-reactive metals; in other words it does not easily lose electrons and moreover very readily accepts further electron charge. Fortunately it is only slightly air-sensitive (for instance, superior to iron in this respect). Both crystals and layers, when pure, lose their shiny surface slowly. The tarnishing is noticeable after 2–3 weeks and the hydrolysis product is easily polished off. From i.r. evidence,[21] it appears that when $(SN)_x$ layers are exposed to moisture, hydrolysis occurs almost entirely at chain ends. Perhaps, when explosion occurs on severe percussion, decomposition progresses similarly along the chains; this would explain why chain terminal $Me_3Si$ groups reduce the shock sensitivity of powdered $(SN)_x$.[18] It is important to note that all physical and chemical properties (e.g. air stability) are much dependent upon polymer purity and the state of subdivision.

Proposed applications (patents are largely from the United States and Japan) include its use in electroconductive resins,[6] battery electrodes,[7] $(SN)_x$ surface modified electrodes for electrocatalysis (e.g. direct reduction of acetylene)[8] and as ion-sensitive electrodes (e.g. for $Ag^+$),[9] in Schottky barrier devices,[10,11] integrated optical systems,[12] explosion initiators,[13] non-linear optical films,[23] in the exterior magnetic shielding of electrical equipment (using $Br_2$-doped $(SN)_x$)[24] and for information recording.[14] The most developed applications appear to be those using $(SN)_x$ layers, e.g. as electrodes. There are several patents on high output light-weight batteries which use $(SN)_x$ for one or both electrodes.[7] We regularly use a polished $(SN)_x$ crystal micro-electrode for electrochemical studies in acetonitrile and liquid $SO_2$.[3] Interestingly, in cyclic voltammetric studies the pure $(SN)_x$

single crystal electrode behaved like platinum. In acetonitrile with suitable support electrolyte no electrode process was observed between +3·5 and −2·5 V (versus SCE).[3]

Perhaps the most impressive applications of $(SN)_x$ are in electronic (Schottky barrier) devices, for which the most effective metal component is usually gold. There is a dramatic improvement when the gold is replaced by $(SN)_x$: a 40% increase in quantum efficiency in the case of gallium arsenide solar cells[10] and a hundredfold increase in the high energy blue radiation from LEDs based on zinc sulphide.[11] Since devices require a fraction of 1 $\mu$m of (high electronegativity) metal, the cost of even the classical polymer, approximately 50 pairs $mg^{-1}$, is not prohibitive as 1 mg will cover over 40 $cm^2$ at 0·1 $\mu$m thickness (assuming no losses during the coating process). Other uses will no doubt emerge from a wider examination of physical and chemical properties; chemical studies of $(SN)_x$ are limited to its use as a source of $S_3N_3$ radicals in the synthesis of [Ph CNSSN] [SNSNSN].[25] Shorter chain polythiazenes have been patented as high sensitivity photoconductors for use in electrophotographic plates, and as pigments for electrophoretic imaging.[15]

Since the classical preparation is inconvenient, expensive, a trifle dangerous (highly purified $S_4N_4$ is explosive) and tedious (the polymerisation step takes 6–8 weeks), alternative syntheses have been much studied. There are about a dozen new preparations worldwide, half of them from Durham (three are patented).[16]

The route that is probably most amenable to scale-up is the azide route (Fig. 2); the polymer precipitates from solution as a black powder. It is air-sensitive and should be handled in a glove-box. The first method[17] is particularly simple but the purity of the latter product (~95%) is much superior. Exhaustive extraction with liquid $SO_2$ removes most contaminants[18] though not NaCl. Powdered $(SN)_x$ can be obtained in >99% purity by chemical or electrochemical reduction of $S_5N_5^+$ salts.[19]

Probably the most suitable routes to $(SN)_x$ for devices are polymerisation of $S_2N_2$ or electrochemical reduction[1] of S/N cations, especially

$$S_2Cl_2 + NaN_3 \xrightarrow[CH_3CN]{} NaCl + (SN)_x + N_2 \qquad (1)$$

$$(NSCl)_3 + Me_3SiN_3 \longrightarrow Me_3SiCl + (SN)_x \qquad (2)$$

Fig. 2.

$$NH_4Cl/S_2Cl_2 \xrightarrow{\text{reflux}} S_3N_2Cl_2 \xrightarrow{Cl_2} (NSCl)_3$$

$$(NSCl)_3 \xrightarrow[CH_3NO_2]{Fe} S_5N_5FeCl_4 \xrightarrow{CsF} S_5N_5Cl$$

**Fig. 3.**

$S_5N_5^+$. Such $S_5N_5^+$ salts can be prepared[19] without the use of explosive intermediates ($S_4N_4$ and $S_2N_2$ in the polymerisation method), for example as shown in Fig. 3.

The electrochemical method has three main attractions:

—high purity product
—safety (the intermediate $S_5N_5^+$ salts are not explosive and can be stored indefinitely)
—simplicity; the $(SN)_x$ can be deposited in a controlled manner directly on the device or other conducting or semiconducting substrate.

It does not seem that any problems of synthesis have inhibited the flow of patents from the United States and Japan. Is this another area of industrial value hardly recognised in Europe?

Looking to the future, it will no doubt be possible to streamline preparations of $(SN)_x$ and much reduce costs. This will then encourage chemical, catalytic and further physical studies. The synthesis of 'second generation' polythiazenes is now under way though only a few systems (conductivities up to $10^{-4}\,S\,cm^{-1}$) have so far been studied, with benzene, pyridine or thiophene spacer groups between short $S_xN_y$ chains.[20] An alternative approach is to convert some of the chain sulphur atoms into three-coordinate $S^+$, thereby enabling attachment of side groups at sulphur and/or stronger cross-linking between chains. Such enhanced interaction between chains perhaps occurs in the polymer $(S_5N_5^+AsF_6^-)_x$.[26] We have recently prepared this compound electrochemically as highly conductive copper-coloured crystals.[27]

In the last few years, the research emphasis in the development of new, potentially conductive thiazenes has been shifting from delocalised S/N and C/S/N chains to delocalised C/S/N rings (ref. 28). In such C/S/N systems, one can preserve the secondary interactions (e.g. S · · · S) which are so important in facilitating inter-chain or inter-ring conductivity. Cowan *et al.*[29] have produced an extremely useful discussion of the criteria to bear in mind when designing the molecular components of organic conductors.

Our studies of polythiazenes are part of an expanding research programme by 12 members of Durham chemistry staff (within or in association with the Interdisciplinary Research Centre for Science and Technology) on the preparation and characterisation of new polymer systems and conducting materials.[22]

## References


1. Labes, M. M., Love, P. & Nichols, L. F., *Chem. Revs,* **79** (1979) 1, and references therein.
2. Garfield, E., *Current Contents,* **16** (1976) 6.
3. Banister, A. J. & Hauptman, Z. V., unpublished work.
4. Lau, W. M., Westwood, N. P. C. & Palmer, M. H., *J. Am. Chem. Soc.,* **108** (1986) 3229.
5. Scranton, R. A., *et al., J. Appl. Phys.,* **48** (1977) 3838A *J. Vac. Sci. Technol.,* **14** (1977) 973.
6. Nippon Denso K. K., Jap. Patent 59 129 261, 25 July 1984; *Chem. Abs.,* **102** (1985) 46994e.
7. Asahi Chem. Ind. Co. See *Chem. Abs.,* **99** (1983) 79004e and 79008i.
8. Rubinson, J. F., Behymer, T. D. & Mark Jr, H. B., *J. Am. Chem. Soc.,* **104** (1982) 1224.
9. Nowak, R. J., Kutner, W., Rubinson, J. F., Voulgaropoulos, A., Mark, H. B., Jr. & MacDiarmid, A. G., *J. Electrochem. Soc.,* **128** (1981) 1927.
10. Cohen, M. J. & Harris, J. S., US Patent 4 227 943, 14 October 1980, and *Appl. Phys. Letters,* **33** (1978) 812; Sharp KK, Jap. Patent 55 160 475, 13 December 1980.
11. Thomas, A. E., Woods, J. & Hauptman, Z. V., *J. Phys. D: Appl. Physics,* **16** (1983) 1123.
12. University of Pennsylvania, US Patent 4 016 331, 26 January 1976; *Chem. Abs.,* **86** (1977) 172655j.
13. US Sec. of the Army, US Patent 4 206 705, 10 June 1980.
14. Sony Corp., Jap Patents 55 108 944, 21 August 1980, and 87 013 730, 23 August 1987.
15. Grushkin, B., US Patents 3 615 409 and 3 616 393, 1971; *Chem. Abs.,* **76** (1972) 40262 and 40283.
16. Banister, A. J., Fielder, A. J., Hauptman, Z. V., Hey, R. G. & Smith, N. R. M., British Patent Appls 79/24 580, 13 July 1979, and 83/22 621, 23 August 1983; *Chem. Abs.,* **95** (1981). P83144k, and **102** (1985) P187455s.
17. Kennett, F. A., Maclean, G. K., Passmore, J. & Rao, M. N. S., *J. Chem. Soc. Dalton Trans.* (1982) 851.
18. Banister, A. J. *et al., J. Chem. Soc. Dalton Trans.* (1986) 2371.

19. Banister, A. J., Hauptman, Z. V., Kendrick, A. G. & Small, R. W. H., *J. Chem. Soc. Dalton Trans.* (1987) 915.
20. Wolmershäuser, G., Fuhrmann, J., Jolter, R., Wilhelm, T. & Scherer, O. J., *Mol. Cryst. Liq. Cryst.*, **118** (1985) 435; Wolmershauser, G., Jolter, R. & Wilhelm, T. *J. de Physique,* **44** (Suppl.) (1983) C3.729; Chien, J. C. W. & Zhou, M.-Y., *J. Polym. Sci., Pt A,* **24** (1986) 2947.
21. Ancelin, H., Hauptman, Z. V., Banister, A. J. & Yarwood, J., *J. Polym. Sci., Pt. B, Polym. PHys.*, **28** (1990) 1611.
22. Further details of the types of work conducted within the IRC can be obtained from Professor W. J. Feast (Associate Director) at the University of Durham.
23. Yoshimura, T. & Kubota, Y., Jap. Patent 1 267 529, A2, 25 October 1989; *Chem. Abs.*, **112** (1990) 159497v.
24. Mennecier, P., French Patent 2 619 655. A1, 24 February 1989; *Chem. Abs.*, **111** (1989) 145161y.
25. Bannister, A. J., Hansford, M. I., Hauptman, Z. V., Luke, A. W., Wait, S. T., Clegg, W. & Jørgensen, K. A., *J. Chem. Soc. Dalton Trans,* (1990) 2793.
26. Mews, R., *Angew. Chem. Internat. Edit.*, **15** (1976) 691.
27. Banister, A. J., Hauptman, Z. V. & Wait, S. T., unpublished work.
28. Oakley, R. T., Cyclic and heterocyclic thiazenes. In *Progress in Inorganic Chemistry,* ed. S. J. Lippard, Vol. 36. Wiley, New York, 1988, p. 299.
29. Cowan, D. O., Kini, A., Chiang, L. Y., Lerstrup, K., Talham, D. R., Poehler, T. O. & Bloch, A. N., *Mol. Cryst. Liq. Cryst.*, **86** (1982) 1.

# Absorbent polymers

**D. Marshall**

*Allied Colloids Limited, Bradford, UK*

Working for a company that for the last thirty years has endeavoured to perfect the production of water-soluble polymers, it came as quite a shock to realise that crosslinked swellable polymers could be considered as products in their own right. Before this time swelling, non-soluble polymers were a problem to be eliminated in the production of soluble types.

The dedication of production plant to product absorbent products was the next essential step to avoid cross-contamination of polymers even though the chemistry of the solubles and swellables can be very close.

The initial interest in absorbent polymers was directed towards the agricultural and horticultural outlets where high water capacity in the polymer was considered essential. It was found to be even more important to ensure capability of absorbent polymers to absorb aqueous fluids containing dissolved salts, the most aggressive system with regard to salt concentration being encountered in phase change materials (PCM) for heat storage systems.

Before describing several practical outlets for absorbent polymers it is appropriate at this stage to consider the chemical structures of such materials with particular emphasis on the types produced by Allied Colloids (Fig. 1).

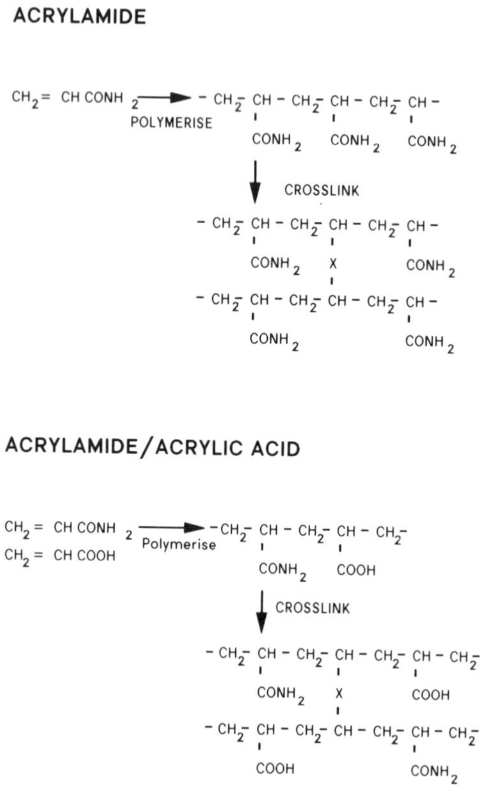


**Fig. 1.**

Other polymers with absorbent properties include the following chemical types:

—crosslinked starch-grafted polyacrylates
—crosslinked cellulose derivatives (e.g. CMC)
—crosslinked maleic isobutylene copolymers

The physical forms in which the polymer types are available can vary from irregular granules, spherical beads, fibres and dispersions in non-aqueous media.

Figure 2(a) shows a typical granular polymer with a particle size range from 45 to 800 $\mu$m, (b) shows a bead polymer with similar particle size range, (c) shows a fibrous absorbent having a denier of

ACRYLIC ACID

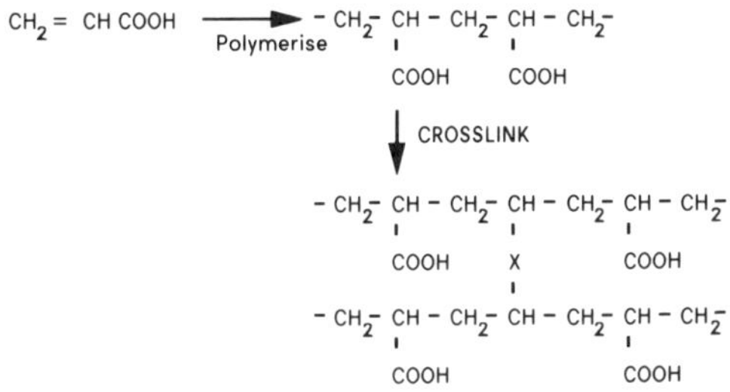


ACRYLONITRILE

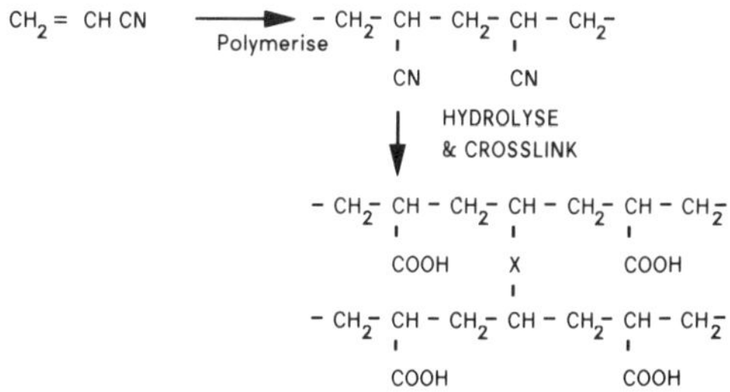


**Fig. 1.** —*contd.*

about 18 and (d) shows a liquid polymer dispersion (LPD) with particle size range 0·2–2·0 $\mu$m.

The granular polymer is produced by bulk polymerisation giving rise to large gel structures. These structures are cut and minced prior to drying on a fluid bed drier. After drying the particles are reduced in size by crushing or grinding followed by classification and any further treatments.

The bead polymer is produced by polymerising droplets of monomer suspended in a solvent. After the reaction is completed the solvent is stripped off and the bead dried to give uniform spheres of polymer. This method produces products that are free flowing and with very low free monomer levels.

The fibrous absorbent polymers are produced from a polymer dope

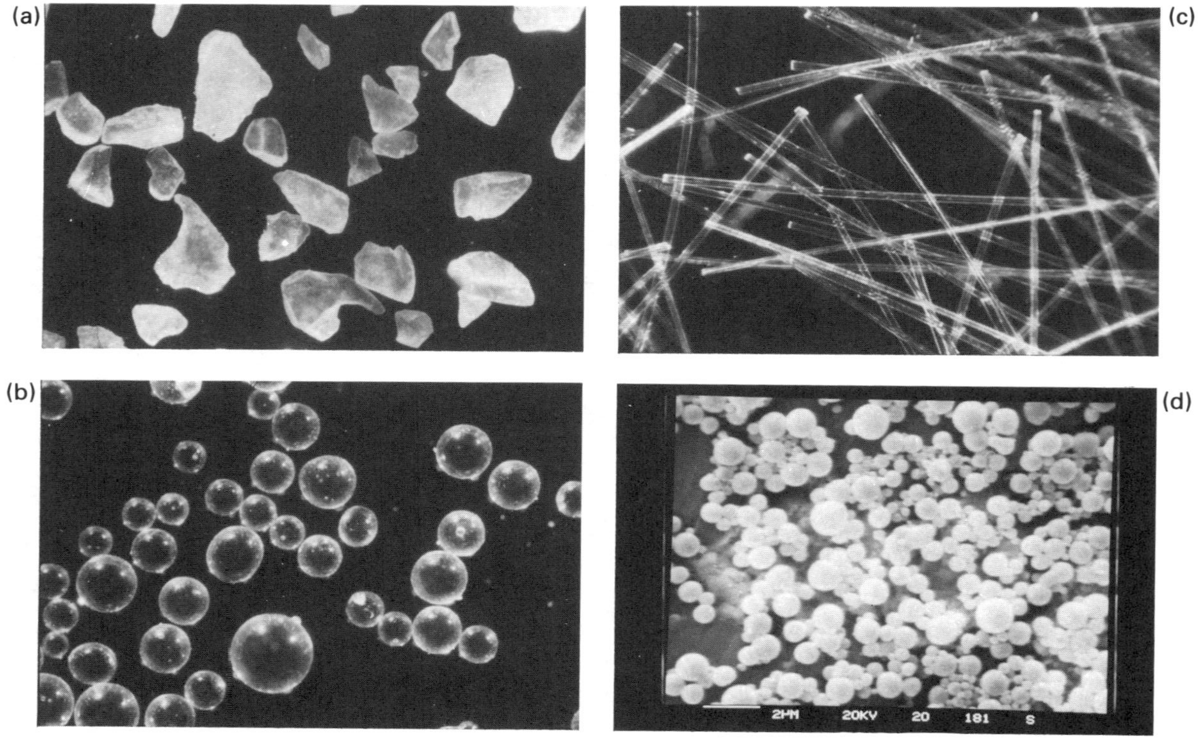


**Fig. 2.** (a) Granular polymer, (b) bead polymer, (c) fibrous polymer and (d) polymer dispersion.

which is passed through a spinneret under pressure. The dope composition is such that on post heat treatment the polymer reacts to form an absorbent matrix.

Surface treatments are applied to counter static and also to promote good wetting of the fibre.

The liquid polymer dispersions are made by the polymerisation of monomers emulsified into a solvent. The solvent is extracted and replaced by the non-aqueous carrier medium. Polymer contents up to 70 wt % can be achieved by this technique.

As a general rule absorbent polymers are tailored to meet a market requirement as no single product will fulfill all application needs.

Typical of such requirements are:

- particle size and particle size distribution
- capacity of polymer to absorb fluid
- retention of fluid in the polymer
- release of fluid from the polymer
- rate of fluid uptake
- persistence of fluid holding capacity over long periods —weeks, months, years
- low toxicity
- cost performance in a process

The question, 'What is an absorbent polymer?', was answered comprehensively by Dr F. Masuda in a recent paper[1] and in essence described the two main functions for aqueous fluid uptake: briefly, the degree of crosslinking and the extent of ionic character of the polymer.

It is the ionic nature of absorbent polymers that makes them sensitive in terms of capacity to dissolved salts in aqueous fluids. This property was made clear in the first place when polymers were made for the agricultural/horticultural markets.

## Agricultural/horticultural markets for absobent polymers

It is almost twenty years since absorbent polymers were developed for use in agriculture as water/nutrient holding products.[2] Such polymers were based on hydrolysed acrylonitrile starch graft copolymers and have since opened up a new area of technology into many different markets.

Our primary interest in water absorbent polymers was derived from technology based on high molecular weight polyacrylamides.

For desert regions the major constraint on plant growth is the scarcity of water. Sand/soil mixtures are extremely porous and water rapidly drains from the root zone, and even with frequent watering the water content of the substrate may be insufficient to support growth. Evaporation losses are high and salt content is also high in many desert sands.

Such problems can be overcome by introducing a water imbibing polymer to maintain water in the root zone and prevent poor crop emergence. The water imbibed by the polymer must be available to the plant without deterioration in its performance.

The presence of salts should ideally not affect the polymers absorption capabilities; neither should the salts accumulate or be rejected by the absorbent polymer.

Alcosorb AB1 and Alcosorb AB3 were developed to meet specific requirements. Alcosorb AB1 absorbs 30–50 times its own weight of pure water and is only slightly affected by salts (Fig. 3). Alcosorb AB3 absorbs over 400 times its own weight of pure water and is salt sensitive depending on ionic concentration of the salt solution (Fig. 3).

The former product is virtually non-ionic and this results in an extremely good salt tolerance. The latter product, although it has a much higher capacity for pure water, is sensitive to salts. The degree of ionic character in the polymer is around 30%. It must be remembered that although the Alcosorb AB1 has a much poorer capacity for pure water the benefit in high salt concentrations can be significant.

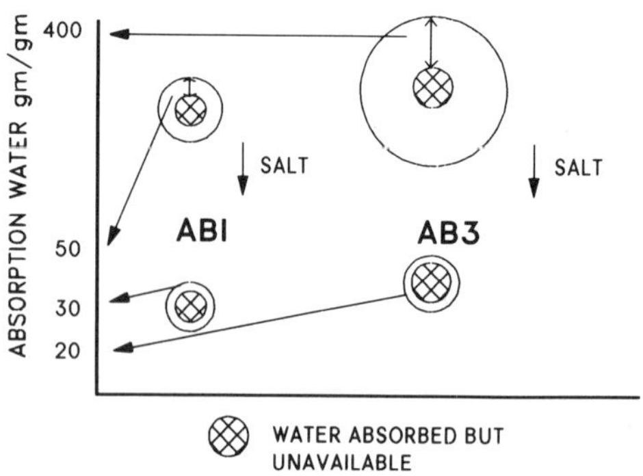


**Fig. 3.** Water absorption and salt tolerance of Alcosorb AB polymers.

In the market the high swelling grade is the one of choice because of the weight per weight absorption of fluid, even at modest salt concentration. Both products can be used successfully in the following areas:

- pot plants
- blended growth media
- seedling propagation systems
- establishment and maintenance of grass on lawns, embankments, sports areas
- seed coatings
- planting and transporting of trees
- transporting of orchids

Nutrients dissolved in the water are absorbed in the polymer and are freely available to the plant as the aqueous fluid is extracted (Table 1).

The results illustrated in Table 1 clearly show there is no significant accumulation or rejection of salt by the polymer. Similar results are found with the higher swell polymer Alcosorb AB3.

One of the outstanding benefits of absorbent polymers in this area is the increase in field capacity. This can easily be demonstrated using sharp sand/polymer blends (Fig. 4).

Figures 5 and 6 illustrate evaporation of water from maximum field capacity and evaporation from sand, sand/polymer mixes starting from the same water addition respectively. In both cases the influence of polymer can be clearly seen and this results in extension of wilt time of the plant (Table 2).

**Table 1.** Free flow of ions

| Salt | Concentration in supernatant (%) | |
|---|---|---|
| | Before polymer[a] addition | After polymer[a] addition |
| NaCl | 1·01 | 0·97 |
| NaCl | 2·02 | 2·22 |
| $MgCl_2$ | 1·02 | 0·98 |
| $MgCl_2$ | 1·97 | 1·97 |
| $CaCl_2$ | 1·01 | 1·01 |
| $CaCl_2$ | 2·03 | 2·02 |

[a] Polymer: Alcosorb AB1.

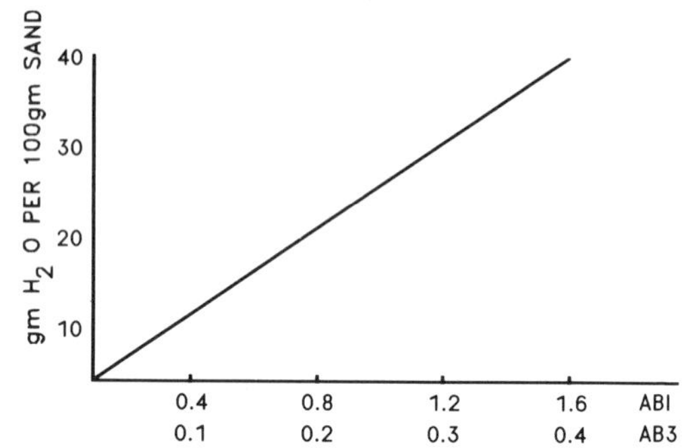


**Fig. 4.** Field capacity improvement.

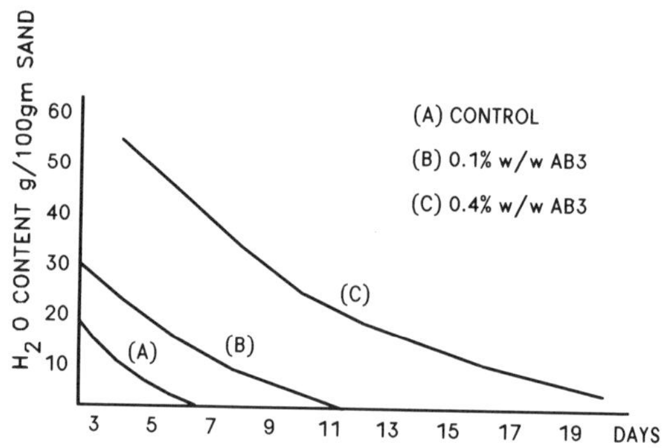


**Fig. 5.** Effect of polymers on rate of water loss with time.

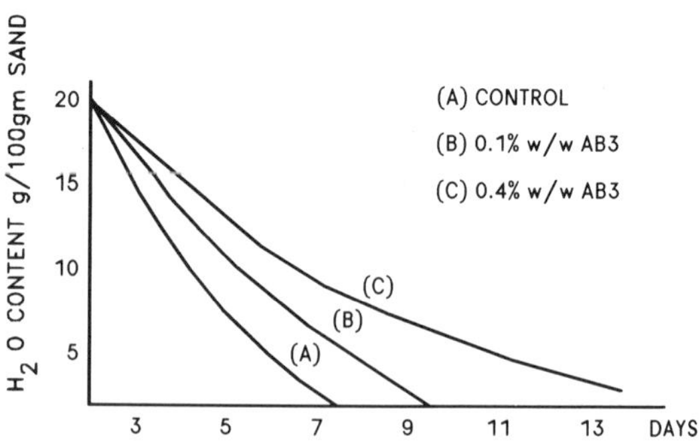


**Fig. 6.** Effect of polymers on rate of water loss from sand.

**Table 2.** Emergence and wilt time of seedlings

| Alcosorb AB3 Dosage (% w/w) | Initial water level g/100 g sand | Emergence | | Wilt time: days since last water |
|---|---|---|---|---|
| | | Day 5 | Day 8 | |
| 0 | 21·9 | 10 | 40 | 4 |
| 0·05 | 21·9 | 78 | 98 | 6 |
| 0·05 | 23·0 | 78 | 100 | 7 |
| 0·10 | 21·9 | 85 | 98 | 7 |
| 0·10 | 26·5 | 75 | 95 | 7 |
| 0·20 | 21·9 | 85 | 98 | 8 |
| 0·20 | 32·0 | 98 | 100 | 9 |
| 0·30 | 21·9 | 58 | 98 | 9 |
| 0·30 | 40·0 | 88 | 98 | 9 |
| 0·40 | 21·9 | 63 | 95 | 9 |
| 0·40 | 46·0 | 95 | 98 | 9 |

One set of polymer containing pots was watered to the field capacity of the control, the others until drainage.

Control emergence was poor. In contrast all polymer treatments showed excellent emergence, reflecting the reduced water evaporation in the presence of Alcosorb.

The benefit of very low dosages (0·05–0·20%) is immediately apparent. There is no gain in increasing polymer levels above 0·20% w/w.

There are several branded products on the market today including Alcosorb—products such as:

- Terrasorb
- Aquasorb
- Hydrosource
- Agrosoke
- Broadleaf

with different grades under these proprietary names.

The market for absorbent polymers worldwide for agricultural/horticultural use is large but somewhat hard to define, with estimates ranging between 5000 and 25 000 tonnes.

In general the chemistry of such absorbent polymers is rooted in acrylamide and copolymers of acrylamide, although starch grafted acrylates are also used.

## Use of absorbent polymers to improve handleability of mineral slurries

There is nothing new in the use of high molecular weight water soluble polymers for the conditioning of mineral slurries by flocculation; this allows easier filtration and liquid solids separation.

The use of absorbent polymers is somewhat novel in that the water content of the slurry is not separated but remains in the system as a gelled entity. A good example of this technique which is finding a very important market is in the coal industry. Modern methods of mechanical coal mining using high output shearers have resulted in an increase in the amount of fine coal produced from 10 to 25% of total run-of-mine production.

After removal of the shale in the washery, fines are often recovered by flotation and dewatered by thickening and vacuum filtration.

It is very difficult to remove water from very fine particles of coal; even after vacuum filtration some fine coal remains very wet and unsuitable for blending with fuel for utilities. This fine coal has been stockpiled—a valuable but unusable source of energy.

The addition of an absorbent polymer specifically designed for this outlet results in changing the fine coal mass from a sticky dilatant liquid into a free flowing crumbly product which is easy to handle and to blend with regular utility fuel.

The normal method for measuring the handleability of a blended coal is to pass the product through a Durham cone which measures the ability of the coal blend to flow through an orifice.

The results of using absorbent polymer show that handleability is greatly improved and blends of fine and larger coals can flow better than the larger coal alone (Table 3).

**Table 3.** Flow improvement in coal fines/blends

| Ratio of washed smalls to filter cake | Untreated filter cake (kg/s) | Alcosorb treated filter cake (kg/s) |
|---|---|---|
| 0 | 1·35 | 1·32 |
| 6:1 | 1·17 | 1·38 |
| 5:1 | 0·87 | 1·46 |
| 4:1 | 0·77 | 1·49 |
| 3:1 | 0·54 | 1·49 |

Typical dosage levels for Alcosorb in coal fines ranges between 0·2 and 2% depending on the water content and salt content of the water.

An example of the coal filter cake is:

water content 20–30%
particle size less than 60 μm
ash content 25%
solids content ~65%

Other applications of absorbent polymer in the mineral processing industry are:

- drying up of wet areas underground to give improved working environments for both men and machinery
- improvement of mineral concentrates stability in shipment without further reduction of moisture levels
- conversion of liquid effluents to gels before disposal
- improved moisture retention in mulches used for revegetation of mine sites

The market for absorbent polymers in the mining industry is into five figure tonnage in coal alone. The future looks very promising for this area of activity.

## Heat storage systems with particular reference to phase change materials[3]

The ability to store energy—chemically, physically or thermally—allows it to be used in more efficient ways. It is possible using thermal storage to take advantage of cheaper energy forms available in out-of-phase usage patterns.

Obvious examples of thermal energy storage systems are off-peak storage radiators, domestic hot water cylinders and the fuel rods of a nuclear fission reactor. The less obvious include the houses we live in, which if properly designed can be operated to reduce energy consumption by minimising day-time overheating and night-time overcooling.

Examples of processes that are inefficient because they lack a thermal storage component are:

- greenhouses—vented by day to avoid overheating—heated at night

- houses—constructed in low mass materials but inadequately designed for the use of natural sunlight
- in general, all batch processes where energy is rejected which could be used for preheating in the following cycle

The commonest way of storing energy is to heat a quantity of cheap available material to whatever temperature can be reached by the energy source. The stored thermal energy is preserved as effectively as possible by insulation until the time when the energy is needed.

The storage materials may be liquid (domestic hot water systems) or solid (night storage radiators) and the amount of energy recovered ($Q$) is determined by the mass ($M$) of material, its specific heat per unit mass ($C$), the temperature range of the material during the process ($T$) and an efficiency factor ($n$) which accounts for natural losses from imperfect insulation:

$$Q = M \times C \times T \times n$$

The temperature range $T$ is dictated by practical considerations such as the characteristic of the energy source, the required temperature of delivery from storage and the balance between acceptable insulation costs and cycle efficiency. Unless the storage temperature is very high compared with the delivery temperature or unless special controls are used the delivery temperature will fall progressively with discharge.

In order to maximise $Q$ there are conflicting trade-offs in storage unit sizes, insulation levels, performance and cost. Some of these limitations do not apply to the same extent to the newer class of phase change storage materials (PCMs). Such systems change from solid to liquid and back again during a complete cycle.

The values of density and specific heat are similar to conventional materials. It is the melting/freezing process that involves a very large extra amount of energy—latent heat because it does not involve a measurable change in the physical temperature of the PCM.

The active materials in PCMs are hydrates of commonly available salts such as:

| | |
|---|---|
| sodium sulphate | 10 molecules water |
| calcium chloride | 6 molecules water |
| sodium pyrophosphate | 10 molecules water |

which are all solids below their transition temperatures because the water molecules are bound in the crystal form. Above the transition temperatures these crystalline hydrate forms cannot exist and the energy used to detach the water molecules is stored as latent heat.

Serious problems occur when attempts are made to recover that stored energy because it will only release in full if all the water can recombine with the salt crystal within the time of the discharge cycle. It is normally required that the energy is released by the PCM at or near the transition temperature. For example, ice is an ideal PCM because it melts completely (congruently) to give a liquid of the same composition and therefore can only produce 100% ice again as discharge.

A salt hydrate is a compound between salt and water and unless the water can dissolve all the salt when melted (congruent) some of the salt will settle out and remain on the base of the container. This can prevent or slow down the reformation of the hydrate and restrict total recovery of the stored energy.

Sodium pyrophosphate decahydrate is a good example of this behaviour which has hitherto prevented its use (Fig. 7).

As indicated in Fig. 8 the net energy turned over per cycle would decrease gradually until about 50% of the originally stored energy was available.

The use of absorbent polymers can overcome the problem of incongruency which causes the inefficient cycling of PCMs.

In very simple terms the presence of a suitable crosslinked hydrophilic polymer can prevent the settlement of anhydrous salt upon charging of the system (Fig. 9).

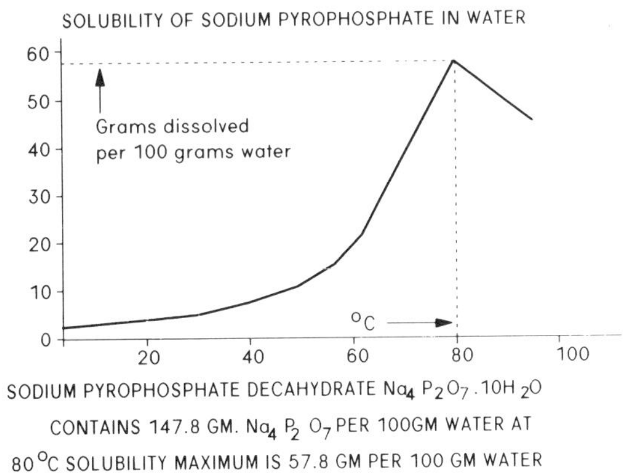


**Fig. 7.** Incongruent melting of salt hydrate.

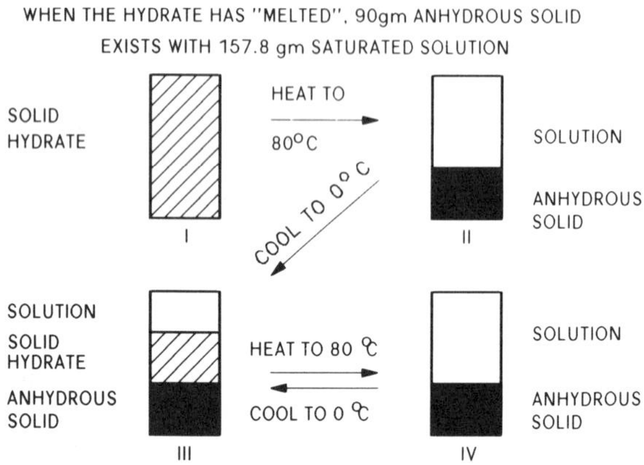


**Fig. 8.** Effect of thermal cycling on unstabilised hydrate.

The presence of the absorbent polymer maintains the anhydrous salt in suspension over hundreds of cycles, allowing the salt hydrate to reform almost completely on discharge.

Depending on the salt type chosen for the PCM the polymer can be varied in composition to suit the particular chemistry. For example, cationic absorbent polymers were developed for calcium salts used in PCMs, as such polymers were more stable in the presence of divalent cations.

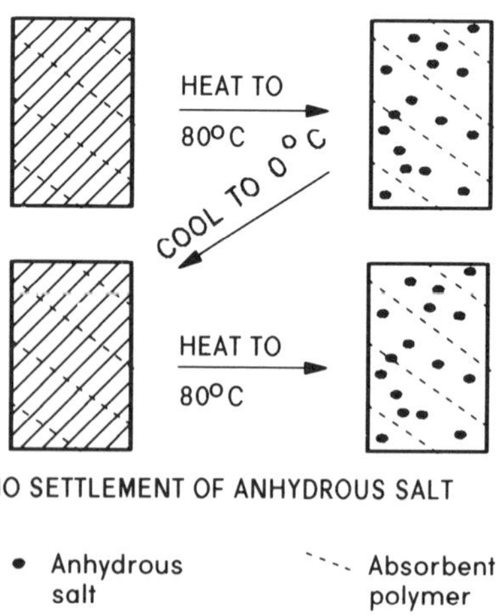


**Fig. 9.** PCM with absorbent polymer.

The application of polymers into PCM systems can be varied. Originally the soluble polymer was crosslinked *in situ* by heating through one or two cycles. Problems arose from the very high viscosity of the system and storage system containers were very difficult to fill.

*In situ* polymerisation was also an alternative such that the filling was easier—no viscosity contribution from the monomers but polymerisation control was difficult.

The ideal method is to use a crosslinked absorbent polymer in a relatively fine particulate form, for example, all less than 100 $\mu$m, and blend this in the appropriate proportion with the salt hydrate with thorough mixing.

Polymers are now being used in PCMs with transition temperatures of 10, 21, 48, 58 and 70°C. The 70°C system is particularly interesting as a domestic installation linked with a conventional wet system of radiators.

Interest in PCMs has fluctuated over the past ten years with the savings in basic fuel prices, but is on the upcurve at present. Estimated polymer tonnage is very difficult but is in the hundred of tonnes per year.

## Absorbent polymers in consumer disposables

Manufacturers of hygienic disposables such as adult incontinence diapers, baby diapers and sanitary pads have come to accept over the past four or five years the use and usefulness of absorbent polymers normally called superabsorbents..

Initially superabsorbents were merely added in small quantities to existing constructions with little regard for positioning in the diaper or an in-depth knowledge of the polymer type and performance. This resulted in showing little if any benefit to the consumer.

It became clear at a very early stage in polymer usage that to obtain the true benefits from superabsorbents, diaper design had to be changed to accommodate this radical development.

The most significant benefits to be gained by using superabsorbent polymers are:

- drier diapers in use
- slimmer diapers
- better fitting diapers

Drier diapers can reduce problems such as diaper rash and are generally more comfortable to wear. The superior dryness is derived from the polymer having excellent fluid retention properties compared with the traditional cellulosic fluff pulp.

Slimmer diapers result in the partial replacement of the bulky cellulosic fluff pulp by the polymer. This leads to a closer fitting diaper hugging the body contours.

In the early development of superabsorbent polymers for baby diapers, manufacturers of diapers requested polymers with the highest possible fluid uptake. Polymers with 80–100 g/g of fluid capacity were not uncommon.

It was soon realised that such polymers, although having high free swell capacities, were not performing well in the diaper because of poor fluid retention under pressure. This led to release of fluid back through the outer cover of the diaper and on to the skin.

New polymers were developed with lower total capacities, say 40–50 g/g, but with much better retention under pressure properties. More polymer per unit disposable has to be used with this polymer type but the improvement in performance is dramatic.

Today the following parameters for superabsorbents are thought to be the most important to provide good performance in the finished disposable:

- high fluid retention under pressure
- high fluid uptake under pressure
- excellent fluid transport through the polymer
- high gel strength of swollen polymer
- controlled rate of fluid uptake
- particle size distribution with minimum of fine particles
- low free monomer content
- toxicologically acceptable

The majority of superabsorbents used today in disposable health care products are based on crosslinked polyacrylates. Although many grades from several suppliers are available worldwide the chemistry of these polymers is very similar.

Variations and improvements are always being developed to meet special needs and the total market worldwide is in excess of 100 000 tonnes.

The following companies with generic names for the superabsorbent polymers are the best known—alphabetically:

| | |
|---|---|
| Allied Colloids | Salsorb |
| Chemdal | Aridall |
| Dow | Drytech |
| Hoechst Celanese | Sanwet |
| Nippon Shokubia | Aqualic |
| Sanyo | Sanwet |
| Stockhausen | Favor |

All the above producers provide excellent products for disposables and the key to providing the best product is knowing the construction of the disposable and where best to site the polymer and of course which grade of polymer to use.

Ultimately the performance of the diaper or adult incontinence unit in use is the only benchmark that determines the choice of polymer.

A new superabsorbent polymer in fibre form is being developed based on polyacrylate chemistry and will be available for trials in the near future.

The attraction of fibre is, of course, the physical form for blending with other fibres used in disposable goods. Although powder handling is not difficult, given the systems developed over the past few years there is less chance of polymer fall-out from core constructions with fibre. To compete with granular polymers the fibrous absorbents will have to have a comparable capacity rating and not least of all, in certain areas, will have to be cost competitive.

In summary, the existing business for absorbent polymers is very large and likely to increase over the next few years to approach two hundred thousand tonnes per year. Whilst there are many suppliers in this market it is thought that the survivors will be the manufacturers who are basic in the monomers used for such polymers or who have the knowledge to innovate and develop polymers for this speciality area.

## References


1. Masuda, Dr F., Pira Fibramerics Program, Products and Technologies Conference, London, UK, December 1987.
2. *Chemical Week,* p. 24, 21 July 1974.
3. Page, J. K. R., Calor alternative energy technical review. Calortherm Phase Change Storage Systems, 1984.

# High purity organometallic compounds for the preparation of compound semiconductors

**A. B. Leese & A. C. Jones**

*Epichem Limited, Bromborough, UK*

## Introduction

It is possible by combining elements of the Groups III–V, II–VI to produce alloys that have a variety of semiconductor and optoelectronic qualities, similar but in a number of ways far superior to those produced with silicon.

All the organometallic chemicals that we manufacture and purify are used in these applications and have very demanding requirements placed upon them.

To identify with the customers' problems it is necessary to understand the technology they are involved in. The particular expertise we have developed involves understanding the solid state physics of the various device structures with particular reference to the effects of impurities upon the electrical properties and relating this back to the chemistry of producing and purifying the organometallic precursors.

I would like, therefore, to firstly give a brief introduction into the way our chemicals are used and to relate this to the novel chemistry we have employed to obtain extremely high purity and consistent product.

## The mechanism of producing semiconductor structures

The fundamental chemistry behind the preparation of semiconductor devices is to start with a substrate of a material such as silicon and to grow thin layers of lattice matched crystalline material which have modified properties to cause electrons to behave in various ways.

The traditional method for forming thin layers in the case of the silicon industry is by using chemical vapour deposition. This enables large areas to be produced and is the basis for low cost silicon integrated circuits. In the case of silicon, the typical reaction would be:

$$\underset{\text{Silane}}{SiH_4} \xrightarrow[\text{1100°C}]{\text{Silicon substrate}} Si + 2H_2$$

By introducing small amounts ($10^{-6}$ mol %) of impurities such as arsine, phosphine or diborane into the gas stream the electrical properties can be modified to be either *p*-type (electron acceptors) or *n*-type (electron donors).

## Compound semiconductors and metal organic vapour phase epitaxy (MOVPE)

As indicated earlier, it is possible by combining various elements to produce a whole range of materials which are isoelectronic with silicon and can be modified to produce various semiconductor and optoelectronic properties. To achieve these results it is necessary again to grow various thin layered materials with modified properties.

Several techniques can be employed to do this. They include:

(1) chloride or hydride vapour phase epitaxy
(2) liquid phase epitaxy
(3) molecular beam epitaxy
(4) metal organic vapour phase epitaxy.

The latter technique is growing in popularity since it allows complex thin layer structures to be grown, and because of its similarity to the processes used in silicon epitaxy it lends itself to large area mass production techniques.

The original work on this method was carried out by Manasevit in 1968.[1] He showed that it was possible to grow epitaxial layers of GaAs by passing the vapour of trimethyl gallium together with arsine over a gallium arsenide substrate heated to 700°C:[2]

$$Ga(CH_3) + AsH_3 \xrightarrow[GaAs]{700°C} GaAS + 3CH_4$$

He then went on to show that a wide range of II–VI, IV–VI and III–V compounds could be grown using this very versatile technology (see Table 1).

A typical layout of a MOVPE reactor is shown in Fig. 1.[3]

**Table 1**

| Compounds | Reactants |
|---|---|
| Group III–V | |
| Binary | |
| GaAs | TMG, $AsH_3$ |
| GaP | TMG, $PH_3$ |
| AlN | TMA, $NH_3$ |
| GaN | TMG, $NH_3$ |
| InP | TMI, $PH_3$ |
| InAs | TMI, $AsH_3$ |
| InSb | TMI, TMSb |
| Tertiary | |
| GaAsP | TMG, $AsH_3$, $PH_3$ |
| GaAlAs | TMG, $AsH_3$, TMA |
| GaInAs | TMG, TMI, $AsH_3$ |
| GaAsSb | TMG, $AsH_3$, $SbH_3$ |
| InAsP | TMI, $AsH_3$, $PH_3$ |
| InAsSb | TMI, $AsH_3$, TeSb |
| Quaternary | |
| GaInAsP | TMG, TMI, $AsH_3$, $PH_3$ |
| Group II–VI | |
| Binary | |
| ZnS | DEZ, $H_2S$ |
| ZnSe | DEZ, $H_2Se$ |
| ZnTe | DEZ, DMTe |
| CdSe | DMC, $H_2Se$ |
| CdS | DMC, $H_2S$ |
| CdTe | DMC, DMTe |
| Tertiary | |
| CdHgTe | DMC, Hg, DET |

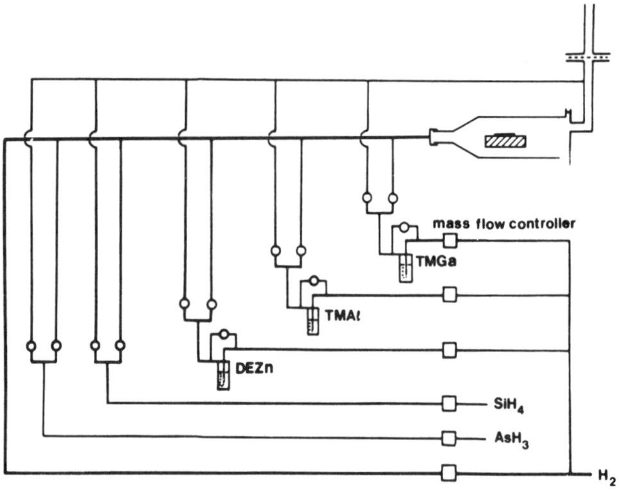


Fig. 1.

# Applications

## Semiconductors

The use of gallium arsenide to produce integrated circuits or discrete devices in place of silicon gives two significant advantages:

(1) they operate at 8 times the speed
(2) they are more resistant to radiation

The disadvantage at present is the very high cost compared with silicon. Uses in these areas have therefore concentrated on space and military applications.

## Optoelectronics

These areas are the ones that have generated the most commercial interest. As alloy semiconductors have a direct band gap, they are much more efficient emitters of photons than silicon which has an indirect band structure.

By varying the composition of the alloys, the wavelength of emission can be varied over a wide range. A good example of this is given in Fig. 2 for the Ga–In–As–P system.[4]

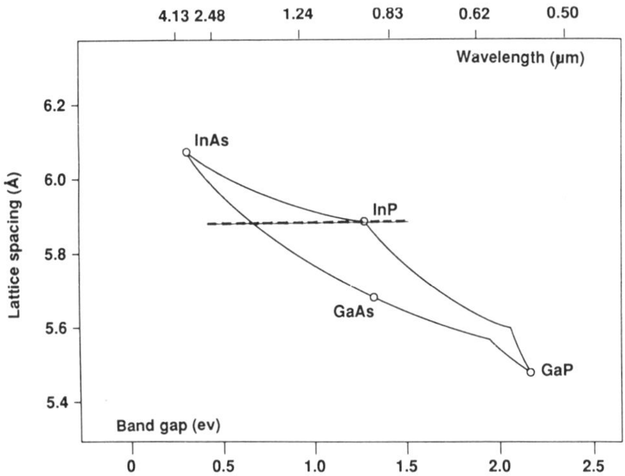


**Fig. 2.** Variation of band gap (and λ) with lattice spacing for the Ga–In–As–P system.

Applications of this property cover a wide range of areas, examples of which are:

(a) solar cells for space applications (GaAlAs)
(b) photocathodes for night vision (GaAlAs)
(c) solid state lasers for compact disc reading (GaAlAs)
(d) emitters and detectors for optical fibre systems (InGaAsP)
(e) infrared detectors and range finders (CdHgTe).

## Requirements of the source materials

To produce efficient semiconductor devices using MOVPE it is necessary for the grown layers to have very low levels of impurities present. These impurities, such as Zn and Si, have been shown to be introduced into the layers from the precursor materials, in particular the organometallic sources. Often the impurity elements are closely associated with the basic metals and tend to have similar chemical properties so they inevitably appear in the final product in levels which can cause problems, i.e. 5–10 ppm.

The problem of impure organometallic sources was identified by the UK Ministry of Defence who initiated research programmes at

Liverpool University and Queen Mary College to tackle the problem. It was this initial work which showed the way and led to the development of a whole range of adduct purification techniques used by Epichem to produce ultra high purity alkyls.

## Adduct purification

The principles of adduct purification are summarised in Fig. 3. A variety of Lewis bases (L) can be used; the exact nature of L is determined by the specific chemistry of the metalorganic substance to be purified. The bidentate phosphorus ligand 1,2-bis(diphenylphosphino)ethane (DIPHOS) has proved particularly suitable for the purification of trimethylindium. The aims of adduct purification are to obtain an ultra pure product and to render final organometallic product purity effectively independent of batch-to-batch variation in starting material purity, thus achieving a high level of consistency.

An important feature of adduct purification processes is that generally both a chemical purification stage (adduct purification) and a physical purification stage (distillation) are involved. In early work, it was found that non-metallic impurities such as methyl iodide and short chain ($C_2$–$C_4$) hydrocarbons were difficult or even impossible to remove from TMG using conventional fractional distillation tech-

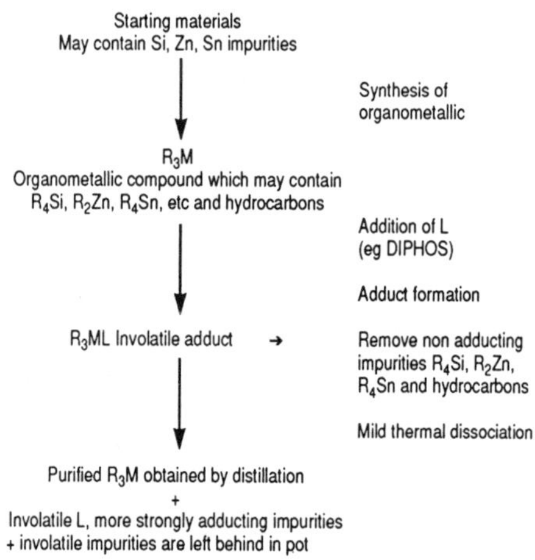


**Fig. 3.** Principles of adduct purification.

niques. We felt that chemical purification techniques would prove to be more suitable. We found that we were able to form a relatively involatile adduct between the impure TMG and an appropriate Lewis base. After removal of the non-adducting methyl iodide and $C_2$–$C_4$ hydrocarbons *in vacuo,* the adduct was then thermally dissociated to yield the purified TMG. It has since been demonstrated that adduct purification is also effective in removing a range of metallic microcontaminants (e.g. Si, Zn and Sn alkyls) from a number of metalorganic MOVPE precursors. By reference to TMG and TMA we shall further demonstrate the efficiency of the process.

## Analysis

The organometallic compounds are generally pyrophoric liquids or solids; hence before analysis they traditionally needed to be decomposed. We did in fact employ a closed system decomposition technique to sample the organometallics for subsequent analysis using inductively coupled plasma emission spectroscopy (ICP). Due to the low levels of detection required (sub-parts per million for most elements), this technique was only partially successful in retaining volatile metallic impurities in the organometallic. We subsequently developed a proprietary technique of direct injection, in which the neat organometallic is directly aspirated into the plasma source using a closed oxygen-free inlet system. Aspiration rates are typically 1 g $min^{-1}$.

**Table 2.** Metals analysis using inductively coupled plasma emission spectroscopy

| Elements analysed for (with detection limit, ppm by weight metals basis) | | | | | | | | | |
|---|---|---|---|---|---|---|---|---|---|
| Ag | <0·4 | Cd | <0·02 | In | <0·5 | Ni | <0·05 | Si | <0·03 |
| Al | <0·5 | Co | <0·4 | K | <1·0 | P | <0·5 | Sn | <1·0 |
| As | <0·5 | Cr | <0·4 | La | <0·4 | Pb | <1·0 | Sr | <0·1 |
| Au | <0·5 | Cu | <0·05 | Li | <0·4 | Pd | <0·5 | Tb | <0·5 |
| B | <0·4 | Fe | <0·1 | Mg | <0·01 | Pt | <0·5 | Ti | <0·2 |
| Ba | <0·1 | Ga | <0·5 | Mn | <0·02 | Rh | <0·5 | U | <0·5 |
| Be | <0·02 | Ge | <0·5 | Mo | <0·5 | S | <1·0 | V | <0·5 |
| Bi | <0·5 | Hg | <0·5 | Na | <0·5 | Sb | <1·0 | W | <0·5 |
| Ca | <0·02 | I | <0·0 | Nb | <0·5 | Se | <1·0 | Y | <0·02 |
| | | | | | | | | Zn | <0·2 |

Wherever possible, volatile organometallic compounds (e.g. $SiR_3$, $ZnR_2$ where $R = CH_3$) are used as calibration standards. Our current lower detection limits are shown in Table 2.

The other main analytical technique used is mass spectrometry which is used to detect hydrocarbon impurities.

## Correlation with electrical properties of grown layers

No analytical technique can be viewed in isolation, especially in an area demanding such low levels of impurity. We have therefore carried out much work in conjunction with various companies and research establishments in the United Kingdom to ensure correlation between the levels of impurity we find using ICP and the electrical properties of the grown layers.

## Examples

### Trimethylgallium

The reagents most commonly used for the growth of GaAs by MOVPE is $Me_3Ga$ in combination with $AsH_3$ (either in pure form or as a dilute mixture in $H_2$). There are a number of published routes to $R_3Ga$ compounds, as outlined below.

#### *Synthesis of $R_3Ga$*

For $Me_3Ga$, the routes available may be summarised thus:

(a) $2Ga + 3Me_2Hg \rightarrow 2Me_3Ga + 3Hg$[5]
(b) $Ga_2Mg_6 + 8MeI \rightarrow 2Me_3Ga + 2MeMgI + 3MgI_2$[6,7]
(c) electrolysis of $Na[AlMe_3F]$ with a sacrificial Ga anode[8]
(d) electrolysis of MeMgX or $Me_2Mg$ with a sacrificial Ga anode[9]
(e) $2GaCl_3 + 3Me_2Zn \rightarrow 2Me_3Ga + 3ZnCl_2$[10]
(f) $2GaCl_3 + 6MeMgCl \rightarrow 2Me_3Ga + 6MgCl_2$[11]
(g) $GaCl_3 + 3MeLi \rightarrow Me_3Ga + 3LiCl$[12]
(h) $GaCl_3 + 3Me_3Al \rightarrow Me_3Ga + 3Me_2AlCl$[13]

Only a few of these routes have proved suitable for the production of useful amounts of pure $R_3Ga$ product. For instance, routes (a) to (e) can lead to a product heavily contaminated with Hg or Zn. Route (b) involves prior synthesis of a homogeneous Ga/Mg alloy, whilst the electrolytic routes are difficult to scale up. Routes (f), (g) and (h) involving the alkylation of $GaCl_3$ with RMgX, RLi or $R_3Al$ are generally the preferred routes to high purity $R_3Ga$, although there may be a difficulty in separating the $R_3Ga$ product from the reaction mechanism when RMgX or RLi are used.

Even though methods (f), (g) and (h) lead to good yields of organometallic, the final product may still contain trace impurities derived from the starting materials. Typically, silicon and zinc levels may be in the 5–15 ppm range—totally unacceptable for good quality layer growth. By adduct purifying the trimethylgallium, the levels of impurities may be reduced significantly and greatly improve the quality of the layers (Table 3).

## Trimethylaluminium

AlGaAs has conventionally been grown in MOVPE using $Me_3Al$ together with $Me_3Ga$ and AsH,[14] whilst at lower pressures, $Me_3Al$ has sometimes been replaced with $Et_3Al$.[14,15]

### *Synthetis of $R_3Al$*

$R_3Al$ compounds have achieved great industrial importance due to their large scale use as co-catalysts for leading to isolactic (or

**Table 3.** Effect of reducing impurity levels of adduct purification on electrical properties

| TMG sample | ICP analysis[a] (ppm w.r.t. Ga) | Carrier concentration at 77 K ($cm^{-3}$) | Mobility at 77 K ($cm^2\,V^{-1}\,S^{-1}$) |
|---|---|---|---|
| 1 | Si 12, Zn 6 | $3{\cdot}2 \times 10^{15}$ | 22 711 |
| 2 | Si 5, Zn 3 | $1{\cdot}1 \times 10^{15}$ | 37 000 |
| 3 | Si 1·2, Zn 2 | $1{\cdot}1 \times 10^{15}$ | 42 600 |
| 4 | Si 0·03, Zn 0·2 | $1{\cdot}0 \times 10^{14}$ | 127 000 |

[a] All other elements were not detected.
Growth temperature 690°C, layer thickness 12 m, V/III ratio = 40 : 1.

stereoregular) polymers.[16] There are a number of routes available for the synthesis of $R_3Al$ compounds as shown below:

(a) $2Al + 3Me_2Hg \rightarrow 2Me_3Al$[17]
(b) $Me_3Al_2X_2 + 3Na \rightarrow Me_3Al + Al + NaX$[18]
(c) $Al_2Mg_3 + 6MeCl \rightarrow 2Me_3Al + 2MgCl_2$[19]
(d) $Al + H_2 + 3(CH_2{=}CH_2) \rightarrow Et_3Al$[18]

Routes (c) and (d) have generally proved the most satisfying for the production of large quantities of $R_3Al$ compounds, though the crude product is generally heavily contaminated with both silicon and hydrocarbon impurities. Until the development of adduct purifying techniques the electrical properties of AlGaAs were greatly influenced by the levels of Si in the metalorganic precursor as described below.

### *Adduct Purification of $Me_3Al$*

Conventional distillation methods have proved ineffective in removing Si or hydrocarbon contaminants from $Me_3Al$; for example, TMAL source 1 was found to contain 14 ppm Si even after fractional distillation. However, adduct purification of this source reduced the Si level to <1 ppm. The effect of this reduction in levels of silicon can be seen from the results in Table 4.

## Trimethylindium

This product has particular problems associated with it. It is a pyrophoric solid and is particularly difficult to purify to the levels necessary for good layer growth of indium phosphide.

Work on this area was carried out under the Joint Opto Electronic Research Scheme at a UK Department of Trade and Industry sponsored programme. In conjunction with Queen Mary College,

**Table 4**

| TMA source | ICP analysis (ppm w.r.t. Al) | S t charge density ($cm^{-2}$) | Mobility 77 K ($cm^2\,V^{-1}\,S^{-1}$) |
|---|---|---|---|
| Non-purified | 14 | $1{\cdot}0 \times 10^{12}$ | 6 000 |
| Adduct purified | 0·7 | $5{\cdot}7 \times 10^{11}$ | 80 000 |

Plessey and STC we were able to develop an adduct purification technique based upon a bidentate phosphorus ligand—1,2 bis(diphenylphosphino)-ethane. This has enabled levels of purity to be reached that were previously unobtainable. Typically, trimethylindium purified by this technique will have no detectable impurities, including silicon levels as low as 30 ppb.

Using products produced by this route, workers at STC Harlow and the University of Illinois have obtained record levels of purity in indium phosphide layers with mobilities at 77 K of 298 000 $cm^2 V^{-1} S^{-1}$—some two times the previous maximum.

## Conclusions

By close collaboration with customers and research establishments in the UK we have been able to develop new and novel chemical purification techniques that have greatly assisted in the development of MOVPE as a production technique for a new generation of semiconductor materials.

## References


1. Manasevit, H. M., *Appl. Phys. Lett.*, **12** (1968) 56.
2. Manasevit, H. M., *J. Cryst. Growth,* **55** (1981) 1.
3. Gerrard, Neil D., *Chemtronics,* **3** (1988).
4. Moss, Rodney H., *Chemistry in Britain,* **9** (1983).
5. Coates, G. E., *J. Chem. Soc.* (1951) 2003.
6. Bregadze, V. I., Golubinskaya, L. M., Tonegan, L. G. & Gribov, B. G., *Dokl. Akad. Nauk. SSSR,* **212** (1973) 880.
7. Jones, A. C., Cole-Hamilton, D. J., Holliday, A. K. & Ahmed, M. M., *J. Chem. Soc., Dalton Trans* (1983) 1047.
8. Ziegler, K., UK Patent 814 609 1959.
9. Mullin, J. B., Holliday, A. K., Cole-Hamilton, D. J., Jones, A. C. & Gerrard, N. D., UK Patent GB 2 125 795 and US Paent 4 464 233, 1985.
10. Wiberg, E., Johannsen, T. & Stecher, O., *Z. Anorg. Allg. Chem.,* **251** (1943) 114.
11. Jones, A. C., Holliday, A. K., Cole-Hamilton, D. J., Ahmed, M. M. & Gerrard, N. D., *J. Cryst, Growth,* **68** (1984) 1.
12. Kover, R. A., Derr, H., Brandau, D., & Calloway, J. O., *Inorg. Chem.,* **14** (1975) 2809.

13. Gaines, D. F., Borlis, J. & Fody, E. P., *Inorg. Synth.*, **15** (1974) 203.
14. Keuch, T. F., Vouhoff, E. Kuan, T. S., Deline, V. & Potenski, R., *J. Cryst. Growth,* **77** (1986) 257.
15. Kobayashi, N. & Makimoto, T., *Jap. J. Appl. Phys.*, **24** (10) (1985) L824.
16. For a comprehensive review, see: Ziegler, K., *Organometallic Chemistry,* ed. H. Zeiss. Perrall, New York, 1960, Ch. 5.
17. Buckton, G. B. & Odling, W., *Proc. Roy. Soc.,* **14** (1865) 19.
18. Ziegler, K., *Angew. Chem.,* **64** (1952) 323, 330.
19. Von Grosse, A., *J. Org. Chem.,* **5** (1940) 106.

# Recent advances in radiation curing

**R. R. Kemmerer & M. A. Johnson**

*Radcure Specialties SA, Drogness, Belgium*

## Introduction

One of the fastest growing segments of speciality chemicals is raw materials for radiation curing. This growth is due to a combination of two factors:

(i) expansion of established applications (overprint varnishes, inks, etc.) due to the oft-cited advantages over conventional conversion processes and
(ii) development of a number of novel end uses.

In the final analysis, unique performance of radiation cured systems vs other technologies is the key to growth.

Radiation curing is polymerization (crosslinking) of a chemical system which is initiated by incident radiation. Though most radiation cured products are coatings, inks, adhesives, etc., which are cured as thin (ca. 1–2 mils) films on various substrates (paper, metal, or plastic), examples of thick films and three-dimensional products cured by ultraviolet (UV) radiation have been recently reported. Almost all industrial radiation cured processes are by UV or electron beam (EB). While most radiation curing is by free-radical polymerization of acrylate-functional oligomers and monomers, methacrylate-functional materials are also used. Some UV curing involves cationic mechanisms of epoxy resins as well.

The commercial process involves (i) formulation, compounding, storage, and handling of these highly reactive materials, (ii) coating and handling of substrates—some of which may be sensitive and/or expensive—and (iii) converting the coating, etc. with a radiation processor. Product performance is primarily dependent on coating raw materials.

There are increasing numbers of novel applications that open up new manufacturing opportunities and generate new requirements for suppliers of raw materials. The purpose of this paper is to highlight some of the newest applications for radiation curing chemistry which offer promise for the future in light of the most successful commercial applications to date.

## Historical background

Compared to that of most paint and coatings technologies, the history of UV/EB curing is short. Although theoretical discussion of radiation cured polymers were published in the early 1950s,[1] the first commercial applications of this technology to coatings was in the late 1960s in Europe. These were UV cured wood coatings. A UV cured particle board filler process was commercialized in the United States shortly thereafter, and both processes are widely employed today. Electron beam cure was utilized by Ford in the United States for automotive interior coats in the early 1970s.

The growth rate for radiation curable materials is estimated at 20% per year. Some of the newer types of applications being pursued today (curing three-dimensional parts, magnetic tape binders, fibre optics, conductive coatings, structural adhesives and binders, etc.) bear little resemblance to the relatively simple wood fillers and topcoats developed in the 1960s.

## Advantages and disadvantages

The most frequently cited advantages for UV/EB curable processes compared to conventional methods include coatings with ultra smooth surfaces, low energy requirements, rapid throughput or high productivity, reduced solvent emissions, reduced space requirements, and unique performance. The latter category includes, for example, the ability to coat heat-sensitive substrates and speciality applications. An example of these applications is the manufacture of magnetic storage tapes and disks with maximum packing density by instantaneous curing of the binder while the magnetic particles are oriented by magnetic flux.[2] Another is three-dimensional curing of prototype tools via computer aided design.[3]

Some of the problems or perceived disadvantages for this technology pose challenges to the industry, some of which are already being met. First, ultra smooth surfaces are advantageous if high gloss is desired, but may be a disadvantage if flat finish is required. Reducing gloss to low levels is inherently more of a problem with radiation cured coatings than with conventional solvent-based systems. Raw material costs are generally higher, even taking into account solvent losses of conventional coatings and adhesives. The highly reactive acrylate monomers and some oligomers used in radiation curing may cause skin irritation. The rapid cure via UV/EB radiation may result in film shrinkage with poor adhesion to some substrates, notably some metals and some plastics. Finally, as most radiation cured coatings and adhesive formulations are 100% reactive, high solids formulations often have high viscosity which may lead to difficulty in some coating processes.

## The radiation curing process

### Equipment

#### *UV*

Medium pressure mercury vapour (MPM) lamps with outputs of 200 watts/in became available in 1970, and this development was key to the commercialization of UV radiation cure in the United States. In 1974 electrodeless lamps powered by microwave were introduced with higher UV output, and more flexible spectral output through use of different bulb compositions. Power up to 300 watts/in became available and curing efficiency was enhanced. Most recently, commercial mercury spectral lamps with power outputs up to 400 watts/in have become available. Pulsed xenon lamps offer very high output in short bursts or pulses, which are useful for certain applications. Most commercial systems today, however, are of the MPM or electrodeless type.

### *EB*

The original Ford electron beam units were of the scanning beam design. In 1971, electron accelerators with linear cathodes which did not need to be scanned were introduced.[4] This so-called electron curtain process was commercially developed by Energy Sciences International and Radiation Polymer Company, and represents the current state-of-the-art. Capital outlay remains a major factor in the decision to utilize electron beam curing for a commercial process.

## Chemistry of UV/EB Radiation Curing

Radiation curable coatings, adhesives, and other products are comprised of four basic types of materials: acrylated oligomers or prepolymers, acrylated monomers, additives, and photoinitiators. The basic film forming properties are derived from the oligomeric portion. Monomer can be either monofunctional acrylates (or methacrylates) or multifunctional acrylates. The primary function of monomers is to reduce viscosity of the formulation. Additives are used for a variety of reasons: surfactants to improve wetting or flow, pigment to provide colour or opacity, fillers for cost reduction or plasticization, slip aids, and numerous others. Photoinitiators are required for UV cured systems but not for EB cured systems. These will be discussed, with the other components, in the section entitled ‘Raw materials’ below.

Commercial applications of radiation cured chemistry is predominately free radical polymerization. The initiating species (‘radical’) (Fig. 1) can be generated by incident UV light or, in the case of E-beam, high energy electrons. The most commercially important class of materials are acrylate functional, although methacrylate, vinyl, and allyl functionalities are seen in some applications. Thiol/ene chemistry, involving the addition of the SH group across a double bond, is also a free radical process.

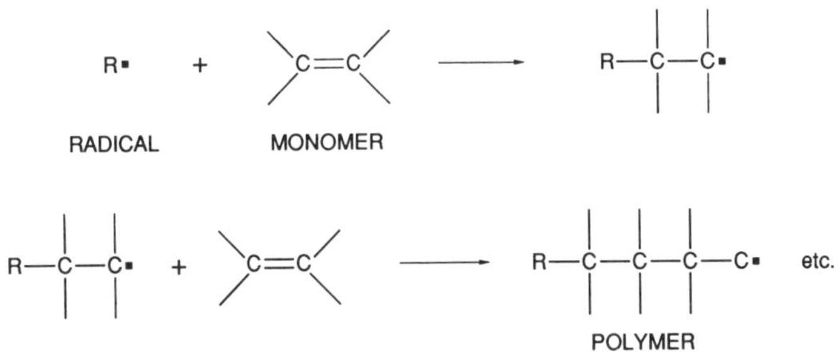


**Fig. 1.** Basic free radical polymerization reaction.

## Raw Materials

### *Acrylated oligomers*

The basic film forming properties of the UV/EB formulations are obtained from the acrylated oligomer or prepolymer. Three types of oligomer dominate the industry: acrylated epoxies, acrylated urethanes, and acrylated polyesters.

*Acrylated epoxies.* The diacrylate of the diglycidyl ether of bisphenol A (Fig. 2) is the single largest volume oligomer of this class used today. Epoxy acrylates are chosen for relatively fast cure, lower cost, hardness, and chemical resistance. A variety of acrylated epoxies other than of the structure shown in Fig. 2 exist, having improved flexibility, certain desired additional chemical functionality, etc.

A notable feature of this class of compounds is the hydroxyl functionality shown in Fig. 2 which results in high viscosity through hydrogen bonding. Considerable interest has been shown in reducing viscosity of this type of material for ease of handling. A recent development is the use of small amounts of selected salts, and water, which dramatically reduces the viscosity of this resin class.[5] Another recent advancement is commercialization of rubber-modified acrylated

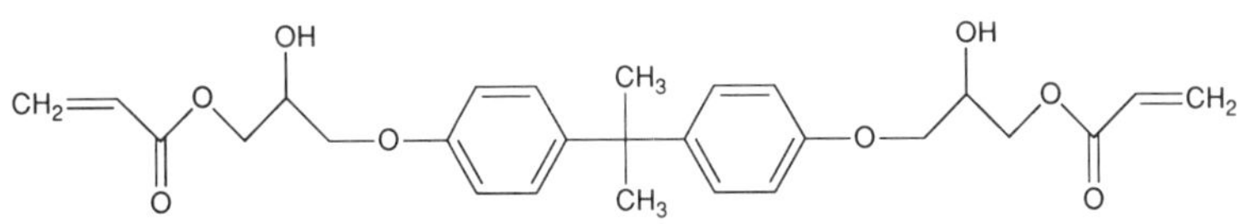


**Fig. 2.** Acrylated epoxy oligomer—bisphenol A based.

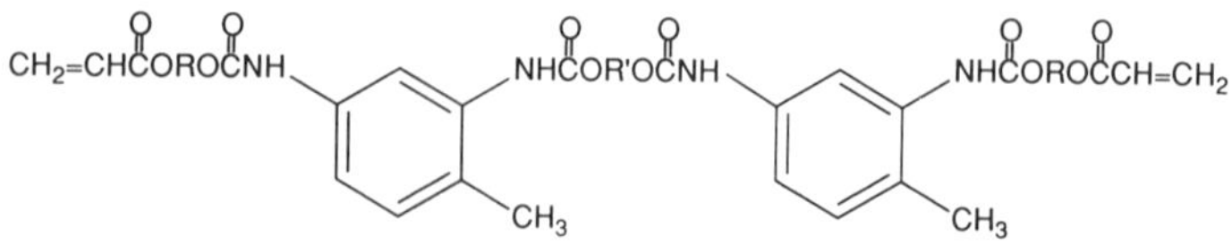


**Fig. 3.** Acrylated urethane oligomer—TDI based.

epoxies, which offer improved mechanical and thermal shock resistance.[6] It has been shown that variations of type and amount of rubber in these compounds will change tensile, electrical, and thermal properties of cured films containing these rubber modified acrylated epoxies.

*Acrylated urethanes.* Acrylated urethanes are desired for tough, flexible applications, with sometimes superior adhesion properties. They are divided most broadly into aliphatic and aromatic (Fig. 3) types. Although slightly more expensive, the aliphatic types are desired for the special resistance to yellowing on prolonged exposure to sunlight. Some of these materials are also of extremely high viscosity while other urethane acrylates are lower in viscosity, depending on composition.

*Acrylated polyesters.* The wood fillers and wood coatings discussed above in the first UV cure applications were unsaturated polyesters. Acrylated versions followed, produced by direct esterification or transesterification of residual hydroxyls (Fig. 4) in Europe. Wood fillers and wood coatings are still important applications for these materials. Properties of these resins can vary significantly depending on the raw materials used. These oligomers typically are lower

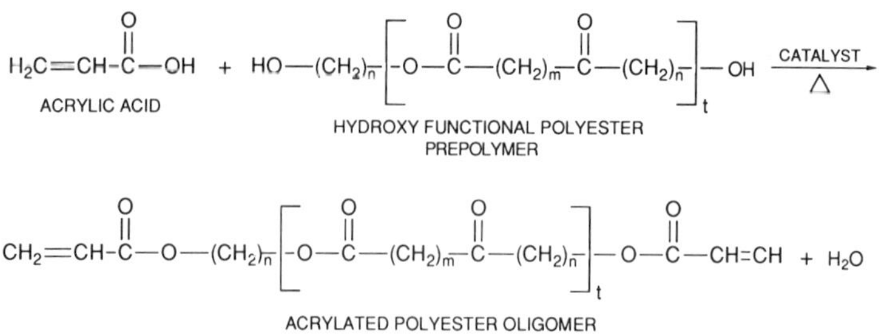


**Fig. 4.**

viscosity and thus may offer a handling advantage. Other applications are for coatings of flexible substrates, laminating adhesives, magnetic film binders, and inks.

### *Acrylated monomers*

Monomers are used in UV/EB formulations to reduce viscosity of the formulation. Although the basic film-forming properties of a formulation are supplied by the oligomer, monomers also exert a strong effect on the final properties of the cured product. Thus, in addition to viscosity reduction, multifunctional monomers (MFMs) provide crosslinking in the cured homopolymer, resulting in increased hardness and chemical resistance, and, in many cases, improved tensile properties. Some typical MFMs are:

glycerylpropoxy triacrylate
1,6-hexanediol diacrylate
pentaerythritol triacrylate
tetraethyleneglycol diacrylate
trimethylolpropane ethoxy triacrylate
trimethylolpropane triacrylate
trimethylolpropane trimethacrylate
tripropyleneglycol diacrylate

Monofunctional monomers do not crosslink, but they cure into the body of the film. While their main purpose in the formulation is viscosity reduction, sometimes desirable effects on the finished product are achieved by the proper choice of a monofunctional monomer—e.g. a harder or softer film, etc. Some examples of commercial monofunctional monomers are:

beta-carboxyethyl acrylate
2-(2-ethoxyethoxy)ethyl acrylate
2-ethylhexyl acrylate
isobornyl acrylate
*N*-vinyl-2-pyrrolidone
2-phenoxyethyl acrylate
octyl/decyl acrylate
styrene
tetrahydrofurfuryl acrylate

Recent developments include the availability of alkoxylated (usually ethylene oxide or propylene oxide) multifunctional monomers which are proposed for lower skin and eye irritation. Also, lower odour MFMs are in demand for some applications, and commercial products are now available.

### *Additives*

Generally, additives are non-reactive, and the choice of additives for formulating UV cured coatings and inks parallels the conventional art. For example, a variety of wetting and flow control agents are available from manufacturers of these speciality materials. Recent advances in this area, however, include functionalized additives. For example, (i) reactive surfactants and (ii) acrylate functional silicone oils are now commercially available.

Colouring additives are compounded especially for radiation cured systems. Generally these are dispersions of inorganic pigments in reactive monomer. Pigments can interfere with UV curing, but electron beam radiation will pass through pigmented systems, up to the penetration limits of the incident radiation (about 0·016 in, using 200 kV, depending on formulation density).

### *Photoinitiators*

Photoinitiators absorb UV light and generate free radicals, which in turn initiate the free radical polymerization process. A variety of photoinitiator systems are commercially available today, and selection depends on cost, UV equipment available, absorption characteristics of the photoinitiator and that of the formulated system—including pigmentation requirements, film thickness and through-cure, potential for chemical interference, exposure to weather, toxicity, and handling (liquid vs solid) requirements.

The two most common types of photoinitiator systems are photocleavage (Fig. 5) and photoabstraction (Fig. 6). These two $\alpha$-substituted acetophenone types are especially desirable in applications requiring good weathering, i.e. non-yellowing. Two examples are shown in Fig. 7. The first product (Irgacure® 184) is a solid at room temperature, while the second, chemically similar, material, Darocur® 1173, is a liquid.

Photoabstraction photoinitiator systems consist of a photosensitizer and a co-initiator (Fig. 8). The co-initiator is typically a tertiary

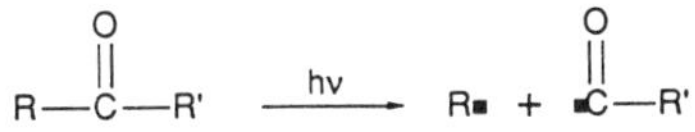

INTRAMOLECULAR PHOTO-FRAGMENTATION

SUBSEQUENT INTERNAL REARRANGEMENTS MAY OCCUR

**Fig. 5.** Photocleavage initiators.

$$RX \xrightarrow{h\nu} RX^{*}$$

$$RX^{*} + R'H \longrightarrow R'\bullet + \bullet RXH$$

INTERMOLECULAR HYDROGEN ATOM ABSTRACTION

**Fig. 6.** Photoabstraction initiators.

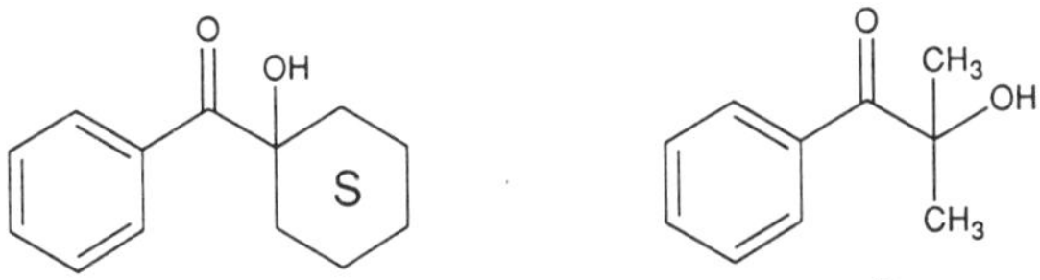

Irgacure ® 184 Darocur ® 1173

**Fig. 7.** Common photoinitiators (cleavage).

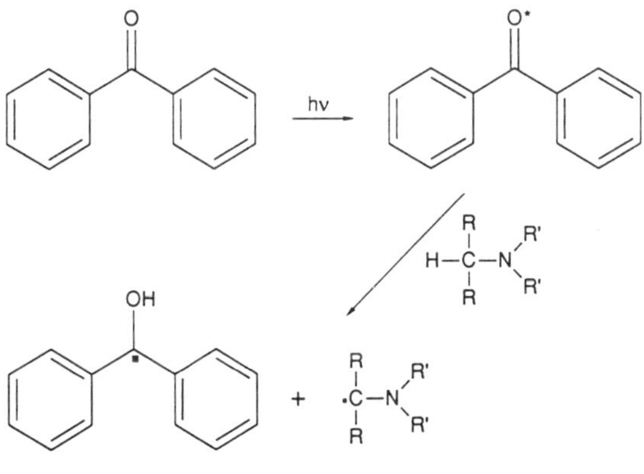

**Fig. 8.** Photoabstraction reaction.

• QUENCHING REACTION

$PI^{*} + O_2 \longrightarrow PI + O_2^{*}$

• SCAVENGING REACTION

$R\cdot + O_2 \longrightarrow ROO\cdot$

STRONG PROPAGATING RADICAL — WEAK PROPAGATING RADICAL

**Fig. 9.** Oxygen inhibition reactions.

amine, which is the proton source in the hydrogen ion abstraction process. Benzophenone is a commonly used photosensitizer, and this material is relatively inexpensive compared to many others, including the two other photoinitiators described above. Another benefit of this system is the mitigation of oxygen inhibition.

Oxygen inhibits free-radical reaction (i) through quenching and (ii) through scavenging of free radicals, by generation of peroxide radicals (Fig. 9), which are weak propagating radicals. The amine co-initiator interferes with oxygen inhibition by consuming oxygen in a chain process.

## The Impact of Regulatory Activity

The Toxic Substances Control Act of 1977 (TSCA) gave the EPA broad authorization to regulate new and existing chemicals.[8] This law affects both manufacturers and processors of industrial chemicals. Regulatory action impacts raw material costs, work place practices, and documentation. This regulatory activity has severely restricted the development and commercialization of new acrylates and methacrylates.[9]

The Speciality Acrylates and Methacrylates Program (SAM) was organized through the Chemical Manufacturers Association in 1987 to overcome this problem. SAM proposes to work with the EPA to formulate regulatory procedures that do not unnecessarily impede development and commercialization of new reactive chemicals, while providing adequate protection to health and the environment. This work includes development of a testing programme to better answer scientific/toxicology questions, while allowing industry to continue to introduce new acrylates and methacrylates.

## Other recent advances

Adhesion of radiation curable coatings to difficult substrates such as certain metals and plastics remains one of the greatest challenges of this industry. Many formulations tend to shrink on curing, and the rapidity of cure compounds the difficulty. Considerable efforts to provide oligomers with good adhesion to difficult substrates have had only limited success.[10] A recent development is the use of radiation curable monomers which expand on curing.[11] However, these are extremely high cost materials, such that they offer no practical solution to the problem for the general needs of the industry.

Pretreatment of substrate is widely used to improve adhesion. Examples include phosphatized steel and anodized aluminium, fluorinated[12] and corona-treated polyester.

Recently, also, there has been a resurgence of interest in water dispersible and water reducible radiation curable systems. The principal advantages of such systems is viscosity reduction without use of solvent or monomer. Volatile organic solvents are a source of unwanted exhaust emissions. Monomer addition may result in possible undesired changes in the coating film. Oligomer suppliers are offering some water dispersible oligomers on a limited basis. Photoinitiator suppliers have also responded to this need with work on photoinitiator systems for use in water dispersible systems.[13]

Other recent developments in raw materials for radiation curing are the so-called hybrid and dual-cure systems. In the latter system, partial cure is effected by free-radical polymerization, and heat or catalysed epoxide cure completes the process. This technology may offer advantages for various applications[14] from electronics to coatings with enhanced adhesion.

## Conclusion

Radiation curing has made impressive progress in the scant two decades since first commercialization. Growth of this industry continues at a rapid pace, and this is expected to continue. Regulatory pressures have contributed in some ways to the growth of this industry (requirements for emissions controls), and has hindered growth in other ways (TSCA regulation of acrylates and methacrylates). The advantages of reduced energy requirements, rapid throughput, and reduced space requirements seem to be universal benefits which will continue to stimulate growth of this technology. Still other advantages, such as the ability to coat sensitive substrates, and the many specialized applications offer additional promise of a bright future for radiation curing.

## References


1. See: Christmas, B. K., Kemmerer, R. & Kosnik, F., UV/EB curing technology: a short history. In *HIgh Performance Polymers: Their Origin and Development,* ed. R. B. Seymour & G. S. Kirshenbaum. Elsevier, New York, 1986, 331–7, and references cited therein.
2. Brodriguez, T., US Patent 4 717 623, January 1988.
3. *Wall Street Journal,* p. 1, 31 July 1987; *Chemical Week,* pp. 24–5, 9 September 1987.
4. Wasselle, L. A., Electron beam curing of polymers. Stanford Research Institute Report, July 1977.
5. Kemmerer, R. R., Christmas, B. K., Willard, F. K. & Scott, B. D., RadTech '88–North American Conference Papers, Radtech International, Northbrook, Ill., April 1988, pp. 290–308.
6. Johnson, M. A. & McConnell, J. A. *ibid,* pp. 264–67.
7. Bartholomew, R. F. & Davidson, R. S., *J. Chem. Soc.* (*C*) (1971) 2342.
8. See, for example, Plamondon, J. & Keener, R., Radcure '84 Conference Proceedings, Section 14, 1984, pp. 22–34.
9. Nelson, P. C. & Moran, E. J., Regulatory restriction on commercializing new chemicals. RadTech '88—North American Conference Papers, Radtech International, Northbrook, Ill, April 1988, pp. 120–6.
10. Christmas, B. K. & Zey, E. G., *Speciality Chemicals* (February 1988) 24–6; (August 1988).
11. Bluestein, C., *Cure Letter,* **5**(4) (April 1988) 2.
12. Brown, W. H. & Anderson, D. G., US Patent 4 717 623, January 1988.
13. Koehler, M. & Ohngemach, J. RadTech '88—North American Conference Papers, Radtech International, Northbrook, Ill., April 1988 pp. 150–9; Bassi, G. L., Broggi, F. & Revelli, A., *ibid,* pp. 160–6.
14. Christmas, B. K., Kemmerer, R. R. & Willard, F. K., RadTech '88—North American Conference Papers, Radtech International, Northbrook, Ill., September 1988, pp. 274–89.

# Trends in low viscosity acrylate resins for radiation cured formulations

**A. Howell & D. A. Cornforth**

*Anchor Chemical (UK) Limited, Manchester, UK*

Radiation curing techniques have developed considerably since the process became commercially viable in the early 1970s. Raw material development has also advanced during this time, partly as a result of new technology, but more importantly due to restrictions put on many materials relating to health and safety aspects.

Present day formulations still rely heavily on the principle that a system comprises oligomers, monomers, photoinitiators and additives. In many cases the oligomer components are comparable to those available 10–15 years ago; these include examples such as epoxy acrylates and urethane acrylates. Traditionally, these types of material are very viscous and require dilution with reactive monomers to achieve the desired viscosity and reactivity.

A number of monomers previously utilised have been eliminated due to factors such as toxicity, odour and skin irritation. Although monomers used at present perform very well, in certain areas high levels of safer monomers are required to achieve workable viscosities, and this can detract from the film properties imparted by the base oligomer.

New opportunities are now available to examine novel systems specifically designed to use either a minimum amount of monomer or eliminate the use of monomers completely. These include low viscosity monomer free oligomers, water thinnable oligomers and in special cases oligomers containing volatile, non-reactive solvents.

This paper is designed to give an insight into these novel materials and stimulate ideas as to where these products can be used.

In the United States and Japan, the markets and the requirements for raw materials differ considerably, as does the balance between UV and EB Curing techniques. These markets however, along with Europe, share the same growth potential over the next few years.

## European market—1989

- epoxy acrylate consumption 3500 tonnes
- urethane acrylate consumption 1500 tonnes
- polyester acrylate consumption 2500 tonnes
- monomer consumption 9500 tonnes
- estimated growth potential over next 3–5 years 8–12% by volume

For 1989, the European market analysis indicated that 3500 tonnes of epoxy acrylate was consumed; the bulk of this material was used on either wood or paper substrates. Urethane acrylate usage was less than half that for the epoxy acrylate whereas polyester acrylate consumption was mid-way between the two.

Monomer consumption was greater than all three of the oligomer types combined, at 9500 tonnes. The bulk of this usage was made up of materials such as TPGDA, TMPTA, HDDA and alkoxylated trifunctional monomers.

## Why low viscosity systems?

Limitations on monomer developments:

- raw materials
- European legislation on monomers and polymers
- costs involved

Although it would be presumptuous to say that raw material sources have been exhausted, the scope for innovative monomer development is limited, even though an infinite number of ethoxylated or propoxylated polyols can still be promoted as novel materials.

Concerns in the United Kingdom relating to skin irritancy and the SBPIM guidelines mean that many monomers perfectly acceptable a few years ago cannot now be used in radiation curable inks. This has had a knock-on effect in the varnish industry. The trend to toxicologically safe products will continue to restrict the use of existing monomers.

European legislation which sets guidelines for what constitutes a monomer or polymer, makes it more attractive to have a material classified as the latter rather than the former. As an estimate, it is likely to cost in excess of $60 000 to test a new development monomer satisfactorily. This is mandatory on products that are classified as new monomers by the European Commission.

In the case of polymers, this is not mandatory though it is still essential that tests be carried out to ensure that these materials are safe.

## Options open to resin manufacturers and formulators

- monomer diluents
- volatile solvents
- low viscosity monomer-free oligomers
- water thinnable oligomers

Manufacturers of resins and monomers have certain options and avenues of approach to ensure their products are received enthusiastically. To attain low viscosities, simple measures can be taken such as diluting with monomers or even organic non-reactive solvents. A novel approach is to look at specifically designed monomer-free oligomers or, even more revolutionary, the manufacture of water thinnable oligomers.

## Historical developments

| Monofunctional | Difunctional | Polyfunctional |
|---|---|---|
| HEA | BDDA | TMPTA |
| EOEOEA | NPGDA | PETA |
| PEA | HDDA | EOTMPTA |
| EHA | TPGDA | GPTA |
| IDA | Alkoxylated BPADA | EO PETA |
| IBA | Alkoxylated NPGDA | |

Traditionally, monofunctional monomers were classed as having unacceptable odours and high skin irritation; this is misleading as many of the newer types of monofunctional diluents are considerably safer than their predecessors.

Monomers such as phenoxy ethyl acrylate (PEA), ethyl hexyl acrylate (EHA) and ethoxy ethoxy ethyl acrylate (EOEOEA), though very efficient diluents, all have problems with odour, skin irritation and can be potentially toxic.

Monomers such as isodecyl acrylate (IDA) and isobornyl acrylate (IBA) are more acceptable monomers for use in today's applications. IDA is a good example or a relatively inexpensive monomer with good dilution characteristics yet incapable of improving the cross-linked density of cured formulations. Though not as fast to cure as polyfunctional monomers it can give better flexibility and adhesion on many difficult substrates.

There has been a number of diacrylate monomers introduced onto the market over the years. Many of the earlier ones also had problems with odour and skin irritation or were classed as carcinogens. Butane diol diacrylate (BDDA) and neopentyl glycol diacrylate (NPGDA) were both ruled out on a combination of these factors.

Hexane diol diacrylate (HDDA), on the other hand, is an excellent monomer, especially on difficult substrates such as plasticised PVC, with good viscosity cutting characteristics and fast cure speed. However, its high Draize rating severely restricts its use. HDDA is still used extensively in some areas of application in Europe but in the United Kingdom it cannot be used in inks and it has an uncertain future.

Tripropylene glycol diacrylate (TPGDA) has been available for many years; it has a relatively low Draize rating and performs well. It is one of the few monomers that continues to meet current health and safety requirements.

Alkoxylation has been an option open to produce safer monomers and is used with both neopentyl glycol and Bisphenol A. However, the poor viscosity cutting characteristics of the latter coupled with its aromatic nature, restricts the use of this product in many new applications.

Polyfunctional materials are generally used for their speed of cure and ability to enhance the cross-link density. Many problems were apparent with trimethylol propane triacrylate (TMPTA) and the tri- and tetracrylate esters of pentaerythritol (PETA). However, both are still used in some areas with large tonnages of TMPTA, used especially outside the United Kingdom.

Alkoxylated aliphatic monomers are now far more common and have more acceptable Draize ratings. Many can also be used without problems in the ink industry, especially in litho inks where they are ideal for achieving the desired water balance.

Varying degrees of ethoxylation and propoxylation on trimethylol propane, glycerol and pentaerythritol (EOTMPTA, GPTA, EOPETA) ensure that many options are open to the monomer manufacturer. Conversely, the action of the monomer as an efficient diluent can be severely diminished by higher degrees of alkoxylation and again this can detract significantly from the excellent properties imparted by the base oligomers.

## Problems with monomer diluents

- odour
- skin irritancy
- carcinogenic?
- high film shrinkage
- loss of adhesion
- variable cure speeds
- change in overall film properties

To summarize, skin irritation and toxicology must be acceptable before monomers can even be considered for use in this technology. Low odour and low taint are becoming increasingly important, demanding that high odour products be avoided.

On cure monomers generally have quite high shrinkage values, resulting in poorer adhesion; they can also have an adverse effect on their cure speeds depending on their functionality and molecular weight.

In future, no doubt many new monomers will become available. The evidence suggests that the majority of these novel or specialised monomers may be expensive and this will prohibit their use in many areas of industry where competition is fierce.

New ways have to be found to overcome the present restrictions.

## Use of volatile organic solvents

- rapid viscosity reduction
- may not require removal stage
- oligomer properties emphasised
- high MWT epoxies/novolacs can be used
- oligomers can be supplied in solvent

The use of non-reactive volatile organic solvents as diluents is found in special application areas. Figure 1 shows the rapid viscosity reduction found when MEK is compared to TPGDA in combination with an epoxy acrylate resin. Although the effects on viscosity reduction are excellent, the negative aspects are more important. These include fire hazards, an extra solvent removal stage, problems with solvent emissions and odour.

In cases where coatings are applied by spray, where low viscosities are required of less than one poise, small amounts of solvent can be used in preference to the use of large proportions of reactive monomers.

A major advantage in the use of non-reactive solvents is that an oligomer, tailor-made for a specific application, can be applied to a

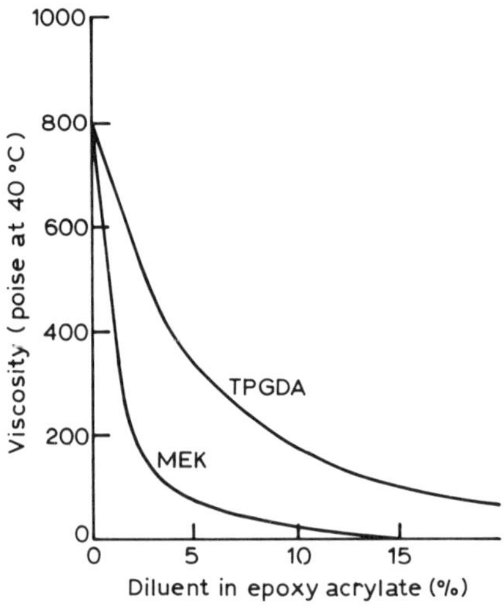


**Fig. 1.**

substrate and after solvent removal impart exactly the properties that the resin has been designed to offer. This is especially applicable to high molecular weight solid acrylate esters of epoxy resins and epoxy novalacs. With these types of material it is common to supply the oligomers in a solvent, though glycol ethers would normally be used rather than MEK. Examples of this are in developable etch resists and solder resists for the electronics industry.

## Developable etch and solder resists

- usually supplied at 50–70% solids
- glycol ether solvent
- screen printed coating
- touch dry uncured film after solvent removal
- good definition exposure from photomask
- uncured resin developable by solvent or base
- may remain intact (solder) or solvent removed (etch) resists

These are good examples where monomer-free coatings are applied at normally 50–70% solids. The glycol ether solvents can be changed to give the required evaporation rate. The systems are normally screen printed and, after solvent removal, touch dry films are produced when completely uncured.

Following exposure through a photomask, the unexposed areas can be removed by either solvent or base and the cured resin forms the basis of a solder resist. It can be removed at a later stage by an aggressive solvent in the case of an etch resist.

In less specialised areas where large quantities of materials are utilised, this type of approach cannot be used. Solvents are not viable and other ways have to be found to produce monomer-free oligomers. In general terms the bulk oligomers are usually of the three following types.

## Conventional oligomers (Table 1)

The standard epoxy acrylate is a well known and established raw material. In its undiluted form it is extremely viscous although it is soluble in most monomers and the rate of viscosity reduction is very

**Table 1**

| | Epoxy acrylate | Urethane acrylates | Polyester acrylates |
|---|---|---|---|
| Viscosity | High | High | Variable |
| Dilution with monomers | Easy | Easy | Easy |
| Viscosity cutback | Good | Fair | Good |
| Cure speed | Very fast | Variable | Variable |
| Flexibility | Poor | Good | Variable |
| Chemical resistance | Excellent | Good | Good |

rapid. The resin itself shows fast cure and has excellent chemical resistance though the cured oligomer is very brittle.

With urethane acrylates many different types are available. They are all high viscosity but can be diluted with most monomers; however, the viscosity reduction rate is usually inferior to that of epoxy acrylates due to the chemical nature of the oligomer. Cure speeds vary considerably but chemical resistance and flexibility are generally good.

Even more variables are possible with polyester acrylates, and they can be tailor-made to suit an application. Many materials have been examined in the past to see whether they can be compared to epoxy acrylates on formulation, but combinations of cure speed, chemical resistance and overall performance are difficult to match.

In designing a suitable yet novel alternative, certain criteria had to be kept in mind.

## Monomer-free low viscosity epoxy acrylates

- toxicologically acceptable
- easy handling
- fast cure speeds
- excellent gloss
- lower shrinkage on cure
- better adhesion to substrates
- reduced solvent levels in spray coatings

The materials have to be acceptable toxicologically. This also covers areas such as odour before and after cure in an attempt to produce low odour, taint-free films. Ease of handling is also a major consideration; an easy flowing monomer-free oligomer has excellent potential if it also conforms to the fast cure speeds and good gloss coatings associated with epoxy acrylates. A further advantage is the promise of lower shrinkage on cure and hence most likely better adhesion. If spray coatings are required, it is still likely that volatile solvents may be needed but only in very low concentrations.

# Low viscosity oligomers—development

A series of resins has been examined to produce a good low viscosity material capable of comparing favourably with commercially available epoxy acrylates (Table 2).

Resins 1 to 4 are all approximately of the same molecular weights. The variation in viscosity of the undiluted resins is, however, significant. Resin 1 is a standard aromatic epoxy acrylate; it has a viscosity in excess of 1000 poise at ambient temperature and shows excellent speed of cure and resistance to MEK. Unfortunately, due to the brittle nature of the cured film the tensile strength, modulus and elongation cannot be measured easily.

Resin 2 was manufactured to try to overcome many of the restrictions found with the Resin 1 type. This product was, however, much slower to cure and the chemical resistance was inferior. Although the modulus showed this oligomer to have good flexibility,

**Table 2**

| | Resin 1 | Resin 2 | Resin 3 | Resin 4 |
|---|---|---|---|---|
| Molecular weight | 530 | 520 | 450 | 470 |
| Viscosity (poise at 25°C) | 1000+ | 12 | 7 | 25 |
| Cure speed (ft min$^{-1}$) | 100+ | 40 | 60 | 90 |
| Chemical resistance (MEK rubs) | 200+ | 120 | 200+ | 200+ |
| Tensile strength (N mm$^{-2}$) | — | 1 | 2 | 9·3 |
| Tensile modulus (N mm$^{-2}$) | — | 32 | 57 | 223 |
| Elongation (%) | — | 4 | 3 | 7·5 |

it had poor properties, highlighted by a negligible tensile strength and elongation. The resin was of poor colour with an acrid odour. Although many of the restrictions from the standard epoxy had been overcome, they were replaced by others.

Resin 3 was aimed at improving the product colour and residual odour; this was successful yet the tensile strength and elongation were still below par. The modulus figure showed a further improvement on flexibility with the chemical resistance improving.

Finally, Resin 4 has all the best facets of the resins under study. Cure speed is around 90 ft $min^{-1}$ and the cured film has excellent chemical resistance. It has a useful tensile strength of 9·3 N $mm^{-2}$, good flexibility and a good elongation at 7·5%, not normally associated with an epoxy acrylate. The product has a colour value of 4–6 Gardner and a low viscosity at 25 poise on the monomer-free form. The product has a much better odour than the lower viscosity resins and promises to be very useful in a number of areas, especially for low odour and low taint coatings.

## Commercially available resins

In Table 3, a comparison is made of the performance of a standard epoxy acrylate and a polyester acrylate specifically designed for use in overprint varnishes, with the low viscosity monomer-free oligomer.

**Table 3**

| | A | B | C |
|---|---|---|---|
| Epoxy acrylate A | 50 | — | — |
| Low viscosity epoxy acrylate B | — | 80 | — |
| Polyester acrylate C | — | — | 90 |
| TPGDA | 50 | 20 | 10 |
| Photoinitiators | 5 | 5 | 5 |
| Resin viscosity (poise at 60°C) | 50 | — | — |
| Resin viscosity (poise at 25°C) | — | 25 | 11·8 |
| Formulation viscosity (poise at 25°C) | 6·3 | 7·2 | 5·4 |
| Cure speed (ft $min^{-1}$) | 80 | 100 | 90 |
| Flexibility (180° crease) | Poor | Good | Good |
| Chemical resistance (MEK rubs) | 200+ | 185 | 168 |
| Gloss level (60° glossmeter) | 96·4 | 95·4 | 100·6 |

Formulation A is comparable to many simple overprint varnishes available without the added flow aids, slip aids, etc. To achieve a viscosity of 5–8 poise, the monomer level has to be at 50%.

The polyester acrylate based formulation C had a low initial viscosity. The level of monomer required was also low and its performance was good. Cure speed at 90 ft $min^{-1}$ was better than the epoxy acrylate at 50% dilution. The gloss level was the best of the three examples but the resistance to MEK was the poorest.

The use of this polyester acrylate to emulate the performance of an epoxy acrylate has the major drawback that it is available commercially but is considerably more expensive than the standard epoxy acrylate or low viscosity epoxy acrylate.

Formulation B using the new low viscosity resin has a maximum of 20% monomer added, cure speed and performance are good and the added flexibility is a bonus; adhesion is also marginally better. The base resin is easy to use initially and does not require warming prior to formulation; also wastage is minimal. The gloss level reached was marginally worse than the standard epoxy acrylate. This product, however, excelled on flexibility, adhesion and cure speed.

## Water thinnable resins

- an inexpensive yet efficient diluent
- performance comparable to undiluted oligomers
- excellent gloss
- good chemical resistance
- versatility in roller coating/screen printing
- better clean-up procedures
- reduced ecological problems

Another way of eliminating monomers from formulations is to use a cheap, efficient, colourless and odourless diluent—water. The concept of water as a diluent is not new and although litho processes would not be responsive to water as a diluent, many other processes would be.

The main advantage of water thinnable systems is that the oligomers can be tailor-made to suit special applications. This class of product offers high reactivity and the materials can be cured either undiluted or with variable amounts of water added as a diluent. They perform well as overprint varnishes, exhibiting high gloss levels plus good water and chemical resistance. Materials are available that can be used in other low viscosity or higher viscosity applications.

An added advantage is that these resins can be used in water-free formulations, where the affinity of these products for water means they can be used to ensure easy clean-up of machinery a ıd eliminate the use of organic solvents in this procedure.

## Solubilising UV curable resins gives scope for different applications

Obviously, requirements vary considerably (see Table 4). Taking the two extremes; a normal type of overprint varnish uses a low viscosity in the region of 3–7 poise, low water uptake in the oligomer is advisable and the prime requisites are fast cure, high gloss and good resistance to water and MEK. These can be achieved easily.

Conversely, for screen printing the requirements are very different. A moderate or high water uptake enables a lower film weight to be coated. This enables the screen to move into process printing. Fast cure speeds are not as important, and gloss, chemical resistance and water resistance, although important, are not as critical as in overprint varnishes.

## Commercially available water thinnable UV curing resins

In Table 5, the first product, Resin X, is an example of a water thinnable oligomer. At ambient temperature it flows easily and contains no

**Table 4**

| Overprint varnish | Screen print varnish |
|---|---|
| Low water uptake | Moderate–high water uptake |
| Fast cure speed | Intermediate cure speeds |
| High gloss | Variable |
| Good chemical resistance | —Gloss |
| Good water resistance | —Chemical resistance |
| | —Water resistance |

**Table 5**

| | Resin X | Resin Y |
|---|---|---|
| Initial viscosity (poise at 25°C) | 37 | 550 |
| Maximum water uptake (%) | 25 | 60 |
| Minimum viscosity (poise at 25°C) | 1·5 | 0·5 |
| Cure speed (ft $min^{-1}$) | 100 | 80 |
| Water resistance (double rubs) | 200+ | 90 |
| MEK resistance (double rubs) | 200+ | 200+ |

water or added monomers. It can be diluted to approximately 75% solids, at which stage the homogeneous solution will have a viscosity of 1·5 poise. A typical cure speed on a low viscosity varnish will be 100 ft $min^{-1}$ and it exhibits excellent resistance to both MEK and water.

Resin Y is designed for use in areas such as screen printing. It has a high initial viscosity at 550 poise yet will readily dilute to a minimum of 40% solids, at which stage the viscosity will be less than one poise. The cure speed of a typical varnish at, say, 20–30 poise is 80 ft $min^{-1}$; resistance to MEK is excellent and to water it is still good.

## Resin X in a roller coating varnish formulation

Figure 2 shows that Resin X can be diluted with conventional monomers such as TPGDA as well as water. In the case where the formulation is used for better clean-up facilities then Resin X is compatible with a host of conventionally used additives.

Water, when used as a diluent, shows a dramatic viscosity cutting effect. Water is easily incorporated and the formulation shown with 20% added has a viscosity in the range of 3–5 poise. Irgacure 184 and benzophenone have been found to perform as well as any photoinitiators in this type of system and a level of 2·5% of each on the total solids is recommended as a good starting formulation.

When coated onto litho printed board and cured with one medium pressure 80 W $cm^{-1}$ mercury lamp, a cure speed of 100 ft $min^{-1}$ is typical. The cured film produces excellent gloss and adhesion. The flexibility as characterised by the 180° crease test is very good. This

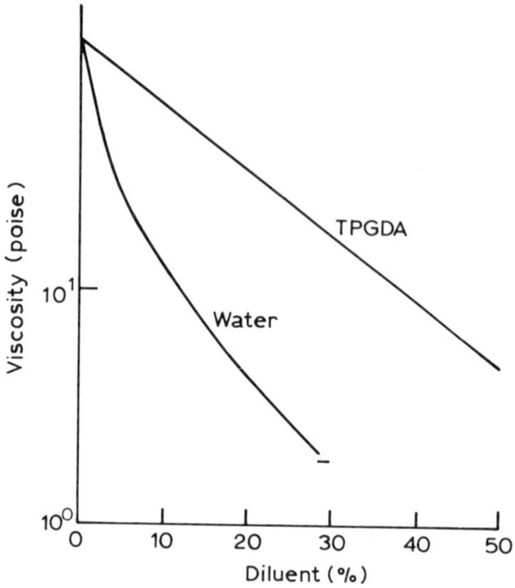


| | |
|---|---|
| Resin X | 80 |
| Water | 20 |
| Irgacure 184 | 2 |
| Benzophenone | 2 |
| Coat Thickness ($\mu$m) | 6 |
| Cure speed (ft min$^{-1}$) | 100 |
| Adhesion on board (%) | 100 |
| 180° crease | Good |
| Water resistance (double rubs) | 200+ |

**Fig. 2.**

would be impossible to achieve on a standard epoxy acrylate without the use of modifying monomers or flexibilisers. Chemical resistance to both water and MEK is excellent.

Resin X is designed for use as a low viscosity water thinnable oligomer primarily as a varnish on paper and board substrates. Alongside this, there are possibilities for its use in the wood coating industry. The potential for this material in inks has not yet been fully studied.

## Resin Y in a screen print varnish formulation

Resin Y shows the same viscosity cutting trends as Resin X. With water this is very rapid and Resin Y can take up much more water, bringing the solids content down to 40%, and thereby producing a very low viscosity. Again this material is compatible with reactive diluents such as TPGDA but the viscosity cutting is less efficient.

The formulation shown was designed as a screen print varnish for paper and board. A product viscosity of 24 poise was chosen and the use of a small amount of a standard epoxy acrylate plus a synergist gave a faster cure speed at 120 ft min$^{-1}$. Adhesion to board was good

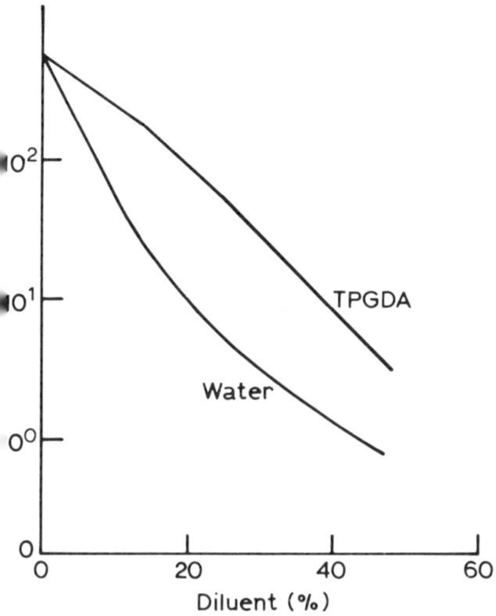


| | |
|---|---|
| Resin Y | 62·5 |
| Water | 12·5 |
| Amine synergist | 10·0 |
| Epoxy acrylate | 10·0 |
| Irgacure 184 | 2·5 |
| Benzophenone | 2·5 |
| Coat thickness ($\mu$m) | 25 |
| Viscosity (poise at 25°C) | 24 |
| Cure speed (ft min$^{-1}$) | 120 |
| Gloss | Excellent |
| Water resistance (double rubs) | 79 |
| Adhesion to board | Good |

**Fig. 3.**

and gloss was excellent. Water resistance was also good, though not as high as with the lower viscosity type of materials such as in the cure of Resin X.

Resin Y has been designed for different application areas than those intended for Resin X. In screen print applications the advantages are obvious and variable viscosity can be manipulated effectively to suit requirements. It is anticipated that the use of this material will again be best rewarded on paper, board and possibly wood.

In the ink area it has potential as a water-based screen ink vehicle, though this has yet to be optimised. At the other end of the scale it could be useful as a flexo ink vehicle in an effort to eliminate the use of volatile organic solvents and their environmental problems.

## Conclusions

There is a high level of interest in low toxicity, low viscosity, radiation curable materials.

- Low viscosity oligomers are now commercially available, offering fast cure, excellent film properties and low skin irritation

- Water-based UV curable oligomers are available, eliminating the need for monomers and solvents in formulations
- Water thinnable systems may be used to eliminate the use of organic solvents for washdown procedures

There is a high level of interest in low viscosity, low toxicity, radiation curable resins. They offer many advantages over conventional systems currently available.

With low viscosity systems, whether water thinnable or specially designed low viscosity epoxy acrylates, they offer excellent performance and versatility.

Elimination, or the reduction in the use of monomers, offers a new concept. Areas of interest are likely to be mainly in the paper and board industry with wood coatings also a possibility.

# Index